Physics One

The Alpha Edition

George D. J. Phillies

Copyright 2020 by George D. J. Phillies

Cover artwork by Cedar Sanderson, Sanderley Studios

About the Cover

The two gentlemen on the cover are Isaac Newton (1642-1746) and Josiah Willard Gibbs (1839-1903), the two greatest scientific geniuses of the Second Millennium.

Isaac Newton, Lucasian Professor of Mathematics at the University of Cambridge, created the differential calculus, Newtonian mechanics, and explained the motions of the planets, as revealed in his book *Philosophiæ Naturalis Principia Mathematica*. He made major contributions to our understanding of light, color, and the motion of fluids, and made other substantial contributions to mathematics. His attempts to reform our understanding of classical alchemy were less successful; his studies of biblical chronology and theology he wisely left unpublished.

Josiah Willard Gibbs, Professor of Mathematical Physics at Yale University, was a quiet, private man, who left behind few records for erstwhile bibliographers. We may contrast with Newton, whose manuscript writings total perhaps ten million words. Gibbs took the laws of thermodynamics that governed steam engines, and wrote down in one vast two-part paper of 300 pages and 700 equations much of modern chemical thermodynamics. He then created modern vector calculus. A series of his papers on physical optics showed that those phenomena were explained by Maxwell's equations, validating Maxwell's theory of electrodynamics. Langley consulted with him on aircraft design. Finally, his 1902 volume *Elementary Principles in Statistical Mechanics* set down from first principles the science of statistical mechanics in its modern classical form.

This is the Alpha Edition of Physics One. While considerable effort has been invested in eliminating typographic errors, incorrect historical observations, and the like, one of the reasons this book costs $20 or so rather than $300 is that I am not backed by a huge staff of proofreaders, editors, etc. Errors doubtless remain. If you find any, please advise at phillies@4liberty.net. I anticipate by and by releasing a Beta edition, hopefully with fewer errors.

Dedication
To Professors Anthony Philip French and Albert Gordon Hill,
who started me toward becoming the physicist I had always wanted to be,
and Professors Rainer Weiss, H. Eugene Stanley, and Daniel Kivelson,
who provided critical support as my career advanced.

Contents

i	**Preface I: For Faculty**	v
ii	**Preface II: Students! Read Me!**	vii
	ii.1 Studying	vii
	ii.2 Organization of the Book	ix
1	**Measurements**	3
	1.1 Coordinate systems	3
	1.2 Dimensions and Units	3
	1.3 Dimensional analysis	5
	1.4 Error	6
	1.5 Significant Figures	7
	1.6 Discussion	9
	1.7 Worked Problems	9
	1.8 Homework Problems	10
	1.9 Solutions to Worked Problems	11
2	**Vectors**	13
	2.1 Introduction	13
	2.2 Vector Multiplication	17
	2.3 The Curie Principle	20
	2.4 Discussion	20
	2.5 Worked Problem	21
	2.6 Homework	21
	2.7 Solution to the Worked Problem	21
3	**Calculus; Motion at Constant Acceleration**	23
	3.1 Introduction	23
	3.2 The Standard Functions	23
	3.3 Motion at Constant Acceleration	25
	3.4 Discussion	30
	3.5 Worked Problems	31
	3.6 Homework	31
	3.7 Solutions to the Worked Problems	35
4	**Chapter 4. Averages, Average Velocity**	37
	4.1 Averages	37
	4.2 Discussion	39
	4.3 Worked Problems	39
	4.4 Homework	40
	4.5 Solutions to the Worked Problems	42

5 Derivatives of Vectors, Circular Motion   45
    5.1 Introduction 45
    5.2 Derivatives of Vectors: Cartesian Coordinates 45
    5.3 Circular Motion 46
    5.4 Ballistic Motion 48
    5.5 Discussion 50
    5.6 Worked Problems 50
    5.7 Homework 50
    5.8 Solutions to the Worked Problems 53

6 Newton's Laws of Motion   55
    6.1 Introduction 55
    6.2 Newton's Laws 55
    6.3 Applications of Newton's Laws 59
    6.4 Discussion 63
    6.5 Worked Problems 63
    6.6 Homework 64
    6.7 Solutions to the Worked Problems 69

7 Inertial and Non-Inertial Reference Frames   73

8 Applications of Newton's Laws of Motion   77
    8.1 Introduction 77
    8.2 The Rocket Car on a Hill 77
    8.3 Coupled Masses 80
    8.4 The Hanging Mass 82
    8.5 Worked Problems 84
    8.6 Homework 85
    8.7 Solutions to the Worked Problems 90

9 Tribology   95
    9.1 Introduction 95
    9.2 Kinetic Friction 96
    9.3 Static Friction 98
    9.4 Coupled Masses with Friction 99
    9.5 Rolling Friction and the Tractive Force 102
    9.6 Friction in Fluids 102
    9.7 Discussion 104
    9.8 Worked Problems 104
    9.9 Homework 104
    9.10 Solutions to the Worked Problems 108

10 Examination 1   113
    10.1 Solutions to Examination I 115

11 Springs   119
    11.1 Introduction 119
    11.2 Forces On and By Springs 120
    11.3 Hooke's Law Springs 121
    11.4 Spring Attached to Wall 124
    11.5 Discussion 125
    11.6 Worked Problems 126
    11.7 Homework 126
    11.8 Solutions to the Worked Problems 127

CONTENTS   v

12 Momentum, Conservation, Collisions   131
  12.1 The Center of Mass 131
  12.2 Motion of a Group of Bodies 132
  12.3 Collisions 133
  12.4 Discussion 135
  12.5 Worked Problems 135
  12.6 Homework 136
  12.7 Problem Solutions 137

13 Work, Kinetic Energy, and the Work-Energy Theorem   141
  13.1 Work 141
  13.2 The Work-Energy Theorem 142
  13.3 Power 144
  13.4 Examples of the Work-Energy Theorem 145
  13.5 Worked Problems 149
  13.6 Homework 150
  13.7 Solutions to the Worked Problems 153

14 Energy   157
  14.1 Introduction 157
  14.2 Stability 159
  14.3 Gravitational Potential Energy of an Extended Body 160
  14.4 I Threw a Rock Into the Air 161
  14.5 The Radical Roller Coaster 162
  14.6 Two Masses Connected by a Wire 163
  14.7 Potential Energy of a Spring 164
  14.8 Discussion 165
  14.9 Worked Problems 165
  14.10 Homework 166
  14.11 Solutions to the Worked Problems 170

15 Energy Conservation, Collisions, and Friction   175
  15.1 Introduction 175
  15.2 Examples of Collisions 176
  15.3 Discussion 179
  15.4 Worked Problems 180
  15.5 Homework 180
  15.6 Problem Solutions 182

16 Examination 2   183
  16.1 Examination 2 Solutions 185

17 Descriptions of Rotation   187
  17.1 Rotation 187
  17.2 Vector Products 187
  17.3 Circular Motion 189
  17.4 Circular Motion in an Arbitrary Plane 191
  17.5 Discussion 193
  17.6 Worked Problems 193
  17.7 Homework 193
  17.8 Solutions to the Worked Problems 194

18 Motion in Cylindrical Polar Coordinates — 197
- 18.1 Introduction . . . 197
- 18.2 Worked Problems . . . 201
- 18.3 Homework . . . 201
- 18.4 Solutions to the Worked Problems . . . 202

19 Angular Momentum — 205
- 19.1 Introduction . . . 205
- 19.2 Angular Momentum . . . 205
- 19.3 Torque . . . 206
- 19.4 Conservation of Angular Momentum . . . 207
- 19.5 Displacement of the Origin . . . 208
- 19.6 ω and L . . . 209
- 19.7 Discussion . . . 211
- 19.8 Worked Problems . . . 211
- 19.9 Homework . . . 212
- 19.10 Problem Solutions . . . 214

20 Moment of Inertia — 217
- 20.1 Introduction . . . 217
- 20.2 Rigid Body Motion; Rolling Motion Without Slip . . . 217
- 20.3 Composite Rotating Systems . . . 219
- 20.4 Discussion . . . 220
- 20.5 Worked Problems . . . 220
- 20.6 Homework . . . 220
- 20.7 Solutions to the Worked Problems . . . 221

21 Rigid Body Rotation — 223
- 21.1 Introduction . . . 223
- 21.2 Example: The Physical Pendulum . . . 223
- 21.3 Koenig's Theorem . . . 226
- 21.4 The Rolling Cylinder . . . 226
- 21.5 Worked Problems . . . 227
- 21.6 Homework . . . 228
- 21.7 Solution to the Worked Problem . . . 230

22 Torque Diagrams; Pendulums — 233
- 22.1 The Torque Diagram . . . 233
- 22.2 The Simple Pendulum . . . 236
- 22.3 Pendulum with Extended Bob . . . 237
- 22.4 Pendulum Motion: A Solution . . . 239
- 22.5 Period of a Pendulum . . . 240
- 22.6 Discussion . . . 240
- 22.7 Worked Problems . . . 241
- 22.8 Homework . . . 241
- 22.9 Solutions to the Worked Problems . . . 242

23 Coupled Motion Including Rotation — 247
- 23.1 Introduction . . . 247
- 23.2 One Mass and a Wheel . . . 247
- 23.3 Two Masses and a Wheel . . . 248
- 23.4 Discussion . . . 252
- 23.5 Worked Problems . . . 252
- 23.6 Homework . . . 253
- 23.7 Problem Solutions . . . 256

CONTENTS

24 Statics — **259**
- 24.1 Simple Statics Problem . . . 260
- 24.2 Ladder on a Wall with Friction . . . 261
- 24.3 The Hinged Flagpole . . . 262
- 24.4 Discussion . . . 263
- 24.5 Worked Problems . . . 264
- 24.6 Homework . . . 265
- 24.7 Solutions to the Worked Problems . . . 268

25 Gravity — **273**
- 25.1 Introduction . . . 273
- 25.2 Newtonian Gravity . . . 275
- 25.3 Escape Velocity . . . 278
- 25.4 Weightlessness . . . 279
- 25.5 Discussion . . . 279
- 25.6 Worked Problems . . . 280
- 25.7 Homework . . . 280
- 25.8 Solutions to the Worked Problems . . . 283

26 Planetary Orbits — **287**
- 26.1 Discussion . . . 290
- 26.2 Worked Problem . . . 290
- 26.3 Homework . . . 290
- 26.4 Solution to the Worked Problem . . . 291

27 Examination 3 — **293**
- 27.1 Examination 3 Solutions . . . 295

28 A Sequence of Experiments — **301**

29 Experiment One: Measurements are Imprecise — **305**
- 29.1 Experimental . . . 305
- 29.2 Data Analysis . . . 306
- 29.3 Making Graphs . . . 308
- 29.4 The Search for Systematic Error . . . 310
- 29.5 Laboratory Report . . . 310

30 Experiment Two: The Search for Best Technique — **313**
- 30.1 Experimental . . . 313
- 30.2 Data Analysis . . . 314
- 30.3 Significant Figures . . . 315

31 Experiment Three: Properties of the Pendulum — **317**
- 31.1 Data Fitting . . . 317

32 Experiment Four: The Physical Pendulum — **321**

33 Experiment Five: The Atwood Machine — **323**

34 Experiment Six: The Static Force Diagram — **325**

35 Harmonic Motion — **329**
- 35.1 Note . . . 330

36 Complex Numbers — 331
- 36.1 Arithmetic with Complex Numbers … 332
- 36.2 The Euler Identity … 333
- 36.3 Note … 334
- 36.4 Homework … 334

37 Harmonic Oscillation — 337
- 37.1 A Mass on a Spring … 337
- 37.2 Canonical Forms for a Harmonic Oscillator … 339
- 37.3 The Pendulum … 340
- 37.4 Complex Variable Method … 341
- 37.5 Energy of a Pendulum … 342
- 37.6 Pendulums: Torque Approach … 343
- 37.7 Physical Pendulum: Energy Approach … 345
- 37.8 Calculating Constants of Integration … 346
- 37.9 Homework … 347

38 Harmonic Motion, Damped or Driven — 351
- 38.1 Harmonic Oscillation with Damping … 351
- 38.2 Energy Storage–The Quality Parameter … 354
- 38.3 Harmonic Oscillation with an External Driving Force … 355
- 38.4 Discussion … 356
- 38.5 Homework … 357

39 The Damped, Driven Harmonic Oscillator — 359
- 39.1 Basic Calculation … 359
- 39.2 Form of the Amplitude Curve … 362
- 39.3 Power Absorption … 365
- 39.4 Homework … 369

40 Coupled Harmonic Oscillators — 373
- 40.1 Mathematical Interlude … 373
- 40.2 An Example … 377
- 40.3 Discussion … 379
- 40.4 Homework … 379

41 The Double Pendulum — 381
- 41.1 Discussion … 384
- 41.2 Homework … 384

42 The Oscillating String – Standing Waves — 387
- 42.1 Normal Modes of a String … 387
- 42.2 Energy in a Vibrating String … 392
- 42.3 Homework … 395

43 The Oscillating String – Traveling Waves — 397
- 43.1 Introduction … 397
- 43.2 Travelling Waves … 397
- 43.3 Terminology … 399
- 43.4 Longitudinal and Transverse Velocities … 400
- 43.5 Homework … 402

44 About the Author — 405

Chapter i

Preface I: For Faculty

and For People Doing independent Study

Physics One, The Alpha Edition is based on the decades I spent teaching this course as a college professor. This book represents a radical change in how we teach physics. Why is this book different from many other introductory mechanics books? Most important, students can probably afford to buy it. In some states, this book costs less than the sales tax on some of its competitors. That's possible because this volume is independently published, using my figures and editing skills. My illustrations are the sort of line drawings I would put on the blackboard during lecture.

Physics One presents calculus-based, college-level physics. By calculus-based, I mean that students must consistently use calculus in homework. They should also need calculus in examinations, but that is under your control, not mine. By college-level, I mean that for the most part the course is based on symbolic, not numerical, reasoning.

Why is this *The Alpha Edition*? I spent considerable time proofreading and fact checking. However, this volume costs so little because it is not being used to support a vast horde of artists, photographers, editors, proofreaders, indexers, marketeers, salesmen, Vice Presidents, expense accounts, corner offices[1], consultants, and other doubtless nice people. Correspondingly, despite my best efforts, there are doubtless remaining typographic errors, infelicities of phrase, and other mistakes I failed to remove. I would much appreciate your calling these to my attention, at phillies@4liberty.net, following which at some future date I shall be replacing *Physics One, The Alpha Edition* with *Physics One, The Beta Edition*.

You will hear colleagues say that many freshmen are not yet reasoning on the symbolic level, the level where they would solve problems using algebra and calculus rather than plugging numbers into their pocket calculators. To that I say students came to the university to learn new things. For those students, symbolic reasoning is one of those things that they need to learn, and that you and your colleagues need to teach them. To help students learn symbolic reasoning, the book presents a lot of algebraic calculations, in which I take very small steps from line to line, so students can follow the steps for themselves. I try to emphasize why steps are being taken. In addition, I provide almost no numerical problems, only algebraic and calculus-based problems. For each Chapter in Part I, I present a set of self-test questions for students, with solutions provided at roughly the level of detail I would have presented the solutions myself in a conference class. These solutions present new material not in the rest of the chapter.

Physics One has as its foundation one rule and two equations.

The one rule is: Physics is based on *praxis*, learning to get the right answers for the right reasons. A wrong method that happens to give the right answer is wrong. Read Section 3.3 Motion at Constant Acceleration to see where an alarming number of students have learned wrong methods.

The two equations are

$$\bm{F} = \frac{d\bm{p}}{dt} \tag{i.1}$$

and

$$\text{PHYSICS} - \text{CALCULUS} = \text{NONSENSE}. \tag{i.2}$$

The first equation is Newton's Second Law of Motion. Your students may have seen it written differently. Don't worry if they don't recognize my notation. After we finish the first few chapters, they will understand the symbols I just used.

The second equation is the main reason this book is radical. I say: To understand physics, you must use calculus. This book assumes that a student can take integrals and derivatives of a few standard functions: sin, cos, exp, log, polynomials, and powers, and is familiar with the product and chain rules.

The second equation implies that students have to some extent studied calculus before they pass far into this text. Many High Schools now cap their math sequence with calculus, so that your students may well have seen the needed derivatives and integrals (and probably a considerable number of less useful integrals and derivatives). At a few colleges, students may be taught with the Kingsbury approach, in which students learn how to take derivatives and integrals of, e.g., polynomials, first, and turn to limits and analysis later; those students with some coordination may know enough calculus by the time they need it. If you are studying physics independently, you need to learn some calculus first.

Corresponding to the expectation that students learn and use symbolic reasoning and calculus in homework and exams is the need for student homework to be graded so as to learn how to use symbolic reasoning and calculus. That's quite different from homework sets in which students only need to supply the right answer. Serious attention to grader performance will be needed.

Part I of the text develops basic mechanics. The experiments in Part II of the text are usefully done in parallel with the theoretical development in Part I. Part III develops harmonic oscillators; those can be challenging to study with experiments unless there is an appreciable equipment budget.

The course follows the lectures and recitations sections I taught for many years at the Worcester Polytechnic Institute. Our academic schedule was atypical. Terms ran seven weeks, a week including three lectures, two recitation sections, and two hours of laboratory. Three of the lectures were used for exams. Homework was assigned to be collected in every lecture including the first, and was returned in the next recitation section or lecture. Students took three courses at a time; they were expected to spend 10-15 hours a week outside of class studying and doing homework. Approximately speaking, each chapter in Part I corresponds to a lecture. The Lab Exercises in Part II represent a novel approach based on using minimum equipment. The Chapters in Part III on harmonic motion, are organized by topic, and represent multiple lectures each.

Finally, I found a path to involve all students in answering non-rhetorical questions (you can now duplicate my scheme electronically, but my method, like my book, costs much less.) I would pose a question, and list on the board several answers, labelled A-B-C-D. Students had each been issued a large sheet of paper labeled A-B-C-D in its four corners. I gave ample time to think and then told them all to pick up the paper by the corner matching their answer. They all answered at once, so they all had to think. Yes, my "obvious" answer was usually the wrong answer. *Cognitive dissonance* is an enormously powerful teaching tool.

A closing thought on homework problems: Over the term, students should work a large number of homework problems. To simplify grading, I used statistical sampling, grading one of the three or four problems on each homework set. The Grading for 20 homework sets was 6 - solution process and final answer are correct, 4 - solution process is correct but there are modest errors, 2 - solution process is wrong but all problems had serious attempts, and 0 - homework was wrong, seriously incomplete, or missing. This process generates a grade. It keeps things simple enough that there is no common excuse for a grader not to finish everything immediately, and eliminates more or less all disagreements about how many points an answer is worth.

Finally, some words of thanks: Word processing and formatting were done with LaTeX, namely WinEdt overlaying MikTex, including various American Mathematical Society packages. The drawings, generated with Inkscape, are very close to what I would have put on the blackboard. I should specifically thank Cedar Sanderson for calling my attention to Inkscape, George Grätzer for his series of LaTeX volumes especially *More Math Into LaTeX*, long-time friend Rich Moore for suggesting that I rewrite the introduction, and above all generations of students whose questions, comments, and especially mistakes did much to improve this work.

[1] OK, I confess. My working office, upper floor of my home, has picture windows on two sides.

Chapter ii

Preface II: Students! Read Me!

This book had readers before it was released. Talking about this Preface, I was given the feedback 'I wish I'd been told all this when I was an undergraduate.' I've tried to make it sound as friendly as possible, but that's a bit challenging.

Physics One, The Alpha Edition is based on the many years I spent teaching this course as a college professor. This book represents a radical change in how we teach physics. Why is this book different from many other introductory mechanics books? Most important, you can probably afford to buy it. In some states, this book costs less than the sales tax on some of its competitors. That's possible because this volume is independently published, using my figures and editing skills. My illustrations are the sort of line drawings I would put on the blackboard during lecture.

Physics One presents calculus-based, college-level physics. By calculus-based, I mean that you must consistently use calculus in homework. Your should also need calculus in examinations, but that is under your professor's control, not mine. By college-level, I mean that for the most part the course is based on symbolic, not numerical, reasoning.

Why is this *The Alpha Edition*? I spent considerable time proofreading and fact checking. However, this volume costs so little because it is not being used to support a vast horde of artists, photographers, editors, proofreaders, indexers, marketeers, salesmen, Vice Presidents, expense accounts, corner offices[1], consultants, and other doubtless nice people. Correspondingly, despite my best efforts, there are doubtless remaining typographic errors, infelicities of phrase, and other mistakes I failed to remove. I would much appreciate your calling these to my attention, at phillies@4liberty.net, following which at some future date I shall be replacing *Physics One, The Alpha Edition* with *Physics One, The Beta Edition*.

What is physics? Physics is the study of all natural phenomena. The objective of physics is to reduce all natural phenomena to a small number of principles and elementary objects. Physics explains all natural phenomena in terms of matter and motion, in terms of particles, fields, and forms. The original difference between physics (and its sister sciences, and engineering) and all other natural philosophies arises from two understandings: (1) The natural world can be described quantitatively, notably by analytic geometry and calculus, and (2) when you can make accurate predictions about the world, you can also calculate how accurate your predictions might be. Physics covers a vast number of particular topics, some of which have been given openings in this book. To find those openings, consult 'advanced topics' in the Index.

ii.1 Studying

To learn physics, there is no substitute for spending time studying. Study your notes. Study your book. Study the solutions to the Worked Problems, after you have tried solving the problems yourself. Study means questioning what you are reading. If there are a series of equations, do the algebra for yourself and see if you get the same answer that I did. Some people learn more from reading. Some people learn more from working problems. You have to learn what works for you. If you hit something that you do not understand, try looking at a different physics book. See what you can find on the internet.

For many of you, doing homework is how you learn physics. You should view homework as a self-test. The best way to do homework is to read the chapter and your lecture notes carefully, perhaps taking notes on the text, work through the calculations in your notes, and then do the homework *without* looking back into the chapter. If you have learned the material, you usually will not need to look up procedures. I say "For many of you" because for some of us homework problems are a nuisance that must be overcome before serious studying takes place.

In doing homework, there are some helpful tricks:
(0) Use lined paper.
(1) Write in pencil on only one side of the page.
(2) Write equations in a single vertical column.
(3) Present your calculations in a logical order. If the equations are written in random order like a section of computer spaghetti code, they are wrong. If your solutions are done properly, anyone has but to look at the last line of your solution to find your answer, and read down the column to see how you reached your answer.
(4) In a large calculation, do the work on scratch paper and then write up a final solution. That's not different than doing a paper in English, in which you do a rough draft, line edit it, perhaps do several rough drafts, and then present a final draft.
(5) If the first character in an equation is an equals sign, the equation is wrong.
(6) Write large enough that your handwriting can be read. For example, derivatives should generally spread vertically over two lines.
(7) Work problems using symbols, and plug in numbers toward the end.

You will not always solve every problem immediately. If you can't solve a problem, at some point go do something else. Let your subconscious work. Return to the problem later. If you always run for help after 15 minutes, you will not learn the perseverance and problem solving skills you will need when you are graduated.

Reading a physics text is not the same as reading a history book. Having read the words does not count. To read a physics text, you need to go through line by line, consider what is being said, and do the math steps for yourself, so that you come to understand the processes being used in the calculations. 'Memorizing the derivations' misses the point.

In listening to lecture, take notes. Before you go to sleep that night, recopy them, inserting all the details you remember but didn't write down. That's using short-term memory. The next day, your short-term memory will have faded, and your original notes will be much less clear.

You understand physics if you can solve problems that are not exactly the same as the problems you saw solved in the book or worked in the homework. A person who tells you 'I understand the material, but I can't work problems' does not understand two things. First, he does not understand the material. Second, he does not understand the word 'understand'.

What about group study? Actual pedagogical research due to Kingsbury shows that the only effective group for group study contains exactly two people. They each solve the same two problems. They take turns showing solutions. One talks; one knows no one else is listening. Both therefore pay careful attention. What do you say when it is your turn to talk? You explain each step, why you took it, and why the step is correct. The other person interrupts you and forces you to defend your thinking. This approach to learning is the Socratic Dialogue. It is two-and-a-half millennia old, and still works. Group discussions between people who each tried to solve the same problem can be fruitful. Group discussions in which you listen, and someone else announces answers to problems you did not try to solve, are like lying in bed watching an exercise video: It may be fun. It's not exercise. Practical advice: For some people, group work is a waste of time.

Some of my colleagues think that we should use concrete problem questions. Accordingly, I supply an extended series of these, based on your being the physics consultant for a motion picture director. The director is a firm believer in using stunts, explosives, car chases, and tear-jerking maudlin romance scenes as a replacement for plot, characterization, dialogue, acting ability, and cinematographic technique. Alas, he does not have a CGI budget, so all these stunts are going to be done with the real actors and functional props. The accuracy of your predicted results is significant. On the other hand, he pays well.

Memorization is an extremely valuable tool for putting large bodies of fact into a useful order in your mind. No matter what your major, material in core courses is material you need to know *without looking it up*. If someone tells you 'I don't need to memorize things, because I can always look them up', challenge them to learn a topic, using a text written in a foreign language that they do not speak or read. They may use a bilingual dictionary. Korean is a fine choice of language; the Hangul script is by rational design highly

ii.2 Organization of the Book

A brief description of the rest of the course: The first five Chapters cover much of the mathematical language you need in order to understand the course. Chapter Six presents Newton's Laws of Motion and a few applications. Everything in the book after Chapter Six is an application or special case of Newton's Laws.

The text in each Chapter follows very closely the lectures I gave when I taught this course, down to the historical anecdotes and occasional joke. Most chapters in Part I correspond to a single lecture. As I taught the course, for each lecture there was homework, collected at the start of the next lecture, so I assigned three homework sets a week, returned more-or-less at the following lecture. I can't arrange that for you, so each Chapter in Part I includes at the end several *Worked Problems* and their solutions, roughly as I would have presented solutions in class.

Each Chapter in Part I has Worked Problems. *The Worked Problems* were a Homework set. Following the Worked Problems are the regular Homework Problems, and following the regular Homework Problems I supply my complete *Solutions to the Worked Problems*, more or less as I presented them in class. The solutions include new material worth knowing, material not covered in the rest of the chapter. As a study technique, after studying the text and your notes, try the Worked Problems yourself, check your answers, then compare with my answers. I spell out the process of reaching each solution in great detail. A complete, correct solution worth full credit could have been much shorter than my solution. My solutions are written out to help you if you are unsure or confused, even if someone who understands what is going on wishes I would give a much shorter solution.

The Sample Examinations: I have included in Part I of the book three sample examinations, spaced roughly after each third of Part I. Why are there sample examinations? When you think you have studied enough to take the exam, you should find a quiet place, set out a pad of paper and writing implements, and do the exam yourself under exam conditions, meaning in particular that you have a timer running. When you're done, compare your solutions with the textbook solutions, and see how you did. I hope that you did well. The purpose of the sample exam is to give you a self test as to whether you've actually studied enough. The Sample Exams are exams I actually used once upon a time; students had 50 minutes to complete them. Note that I used university rather than high school grading standards. In real work, you do a calculation, set it aside for a day and then check if it is right. You can't do that on an exam. I therefore wrote exams so that the pass line was around 45-50, the C-B line was around 65, and the B-A line was in the mid-80s, so that minor errors in work done in haste would not perturb letter grade outcomes. When you work a trial exam, don't be surprised if you don't get everything right.

Part II of the Book gives a series of *Laboratory Exercises*. The purpose of the experimental sequence is to show you how real experiments are done. Warning: That's not similar to lab experiments you are likely to have seen in other courses. You need to think about what you are doing. You are pointed in a direction. You are not told exactly what to do. There is a significant emphasis on the notion of error in making measurements. An important part of experimental work is to realize that there may well be several different paths to the same result, and that experiments may be needed before you decide which experimental approach gives the best results.

The experiments are designed so that if you have school supplies, a computer, and a few household tools, you can carry them out yourself. The services of a children's toy construction set, an accurate scale, and a high-precision stopwatch will be helpful. Some experiments will be easier for several people working together. Real experiments have random errors, because measurements are imperfect. We therefore begin with experiments that demonstrate random error and statistical analysis. We then consider designing an experiment; we consider different methods for measuring the period of a pendulum. We ask how the period of a pendulum depends on the pendulum's properties. I then ask you to construct the device that was the first quantitative experimental test of Newton's Laws of Motion, the Atwood Machine. Can you do as well as Atwood did? The final laboratory is to build the experimental equivalent of a force diagram for a stationary object.

Part III of the book does a detailed analysis of *harmonic oscillators*. The work is somewhat more complicated than the physics shown in Part I. It's a supplement, based on a freshman course that I taught on harmonic oscillators. It's shorter than Part I, because the front third of the harmonic oscillator course had to cover roughly the front half of Part I of the text.

Finally, some words of thanks: Word processing and formatting were done with LaTeX, namely WinEdt overlaying MikTex, including various American Mathematical Society packages. The drawings, generated with Inkscape, are very close to what I would have put on the blackboard. I should specifically thank Cedar Sanderson for calling my attention to Inkscape, George Grätzer for his series of LaTeX volumes especially *More Math Into LaTeX*, long-time friend Rich Moore for suggesting that I rewrite the introduction, and above all generations of students whose questions, comments, and especially mistakes did much to improve this work.

[1] OK, I confess. My working office, upper floor in my home, has picture windows on two sides.

Part One

The Physics of Isaac Newton

This page reserved for your notes.

Chapter 1

Measurements

In this chapter, we briefly discuss coordinate systems, dimensions and units, dimensional analysis, experimental errors and significant figures, and close with some remarks on approaches to learning physics.

1.1 Coordinate systems

In most of this course we will use Cartesian coordinates. You should by now have studied analytic geometry, in which points on a plane are identified with their x and y coordinates. However, physical space is three-dimensional, so we need three coordinates, namely x, y, and z. The x-, y-, and z-coordinates are perpendicular to each other. Here's a sketch of Cartesian coordinates, and a particle at $\bm{r}(t)$ moving with respect to the Cartesian coordinate axes. Many of you are aware of other coordinate systems, e.g., cylindrical coordinates or spherical polar coordinates. Rewriting mechanics equations from Cartesian to other coordinate systems is, except for a few special cases, not trivial.

There is also a time coordinate t. Objects move through space, so the x, y, and z coordinates of an object are each a function of time, written $x(t)$, $y(t)$, and $z(t)$. That last statement is a bit more subtle than it sounds:

First, almost all real objects are not points; they have physical extent, leading to a question: In an object that is not a point, which point in the object is to be identified as the object's location? There turns out to be a reasonable answer. [Aside: As of this writing, no measurement has ever found that the electron is not a mathematical point. If the electron has a non-zero physical extent, we have not yet found it.]

Second, the statement that anything is a function of time embodies an assumption about the nature of time, namely that all observers can come to an agreement (up to time zones, corrections for speed-of-light signal delays, and the like) as to what time it is, at each point in space. Special relativity shows that this claim is false. If two observers are moving with respect to each other, and one says that a line of clocks is properly synchronized so that they all agree with each other, the other observer will find that the clocks are not synchronized and do not all agree with each other. We will not treat special relativity farther in this course.

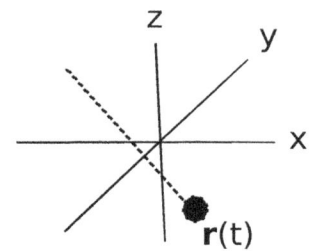

Figure 1.1: Three-dimensional Cartesian coordinates (solid lines) and the trajectory (dashed line) of a moving particle (circle).

1.2 Dimensions and Units

In the physical sciences, we introduce large numbers of different physical quantities, such as length, force, and energy. When we do a measurement of a physical quantity, we report a number, but associated with the number is a unit, so that the distance between two points is so many meters or so many miles or so many light years. If you want, you can interchange units, so instead of reporting a distance in meters you can report the same physical distance in miles. The units have changed, the number has changed, but the distance remains the same. For example, instead of saying the distance between two points is exactly one mile, you could say that the distance between the same two points is exactly 5,280 feet. Instead of saying the distance is one mile, you could also report the same distance in meters.

In most of this course, we will use International System (SI) units, meter-kilogram-second (MKS) units, or centimeter-gram-second-electrostatic (cgs-esu) units. (For very round number approximations, I will sometimes invoke English units.) Why do we use SI or cgs-esu units? We use them because they are the international system of units used in all of science. Of course, Americans largely use English units (pound-foot-second) in daily life. However, when SI units were becoming the universal units of science, science was almost entirely European. Before World War I, there were a very small number of great American scientists, such as Josiah Willard Gibbs and Albert Abraham Michelson, but until World War I ended science was almost entirely European. European units (SI units were originally French Imperial units) therefore dominated and are still totally dominant in science.

In describing a physical quantity, underlying all these different units are the *dimensions* of physical quantities. The word *dimension* is here being given a new meaning, which is not the same as its meaning in the sentence *space has three dimensions*. There are vast numbers of different units, but all physical quantities have the same three dimensions, namely length ℓ, mass m, and time t. No matter the physical quantity, its dimensions can be written as a product of length, mass, and time, each raised to some power, namely

$$\ell^a \cdot m^b \cdot t^c. \tag{1.1}$$

Here a, b, and c are simple numbers, the powers to which ℓ, m, and t are raised. Note that in this expression m stands for the dimension mass, but in discussing units of length the m stands for meters.

Units have dimensions, but units are not dimensions. The distance between the ends of a meter stick may be one meter, but the dimensions of that distance are

$$\ell^1 \cdot m^0 \cdot t^0 \equiv \ell. \tag{1.2}$$

(I just used a math result: $x^0 = 1$ for any x except $x = 0$.) It is wrong to say that the dimensions of a distance are meters; the dimension of a distance is length.

As a second example, consider a car driving down a highway. Its dashboard display reports a speed in miles per hour, a length divided by a time. The units of the speed are miles divided by hours. Miles are a length, hours are a time. The dimensions of speed are then $\ell \cdot m^0 \cdot t^{-1}$.

Some units have names. For the last century, there has been a fad of giving names to units, generally names of famous dead scientists. For example, in SI units the unit of energy is the Joule (often pronounced to rhyme with Yule, though James Prescott Joule may well have pronounced it jowl.) However, all these named units have dimensions. As examples of non-SI units:

The furlong, a traditional unit of length, has dimensions ℓ^1.
The pipee, a French pre-Revolutionary unit of area, has dimensions ℓ^2.
The hogshead, a traditional English unit of volume, has dimensions ℓ^3.

There are also named SI units. Using the abbreviations kg, m, and s for kilogram, meter, and second: The Newton, a unit of force, is 1 kilogram-meter per second per second, so its dimensions are $m^1 \cdot \ell^1 \cdot t^{-2}$. The Joule, a unit of energy, is one kilogram meter-squared per second-squared. It has units force times distance, so its dimensions are $m^1 \cdot \ell^2 \cdot t^{-2}$. A unit of torque is the Newton-meter, 1 kilogram meter-squared per second-squared, which has units force times distance, so its dimensions are $m^1 \cdot \ell^2 \cdot t^{-2}$.

Point of confusion: Energy and torque have the same dimensions, but it is incorrect to report torque in Joules. Energy and torque are not the same thing. Why are they not the same. The trick is the word 'times'. When I wrote 'force times distance', first for energy and then for torque, the word 'times' had two completely different meanings, one in the definition of energy and another in the definition of torque. In this course, the word 'times' turns out to have not one or two but four distinct meanings, none of which are captured completely by the "·" character in $\ell^a \cdot m^b \cdot t^c$ As will become clear, energy is a scalar, and torque is a pseudovector. We'll discuss that more when we consider vectors.

Second point of confusion: The letter m is being used here with two different meanings. In giving units, m stands for meter. In giving dimensions, m stands for mass.

1.3. DIMENSIONAL ANALYSIS

How do algebraic symbols combine? Consider quantities a and b. The product ab has the dimensions of the product of the individual dimensions. For example, if v is a speed and t is a time, vt has dimensions (length/time) (time) or length. If a and b have the same dimensions, then $a + b$ has the dimensions of a or of b. If a and b have different dimensions, the sum $a + b$ is mathematically invalid.

Interesting question. Consider the angle θ. Note the figure.

What are the dimensions of θ? If you think back to calculus, you will recall that an angle θ was defined in terms of a circle. The circle had radius r. The angle θ spanned a length s of the circle's circumference, with θ being defined by $\theta = s/r$. θ thus has dimensions length/length, these being the dimension unity, same as the dimensions of the number 1. For example, if θ spans half of a circle, $s = \pi r$ and $\theta = \pi r/r$, meaning for a half circle that $\theta = \pi$. θ has units radians, but the dimensions of radians are unity ($\ell^0 m^0 t^0$).

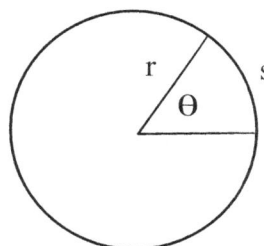

Figure 1.2: The angle θ, defined in a circle having radius r by the segment s of the circle's perimeter that the angle θ spans.

[Aside: As is not always emphasized by mathematicians, all those formulas you learned in calculus class for derivatives and integrals of trig functions are only true if θ is in radians. If θ is in degrees or grads or brads or mils, the calculus formulas you know don't give the right answers. If you are accustomed to land or water navigation, note that in navigation $\theta = 0$ naturally points to the top not the side of the page, and increasing θ runs clockwise, not counterclockwise. In calculus, in the x-y plane increasing θ runs counterclockwise, now clockwise. In the x-z plane, calculus θ increases in the clockwise, not the counterclockwise, direction. The book has a number of these asides. They discuss matters worth knowing, but not part of the regular course.]

Suppose I have an equation, e.g.
$$F_x = m\frac{d^2 x}{dt^2}. \tag{1.3}$$

That's a special case of Newton's law; it describes motion in one dimension. Here F_x is the total force on mass m in the x-direction, and dx^2/dt^2 is the x-component of the mass's acceleration.

It is totally wrong to write units inside an equation. If you did this horrible wrong thing, you might write
$$F_x(N) = m(kg)\frac{d^2 x(m)}{dt^2(s^2)}, \tag{1.4}$$

where in this abomination the extra letters N and kg, one of the two letters m **but not the other one**, and s stand for Newtons, kilograms, meters, and seconds, respectively. The other m stands for the mass of the object. Why is this equation worse than wrong? For starters, equation 1.3 is not only true in SI units, it is true in any other *consistent set of units*. For example, the equation is true if the force, mass, distance, and time are measured in dynes, grams, centimeters and seconds.

In some systems of units, in particular SI units, electrical charge is a special case, effectively a fourth dimension in the same sense that mass, length, and time are dimensions.

1.3 Dimensional analysis

Physical equations have an important property. If an equation is correct, the dimensions of the two sides of the equation are the same. (That statement does not work the other way. Even if the two sides of an equation have the same dimensions, the equation may still be wrong.)

Let's consider an example, for motion at constant velocity. In one dimension, we may write
$$x(t) = x_0 + vt. \tag{1.5}$$

Here v is the velocity in the x direction and x_0 is a constant position.

What are the dimensions of the terms in this equation? Replacing each algebraic character with its dimensions, we find
$$\text{length} = \text{length} + (\text{length/time})\text{time}. \tag{1.6}$$

Every term on that line, including $x(t)$, x_0, and vt, has the same dimensions, namely length. The left- and right-hand sides of that equation have the same dimensions, so that equation could possibly be true.

One the other hand, suppose someone wrote

$$x(t) = x_0 + vt^2. \tag{1.7}$$

Could this equation possible be true? On the right-hand-side of the equation, the dimensions are length + length/time*(time)^2. The right-hand-side of the equation tries to add two terms that have different dimensions. That's not allowed. Equation 1.7 can't possibly be true.

1.4 Error

In fair part, we will be doing physics calculations using symbols, not numbers. However, some physics calculations incorporate numbers from physical measurements. Most physical measurements have some level of experimental error, so that repeated measurements of the same quantity give slightly different answers. The differences between values from repeated measurements are an estimator of the error in the measurement. We call the differences between the measurements the *error*, but the word error does not mean that the measurements are defective, or that the experimenter must have made a mistake. The word error, as applied to measurements, instead implies that there is a limit to the accuracy of the measurements.

There are four sorts of error, namely random, systematic, conceptual, and quantum. Random errors correspond to unpredictable variations from measurement to measurement in repeated measurements of a given quantity, measurements being made under identical conditions. The random error is the size of the variation from measurement to measurement. A systematic error is an error that always occurs in the same direction. For example, if you weigh yourself on a bathroom scale while holding a pair of bricks, your measured weight will always be wrong by the weight of the bricks. That's systematic error. It could be said that the systematic errors mean that the measurement is defective. Conceptual errors arise when one attempts to transform, for example, the reading on a meter, into an interesting physical parameter, while using a theoretical model that is invalid in the system of interest. Quantum error refers to measurements of the same quantity that change from measurement to measurement because of quantum effects.

How large are random errors? A bathroom scale may give your weight to within a pound, an error of a fraction of a percent. The Large Interferometric Gravitational Observatory (LIGO) detects gravity waves by comparing two lengths, with an accuracy of one part in 10^{21} or so (it keeps improving). How accurate is one part in 10^{21}? The distance from here to the moon is $3.8 * 10^8$ m. If you measured that distance with LIGO's accuracy, your measurement would have an error of 10^{-13} m. That's about a thousandth of the diameter of an atom.

There is an extremely extensive theory in statistics on how random errors typically behave, and how to include them in an analysis of an experiment. You are not reading a textbook of statistics, so here I am going to supply only a few rules of thumb that are adequate as a first approximation.

Suppose we measure some quantity q. If we repeat the measurement many times, we find that the measurements have some average value a. We also find that the measurements are scattered around the average within some range b, so that *most* measurements are in a range $a \pm b$. Note that I said *most* measurements, not *all* measurements. If you do an experiment enough times, occasionally you will find an *outlier*, a point that is far away from the average. At least some outliers are supposed to be there. If you make many measurements but see no outliers, something is wrong, or something strange is going on. Deciding what to do about outliers is a significant challenge. They're one of the reasons that experiment is an art as well as a science. We will skip treating outliers for now.

Suppose I now make a series of N measurements and take their sum S. In the experimental section of the text, Part Two, you will actually do this. If I repeat the N measurements many times, the average value of S will be approximately Na. However, I can also find the scatter in S around its average value. If the scatter in the result of one measurement is b, as seen in the previous paragraph, the sum of N measurements will usually be in the range $Na \pm \sqrt{N}b$. Note I said 'usually'. Just as there are outliers in the value of individual measurements of q, so also there will be outliers in measurements of S. The sum of N measurements grows linearly with N, but the scatter in the N measurements only grows as the square root of N. The scatter in the sum grows more slowly than the sum itself because the scatter around the average is sometimes positive

1.5. SIGNIFICANT FIGURES

and sometimes negative. The positive and negative outcomes tend to cancel. They do not cancel exactly, but they tend to cancel.

Instead of taking the total S of N measurements, we could calculate the average $\bar{a}$ of the N measurements. The average $\bar{a}$ will generally be in a range around the true average a, namely $\bar{a}$ will generally be in a range

$$\frac{1}{N}(Na \pm \sqrt{N}b) = a \pm \frac{b}{\sqrt{N}}. \tag{1.8}$$

around a. As N increases, the size of the scatter in repeated measurements of $\bar{a}$ becomes smaller. Making multiple measurements reduces the random error in your determination of a. By repeating a measurement many times, and taking an average, you can reduce the random scatter, thus improving your experimental accuracy. The scatter in the sum increases with increasing N; the scatter in the average of the terms being summed decreases with increasing N. Making multiple measurements, however, has no effect on *systematic error*. If you weigh yourself on a bathroom scale many times, always holding two bricks, and take an average, your weight will still on the average come out heavy by the weight of those two bricks.

It is much easier to tell that your experiment has random errors than to tell that your experiment has systematic errors. If you make multiple measurements, and look carefully at the data, you will always have some degree of random scatter. By looking, you can calculate roughly how big the random error is. We'll actually do this in the laboratory experiments.

However, if there is a systematic error, you can't easily tell systematic error is there by looking at your measurements, because all measurements are wrong in the same way. One way to detect systematic errors is to do control experiments whose answer should be known. For example, before weighing yourself on a bathroom scale, look at the scale reading when there is no weight on the scale. That reading should be zero. If the reading is not zero, your scale is reporting a systematic error when it reports your weight. Detecting systematic errors is a major challenge in doing accurate experiments.

As an alternative to adding several experimental numbers, you might instead subtract them. If you do that, experimental error becomes much more important. Suppose we take two measurements of a and subtract one from the other. To make life more interesting, a is gradually changing in time so that the two measurements of a will on the average be a and $a + \epsilon$, respectively. In that case the difference between the two measurements becomes

$$a + \epsilon - a \pm \sqrt{2}b = \epsilon \pm \sqrt{2}b. \tag{1.9}$$

The difference ϵ may be quite small, but the error $\pm\sqrt{b}$ in measuring ϵ has not changed. You are seeing what is known as the error in the small difference of two large numbers. If you take the difference between two large numbers, the difference may be quite small, but the error in measuring that difference will not shrink. The two large numbers a and $a + \epsilon$ can more or less cancel each other, but the errors in measuring them do not cancel. As a result, your measurement of ϵ may be much less accurate than your initial measurement of a.

From these general comments on experimental error we reach some rules for handling numbers under scientific circumstances. First, there is a unique *scientific notation* for writing real numbers. An example is $3.05 \cdot 10^7$. The number has been written with one digit before the decimal point and the remaining digits after the decimal point. The number before the decimal point may be a zero; that zero should always be written, so that it is always obvious where the decimal point is. If I had written a number as $.05 \cdot 10^7$, someone could accidentally misread that number as $0.5 \cdot 10^7$, with potential negative consequences.

The number after the decimal point should only be zero if the number before the decimal point is not zero. Thus $0.305 \cdot 10^7$ is a correct form, but $0.05 \cdot 10^7$ is incorrect. The latter number should be written $0.5 \cdot 10^6$.

Aside: Some of you were brought up to write decimal numbers in the form 3.05E7. Except in computer languages, that's simply wrong.

1.5 Significant Figures

We now reached the oversimplified version of accuracy in measurement, the *significant figure*. The lab will be more accurate. What do we mean by 'significant figure'. Consider $3.05 \cdot 10^7$. We count the number of integers after the decimal point, there being two of these, and say that we know $3.05 \cdot 10^7$ to *two significant figures*, about 1 percent or a bit better. The significant figure *approximation* is that in numbers $3.05 \cdot 10^7$ or $3.0512 \cdot 10^7$ the displayed numbers 3.05 and 3.0512 are known exactly, but we have no idea what the next digit of the number is.

Why is this an approximation? For example, $3.05 \cdot 10^7$ might actually be found from repeated experiments to be $(3.053 \pm 0.002) \cdot 10^7$, in which case we have some information on the digit after the 5; it is probably 1, 2, 3, 4, or 5, most probably 3. If we invoke significant figures, $(3.053 \pm 0.002) \cdot 10^7$ becomes $3.05 \cdot 10^7$. Partial knowledge of the final digit, the terminal 3, is replaced with the approximation that we have no idea what lies beyond the terminal digit. Significant figures are a crude approximation to error analysis, so saying that a number has, e.g., three significant figures is itself an approximation.

As a general rule, if you are doing a calculation with a series of numbers that you know to a certain number of significant figures, the final answer should not have many more than that number of significant figures. It is entirely proper, in doing a numerical calculation on a computer or calculator, to keep all the digits the computer can handle, and only round off to an appropriate number of significant figures at the very end. Your calculator may give you a final answer with twelve digits in it, but if your initial numbers were known to two significant figures then your final answer is also known to about two significant figures, not twelve significant figures. Thus, for example, as a final answer 1.00/7.00 is appropriately written 0.14 or 0.143, but 0.1428571438571428 is wrong.

A reasonable way to think of the significant figure approximation is to write $3.05 \cdot 10^7$ as $3.05? \cdot 10^7$. The ? signifies ignorance. That's not stupidity; that's lack of knowledge. If you do arithmetic with a series of numbers, you can imagine ? as an extra integer in addition to the normal integers 0,1,...9, except that if you add, subtract, multiply, or divide anything by ? the answer is ?. I introduce "?" as a way to show you what is going on; you should not use it outside of this course. As an example, consider 2.17? + 3.5?. If you write that out in tabular form, you have

2.17?
3.5?

When you add 7+?, the answer is ?, so the correct sum is 5.6 or 5.7, depending on whether or not you round 2.17 to 2.2 before or after doing the addition. On the other hand, if you were to subtract two numbers, you might have 3.5? - 3.17? = 0.3? You know the value of the difference of those two numbers to only one significant figure.

Let's consider a few more examples of math with significant figures. Here's another addition:

3.97??
$+$2.6341
6.60??

We applied the rules on ? as 1+? =? and 4+? =?.
Here's multiplication:

2.31?
$\times$4.1000
0.231?
9.24?
9.47??

And here's subtraction:

3.17?? ← known to 1%
3.16?? ← known to 1%
0.01?? ← known to one digit, i.e., 100% possible error

As a closing thought, 'significant figures' are a very crude approximate description of accuracy in measurement. It is possible to introduce pointlessly pedantic rules that purport to calculate exactly how many significant figures are present in a measurement, but those rules miss the point, which is that if we are making measurements with an accuracy of, say one percent, we should not be reporting our answer out to eight or twelve digits after the decimal point.

Associated with the notion of significant figures is the notion of *estimation*, making a rough estimate as to the value of a number you do not know. Estimation is a valuable skill in everyday life. I am reminded of the student newspaper that told students that driving to campus rather than taking a bus would cost an average of $14,000 a year in gasoline. Without using a calculator or paper and pencil, estimate roughly how far the alleged average student must have lived from that campus for this statement to be true.

1.6 Discussion

This is a discussion section, not a summary of the chapter. Summaries are things you should write up for yourself after reading the chapter.

A few closing comments that some of you may will annoying. For some, my remarks will be annoying because you already knew them. For others, my remarks will be annoying because they suggest you need to change your opinions.

Some of you will have heard the word *concepts*, for example in the phrase *I understand the concepts, but I can't solve the problems.* In this context, belief in concepts is a dangerous superstition that will keep you from understanding what is going on. Said more succinctly:

Calculation is everything. Concepts are nothing.

There is a more advanced use of the notion of concepts. In the more advanced use, the concept is a handwaving tool used to help you remember how a calculation works. The concept is not the calculation. The concept is words that tell you where you want to go. For example, *Find a minimum of this function* is a compressed reminder about taking derivatives, testing endpoints, and checking for points where the function is not differentiable. If you are familiar with computer programming, in their advanced use concepts are like subroutine calls.

University is not like High School. The purpose of a University is to tell you what you need to learn and understand. Learning is something you do inside yourself. I can tell you what you need to learn, but you have to do the learning yourself. For the most part, you learn by thinking about what you have been told, working homework problems, and studying the textbook. You are not all the same. Some people more or less can only learn physics by solving lots of problems. For those people, I have supplied a set of problems with worked answers. For other people, problems are something to be gotten out of the way so you can get back to studying and understanding the material. You need to work out for yourself where you lie on this problem-solving continuum. "It's less work" is *not* a path to working out what is correct for you.

This is not the only freshman physics book in the world. Sometimes it helps to head off to a library and read another book on the same topic. Different perspectives may clarify your thinking.

Some number of you went to High Schools in which you had to do homework only three or two days a week. If you are in a university course, you should understand that studying is something you do six or seven days a week. If you aren't interested in studying on weekends, you need to learn a key line for your future career. *Sir, would you like fries with that?* comes immediately to mind. Studying a sophisticated manual skill, for example sweeping with a broom, might also be helpful.

1.7 Worked Problems

1. The period T is the time required for a pendulum – some nuts hanging at the end of a string – to swing back and forth. The period T might depend on the gravitational constant g (units meters/(second^2)), the length L of the string, or the mass m of the nuts. Using dimensional analysis, find the combination of g, l, and m that has dimensions (time)^1. (Hints: g, l, or m may be raised to non-integer powers, e.g., $1/3$ or π.) The objective is to confirm $T = Kg^a \ell^b m^c$. When would this dimensional approach fail?

2. Significant Figures. In the following, the numbers are known to the indicated accuracies. Do these by hand, using neat vertical columns. insert a "?" as the digit after the last digit for each number, and remember that ?, when added to, subtracted from, multiplied by, or divided by or into another number is still "?" Report a correct number to the correct number of significant figures as your final answer. I know you can do these with a calculator. That's not the point of the problem. The objective is for

you to see how significant figures materialize in calculations. (a) 8.7-3.29 (b) 2.1+41.32+1.678. (c) 2.2 × 3.784.

3. As an estimate, how many blades of grass (natural or synthetic) are on the nearest athletic field?

1.8 Homework Problems

1. Where is your library? What are some of the other freshman mechanics books in it?

2. See if you can find the Mechanics books by Kleppner and Kolenkow or by Frautschi and co-authors (for which you want the "Advanced Edition"). These texts give alternative readings similar to ours, i.e., at a university rather than a high school level.

3. Report the dimensions of the following quantities: (a) mass, (b) acceleration, (c) (acceleration^2)/force, (d) mass · velocity/volume.

4. Report the dimensions of the following quantities: (a) mass, (b) acceleration, (c) force/(acceleration^2), (d) area/velocity.

5. Report the dimensions of the following quantities: (a) time, (b) time rate-of-change of acceleration, (c) velocity^2/force, (d) area/mass.

6. Dimensional Analysis. You have a block of wood hanging at the bottom end of a spring. At the top end, the spring is attached to the ceiling. The period T is the time required for the block to move up and down as the spring contracts and expands. The period T might depend on the spring constant k (units kilograms/(second^2)) or the mass m of the block of wood. Find the combination of k and m that has dimensions (time)^1. (Hints: The constants may be raised to non-integer powers, e.g., 1/3 or π.) The objective is to confirm $T = K k^a m^b$.

7. Significant Figures. In the following, the numbers are known to the indicated accuracies. Do these by hand, using neat vertical columns. Insert a "?" as the digit after the last digit for each number, and remember that ? when added to, subtracted from, multiplied by, or divided by or into another number is still "?" Report a correct, significant number as your final answer. I know you can do these with a calculator. That's not the point of the problem. The objective is for you to see how significant figures materialize in calculations. (a) 81.7-3.29 (b) 1.9+80.73+2.989. [Do this both by rounding at the end and also by rounding at the beginning.] (c) 1.7 x 2.48.

8. Significant Figures. In the following, the numbers are known to the indicated accuracies. Do these by hand, using neat vertical columns. Insert a "?" as the digit after the last digit for each number, and remember that ? when added to, subtracted from, multiplied by, or divided by or into another number is still "?" Report a correct, significant number as your final answer. I know you can do these with a calculator. That's not the point of the problem. The objective is for you to see how significant figures materialize in calculations. (a) 8.7-3.29. (b) 2.1+41.32+1.678. [Do this both by rounding at the end and also by rounding at the beginning.] (c) 2.2 x 3.784. (d) 5.682/4.2.

9. Significant Figures. In the following, the numbers are known to the indicated accuracies. Do these by hand, using neat vertical columns. Insert a "?" as the digit after the last digit for each number, and remember that ? when added to, subtracted from, multiplied by, or divided by or into another number is still "?" Report a correct, significant number as your final answer. (a) 37.8-2.51. (b) 2.0+37.81+2.6413. [Do this both by rounding at the end and also by rounding at the beginning.] (c)3.1 x 9.15. (d) 9.652/3.1.

10. Estimate how many of the molecules of nitrogen in your lungs, at the moment, at one time passed through the lungs of King George III of England. To keep things simple, ignore mechanisms that move nitrogen into and out of the atmosphere.

11. From the incident power of sunlight on the earth (round number, top of atmosphere, 1360 watts per square meter) estimate the power output of the Sun.

1.9 Solutions to Worked Problems

1. We propose that $g^a L^b m^c \sim T$. If we replace the units with their dimensions, we obtain
$$(\ell t^{-2})^a \ell^b m^c \sim t^1 \ell^0 m^0 \tag{1.10}$$
Collecting powers, we have
$$t^{-2a} \ell^{a+b} m^c \sim t^1 \ell^0 m^0 \tag{1.11}$$
In order for this relationship to be true, ℓ, t, and m must be raised to the same power on each side, giving
$$a + b = 0, \tag{1.12}$$
$$-2a = 1, \tag{1.13}$$
$$c = 0 \tag{1.14}$$
leading to
$$a = -1/2, \tag{1.15}$$
$$b = 1/2, \tag{1.16}$$
$$c = 0 \tag{1.17}$$
and therefore
$$T \sim \left(\frac{L}{g}\right)^{1/2}. \tag{1.18}$$

When would this approach fail? The requirement that powers on the two sides of the equation be the same gives three equations, one each for t, ℓ, and m. But if the unknown quantity is a power of four variables $W^a X^b Y^c Z^d$, we have four unknowns, but only three equations, and cannot solve. The approach also fails if two of the unknown quantities, for example X and Y, have the same dimensions, because then (X/Y) has dimensions unity. We can then insert any number of additional factors X/Y in the solution without perturbing the dimensional analysis.

2. We have
(a)
   8.7?
   −3.29?
   ─────
   5.5?

(b)
   2.1?
   41.32?
   +1.678?
   ───────
   45.0?

(c)
   3.784?
   ×2.2??
   ──────
   ????
   ????
   7568
   7568
   ──────
   8.2?????

3. Of course, a reasonable answer depends on the size of your athletic fields. However, for a place I once taught, they were 100 by 300 yards, or forty million square inches. The lawn – I went out and looked – had twenty blades of grass to the square inch or eight hundred million blades of grass. In round numbers, a billion blades of grass.

This page reserved for your notes.

Chapter 2

Vectors

2.1 Introduction

We now turn to vectors, which are one of the basic mathematical tools needed to describe classical mechanics. As a historical matter, Isaac Newton published his work in 1687, while Josiah Willard Gibbs, the inventor of vectors as a mathematical structure, did not unveil his work until the late nineteenth century[1]. However, vectors are the natural way to describe many things in classical mechanics, so we introduce them here and use them throughout the rest of the book. This is a physics book, not a math book, so in some cases I will not give proofs for mathematical results.

This Chapter has two parts, dealing with vectors and their addition and with the scalar (dot) product of two vectors. The vector (cross) product of two vectors will be discussed in detail later, just before we treat angular momenta. We briefly note the outer (tensor) product of two vectors.

First, a matter of notation. A simple number, or an algebraic symbol that stands for a simple number, is a *scalar*. A typical symbol for a scalar is g, the "gravitational acceleration". There are several standard ways to write a vector, notably the boldface V and the letter symbol with the superposed vector arrow $\vec{V}$. The former notation is common in books, but extremely inconvenient in lectures, because chalkboards do not cooperate in letting you make boldface characters easily. Instead, in lectures and lecture notes one more typically sees $\vec{V}$.

So, what is a vector? Mathematicians started out with a very simpleminded idea of what a vector is, and then refined and generalized the idea. The simplest answer is that a vector is something with a magnitude and a direction. You will sometimes see the word *magnitude* replaced with the word *length*, but "length" is misleading. Many vectors do not have length as their dimension. For example, if a car is headed due east on the interstate at 100 km/h, the magnitude of its velocity vector is 100 km/h and the direction is due east, but 100 km/h does not have units of length. As an improvement on magnitude and direction, we can say that a vector is a quantity that has components. We'll get back in a bit to what components are. Finally, one might say that a vector is an ordered list of numbers or algebraic symbols. By *ordered*, we mean that the lists (m, n, p, q) and (q, m, n, p) are not equal. The same symbols appear in both lists, but the two lists have their characters written down in different sequences.

Returning to the velocity vector of the car, note that the vector has a magnitude and direction, but it does not have a location. The vector would be the same whether the car were speeding through Springfield, Auburn, or Boston. We can, however, generalize the notion of vector by introducing the *vector field*. In a typical vector field, each point in space has associated with it a vector. Almost all of you have actually seen an example of a vector field. Imagine a weather map, on which at different points on the map there are arrows corresponding to wind direction and wind speed. Those wind vectors are pieces of a vector field, because each of them corresponds to a vector, the wind velocity (the wind speed and the wind direction) at a particular weather station. In addition, we will eventually encounter vectors that do have a well-defined location. For example, the position vector, the vector from the origin to the location of a particle, always has its tail at the origin.

Two vectors are equal to each other if they have the same magnitude and the same direction. There are alternative ways of saying the two vectors are equal to each other. These alternative ways are generalizations of 'same magnitude and same direction'.

We will regularly draw vectors. A typical vector appears in Figure 2.1. The vector A points from point a to point b. a is the tail of the vector A, while b is the head of the vector A. The magnitude of the vector A is written A or $|A|$. Those two ways of writing the magnitude of the vector A have exactly the same meaning. You may recognize $|\cdots|$ as the symbol for the magnitude of a real or complex number. The similarity is not a coincidence.

Figure 2.1: A vector A with tail at a and head at b

One can do mathematical processes with vectors. Some processes are easier to draw if the vector is confined to the plane of the blackboard, but vectors aren't always confined to a plane.

We write the sum of two vectors as $A + B$. The sum of two vectors is itself a vector. We represent $A + B$ graphically by translocating one of the two vectors until the head of one vector and the tail of the other vector are at the same point. We can always draw the sum of two vectors as lying in a plane, because the sum diagram has three points: (i) the tail of the first vector, (ii) the head of the first vector and the tail of the second vector, and (iii) the head of the second vector, while three points define a plane.

The vector sum may be drawn as show in Figure 2.2. Vector addition is commutative, meaning that

$$A + B = B + A. \tag{2.1}$$

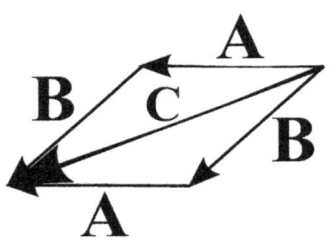

Figure 2.2: Addition of two vectors $A + B$ is commutative. The two vectors A and B define a parallelogram.

We may also introduce C as a third vector, which happens to be the sum of the two vectors A and B. In this case, we can write

$$A + B = C \tag{2.2}$$

and

$$B + A = C. \tag{2.3}$$

Figure 2.2 demonstrates two mathematical facts: (i) The sum of two vectors is commutative, and (ii) Two vectors (in Figure 2.2 the two vectors A and B) define a parallelogram. We return to the latter point when we discuss the vector (cross) product of two vectors.

We may multiply a vector by a scalar. If s is a scalar, then sA is a vector having the same direction as the vector A but having magnitude sA. So for example, if we take a vector A and multiply it by three, $3A$ is a vector that points in the same direction as A but has three times the magnitude, as seen in Figure 2.3.

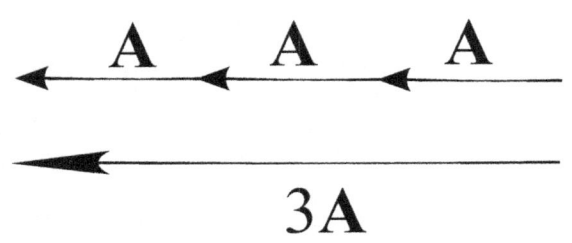

Figure 2.3: Multiplication of a vector by a scalar: $3A$ is equal to the sum $A + A + A$.

Having introduced the notion of multiplying a vector by a scalar, we can now introduce the unit vector. The unit vector for a vector A is written $\hat{A}$. The unit vector $\hat{A}$ points in the same direction as the vector A, but the unit vector has magnitude unity. We may readily calculate the unit vector corresponding to any vector, namely

$$\hat{A} = \frac{A}{|A|}. \tag{2.4}$$

A unit vector is obtained from the regular vector by multiplying the regular vector by one over the regular vector's magnitude (equivalently, the regular vector is divided by its magnitude).

We may represent any vector other than $\mathbf{0}$ in terms of its magnitude and its unit vector, namely

$$A = A\hat{A}. \tag{2.5}$$

In this equation, the magnitude A is always a positive number. The direction of the vector is contained in the unit vector $\hat{A}$.

2.1. INTRODUCTION

What is the unit vector corresponding to the zero vector **0**? You might first ask if there actually is a zero vector, but it's very clear how to construct a zero vector from any other vector:

$$\mathbf{0} = \mathbf{A} - \mathbf{A}, \tag{2.6}$$

which has the property

$$\mathbf{A} + \mathbf{0} = \mathbf{A}. \tag{2.7}$$

The magnitude of **0** is clearly zero; the vector does not go anywhere or point in any particular direction. Division by zero is not allowed in normal mathematics, so from the zero vector **0** you cannot construct a unit vector.

We now turn to representing a vector, in Cartesian coordinates, in terms of its basis vectors. The discussion here refers to a vector that sits in three dimensional space, such as the position vector $\mathbf{R}$, the vector from the origin $\mathcal{O}$ to a point of interest. The basis vectors are a list of unit vectors that have the properties (1) the basis vectors are perpendicular to each other (we will come back to *perpendicular* in a moment) and (2) they are *complete*, *complete* meaning that any vector whatsoever can be written in terms of the basis vectors and a list of scalar *components*. I've now introduced a bunch of terms that have to be explained, but it is easiest to start with the names of the terms.

Figure 2.4: The vector sum of the vectors $\mathbf{A}$ and $-\mathbf{A}$; the two vectors superpose, so that only their heads are clearly distinct. The zero vector, from $\mathbf{0} = \mathbf{A} - \mathbf{A}$, is the (invisible) vector having length zero.

For vectors in three dimensional space using Cartesian coordinates, the basis vectors are a set of three unit vectors that point along the x, y, and z coordinate axes. A picture may help. In Figure 2.5, I drew the three unit vectors as having their tails at the origin, the point where the x, y, and z axes intersect. Vectors do not have locations. Drawing the vectors as seen in the Figure is a point of artistic convenience, used here to emphasize that the three unit vectors are each parallel to one of the three coordinate axes.

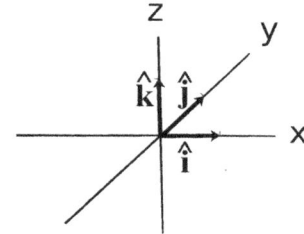

Figure 2.5: The three unit vectors $\hat{\imath}, \hat{\jmath},$ and $\hat{k}$.

The unit vectors have a standard notation. Actually, they have three standard notations. We will stay with the first of these, but in your professional careers you are likely to encounter the other two, so you should be aware of them. For the unit vector parallel to the x-axis we write $\hat{\imath}$, $\hat{x}$, or e_1. For the unit vector parallel to the y-axis we write $\hat{\jmath}$, $\hat{y}$, or e_2. For the unit vector parallel to the z-axis we write $\hat{k}$, $\hat{z}$, or e_3. To represent the three unit vectors, we will use the trio $\hat{\imath}$, $\hat{\jmath}$, and $\hat{k}$. The trio $\hat{x}$, $\hat{y}$, and $\hat{z}$ is self-explanatory. Finally, the trio e_1, e_2, and e_3 is encountered in some older references.

How do the unit vectors enter the description of the general vector $\mathbf{R}$? First, any vector can be written as a sum of three vectors, namely a vector $\mathbf{X}$ parallel to the x-axis, plus a vector $\mathbf{Y}$ parallel to the y-axis, plus a vector $\mathbf{Z}$ parallel to the z-axis, or

$$\mathbf{R} = \mathbf{X} + \mathbf{Y} + \mathbf{Z}, \tag{2.8}$$

as seen here in Figure 2.6. I am calling the full vector $\mathbf{R}$, as if it were a position vector, but $\mathbf{R}$ is just an abstract symbol that could stand equally well for a velocity or a force, position, velocity, and force all being vectors.

The displacements $\mathbf{X}$, $\mathbf{Y}$, and $\mathbf{Z}$ can be taken in any order, because vector addition is commutative. However, from equation 2.5, the three vectors $\mathbf{X}$, $\mathbf{Y}$, and $\mathbf{Z}$ can each be written as the product of a magnitude and a unit vector. By construction, these three vectors are parallel to the three fundamental coordinate axes x, y, and z, so their unit vectors are the unit vectors we just introduced. In other words

$$\hat{\mathbf{X}} = \hat{\imath}, \tag{2.9}$$
$$\hat{\mathbf{Y}} = \hat{\jmath}, \tag{2.10}$$
$$\hat{\mathbf{Z}} = \hat{k}. \tag{2.11}$$

The magnitudes of the three vectors $\boldsymbol{X}$, $\boldsymbol{Y}$, and $\boldsymbol{Z}$ are the trio (X, Y, Z). X, Y, and Z are also the Cartesion coordinates of a point, namely the location of the head of the vector $\boldsymbol{R}$, if the tail of $\boldsymbol{R}$ is located at the origin. Combining equations 2.8 and 2.5, the original vector $\boldsymbol{R}$ may therefore be written as

$$\boldsymbol{R} = X\hat{\boldsymbol{i}} + Y\hat{\boldsymbol{j}} + Z\hat{\boldsymbol{k}}. \tag{2.12}$$

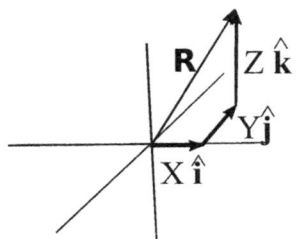

Figure 2.6: A vector as a sum of unit vectors and scalars.

The three magnitudes X, Y, and Z are the *components* of the vector $\boldsymbol{R}$. If the vector $\boldsymbol{R}$ is specified, then its components are determined. If the vector's components X, Y, and Z are given, then equation 2.12 can be used to calculate the vector $\boldsymbol{R}$.

The same vector can also be written

$$\boldsymbol{R} = R_x\hat{\boldsymbol{i}} + R_y\hat{\boldsymbol{j}} + R_z\hat{\boldsymbol{k}}, \tag{2.13}$$

I've introduced a new and very useful bit of notation. The x, y, and z components of the vector are now written R_x, R_y and R_z, respectively. The components are labelled with subscripts x, y, and z. The second notation is especially useful when you are dealing with several different vectors at the same time, namely when you write a component as A_b, the A tells you which vector you are considering while b tells you which component is under consideration.

We could write the vector $\boldsymbol{R}$ in the form of equation 2.12, but we could equally well write

$$\boldsymbol{R} = (X, Y, Z), \tag{2.14}$$

because if we know the three components (X, Y, Z) of the vector we know exactly what the vector $\boldsymbol{R}$ is. The representations $X\hat{\boldsymbol{i}} + Y\hat{\boldsymbol{j}} + Z\hat{\boldsymbol{k}}$ and (X, Y, Z) are *isomorphic*, meaning for each vector in one representation there is exactly one vector in the other representation, and, furthermore, any mathematical statement that is true about $X\hat{\boldsymbol{i}} + Y\hat{\boldsymbol{j}} + Z\hat{\boldsymbol{k}}$ has an exactly corresponding statement that is true about (X, Y, Z), and *vice versa*. (For far more on isomorphic relations, the useful search term is *abstract algebra*).

We have now returned to our original statement that a vector can be viewed as an ordered list of numbers. For a vector in three-dimensional space, a list of three numbers is correct. However, that count can be generalized. Four-component vectors are used in special and general relativity. Many-component vectors are of great practical use in some computer languages.

Let us return to the vector $\boldsymbol{C}$. We wrote $\boldsymbol{C} = \boldsymbol{A} + \boldsymbol{B}$. However, we could equally well write this equation in terms of the components of the three vectors. To do that, we use the systematic notation for the x, y, and z components of the three vectors. The convenient notation is to use x, y, and z as subscripts, so that the x, y, and z components of $\boldsymbol{C}$ are written C_x for the x-component, C_y for the y-component, and C_z for the z-component, and similarly for $\boldsymbol{A}$ and $\boldsymbol{B}$.

We now make a basic statement about vector equations. A vector equation is simply a shorthand way of writing the component equations. When we write $\boldsymbol{C} = \boldsymbol{A} + \boldsymbol{B}$, what we are saying is

$$C_x = A_x + B_x, \tag{2.15}$$
$$C_y = A_y + B_y, \tag{2.16}$$
$$C_z = A_z + B_z. \tag{2.17}$$

The information in the vector equation $\boldsymbol{C} = \boldsymbol{A} + \boldsymbol{B}$ and in the three component equations are exactly the same. Indeed, if one is confronted with a complicated vector equation, it sometimes helps to replace the vector equation with the three component equations and work on the three component equations separately. When we say *component equations* we are referring to the equations that result from taking the scalar products of the underlying vector equation with each of the three Cartesian basis vectors. It may be the case that any or all of the component equations includes vector components lying along several different axes, so that instead of the above equations might find that C_x depends in some way on A_y and B_z.

2.2. VECTOR MULTIPLICATION

How do we know that the equations 2.17 are correct? There is a simple geometric demonstration, seen in Figure 2.7, based on the fact that vector addition is commutative, and that a vector has a magnitude and a direction but not a location. I display the geometric demonstration in two-dimensional space because it's a little easier to see what is going on. An extension to three-dimensional spaces is a homework exercise.

The magnitude (length) of a vector is readily written in terms of the vector's components, *viz*

$$|\mathbf{R}| = \left(R_x^2 + R_y^2 + R_z^2\right)^{1/2}. \tag{2.18}$$

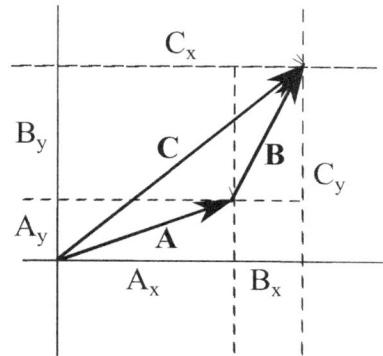

Figure 2.7: Vector component addition, demonstrating that if $\mathbf{C} = \mathbf{A} + \mathbf{B}$ then $C_x = A_x + B_x$ and $C_y = A_y + B_y$.

The scalars (R_x, R_y, R_z) are the components of the original vector $\mathbf{R}$. Equation 2.18 is Euclid's Theorem in three dimensions.

We have now discussed adding vectors. Subtraction and addition are the same process, with the signs of one set of components flipped.

Having said that a vector can be *represented* as a list of its components, there is also a sense in which the vector has an independent real existence. Suppose that you take the Cartesian coordinate system, relocate the origin, and rotate the coordinate axes so that they point in a new set of directions. The components of the vector will all change, in a way determined by the displacement and the representations of the old unit vectors in terms of the new unit vectors. However, the vector itself has not changed. If we started with an automobile going 65 MPH due east, it might have had $\mathbf{v} = 65\hat{\mathbf{E}}$ or $(65, 0, 0)$ as its initial velocity vector, $\hat{\mathbf{E}}$ being the unit vector for east in an east, north, up $(\hat{\mathbf{E}}, \hat{\mathbf{N}}, \hat{\mathbf{k}})$ coordinate system. If we now switch to the northeast, southeast, up $(\hat{\mathbf{NE}}, \hat{\mathbf{SE}}, \hat{\mathbf{k}})$ coordinate system, the car's velocity becomes $\mathbf{v} = 65/\sqrt{2}\hat{\mathbf{NE}} + 65/\sqrt{2}\hat{\mathbf{SE}}$ or $(65/\sqrt{2}, 65/\sqrt{2}, 0)$. The car has not changed its velocity; only the *representation* of the velocity has changed. The figures above showing vectors as arrows being added are all correct as statements about vectors, no matter which coordinate system you choose to use.

2.2 Vector Multiplication

We now turn to vector multiplication. There are actually four sorts of multiplication involving vectors. We have already mentioned multiplying a vector $\mathbf{A}$ by a scalar s to form a product $s\mathbf{A}$. If we have two vectors $\mathbf{A}$ and $\mathbf{B}$, we can form with them three additional, completely different products, namely

- We can form the *inner* product $\mathbf{A} \cdot \mathbf{B}$.

- We can form the *vector* or *cross* product $\mathbf{A} \times \mathbf{B}$.

- We can form the *outer* product $\mathbf{A} \otimes \mathbf{B}$.

The inner product is also called the *scalar* or *dot* product. The outer product is also called the *tensor* or *dyadic* product. The outcome of a scalar (dot) product is a number. The outcome of a vector (cross) product is a vector. The outcome of an outer (tensor) product is a matrix.

here we consider the product of a vector and a scalar and the scalar (dot) product of two vectors. Consider first multiplying a vector $\mathbf{A}$ by a scalar s. The multiplication changes the magnitude of $\mathbf{A}$ but does not change its direction, as given by its unit vector. We say

$$s\mathbf{A} = sA_x\hat{\mathbf{i}} + sA_y\hat{\mathbf{j}} + sA_z\hat{\mathbf{k}}, \tag{2.19}$$

or equivalently

$$s(A_x, A_y, A_z) = (sA_x, sA_y, sA_z). \tag{2.20}$$

If you calculate the magnitude of $\mathbf{A}$ and of $s\mathbf{A}$, you'll find that $|s\mathbf{A}|$ is indeed s times as large as $|\mathbf{A}|$. If you calculate the unit vectors of $\mathbf{A}$ and of $s\mathbf{A}$, you will find that they are the same.

We now turn to the scalar product of two vectors. The product is defined by its effect on the basis vectors. The scalar product of each basis vector with itself is unity. The scalar product of two different basis vectors is zero. For the scalar product, we thus have

$$\hat{i} \cdot \hat{i} = 1, \tag{2.21}$$

$$\hat{j} \cdot \hat{j} = 1, \tag{2.22}$$

$$\hat{k} \cdot \hat{k} = 1, \tag{2.23}$$

$$\hat{i} \cdot \hat{j} = 0, \tag{2.24}$$

$$\hat{i} \cdot \hat{k} = 0, \tag{2.25}$$

$$\hat{j} \cdot \hat{i} = 0, \tag{2.26}$$

$$\hat{j} \cdot \hat{k} = 0, \tag{2.27}$$

$$\hat{k} \cdot \hat{i} = 0, \tag{2.28}$$

$$\hat{k} \cdot \hat{j} = 0. \tag{2.29}$$

The scalar product is commutative and distributive, meaning for three vectors $\boldsymbol{A}$, $\boldsymbol{B}$, and $\boldsymbol{C}$, we may write

$$\boldsymbol{A} \cdot \boldsymbol{B} = \boldsymbol{B} \cdot \boldsymbol{A}, \tag{2.30}$$

for the commutative law, and

$$\boldsymbol{A} \cdot (\boldsymbol{B} + \boldsymbol{C}) = \boldsymbol{A} \cdot \boldsymbol{B} + \boldsymbol{A} \cdot \boldsymbol{C}. \tag{2.31}$$

for the distributive law. These rules apply to any vector, for example the vector $3\hat{i}$.

In taking a scalar product, the vectors should be written in terms of the basis vectors and the scalar constants multiplying them. When the scalar product is taken, the above rules specify what is to be done with the basis vectors. Scalar constants are multiplied out. For example, first using the distributive rule

$$3\hat{i} \cdot (4\hat{i} + 5\hat{k}) = 3 \cdot 4\hat{i} \cdot \hat{i} + 3 \cdot 5\hat{i} \cdot \hat{k}. \tag{2.32}$$

Applying equations 2.29

$$3\hat{i} \cdot (4\hat{i} + 5\hat{k}) = 12 \cdot 1 + 15 \cdot 0 \tag{2.33}$$

so that $3\hat{i} \cdot (4\hat{i} + 5\hat{k}) = 12$. In this example, the scalar product yields a positive number, but in general scalar products can yield either positive or negative numbers. The reason that scalar products can give a negative number as an answer will become clearer when we discuss projections.

If you take the scalar product of two vectors $\boldsymbol{A}$ and $\boldsymbol{B}$, where $\boldsymbol{A} = A_x\hat{i} + A_y\hat{j} + A_z\hat{k}$ and $\boldsymbol{B} = B_x\hat{i} + B_y\hat{j} + B_z\hat{k}$, the distributive law gives you nine terms, six of which are zero, leading to

$$\boldsymbol{A} \cdot \boldsymbol{B} = A_xB_x + A_yB_y + A_zB_z \tag{2.34}$$

The detailed steps are left for homework. As a special case, the dot product can be used to calculate the magnitude of a vector, namely

$$|\boldsymbol{A} \cdot \boldsymbol{A}|^{1/2} = [A_xA_x + A_yA_y + A_zA_z]^{1/2} \equiv |\boldsymbol{A}| \tag{2.35}$$

or

$$|\boldsymbol{A} \cdot \boldsymbol{A}|^{1/2} = A. \tag{2.36}$$

The square root always gives a positive answer, so $|\boldsymbol{A} \cdot \boldsymbol{A}|^{1/2}$ gives the magnitude A of the vector $\boldsymbol{A}$.

Projections are closely related to scalar products. By a projection of a vector $\boldsymbol{B}$ onto a vector $\boldsymbol{A}$, we mean the extent to which the vector $\boldsymbol{B}$ has a component parallel to the vector $\boldsymbol{A}$. For two vectors, we can draw Figure 2.8 to show the projection of $\boldsymbol{B}$ onto $\boldsymbol{A}$.

2.2. VECTOR MULTIPLICATION

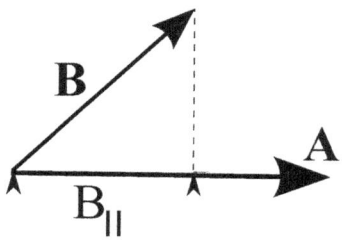

Figure 2.8: Projection $B_\parallel$ of B onto A. Arrows perpendicular to vector A delineate the range of the projection.

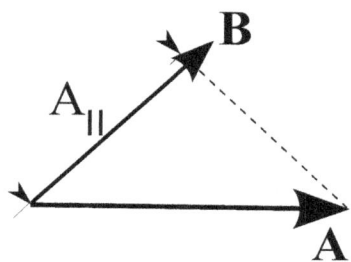

Figure 2.9: Projection $A_\parallel$ of A onto B. Arrows perpendicular to vector B delineate the range of the projection.

In drawing this Figure we used the important rule that a vector has components, but does not have a location. We there moved the vector B until its tail was at the same location as the tail of the vector A. These two vectors, indeed any two vectors, are defined completely by three points, namely the one point at which both of their tails are located and the two points at which their two heads are located. Three points define a plane, so we can always draw two vectors on a sheet of paper or computer screen as shown.

The projection of B onto A is indicated on the figure as $B_\parallel$. It's a scalar. Denoting the angle between B and A as θ, we see that

$$B_\parallel = |B|\cos\theta \equiv B\cos(\theta). \tag{2.37}$$

We can equally construct a quantity $A_\parallel$, shown in Figure 2.9, with

$$A_\parallel = A\cos\theta. \tag{2.38}$$

It does not matter which way you take to be the positive direction for θ, because $\cos(\theta) = \cos(-\theta)$.

Now we come to a math result not derived here. You may look up a proof that

$$\boldsymbol{B}\cdot\boldsymbol{A} = |\boldsymbol{B}||\boldsymbol{A}|\cos\theta \equiv AB\cos\theta, \tag{2.39}$$

which tells us that

$$\boldsymbol{B}\cdot\boldsymbol{A} = AB_\parallel \tag{2.40}$$

and

$$\boldsymbol{B}\cdot\boldsymbol{A} = BA_\parallel. \tag{2.41}$$

Now suppose you reach into those two equations and replace A or B with its unit vector, and remember that the magnitude of a unit vector is unity. In that case, these two equations become

$$\boldsymbol{B}\cdot\hat{\boldsymbol{A}} = B_\parallel \tag{2.42}$$

and

$$\hat{\boldsymbol{B}}\cdot\boldsymbol{A} = A_\parallel. \tag{2.43}$$

When you take the dot product of a unit vector $\hat{\boldsymbol{A}}$ with some other vector B, the result is the projection $B_\parallel$ of B onto A, which is the extend to which the vector B points in the direction of the vector A.

As a specific application, $\hat{\boldsymbol{i}}\cdot\boldsymbol{R} = R_x$, which is a scalar, and similarly for the other Cartesian basis vectors, so that

$$\boldsymbol{R} = (\hat{\boldsymbol{i}}\cdot\boldsymbol{R})\hat{\boldsymbol{i}} + (\hat{\boldsymbol{j}}\cdot\boldsymbol{R})\hat{\boldsymbol{j}} + (\hat{\boldsymbol{k}}\cdot\boldsymbol{R})\hat{\boldsymbol{k}}. \tag{2.44}$$

This application leads directly to the *direction cosines* of a vector. There are three direction cosines, conventionally denoted $\cos(\gamma_1)$, $\cos(\gamma_2)$, and $\cos(\gamma_3)$. The three direction cosines are the cosines of the angles γ_1, γ_2, and γ_3 between the three unit vectors and the vector of interest, so that

$$R_x = R\cos(\gamma_1), \tag{2.45}$$
$$R_y = R\cos(\gamma_2), \tag{2.46}$$
$$R_z = R\cos(\gamma_3). \tag{2.47}$$

Any two vectors define a plane, so each γ_i lies in a plane, the plane formed by its corresponding basis vector and the vector of interest.

We now reach the vector (cross) and tensor products. The vector product of two vectors is itself a vector. The two vectors in the product necessarily define a plane. The product vector is perpendicular to that plane, in a direction to be discussed in a later chapter. The magnitude of the vector product's vector is equal to the area of the parallelogram formed by the two vectors being multiplied. The tensor (outer) product of two vectors is a 3×3 matrix. The components of the matrix are nine numbers M_{ij}, where i and j can each be equal to x, y, or z. For the outer product M of two vectors A and B, we write

$$\boldsymbol{A} \otimes \boldsymbol{B} = \overset{\leftrightarrow}{M} \tag{2.48}$$

with $M_{ij} = A_i B_j$.

2.3 The Curie Principle

The Curie Principle tells us that the quantities on the two sides of an equation must be of the same *order*. The Curie in question is Pierre Curie, Madame Marie Curie's husband and co-discoverer of Radium and Polonium. The quantities on the two sides of an equation must both be scalars, or both be vectors, or both be matrices. If they are vectors or matrices, they must have the same dimensions as each other. For two vectors to be equal, they must have the same number of elements. For two matrices to be equal, they must have the same number of rows and also the same number of columns. Thus, valid equations involving scalars, vectors, or matrices could include

$$A = B \tag{2.49}$$

for two scalars,

$$\boldsymbol{A} = \boldsymbol{B} \tag{2.50}$$

for two vectors, and

$$\overset{\leftrightarrow}{A} = \overset{\leftrightarrow}{B} \tag{2.51}$$

for two matrices. Two vectors or two matrices are equal if their corresponding components are equal.

On the other hand, the equation

$$\boldsymbol{A} = 3 \tag{2.52}$$

cannot possibly be true, no matter what vector is represented by the symbol $\boldsymbol{A}$, because this invalid equation claims that a vector is equal to a scalar.

How does the Curie principle work? At one time, an explanation was difficult to make, but if you are even marginally familiar with computer programming the answer is clear. Suppose we have two vectors (A, B, C) and $(1, 2, 3)$. If we instruct the computer in some generalized way that $(A, B, C) = (1, 2, 3)$, we mean that it should load the memory location corresponding to the variable A with the value 1, load the memory location corresponding to the variable B with the value 2, and load the memory location corresponding to the variable C with the value 3. That would be even clearer if the first vector were written $(A(1), A(2), A(3))$ because then the three components of $\boldsymbol{A}$ would then be loaded with the three components of the second vector.

If you encountered $(A(1), A(2), A(3)) = 3$ you have three components of $\boldsymbol{A}$, labelled 1, 2, 3, but only one number to insert into one of them. What is the poor defenseless computer supposed to do? (Hint: Crash!) That's the Curie principle. If you set two quantities equal to each other, they must each be a scalar, or each be a vector with the same number of components, or each be a matrix with the same dimensions, because otherwise, well, what is supposed to be equal to what?

There is a peculiar exception arising from weak notation. You may encounter $\boldsymbol{V} = 0$, which appears to be setting a vector and a scalar equal to each other. What is meant here by $\boldsymbol{V} = 0$ is $\boldsymbol{V} = \boldsymbol{0}$, where the zero vector $\boldsymbol{0}$ is a vector whose components are all equal to zero.

2.4 Discussion

This chapter has presented a long series of definitions with only a few examples. Definitions are names of things; they are not the things' properties. This chapter presents are series of tools that will be useful in the rest of the book, but the real physics is yet to come. After you use the definitions enough times, they will fit into memory. As an intermediate step, careful notes that you can reference may be of some help.

Note my cursory and obscure reference earlier in the chapter to *abstract algebra*. At a series of point in the book, we will briefly touch on the fringes of a much larger and more complex topic. I will point out that the topic is there, but like Moses and the Promised Land, all you will see here is a brief reference and perhaps a keyword leading to further reading.

[1] Gibbs actually did not unveil vectors, but he did not object when student E. B. Wilson produced the book *Vector Analysis...Founded Upon the Lectures of J. Willard Gibbs*.

2.5 Worked Problem

1. Consider the vectors $\boldsymbol{A} = 2\hat{i} + 5\hat{j} - 1\hat{k}$, $\boldsymbol{B} = -3\hat{i} + 1\hat{j} + 2\hat{k}$, and $\boldsymbol{C} = +4\hat{i} - \hat{j}$. Compute (i) $\boldsymbol{A} + \boldsymbol{B}$, (ii) $\boldsymbol{A} - \boldsymbol{B}$, (iii) $\boldsymbol{A} \cdot \boldsymbol{C}$, (iv) $|\boldsymbol{A}|$, (v) $|\boldsymbol{C}|$, (vi) From (iii), (iv), and (v), find the angle between $\boldsymbol{A}$ and $\boldsymbol{C}$.

2.6 Homework

1. Calculate the magnitudes of $\boldsymbol{A}$ and of $s\boldsymbol{A}$. Is $|s\boldsymbol{A}|$ indeed s times as large as $|\boldsymbol{A}|$?

2. Calculate the unit vectors of $\boldsymbol{A}$ and of $s\boldsymbol{A}$. Are they the same?

3. Prove that $\boldsymbol{A} \cdot \boldsymbol{B} = A_x B_x + A_y B_y + A_z B_z$. Write out the intermediate nine-term step.

4. Starting from equation 2.13, prove $\hat{\boldsymbol{i}} \cdot \boldsymbol{R} = R_x$.

5. Prove that $\boldsymbol{R} = (\hat{\boldsymbol{i}} \cdot \boldsymbol{R})\hat{\boldsymbol{i}} + (\hat{\boldsymbol{j}} \cdot \boldsymbol{R})\hat{\boldsymbol{j}} + (\hat{\boldsymbol{k}} \cdot \boldsymbol{R})\hat{\boldsymbol{k}}$, which may also be written $\hat{\boldsymbol{i}}(\hat{\boldsymbol{i}} \cdot \boldsymbol{R}) + \hat{\boldsymbol{j}}(\hat{\boldsymbol{j}} \cdot \boldsymbol{R}) + \hat{\boldsymbol{k}}(\hat{\boldsymbol{k}} \cdot \boldsymbol{R})$.

6. Consider the vectors $\boldsymbol{A} = 3\hat{i} + 4\hat{j} - 2\hat{k}$, $\boldsymbol{B} = -2\hat{i} + 2\hat{j} + 3\hat{k}$, and $\boldsymbol{C} = +7\hat{i} - 3\hat{j}$. Compute (i) $\boldsymbol{A} + \boldsymbol{B}$, (ii) $\boldsymbol{A} - \boldsymbol{B}$, (iii) $\boldsymbol{A} \cdot \boldsymbol{C}$, (iv) $|\boldsymbol{A}|$, (v) $|\boldsymbol{C}|$, and (vi) from (iii), (iv), and (v), find the angle between $\boldsymbol{A}$ and $\boldsymbol{C}$.

7. Consider the vectors $\boldsymbol{A} = -1\hat{i} + 7\hat{j} - 4\hat{k}$, $\boldsymbol{B} = -6\hat{i} - 9\hat{j} + 8\hat{k}$, and $\boldsymbol{C} = +\hat{i} + \hat{j}$. Compute (i) $\boldsymbol{A} + \boldsymbol{B}$, (ii) $\boldsymbol{A} - \boldsymbol{B}$, (iii) $\boldsymbol{A} \cdot \boldsymbol{C}$, (iv) $|\boldsymbol{A}|$, (v) $|\boldsymbol{C}|$, and (vi) from (iii), (iv), and (v), find the angle between $\boldsymbol{A}$ and $\boldsymbol{C}$.

8. Consider the three vectors $\boldsymbol{A} = -2\hat{i} + 3\hat{j} - 4\hat{k}$, $\boldsymbol{B} = 4\hat{i} - 6\hat{k}$, and $\boldsymbol{C} = \hat{i} + 6\hat{j}$. (i) Compute $\boldsymbol{A} + \boldsymbol{B}$. (ii) What is the y-component of $\boldsymbol{A} + \boldsymbol{B}$? (iii) Compute $\boldsymbol{A} - \boldsymbol{C}$. (iv) Compute $\boldsymbol{A} \cdot \boldsymbol{B}$. (v) If the vectors represent displacements, and an object is moved through $2\boldsymbol{A} + \boldsymbol{B} - \boldsymbol{C}$, how far has the object been moved in the z-direction? (We are using SI units here.) (vi) After the displacement, where is the object with respect to the origin?

9. If $|\boldsymbol{A} + \boldsymbol{B}| = |\boldsymbol{A} - \boldsymbol{B}|$, what is the angle between $\boldsymbol{A}$ and $\boldsymbol{B}$?

10. Find the magnitude, unit vector, and angle with respect to the y-axis of the vector $3\hat{i} - 7\hat{j}$. Find the magnitude, unit vector, and angles with respect to the x-axis, y-axis, and z-axis of the vector $-2\hat{i} - 3\hat{j} + 5\hat{k}$.

2.7 Solution to the Worked Problem

1. (a) $\boldsymbol{A} + \boldsymbol{B} = 2\hat{i} + 5\hat{j} - 1\hat{k} - 3\hat{i} + 1\hat{j} + 2\hat{k} = -\hat{i} + 6\hat{j} + \hat{k}$.

   (b) $\boldsymbol{A} - \boldsymbol{B} = 2\hat{i} + 5\hat{j} - 1\hat{k} + 3\hat{i} - 1\hat{j} - 2\hat{k} = 5\hat{i} + 4\hat{j} - 3\hat{k}$.

   (c) $\boldsymbol{A} \cdot \boldsymbol{C} = 2 \cdot 4 - 5 \cdot 1 = 3$.

   (d) $|\boldsymbol{A}| = (2 \cdot 2 + 5 \cdot 5 + 1 \cdot 1)^{0.5} = 5.477$

   (e) $|\boldsymbol{C}| = (4 \cdot 4 + 1 \cdot 1)^{0.5} = 4.123$

   (f) $\boldsymbol{A} \cdot \boldsymbol{C} = AC\cos(\theta)$ so $\theta = \arccos((\boldsymbol{A} \cdot \boldsymbol{C})/A/C)$. Therefor $\theta = \arccos(3/5.477/4.123) = 1.438$ radians.

This page reserved for your notes.

Chapter 3

Calculus; Motion at Constant Acceleration

3.1 Introduction

At the start of the book, I supplied the fundamental equation
PHYSICS - CALCULUS = NONSENSE

Calculus is not a new tool. It was developed more or less independently by Isaac Newton and Gottfried Wilhelm Leibniz nearly four centuries ago. Newton and Leibniz had a vigorous and somewhat pointless dispute as to who had developed what first. Recently, it became apparent that at about the same time an Austrian monk had also developed the Fundamental Theorem of Calculus. He tragically died, saving a young boy from drowning in a mountain stream, before he could publish his results. It is a curious fact that one of the allowed occupations of Samurai warriors in the Tokugawa period of Japan was abstract mathematics, a skill believed to be as useful as calligraphy or flower arranging. Some historians have made a case that this mathematical school independently developed, using a very different representation of mathematics, the basic ideas of calculus. What you should now have studied is Newtonian calculus, using the more sophisticated notation of Leibniz. As has been said before, Newton was a truly brilliant man, so he didn't worry whether or not his notation was easy to use or prone to introducing errors. Leibniz viewed himself as writing for mere mortals, and therefore thought carefully about how to make his notation easy to understand and unambiguous in employment.

In this chapter, I provide a short refresher on calculus and show a single application, sometimes described as *kinematics* or as *motion at constant acceleration*. As emphasized in the Introduction, this course assumes that you've already had enough calculus to be familiar with the integrals and derivatives of standard functions.

3.2 The Standard Functions

I first mention a few things you should already have heard about. The standard functions of interest are:

- The polynomial $f(x) = a + bx + cx^2 + \ldots$.
- The trigonometric functions $\sin(ax)$ and $\cos(ax)$.
- The exponential e^{ax}, also written $\exp(ax)$.
- The natural logarithm $\log(ax)$, sometimes also written $\ln(ax)$ or $\log_e(ax)$.

For each of these functions, you should be able to take the integral and the derivative. Most of you will also have seen and derived formulas for integrals and derivatives of all sorts of other trig functions, hyperbolic trig functions, and strange functions of polynomials. More or less all of those other formulas are far less useful for this course, so they are not reviewed here.

There is one minor point with which some of you are unfamiliar, namely the use of $\exp(ax)$ as an alternative way to write the exponential e^{ax}. There is a sound reason for introducing $\exp(ax)$ for the exponential. The argument of an exponential can become complicated, in which case writing the argument as a superscript becomes ugly.

There is also a major point which you have surely all seen but whose significance was not always made apparent. That's the constant of integration. For example, if I integrate ax with respect to x, taking the *indefinite integral* in which the limits of integration are not specified, I obtain

$$\int ax \, dx = \frac{ax^2}{2} + x_0 \tag{3.1}$$

and not

$$\int ax \, dx = \frac{ax^2}{2}. \tag{3.2}$$

In this equation, x_0 is a constant, the *constant of integration*. Your calculus preparation may not have stressed why constants of integration are important. Later in the chapter, I will show why constants of integration can be critically important, what they do, and how to determine their values. As another minor point, some of you were shown integrals being written as $\int dx \, ax$ while others will have seen $\int ax \, dx$. These are two ways of writing the same thing. Both forms are correct.

A few of you will have seen the misfortunate misrepresentation

$$\int ax. \tag{3.3}$$

This abomination is nonsense. The reason it is nonsense is that the symbol for integration is $\int d?$, not $\int$. In the correct symbol, the ? stands for the variable being integrated, which must be specified. If the variable is not specified, you have absolutely no idea with respect to what you are taking the integral. It might be x. It might be a. It might be t.

The derivative with respect to t of a function $f(t)$ is often defined

$$\frac{df}{dt} = \lim_{h \to 0} \frac{f(t+h) - f(t)}{(t+h) - t}. \tag{3.4}$$

The Fundamental Theorem of Calculus tells us that the integral is the antiderivative. In particular, for the definite integral of $df(t)/dt$ from a to b, we have

$$\int_a^b \frac{df}{dt} \, dt = f(b) - f(a). \tag{3.5}$$

This form is termed the *definite integral* because the bounds of integrations are specified; the bounds are a and b. What happened to the constant of integration? The subtraction on the right hand side of this equation cancels out the constant of integration in the indefinite integral.

If we take an integral of the function $g(x)$, we may think of $g(x)$ as giving us a smooth curve, which can be plotted as a function of x by making $g(x)$ the distance of the function from the x-axis. The integral $\int g(x) \, dx$ is then the area under the curve. Area is a signed number. It maybe positive or negative. Thus, for example,

$$\int_0^\pi \cos(x) \, dx = 0, \tag{3.6}$$

because

$$\int_0^{\pi/2} \cos(x) \, dx = -\int_{\pi/2}^\pi \cos(x) \, dx. \tag{3.7}$$

The direction of integration is also a signed quantity, so that

$$\int_a^b \frac{df}{dt}\, dt = -\int_b^a \frac{df}{dt}\, dt. \qquad (3.8)$$

In this equation, if you take the integral with respect to t backwards rather than forwards along the t-axis, you get a number with the same magnitude, but the sign is opposite to the sign you get if you take the integral with respect to t forwards along the t-axis.

One point not always emphasized in calculus courses is the dimension – dimension in the sense mass-length-time – of an integral or a derivative. The dimensions that come out of an integral or derivative actually follow immediately from equations 3.4 and 3.5. If we allow that x and t represent a spatial distance and time, then $\frac{d}{dx}$ has dimensions 1/length, while $\frac{d}{dt}$ has dimensions 1/time. For the same reasons $\int dx$ has dimensions length and $\int dt$ has dimensions time.

As an example, consider dx/dt, which gives the speed of an object whose location at time t is $x(t)$. Speed has dimensions length/time. In dx/dt, x has units length, so the 1/time must come from the d/dt. On the other hand, consider $\int \frac{dx}{dt}\, dt$, which has units length. The $\frac{dx}{dt}$ has units length/time, so to make things work $\int dt$ must have units time.

If $f(x)$ and $g(x)$ are both functions of the variable x, then you should recall how to take the derivatives of $f(x) + g(x)$, $f(x)g(x)$, and $(f(x))^n$. You also should know the chain rule, so that if f is a function of g and g is a function of x, there is a process for taking $\frac{df(g(x))}{dx}$.

Finally, there is a standard rule for finding the maximum or minimum of a function $f(x)$, namely you look for the points where $\frac{df(x)}{dx} = 0$. This rule has several failures. In particular, $\frac{df(x)}{dx} = 0$ also locates all of the saddle points of a function. What is a saddle point? Imagine climbing up a hill, and part way up there is a short level stretch of ground before the hill starts climbing again. The short level stretch is a saddle point. The rule also fails at two sorts of points. You can understand these points by imagining the roof of the house. First, if the roof is entirely straight, but tilted, the slope of the roof is not zero anywhere including its two ends, but the low end is the minimum and the high end is the maximum. The derivative rule does not work at endpoints. In addition, if we have a traditional house roof with the peak in the middle, the maximum height of the roof is at the peak, but the derivative is not zero there. In fact, the derivative is not even defined at the peak of the roof, because the slope has a left-hand limit and a right-hand limit, and these two limits are not equal to each other.

The fact that a function is continuous does not mean that it is differentiable. You can have a continuous function that does not have a derivative at a point. Indeed, you can have a function that is continuous everywhere but because of its particular definition does not have a derivative at any point. Almost all physics functions are continuous and differentiable almost but sometimes not quite everywhere.

3.3 Motion at Constant Acceleration

We now turn to a simple use of calculus, namely *motion at constant acceleration*.

For the moment, we will consider motion in one dimension. Real space is three-dimensional; we'll come back to that. We have an object whose position along the x axis is $x(t)$. The object is moving, so its position depends on time, that is, x is indeed $x(t)$. The object then has a position $x(t)$, a velocity $v_x(t)$, and an acceleration $a_x(t)$. x is the position of the object at some time, not the distance through which the object has moved. Note the subscript x. $v_x(t)$ and $a_x(t)$ refer to motions parallel to the x axis. The velocity and acceleration are related to the position x by

$$v_x(t) = \frac{dx(t)}{dt}, \qquad (3.9)$$

$$a_x(t) = \frac{d^2 x(t)}{dt^2}, \qquad (3.10)$$

while the acceleration and the velocity are related to each other through a derivative as

$$a_x(t) = \frac{dv_x}{dt}. \qquad (3.11)$$

By *motion at constant acceleration* I mean that the acceleration has a fixed value A so that we may write for the object an *equation of motion*

$$\frac{d^2x}{dt^2} = A. \tag{3.12}$$

The equation of motion is the equation for the particle's acceleration, with a value inserted for the acceleration. In writing this equation, the value of the acceleration has been given a symbol A that is not the same as the general symbol $a_x(t)$, so that we can tell them apart.

A perhaps-familiar example of motion at constant acceleration is falling motion in the absence of air resistance, as encountered on the surface of the Moon. For a fall at constant acceleration, we conventionally write

$$\frac{d^2z}{dt^2} = -g. \tag{3.13}$$

z is the vertical axis, with z increasing as we go up. Here g describes the acceleration due to gravity. On Earth, to one significant figure, $g = 10$ m/s^2. However, if we were using a different set of units, such as cgs units or English units, g would have a different numerical value, but the equation would still be correct. In one respect this equation is not the way things are normally done. g is an algebraic symbol. Algebraic symbols may be positive or negative, but one usually writes them as though they were positive, and allows negative numerical values to appear at some stage in the solution process. Here we have taken g to be a positive number, the minus sign (needed because objects fall in the downward direction) being inserted explicitly into the equation. That's not literally wrong, but it not the way things are usually done. It's better to let the signs fall out of the solutions, rather than wasting time guessing which sign each variable has, in order to arrange things so that algebraic symbols all stand for positive numbers.

Suppose we want to move from the acceleration to the velocity to the position of an object. We can do this by taking equation 3.12 and integrating it with respect to time. We'll treat the integrals of the two sides of the equation separately, and equate them at the end. The integral with respect to time of the left-hand-side of equation 3.12 is

$$\int_0^T dt\, \frac{d^2x}{dt^2} = v_x(T) - v_x(0). \tag{3.14}$$

One symbol, t, is used for the variable of integration, and a different symbol, T, is used for the time out to which the integral is taken. Nothing in this integral requires $T > 0$. A $T < 0$ is perfectly valid and allows integration backward in time, thus allowing us to calculate where the object came from as well as where it is going. By convention, $v_x(0)$ is written v_{x0}.

Why would you want to integrate backwards in time? One of the most powerful historical dating methods for ancient times uses the date and time of recorded eclipses. A solar eclipse lasts a few minutes and covers only a narrow geographical area. Lunar eclipses are seen everywhere on earth, and appear to last much longer than solar eclipses. One finds a record of an eclipse on a certain date and place in local time, then back calculates from today when an eclipse would have been seen in a particular place at about the right date. Sometimes identifying which eclipse the ancients were describing can be challenging. Once this is done, the exact date for an ancient civilization's calendar has now been determined relative to the dates in our calendar.

We can equally well integrate the right-hand-side of equation 3.12 with respect to time, obtaining

$$\int_0^T A\, dt = AT - 0 \tag{3.15}$$

What is the integral of equation 3.12 with respect to time? We have now done integrals of the two sides of the equation separately. Combining the above results, the integral is

$$v_x(T) - v_{x0} = AT \tag{3.16}$$

or the more familiar

$$v_x(T) = v_{x0} + AT. \tag{3.17}$$

3.3. MOTION AT CONSTANT ACCELERATION

We now have the velocity as a function of v_{x0}, A, and time T. Equation 3.17 is true at all times, so we can once again integrate, this time with respect to the time T, giving

$$\int_0^t dT\, v_x(T) = \int_0^t dT\, (v_{x0} + AT). \tag{3.18}$$

or

$$x(t) - x(0) = v_0 t + \frac{1}{2}At^2 - (v_0 0 + \frac{1}{2}A 0^2) \tag{3.19}$$

On the right-hand-side of this equation, I have shown explicitly the integrated terms at $T = 0$. Those terms are equal to zero in this problem, but in some problems the corresponding terms total to something that is non-zero. By convention the position $x(0)$ as time 0 is written x_0. Finally, in equation 3.17, T is just a symbol for the time. We can replace the letter T with any other letter that is not in use, without changing the meaning of the equation; in particular, we can replace T with t. On rearranging the above equation, we finally have for motion in one dimension at constant acceleration

$$v_x(t) = v_{x0} + At, \tag{3.20}$$

$$x(t) = x_0 + v_{0x} t + \frac{1}{2}At^2. \tag{3.21}$$

We have been very careful not to take any shortcuts in deriving these results from the constant acceleration starting point, equation 3.12, but the derivation is still far shorter than derivations that pretend to dodge calculus with various graphical methods. Furthermore, while you would have to fix a few steps, a slight variation on the above will give you $x(t)$ for an acceleration that depends on time. The key step is to recognize that if $f(t)$ is not a constant, then $\int dt\, f(t) \neq f(t) t$.

Some of you will have seen a peculiar equation $(v(t))^2 - (v(0))^2 = 2AX$ for motion at constant acceleration. Clearly that equation has to be consistent with the equations we just derived, but it's actually not a kinematic equation. This equation is a peculiar way to write the law of conservation of energy, for a particle under the influence of a constant force. *Most important!* In this equation X is not the position of the particle at some time t. Instead, X is the distance that the particle has travelled between times 0 and t. X therefore has a completely different meaning than it does in equation 3.21. In terms of the $x(t)$ of equation 3.21, $X = x(t) - x(0)$. X therefore *does not* tell you where the particle is; X tells you (but only for this very special case) how far the particle has moved.

We now come to the hard part of the discussion. Question: What do the constants x_0 and v_{0x} mean? You may have heard them called *initial* conditions, as though they were the position and velocity at the start of the problem, but the adjective "initial"' is to be charitable misleading. Let's consider a simple example that is actually not so simple.

We advance to mythical California of 70 years ago and the quaint local custom of drag racing. Two cars driven by

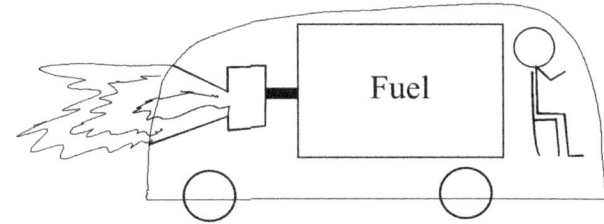

Figure 3.1: The Volksraketenwagon with engine at full thrust.

folks of limited good sense and even more limited regard for the law pull up at a red traffic light and stop. The character in the arrest-me-red sports car shouts at the character driving the apparent period SUV. *Hey, man, want to drag?*, this being an invitation to race the moment the light changes. When the light changes, the sports car takes off down the highway. The other vehicle seems to sit there. However, the other vehicle is not a period SUV. It is a highly modified vehicle, a *Volksraketenwagon* (People's Rocket Wagon), the modification being to replace everything behind the driver with a large liquid-fuel rocket engine hidden by the car's outer hull. The rocket takes ten seconds to power up, but then more or less instantly delivers a considerable constant thrust, to be precise, enough thrust that the Volksraketenwagon accelerates at $A = 100$ m/s^2. (That's ten gravities.) What are the motions of the Volksraketenwagon?

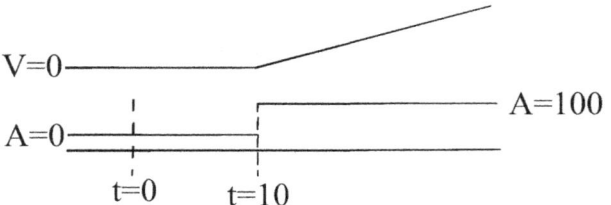

Figure 3.2: Velocity and acceleration of the Volksraketenwagon as functions of time.

A timeline illustrates what happens as time goes on. I've plotted the velocity and the acceleration as functions of time. The velocity curve for $t > 10$ is only qualitative. We are using SI units throughout, so I will not report the units every time. At time $t = 0$ and until $t = 10$ the vehicle is stationary. It is located at $x = 0$ and has $v_x(t) = 0$. At times $t \geq 10$ s, the vehicle accelerates at $A = 100$ m/s^2.

We are now going to advance by the Socratic method. I will ask questions. You get to decide on some answer, and *then* find out if you were right. To make this work, get out a piece of paper, and cover everything on the page below the solid line that follows this paragraph. Then lower the paper until you reach the next solid line. (You may have to flip to the next page to reach the next solid line.)

Question: Can we describe the motion of the vehicle as seen on the entire the time line by inserting one set of numbers for x_0, v_{0x}, and A into the equations

$$v_x(t) = v_{0x} + At, \qquad (3.22)$$

$$x(t) = x_0 + v_{0x}t + \frac{1}{2}At^2? \qquad (3.23)$$

Stop and think. Your answer must be "Yes" or "No". When you are sure you have an answer, lower the paper again to the next line.

The correct answer is "No". Why? Stop and think. When you have an answer, or give up, lower the paper to the next line.

Why are these equations not usable as proposed? The equations describe motion at constant acceleration. However, along the length of the time line the acceleration is not constant. The acceleration has one value for $t < 10$ and a different value for $t > 10$. You need two sets of equations for motion at constant acceleration, one describing times $t < 10$ and the other describing times $t > 10$, and therefore two sets of constants x_0, v_{0x}, and A_x. If you realized the equations were OK if you used two different sets of constants for the two time ranges, so the problem is with the constants and not the equations, you understand well what the question meant.

Now let's work out what the constants are. I'll do this for you for $t < 10$. You should stop when you reach the next line, which is a distance down the page.

We know that at time $t = 0$ the volksraketenwagon is at $x(0) = 0$ and $v_x(0) = 0$, and has $A = 0$. Let's plug those numbers *carefully* into the above two equations. The two equations, with all zeros written explicitly, become

$$0 = v_{0x} + 0 \cdot 0, \qquad (3.24)$$

$$0 = x_0 + v_{0x}0 + \frac{1}{2} 0 \cdot 0^2? \qquad (3.25)$$

In writing these equations, I replaced t with its value zero. I carefully recalled that the velocity and position of the particle are $v_x(t)$ and $x(t)$, as found on the left hand sides of the two equations, and replaced them with their known values, 0 and 0 at time 0. I could equally well have used any other time $t < 10$. It would be wrong to replace x_0 and v_{0x} with 0 and 0; you have to solve for x_0 and v_{0x}. However, if you have seen kinematics in earlier courses, you probably only saw problems in which doing the wrong thing, namely replacing x_0 and v_{0x} with numbers in the problem, did not get you into trouble, so it appeared to be the right thing to do. The problems had been cooked so someone who had no clue what they were doing would

3.3. MOTION AT CONSTANT ACCELERATION

still magically get the right answer. Back at the start of the book I had mentioned *praxis*, getting the right answer for the right reasons; here you are seeing an example of that term. The velocity and position are, however, $v_x(t)$ and $x(t)$, not x_0 and v_{0x}.

Solving first the upper equation for v_{0x}, we get $v_{0x} = 0$. Substituting that solution into the lower equation, we get $0 = x_0$. We have now solved for the constants of integration v_{0x} and x_0. For $t < 10$, both constants are zero. We can now write for the velocity and position of the car at all times $t < 10$

$$v_x(t) = 0, \tag{3.26}$$
$$x(t) = 0. \tag{3.27}$$

Now you get to try the same process. Solve for x_0, v_{x0}, and A for times $t > 10$. So that you all start at the same place, at $t = 10$ and after the rocket engine has fired, $A = 100$ but the car has barely moved, so it is still at location $x(10) \approx 0$ and stationary with $v_x(10) \approx 0$. OK, find equations for $x(t)$ and $v_x(t)$ for all times $t > 10$, meaning you need to find x_0 and v_{0x}. Stop at the solid line following this pragraph, and do not go lower on the page until you have completed your work.

Now move the paper down to the next line and read my answer

What did you find? Were your answers $x_0 = 0$ and $v_{0x} = 0$? After all, 0 and 0 were the initial values for the velocity and position. 0 and 0 are a fairly common pair of answers. Perhaps you found some other pair of numbers. What you should do, before saying you have the answer, is to check if your numbers are right. You do this by plugging them into the equations for $x(t)$ and $v_x(t)$ and solving for $v_x(10)$ and $x(10)$. If you did that, and thought $x_0 = 0$ and $v_{0x} = 0$, you should get

$$v_x(10) = 0 + 100 * 10, \tag{3.28}$$
$$x(10) = 0 + 0 * 10 + \frac{1}{2} 100 * 10^2. \tag{3.29}$$

That is, if your answers for the constants of integration were 0 and 0, then, at the instant the rocket fired, the volksraketenwagon instantly gained a velocity of 1000 m/s and equally instantly transported itself to $x(10) = 5000$ m. That's not what happened. Something clearly went wrong. If you want, try again to find x_0 and v_{0x}.

How do you solve the problem correctly (some of you have already done this) for the three constants. (Three? remember A. I gave you $A = 100$.)

First, for the velocity at $t = 10$, the rocket having already ignited, one has

$$v(t) = v_{0x} + At, \tag{3.30}$$
$$0 = v_{0x} + 100 * 10, \tag{3.31}$$
$$v_{0x} = -1000. \tag{3.32}$$

This solution process has a few inobvious features. First, I always begin with the basic equation into which I will be substituting. That's the first line. I then replace algebraic symbols with numbers, without doing any arithmetic. There is a certain matter of taste involved in solving problems. I could equally well have solved first algebraically for v_{0x}, and then plugged in numbers, again without doing any arithmetic. That would have been equally correct. Then I counted equations and unknowns, to make sure that I have as many equations as there are unknowns. Here I have one unknown, and one equation. That's not a guarantee that the problem is soluble, but if you have fewer equations than you have unknowns, then you probably cannot solve for the unknown you want. Finally, I solve.

Now let's carry out the same process for x_0.

$$x(t) = x_0 + v_{0x}t + \frac{1}{2}At^2, \tag{3.33}$$
$$x_0 = x(t) - v_{0x}t - \frac{1}{2}At^2, \tag{3.34}$$
$$x_0 = 0 - (-1000)10 - 0.5(100)10^2, \tag{3.35}$$
$$x_0 = 5000. \tag{3.36}$$

Why did I solve first for v_{0x} and only then tried to solve for x_0? I looked at the two equations and realized that the equation for $v_x(t)$ contained only one unknown, so it could be solved immediately. The equation for $x(t)$ had two unknowns, one of which would come from the equation for $v_x(t)$, so I should solve for $x(t)$ second.

I again started with the basic equation to be solved. This time I first did a little algebra, so that x_0 was isolated algebraically before I plugged in numbers. Sometimes that step makes life simpler. Sometimes it doesn't. Experience will teach the difference. However, there is a useful anecdote here. Once upon a time, I had a colleague who assigned several students to do the same calculation, back in the days when algebraic calculations could only be done by hand. One of the students was notorious for doing things line by line, always taking only a very small step from one line to the next. The other students, who did several things on each line, were quite critical of him. He was slow! However, in the end, the patient student who always took small steps and checked his work after each step almost always got to the correct answer first. The other students sometimes eventually got to the right answer.

We now have x_0 and v_{x0}, so the equations for $v_x(t)$ and $x(t)$ should be

$$v_x(t) = -1000 + 100t, \tag{3.37}$$

$$x(t) = 5000 - 1000t + \frac{1}{2}100t^2. \tag{3.38}$$

You can check for yourself that these values for x_0 and v_{0x} are correct, namely they show the volksraketen-wagon stationary at the origin at $t = 10$.

According to these equations, at time $t = 0$ the car was at $x = 5000$ m and was travelling at $v_x = -1000$ m/s, that is, the car was travelling at Mach three, three times the speed of sound. Backwards. What does this answer even mean?

Suppose the car always travelled at constant acceleration $A = 100$ m/s^2. In that case, at $t = 0$ it would indeed be 5 km out on front of the red light, headed backward toward the red light at Mach three. However, it would also have a positive acceleration. As a result, over the next 10 seconds its velocity would increase from -1000 m/s to 0 m/s. (Its speed would, by the same argument, be decreasing.) At time $t = 10$, the car would come to a stop, exactly at the red light, the location $x = 0$. It would then head away from the red light again, going in the positive x direction, still accelerating at ten gravities. (Many years ago, when I described in lecture the car coming to a stop exactly at the red light, the description sounded very strange, but you can now watch on the internet something that is very much like the car. One of the commercial rocket launch corporations recovers its first stage rockets by having them return to ground and land vertically. If you find one of these videos, you can see the first stage descend, its rocket firing, and come to a stop just as it reaches the ground.)

We now have an example showing the meaning of the constants of integration x_0 and v_{0x}. x_0 is the position that the object would have had, at $t = 0$, if the object's acceleration had been constant from the times at which the information was given through to time zero. Yes, in the above case that's backward in time from $t = 10$ to $t = 0$. v_{0x} is the velocity that the object would have had, at $t = 0$, if the object's acceleration had been constant from $t = 10$ to as far back in time as $t = 0$. Of course, the acceleration of the car was not constant from $t = 10$ back to time zero, so the $t > 10$ constants of integration in fact do not indicate the position and velocity of the car at $t = 0$.

It may occur to you to try shifting the time origin, so that $t = 10$ becomes $t' = 0$, and solving the problem. That's a special case approach. It fails, for example, if I give you the position of the rocket car at each of three unequally-spaced times. The approach above, sometimes with more algebraic work, is always effective.

Final point on the car problem: $x(t)$ is the position of the car. It is *not* how far the car has travelled. The distance the object has travelled between two times t_1 and t_2 is the object's *displacement*. The displacement between times t_1 and t_2 is $x(t_2) - x(t_1)$.

3.4 Discussion

Consider again what we just did. We have now seen a sketch of a general solution procedure:

Step one. Draw a picture of the situation. Insert in it all of the known quantities, and assign to each of them a symbol.

Step two. Write the basic equations. Identify which variables are unknown. Count that you have as many equations as you have unknowns. If you have fewer equations than you do unknowns, you will usually be unable to solve the equations for the unknowns.

Step three. Perhaps solve algebraically for the unknowns. Sometimes it is better to plug in numbers fairly early on in the calculation.

Step four. Plug in for the known variables and solve for the unknowns.

Aside: I am calling dx/dt the *velocity*. Velocity is actually a vector, of which dx/dt is a component. More on this later. Some authors call dx/dt the *instantaneous velocity*. The word *instantaneous* is redundant, because the definition of a time derivative incorporates the limit $\delta t \to 0$. There is also an *average velocity*, which we will reach in the next chapter.

Finally, remember that my Solutions to the Worked Problems incorporate new material not seen elsewhere in the text.

3.5 Worked Problems

1. Standard notation: $\exp(a) \equiv e^a$. The position of an object as a function of time is given by $x(t) = x_0 t \exp(-at^{1/2})$. Here x_0 and a are numerical constants in the equation for x. In this formula for $x(t)$, x_0 is not the position of the object at $t = 0$, and a is not the acceleration of the object. Discuss the behavior of the velocity as $t \to 0+$ if $a > 0$.

2. A distinguished Faculty Member is loaded into a rocket ship.* The ship takes off from a pit at the bottom of Death Valley, starting at an altitude 100 m below Sea level. At $t = -5$ the engines are ignited. At $t = +5$s, the ship leaves the ground. The vertical acceleration of the ship during the climb is a constant. At $t = 20$ s the ship achieves an altitude of 4000 m above sea level. What is the ship's acceleration? What is its velocity at $t = 30$ s? Solve systematically, beginning with $z = z_0 + v_0 t + 0.5at^2$ and $v = v_0 + at$, not with whatever equation you may have pulled off the internet. Find z_0, v_0, and a. For credit, you must use my time and altitude origins in your calculations. To avoid undue complications, substitute numbers from the beginning. Prove your values for z_0, v_0 and a are correct by showing that they predict correctly the locations and times supplied in the problem. Clue #1) I am fond of this problem, but I keep changing which boundary conditions I supply. A few years ago, 40% of the class got this one wrong. I hope you can do better. Clue #2 You should end up with three equations in three unknowns. You can do systematic elimination to find the unknowns, or you can learn how Mathematica, Maple, or some other computer algebra program will do your work for you. Clue #3) You should have performed the check "prove your values" automatically.

*Yes, one of my former colleagues did go up in the space shuttle.

3.6 Homework

1. It is time for the first test of the atomic train, a steam locomotive in which the burning coal is replaced with a modest-sized block of radium. The engine is started. At some time, the locomotive begins to roll out of the station, maintaining a constant acceleration throughout its travels. After 2 miles, it goes off the rails, making a real mess and producing lawsuits beyond belief. You have been retained as the special physics consultant for a group of litigators. Your objective is to prove exactly where and when the train started.

Fortunately, the railroad tracks are totally straight, and by agreement between the litigating parties they lie along the x axis. Unfortunately the surviving cameras only report the locomotive's position at three moments of time, namely $x = 10$ at time $t = 10$, and later $x = -10$ at $t = 20$, and finally $x = -110$ at $t = 40$. (a) Report the position and velocity of the locomotive as explicit functions of time in the forms

$$x = x_o + v_o t + 0.5 a_x t^2, \tag{3.39}$$
$$v = v_o + a_x t. \tag{3.40}$$

(b) Confirm by appropriate substitutions that your answer agrees with the original information. (c) At the instant the train began to roll forward, where was it located? At what time did it begin to roll forward?

2. Our intrepid astronaut returns from a trip to outer space. At $t = -1$, the spaceship is at an altitude of 100 m above sea level. At $t = +1$, the spaceship is at an altitude of 20 m above sea level. At $t = +2$, the spaceship is at an altitude of 10 m above sea level. Write the altitude and vertical component of the velocity of the spaceship in the forms

$$z = z_0 + v_0 t + 0.5 a t^2, \tag{3.41}$$
$$v_z = v_0 + a t. \tag{3.42}$$

In a successful landing, the spaceship comes to a stop just as it touches the ground. Assuming that the landing was successful, what is the altitude of the ground? What was the altitude of the spaceship at $t = 0$?

3. Another Faculty Member is loaded into a rocket ship. The ship takes off from a mountain, starting at an altitude 1000 m above sea level. At $t = 0$ the engines are ignited. At $t = 7$s, the ship leaves the ground. The vertical acceleration of the ship during the climb is a constant. At $t = 30$ s the ship achieves an altitude of 5000 m above sea level. What is the ship's acceleration? What is its velocity at $t = 30$ s? Solve systematically, beginning with $z = z_0 + v_0 t + 0.5 a t^2$ and $v = v_0 + a t$, not with whatever equation you may have pulled out from the book. Find z_0, v_0, and a. For credit, you must use my time and altitude origins in your calculations. To avoid undue complications, substitute numbers from the beginning. Prove your values for z_0, v_0 and a are correct by showing that they predict correctly the locations and times supplied in the problem.

4. Yet another Faculty member is loaded into a rocket ship, on a secret launch pad in a secret location. At $t = -2$ the ship leaves the ground. The ship's vertical acceleration is constant after take-off. Radar shows that at $t = 2$ the rocket is at an altitude of -50 m relative to sea level. At $t = 10$ the ship has reached an altitude of 200 m. What was the ship's altitude at launch?

5. The rocket ship from the homework is being brought for a landing. It has a constant acceleration until it is brought to a stop, some distance above the ground. At times 1, 2, and 4, respectively, the rocket ship was at altitudes 91, 84, and 76 m, respectively. Write the altitude of the ship in the exact form

$$z(t) = z_0 + v_0 t + 0.5 a t^2, \tag{3.43}$$

including giving numerical values for z_0, v_0, and a.

6. Suppose the acceleration of a particle is given by $\frac{d^2 x}{dt^2} = bt$, where t is the time and b is a constant. That is, the acceleration increases linearly with the time. Find $v_x(t)$ and $x(t)$.

7. Standard notation: $\exp(a) \equiv e^a$. The position of an object as a function of time is given by $x = x_0 t^3 \exp(-a t^{2/3})$. Here x_0 and a are numerical constants in the equation for x. x_0 is not the position of the object at $t = 0$, and a is not the acceleration of the object. Discuss the behavior of the velocity as $t \to 0+$ if $a > 0$.

8. Following equation 3.38, I said that you can check for yourself that you have the right answer here. Do the check. Show your work.

9. The position of an object as a function of time is given by $x = x_0 \exp(a t^{3/2})$. Find the velocity and the acceleration of the object. Discuss the behavior of the velocity and the acceleration as $t \to 0+$.

3.6. HOMEWORK

10. A local driver is proceeding along Salisbury Street in her BMW at an illegal speed of 50 m/s. To simplify the problem, Salisbury Street is taken to run along the x axis, and the car is moving in the $+x$ direction. At $t = -8$, the vehicle's radar, laser, and sonar detectors fire, instantly triggering the brakes. The car slows with constant acceleration a. At $t = -4$, the car's speed is 10 m/s, still in the $+x$ direction. Between $t = -8$ and $t = -4$, the car's velocity in the x-direction may be written $v = v_0 + at$. (i) Show that the acceleration of the car is $a = -10$ m/s^2, and (ii) compute v_0.

11. A truck is approaching an intersection. The truck's motion is filmed and digitized. The supplied coordinates put the center of the intersection at $x = 15$. x increases from left to right. At $t = -20$s, the truck appears on the film at $x = -75$ m with a speed of 15 m/s to the right. The brakes are applied at this time. The magnitude of the resulting acceleration is 2 m/s^2. (a) Give equations for $x(t)$ and $v(t)$ in the seconds after the brakes are applied, using equations in the form

$$x(t) = x_0 + v_0 t + 0.5 a t^2, \qquad (3.44)$$
$$v(t) = v_0 + at. \qquad (3.45)$$

(b) Prove your answers are correct by substituting $t = -20$ in these equations, and showing that you recover the initial position and velocity correctly. (c) How fast is the truck going when it reaches the intersection? (d) At what time does the truck stop?

Clue#1: Some years ago, almost a third of the class got this one wrong. That's much better than the year before, when over half the class was wrong.

Clue#2: Key question: what are the signs of the velocity and acceleration at $t = -20$? Some of you were taught that all numbers are positive, a wrong fact that I will attempt to unteach.

Clue#3: You should always check your answers. You should *automatically* have done the $t = -20$ substitution, without being told, to check your answer.

12. A distinguished faculty member is again loaded into a rocket ship. At $t = 0$, the engines are ignited. At $t = 5$s, the ship leaves its floating launch pad, located exactly at sea level. The vertical acceleration during the climb is a constant. At $t = 25$s the ship attains an altitude of 3000m. What is the ship's acceleration? What is its vertical velocity at $t = 25$ s? Write the ship's altitude as a function of time in the form $z = z_0 + v_0 t + 0.5 a_z t^2$.

13. A distinguished faculty member is once again loaded into a rocket ship. At $t = 0$, the engines are ignited. At $t = 4$s, the ship leaves its floating launch pad, located exactly at sea level. The vertical acceleration during the climb is a constant. At $t = 30$ s the ship attains an altitude of 5000 m. (a) What is the ship's acceleration? (b) Write the ship's altitude as a function of time in the form $z = z_0 + v_0 t + 0.5 a_z t^2$ (you will need to solve for z_0, v_0, and a_z). (c) Write the ship's velocity as a function of time as $v(t) = v_0 + a_z t$. (d) Substitute into your answers for (b) and (c) and confirm that your equations agree with the numbers provided in the problem, e.g., confirm $z(4) = 0$. You should have performed this step automatically. Find the ship's velocity at $t = 30$.

14. You have been hired as a consulting physics detective to find an object dropped from the world's first atomic train. The train was under computer control, giving exact control over the train's acceleration. At $t = 0$ the train was at rest. The train then had an acceleration $d^2x/dt^2 = 0.01$ m/s^2. At time $t = T$, the object was dropped; the engine was thrown into reverse, so that the acceleration of the train became $d^2x/dt^2 = -0.02$ m/s^2. The train came to a stop at a distance $x = L$ from the starting point. The time T is unknown. The value of L has been kept secret from you, because the missing object is the world's only neutrino bomb [1]. [Fortunately, the bomb did not detonate. Unfortunately, neutrino bombs are very small and hard to see, so a physical search of the complete track is impractical.] You are to calculate, in terms of L, the moment T at which the bomb was dropped, and the location along the tracks at which the bomb was dropped. [1] As described in a 4/1/195x Los Alamos memo.

15. A massive pendulum swings back and forth at the end of a long chain. Its horizontal acceleration is measured to be $d^2x/dt^2 = a_0 \sin(\omega t)$. At $t = 0$, the pendulum is located at position $x = s$; at that moment in time it has $v = 0$. [Hint: Therefore, at $t = 0$ the pendulum must not be at the center of its swing.] Find the pendulum's velocity v and its position s as functions of time.

16. It is time for the trial run of the world's fourth atomic train. Because the last three atomic trains were destroyed in the course of past hour examinations, the crew has wisely been replaced with a robot controller. The train exits the yards with an acceleration 0.1 m/s² until it reaches a speed of 10 m/s. It continues at this speed in a straight line in the $+x$ direction until the command is sent to the robot to engage the brakes. At this point, instead of slowing down the train was observed to increase in speed at constant acceleration. When the train reached a speed of 40 m/s, it exited from its tracks and turned into scrap metal. Unfortunately, the instrument recorders on the train mostly failed, but the positions of the train *while it was accelerating* are known at three times. The times and positions are:

$$t = 10, x = 120, \quad (3.46)$$
$$t = 20, x = 230, \quad (3.47)$$
$$t = 40, x = 510. \quad (3.48)$$

i) Write the position and velocity of the train while it was accelerating in the forms

$$x = x_0 + v_0 t + 0.5at^2, \quad (3.49)$$
$$v_x = v_0 + at, \quad (3.50)$$

and supply the needed constants.

ii) At what time did the train begin its final acceleration?

iii) At what time did the train leave the tracks?

17. You are now the world's leading physics detective. Your client [Phillies, et fils, Insurers of MagLevs] brings you to the scene of an ingenious murder on a maglev train. A train carrying a single passenger and a number of crewmen rolled into the train station, going east to west and gradually slowing down. The train's engines provided a constant thrust for the entire period. The crew got off the train as it was rolling to a stop. Unfortunately for the passenger, the train's engine was left engaged, so the train rolled back out of the station, the passenger still on board, and continued to accelerate until the bomb in the passenger compartment detonated. The bomb exploded 2000 m east of the coordinate origin. Unfortunately, the automatic cameras that should have recorded everything were not working very well, so all you have is the location of the train at three times, namely: $x = 2500$ at $t = -50$, $x = 925$ at $t = 400$, and $x = 1125$ at $t = 500$.

Your contract with the insurance company specifies that to get any points you must begin with $x = x_o + v_o t + \frac{1}{2} a_x t^2$ and $v = v_o + at$. Your insurer lets you work numerically rather than symbolically. You must determine (a) the position and velocity of the train as a function of time. (b) the instant at which the bomb exploded.

18. The intrepid Faculty colleague of a prior problem is given another chance to fly into space and return, this time on board a privately-launched spaceship. The launch altitude in central Asia is 2000 m above sea level. At $t = 5$, the engines are ignited. At $t = 10$, the ship leaves the ground. During the climb phase, acceleration is constant. At $t = 100$, the spaceship achieves an altitude of 30000 m above sea level. What is the ship's acceleration? What is the ship's velocity at $t = 100$? Solve systematically, beginning with $x = x_0 + v_0 t + 0.5at^2$, $v_x = v_0 + a_x t$, or corresponding equations for the y and z directions. For credit, you must use my coordinate origins. Do not introduce, e.g., the average acceleration formula or the conservation of energy formula, unless you derive them first.

19. Our intrepid astronaut prepares for a second launch into space. This time, the shuttle has been given an improved engine. The launch occurs from a point 50m above mean sea level. At $t = 3$ seconds, the shuttle clears its launch pad and begins to climb. At $t = 5$, the shuttle has reached an altitude of 110m (above mean sea level). (i) Write the altitude and vertical component of the velocity of the shuttle in

the forms
$$x = x_0 + v_0 t + 0.5at^2 \qquad (3.51)$$
$$v = v_0 + at. \qquad (3.52)$$

(ii) Compute the altitude and vertical component of the shuttle's velocity at $t = 10$ seconds. (iii) Compute v at time $t = 0$. What is the physical meaning of this number?

20. The rocket car discussed in one lecture is travelling in the $+\hat{i}$ direction at cruising speed, 1000 m/s. The car's flight recorder records position, speed, and acceleration at all times. As the car goes down the highway, sensors report that a traffic light dead ahead will turn red, and it will be necessary to stop to avoid running the light. The flight recorder later shows that the retro-rockets were fired at $t = 50$ s, when the car was at $x = 20,000$ m relative to the origin, and at $t = 70$s the speed of the car was down to 500 m/s. The retro-rockets give the car a constant acceleration, and stop firing when the car comes to a rest relative to the pavement. (i) At what time t do the retro-rockets stop firing? (ii) Write the car's position as a function of time, in the form $x = x_o + v_o t + 0.5at^2$, using the coordinate system specified in the problem, for the period while the retro-rockets were firing. Where is the car when it comes to a stop? (iii) In the equation of part (ii) of this problem, what is the physical meaning of the constants x_o and v_o?

3.7 Solutions to the Worked Problems

1. We start with $x(t) = x_o t \exp(-at^{-1/2})$ The velocity $v(t)$ is the first derivative of $x(t)$ with respect to time. Note that I added *with respect to time*. If I had just said "the derivative" I could have meant the derivative with respect to x_o or a. Those are perfectly legitimate derivatives that are actually useful under some circumstances (but not this one). You need to specify with respect to what you are taking the derivative, and I did. Taking the derivative, we have

$$v(t) = x_o \exp(-at^{-1/2}) + x_o t \exp(-at^{-1/2}) \frac{d}{dt}(-at^{-1/2}), \qquad (3.53)$$

where the product rule gave us the two terms and the chain rule led to the final derivative, which simplifies to

$$v(t) = x_o \exp(-at^{-1/2}) + \frac{x_o a}{2} t^{-1/2} \exp(-at^{-1/2}). \qquad (3.54)$$

What is the behavior as $t \to 0$? The term $t^{-1/2}$ diverges. Its exponential, however, goes to zero. An exponential goes to zero more strongly than a polynomial in the same variable diverges, so

$$\lim_{t \to 0} v(t) = 0. \qquad (3.55)$$

2. We should start by collecting all the information given in the problem. The rocket takes off at $t = 5$, at which point it is at altitude $x = -100$ and velocity $v = 0$. At time $t = 20$, it is at $x = 4000$. We are discussing motion at constant acceleration, for which we may write

$$x = x_o + v_o t + \frac{1}{2}at^2, \qquad (3.56)$$
$$v = v_o + at. \qquad (3.57)$$

We now take the known facts and insert them into these equations. I will do that without making any simplifications, so there is no doubt as to where the numbers in the equations came from. This is a good general first step, in that it lets you check exactly where you started.

$$-100 = x_o + 5v_o + \frac{1}{2}a5^2, \qquad (3.58)$$
$$0 = v_o + 5a, \qquad (3.59)$$
$$4000 = x_o + 20v_o + \frac{1}{2}a20^2. \qquad (3.60)$$

The rocket started stationary ($v = 0$) at an altitude $x = -100$, but it would be entirely incorrect to write $x_o = -100$ or $v_o = 0$. x_o and v_o are constants of integration, not the "initial position" or the "initial velocity".

We now count equations and unknowns. There are three equations and three unknowns, so a solution may be possible.

If we subtract the first of these equations from the third, x_o is eliminated from the calculation. It happens that subtraction is a simple step for these particular equations, but in fact subtraction (after multiplication if necessary by a constant) is a highly effective general process for reducing the apparent number of constants. The third equation becomes

$$4100 = 15v_o + 187.5a. \tag{3.61}$$

The first equation has been used to eliminate a constant, namely x_o; the first equation's problem-solving potentialities are thus exhausted. We now use the two remaining equations to eliminate v_o. Multiplying the second equation by 15 and subtracting from the third equation, we find

$$4100 - 0 = 15v_o - 15v_o + 187.5a - 75a, \tag{3.62}$$

which simplifies on dividing by $(187.5 - 75)$ to

$$a = 36.44 \text{m/s}^2. \tag{3.63}$$

We have now found the acceleration. Substituting for a in the second of the original equations, we get $v_o = -182.2$ m/s. v_o is what the velocity would have been at time $t = 0$, before the rocket took off, a hypothetical event that did not happen. Finally, inserting these numbers in the first equation, one finds

$$x_o = -100 - 5(-182.5) - 12.5(36.44), \tag{3.64}$$
$$x_o = 35. \tag{3.65}$$

Therefore, we can write for the position and velocity

$$x = 357 - 182.2t + \frac{1}{2} \cdot 36.44t^2, \tag{3.66}$$
$$v = -182.2 + 36.44t. \tag{3.67}$$

Are we done? No! Always check your results! We still have to check that these equations actually give us the correct initial data. Indeed, if we substitute $t \to 5$, we find $x = -100$ and $v = 0$, while if we substitute $t \to 20$ we find $x = 4000$. Those are indeed the original conditions, so our solution is correct.

We could equally well have used subtraction to eliminate the constant v_o or a. To eliminate a, we could multiply the first equation by $5/12.5$ and subtract the product from the second equation, and separately multiply the first equation by $20^2/5^2$ and subtract the product from the third equation. These two subtractions would give us two equations that depend on the unknowns x_o and v_o, but would be independent of a.

Chapter 4

Chapter 4. Averages, Average Velocity

In this chapter, we consider the nature of averages and the average velocity $\overline{v}$.

4.1 Averages

What we mean by *taking an average*? For many of you, averaging was introduced soon after you learn to do division. For example, you could have been told to calculate the average grade on an exam. That average might have been written

$$\overline{G} = \frac{93 + 13 + 9}{3}, \tag{4.1}$$

where $\overline{G}$ is the average grade, the three grades going into the average were 93, 13, and 9, and the average grade was formed by dividing the sum of the three grades by how many grades there were. The average grade in this equation was computed as an unweighted average. That is, each of the three grades was assigned the same importance in calculating the average.

We might also construct a weighted average of grades. In fact, when I taught this course, the final grade was indeed calculated as a weighted average, namely

$$\overline{G} = \frac{1 \cdot E_1 + 1 \cdot E_2 + 1 \cdot E_3 + 0.8 \cdot H + 0.2 \cdot L}{1 + 1 + 1 + 0.8 + 0.2}. \tag{4.2}$$

In reading this equation for the average grade, E_1, E_2, and E_3 were the grades on the three examinations, H was the total of the homework grades with the total rigged to equal 100, and L was the total of the laboratory grades, also rigged to total 100. The numbers multiplying the various grades, such as the 1 multiplying the grade E_1, are the weights assigned to the different components of the total grade. In this case, the three exams are each assigned the same weight unity, the homework is assigned a total weight of 0.8 (80% of the weight given to a single exam), and the laboratory exercises were assigned a total weight of 0.2 (20% of the weight given to a single exam).

The above equation is an example of a weighted average. It's also possible to write a general formula for a weighted average. That general formula is

$$\overline{G} = \frac{\sum_{i=1}^{m} W_i G_i}{\sum_{i=1}^{m} W_i}. \tag{4.3}$$

In reading this equation, $\overline{G}$ is the average value of the quantity G. The individual values of G that are being averaged are the G_i. In the above equation, the character i is the *index* that identifies each of the G_i that is being included in the average. The sum on i is taken over all quantities being averaged. Implicitly, there is someplace a *complete, non-repeating list* of the quantities being averaged, with each quantity being labeled by its own value of i. W_i is the *weight* or *statistical weight*, the importance being assigned to the quantity G_i. The denominator of the equation, the $\sum_{i=1}^{m} W_i$, is the *normalizing factor*. The normalizing factor acts to divide out the weights that appear in the numerator. Why do we want a normalizing factor? You might reasonably think that the average value of the number 1 should be 1. With the normalizing factor of the denominator in place, you can readily confirm that in this equation the average value of 1 is indeed 1, no matter what the W_i are.

Equation 4.2 is an example of equation 4.3. In the example, the sum on i is written term by term. i goes from 1 to 5. The five W_i are $1.0, 1.0, 1.0, 0.8$, and 0.2, respectively. The five G_i are E_1, E_2, E_3, H, and L, respectively.

Some treatments of averages, particularly in mathematical statistics, assume that all weights have been normalized in advance, so that the normalizing factor $\sum_{i=1}^{m} W_i$ is equal to unity. There is no mathematical obligation to do this. Indeed, in the branch of physics known as *statistical mechanics*, the numerical value of the normalizing factor, the *canonical partition function*, has an exact, fundamental physical meaning that would be lost if the corresponding W_i were pre-normalized to sum to unity.

We may also invoke the general rule for replacing the sum on i with an integral, writing for the unweighted average of the function $g(x)$

$$\overline{g} = \frac{\int dx\, 1 \cdot g(x)}{\int dx\, 1}. \tag{4.4}$$

In this equation, the discrete index i has been replaced with a continuous index x. The list of allowed outcomes $i \in (1, m)$ is replaced with bounds on the integrals. $g(x)$ is a continuous function of x rather than a series of terms with a discrete label i. In this unweighted average, the same weight unity has been assigned to every $g(x)$ included in the average.

The corresponding weighted average is

$$\overline{g} = \frac{\int dx\, W(x)g(x)}{\int dx\, W(x)}, \tag{4.5}$$

in which $W(x)$ is the weighting factor, now written as a function of the continuous variable x.

If we have

$$g(x) = \frac{dG(x)}{dx}, \tag{4.6}$$

we know how to integrate $g(x)$, so we might then write for the unweighted average of $g(x)$

$$\overline{g} = \frac{\int_a^b g(x)\, dx}{\int_a^b dx}. \tag{4.7}$$

In the integrals, the limits of integration have been written explicitly as a and b. We obtain

$$\overline{g} = \frac{G(b) - G(a)}{b - a}. \tag{4.8}$$

As a specific example of this result, suppose that g is the velocity in the x direction, and we average the velocity in the x-direction over a period of time. In this example, we replace g with $v_x(t)$. From the Fundamental Theorem of Calculus, the indefinite integral $\int dt\, dx/dt$ is $x(t)$. Inserting this result into equation 4.7, we find that the average of the velocity between times a and b is

$$\overline{\frac{dx}{dt}} = \frac{\int_a^b \frac{dx}{dt}\, dt}{\int_a^b dt} \tag{4.9}$$

or

$$\overline{\frac{dx}{dt}} = \frac{x(b) - x(a)}{b - a}. \tag{4.10}$$

The average velocity between the times a and b is seen to be the displacement $x(b) - x(a)$ divided by the duration $b - a$ over which the average is taken.

The average velocity is also sometimes written

$$\overline{v_x} = \frac{\Delta x}{\Delta t}. \tag{4.11}$$

4.2. DISCUSSION

in which Δx is the displacement of the object during the time interval Δt. Because we obtained this formula by using the calculus definition of the velocity and then taking the appropriate integrals, it is immediately obvious why $\bar{v}$ is called the average velocity, namely it actually is the outcome of averaging the velocity over some period of time.

There is an interesting example of the average velocity, which takes us back to the observation that the integral is a signed quantity, so that positive areas and negative areas under a curve can cancel. Suppose you run around a tree. For the first half of your run, you run absolutely as fast as possible. For the second half of your run, you walk as slowly as possible until you are back to your starting point. What was your average velocity? That sounds as though it might be difficult to calculate, because the first and second half of your run took very different amounts of time. However, equation 4.11 gives you the answer. You ran in a loop, ending up back at your starting point. If you are back at your starting point, your total displacement $x(b) - x(a)$ over the period of your run is exactly zero. You ran in a circle, more or less, so in the end you didn't get anywhere. Your average velocity was, therefore, zero.

If you look carefully at the steps leading to the formula for the average velocity you will realize that I actually inserted an assumption without stating it. The assumption was that the average of interest is the *time* average, the average taken by assigning to each point in time an equal importance in computing the average. I could also have taken a spatial average, in which I noted that the object is moving through space and therefore at each point in space along its path it has some velocity. If I average over that path, assigning to each point in space an equal weight, I get a different average velocity, the *spatial average velocity*. As a practical matter, it turns out that the time average is usually the average of interest.

4.2 Discussion

Averages are ubiquitous in physics. For example, if we have a sample of radioactive material, a Geiger-Mueller counter reports each time that a radioactive decay is detected. The times at which a signal is detected are totally random and cannot even in principle be predicted. However, we can say on the average how many signals we will receive in a minute.

If I have a container filled with gas atoms, at any given time a certain number of them are pressing against the wall of the container. What pressure do I measure? I have no way to measure where all of the atoms in the container are located. Furthermore the molecules in the gas are moving, so the force that each exerts on some wall of the container fluctuates in time. I have no way to measure the velocity of every gas atom at the same time, so I can't calculate how the atoms will move, causing the pressure to fluctuate as time goes on. What I therefore do in *statistical mechanics* is to say that I will generate a list of all possible locations and positions for all of the gas atoms (this list is very long). For each item on the list, I can calculate the corresponding pressure, giving me a long list of pressures. I then calculate the correctly weighted average of that list of pressures; this is the average pressure found experimentally.

In *quantum mechanics*, we must use an average to calculate the value obtained from an extended series of independent measurements of the same quantity. The average is taken over a list of the allowed *states* of the system.

A polymer is a long, stringy molecule, somewhat resembling a long string of spaghetti. As a chemical molecule, linear polymers are assembled as a line of basic building blocks, with each block in the line linked to the next, somewhat resembling a railroad train. For most synthetic polymers, the processes used to make the polymer molecules are imperfect, so that every polymer molecule does not contain the same number of blocks. *Polymer physics* therefore finds it useful to discuss various average molecular weights, including using as the weight factors the number of molecules with each molecular weight, the mass of the molecules having each molecular weight, and the square of the mass of the molecules having each molecular weight.

You have now read an infinitesimal sampling referencing four branches of modern physics, namely nuclear physics, statistical mechanics, quantum mechanics, and polymer physics.

4.3 Worked Problems

1. I am asked for practical problems. You are working as the *physics consultant* for a motion picture producer. The producer is famous for producing highly profitable films, based on the theory that

pyrotechnics, high explosives, car chases, and incredibly maudlin scenes are a good replacement for acting, plot, characterization, and cinematographic technique. Assume SI units throughout. All numbers are accurate to the nearest hundredth of a unit. In this problem, more or less immediate numerical substitution is appropriate.

In the opening scene of the next motion picture, a film that will show up throughout these homework problems, a truck laden with explosives approaches a railroad crossing. The truck's motion is filmed and digitized by a surveillance camera. The supplied coordinates put the center of the crossing at $x = 40$. x increases from left to right. At $t = -20$ the truck driver notes that a train is approaching the crossing and floors her accelerator. At the time, the camera shows that the truck is at $x = -10$ and has a speed of 20 m/s to the right. The magnitude of the truck's acceleration is 5 m/s^2.

(a) Give equations for $v(t)$ and $x(t)$ of the truck in the seconds after the driver floors the accelerator, writing your answers in the exact forms

$$x = x_0 + v_0 t + 0.5at^2, \qquad (4.12)$$
$$v = v_0 + at. \qquad (4.13)$$

(b) Prove that your answers are correct by substituting $t = -20$ in these equations, and showing that you recover the initial position and velocity correctly.

(c) When did the truck cross the railroad tracks? If you did the calculation correctly, you should have found two answers to the question, one of which is wrong.

(d) The train crosses the intersection at $t = -17$. How long after the truck crossed the rails did the train cross the road?

(e) How fast is the truck going when it reaches the railroad tracks?

[Remarks: This problem only looks easy. When I gave it as a homework problem to freshman, more than half the class got this one wrong. Be aware of signs. You should always check your answers. You should automatically have done the $t = -20$ substitution, without being told, in order to check your answer.

2. One-dimensional motion. The velocity of an object as a function of time is parallel to the x axis, with $v_x = 2t^3 + 7t^2 + \cos(t) - 5$. Find the position, velocity, and acceleration of the object at $t = -1$. (Hint: there will be an unknown constant of integration someplace.) *Derive* a formula for the average acceleration, corresponding to the formula from today's class showing that the average velocity is $\bar{v} = \Delta R/\Delta T$. For this object, find the average velocity and the average acceleration between times -1 and 1.

4.4 Homework

1. A truck is approaching an intersection. The truck's motion is filmed and digitized. The supplied coordinates put the center of the intersection at $x = 15$. x increases from left to right. At $t = -20$s, the truck appears on the film at $x = -75$ m with a speed of 15 m/s to the right. The brakes are applied at this time. The magnitude of the resulting acceleration is 2 m/s^2.

(a) Give equations for $x(t)$ and $v(t)$ in the seconds after the brakes are applied, using equations in the form

$$x(t) = x_0 + v_0 t + 0.5at^2, \qquad (4.14)$$
$$v(t) = v_0 + at. \qquad (4.15)$$

(b) Prove your answers are correct by substituting $t = -20$ in your equations. Showing that you recover the initial position and velocity correctly.

(c) How fast is the truck going when it reaches the intersection?

(d) At what time does the truck stop?

4.4. HOMEWORK

Clue #1: Some years ago, almost a third of the class got this one wrong. That's much better than the year before, when over half the class was wrong.

Clue #2: Key question: what are the signs of the velocity and acceleration at $t = -20$? Some of you were taught that all numbers are positive, a wrong fact that I will attempt to unteach.

Clue #3: You should always check your answers. You should *automatically* have done the $t = -20$ substitution, without being told, to check your answer.

2. An aircraft starts its takeoff run at one end of a 5000 m runway. Its acceleration increases quadratically with time for the first two seconds, during which time its acceleration increases from zero up to a maximum of 3 m/s². After the first two seconds, its acceleration remains at a constant 3 m/s². How far does the aircraft travel down the runway in the first four seconds? What is the aircraft's average speed during the first four seconds of its takeoff run?

3. A car approaches and enters an intersection. The car's motion is filmed and digitized. The supplied coordinates put the center of the intersection at $x = 0$. At $t = +8s$ the car appears on the film at $x = 60$m with a speed of 15 m/s to the left. The brakes are applied at this time. The magnitude of the resulting acceleration is 1 m/s. Give the equations for $x(t)$ and $v_x(t)$ of the car in the forms

$$x(t) = x_0 + v_0 t + 0.5at^2, \tag{4.16}$$
$$v_x(t) = v_0 + at. \tag{4.17}$$

Prove your answers are correct by substituting $t = +8$ in these equations and recovering the original position and velocity.

4. A car approaches and enters an intersection. The car's motion is filmed and digitized. The supplied coordinates put the center of the intersection at $x = 0$. At $t = -8s$ the car appears on the film at $x = -10$m with a speed of 15 m/s to the left. The positive x axis points to the right. The brakes are applied at this time. The magnitude of the resulting acceleration is 1 m/s. Give the equations for $x(t)$ and $v_x(t)$ of the car in the forms

$$x(t) = x_0 + v_0 t + 0.5at^2, \tag{4.18}$$
$$v_x(t) = v_0 + at. \tag{4.19}$$

Prove your answers are correct by substituting $t = -8$ in these equations and recovering the original position and velocity.

5. A bus approaches an intersection. The bus's motion is filmed and recorded by its dashcam. The supplied coordinates put the origin at $x = +15$, the positive x direction being to the right. At $t = -10$, the driver notes that the light has turned yellow and floors the accelerator. At this time, the bus is at $x = 30$ and has a speed of 5 m/s to the left. The magnitude of the bus's acceleration is 2 m/s². (a) Give equations for the bus's position and acceleration, for the times after the accelerator was floored, in the exact forms

$$x(t) = x_0 + v_0 t + 0.5at^2, \tag{4.20}$$
$$v_x(t) = v_0 + at. \tag{4.21}$$

with x_0, v_0 and a replaced with numbers. (b) Prove that your answers are correct by substituting $t = -10$ in these equations, and showing that you recover the initial position and velocity. (c) How fast was the bus moving when it reaches the intersection? (d) The light turns red at $t = -9$. How long after the light turned red did the bus enter the intersection? Clue #1: Beware of signs. Clue #2. Always check your answers. You should automatically have done the $t = 10$ substitution. Clue #3: If you solved on a pocket calculator, and it claimed that a quadratic only has one root, something is seriously wrong.

6. A school bus approaches an intersection. The bus's motion is filmed and recorded by its dashcam. The supplied coordinates put the origin at $x = +10$, the positive x direction being to the right. At

$t = -10$, the brakes are applied. At this time, the bus is at $x = -50$ and has a speed of 25 m/s to the right. The magnitude of the bus's acceleration is 3 m/s².

(a) Give equations for the bus's position and acceleration, for the times after the brakes were applied, in the exact forms

$$x(t) = x_0 + v_0 t + 0.5at^2, \tag{4.22}$$
$$v_x(t) = v_0 + at. \tag{4.23}$$

with x_0, v_0 and a replaced with numbers. (b) Prove that your answers are correct by substituting $t = -10$ in these equations, and showing that you recover the initial position and velocity. (c) How fast is the bus moving when it reaches the intersection? (d) The light turns red at $t = -9$. How long after the light turned red did the bus enter the intersection? (Assume SI units throughout. All numbers are accurate to their nearest hundredth of a unit. More or less immediate numerical substitution is for once appropriate.) Clue #1: Beware of signs. Clue #2. Always check your answers. You should automatically have done the $t = 10$ substitution. Clue #3: If you solved on a pocket calculator, and it claimed that a quadratic only has one root, something is seriously wrong with your calculator.

7. A body has an acceleration $a(t) = a_o \sin(\omega t)$, where a_o and ω are constants. Find the velocity $v_x(t)$ and the position $x(t)$, assuming it started at $x(0) = 0$ and $x_x(t) = 0$.

4.5 Solutions to the Worked Problems

1. We now have a truck, loaded with explosives, being filmed for a motion picture. The vehicle is specified to be moving a constant acceleration (remember, 0 is a constant). You are first directed to write your solutions for the truck's position and velocity in the exact forms:

$$x(t) = x_0 + v_0 t + 0.5at^2, \tag{4.24}$$
$$v(t) = v_0 + at. \tag{4.25}$$

We now gather facts from the problem. We are consistently using SI units, so only the final answers need to have units identified. At $t = -20$, the truck is at $x = -10$ and is travelling at $v(-20) = +20$. The magnitude of the truck's acceleration is $+5$ m/s². "floors the accelerator" is a significant part of the problem; it indicates that the acceleration is $+5$ rather than -5.

(a) We may then write

$$-10 = x_0 + v_0(-20) + 0.5 \cdot 5 \cdot (-20)^2, \tag{4.26}$$
$$20 = v_0 + 5 \cdot (-20) \tag{4.27}$$

The second of these equations solves to give us $v_o = 120$ m/s. Inserting that value into the first of the above two equations tells us $x_o = 1390$ m. We therefore have

$$x(t) = 1390 + 120t + 0.25t^2, \tag{4.28}$$
$$v(t) = 120 + 5t. \tag{4.29}$$

That's the answer for part (a) of the problem.

(b) Part (b) asks: Are these results correct? You should automatically check to see if they are. If you did, you found

$$v(-20) = 120 + 5 \cdot (-20) = 20, \tag{4.30}$$
$$x(-20) = 1390 + 120(-20) + \frac{1}{2} \cdot 5 \cdot (-20)^2 = -10. \tag{4.31}$$

These two results agree with the initial conditions.

4.5. SOLUTIONS TO THE WORKED PROBLEMS

(c) When does the truck cross the tracks? It crosses when $x(t) = 40$, this being when t satisfies

$$40 = 1390 + 120t + 0.25t^2. \tag{4.32}$$

That equation is a quadratic in t. If you solve the quadratic, you find that the solutions are $t = -30$ and $t = -18$. If you did not find two solutions, something is wrong. If you are using a calculator with a "root finder", you should realize that some root finders only find one root, perhaps the wrong one. In this problem, $t = -30$ refers to events before the start of the filming; it is the wrong solution. The right solution is $t = -18$ s.

(d) The train crosses the road at $t = -17$, one second after the truck crosses the tracks.

(e) At $t = -18$, the truck was travelling at $v(t) = 120 + 5(-18) = 30$ m/s.

2. An object's velocity is parallel to the x-axis, with $v_x(t) = 2t^3 + 7t^2 + \cos(t) - 5$. We want the velocity, acceleration, and position at $t = -1$. To reach the latter two quantities we take a derivative and an integral with respect to time, gaining

$$a_x(t) = 6t^2 + 14t - \sin(t), \tag{4.33}$$

$$x(t) = \frac{2}{4}t^4 + \frac{7}{3}t^3 + \sin(t) - 5t + x_o. \tag{4.34}$$

At time $t = -1$ the position, velocity (we are seeing here only the x-component of these quantities), and acceleration are

$$x(-1) = \frac{1}{2}(-1)^4 + \frac{7}{3}(-1)^3 + \sin(-1) - 5(-1) + x_o, \tag{4.35}$$

$$v_x(-1) = 2(-1)^3 + 7(-1)^2 + \cos(-1) - 5, \tag{4.36}$$

$$a_x(-1) = 6(-1)^2 + 14(-1) - \sin(-1). \tag{4.37}$$

You could simplify further.

What is the average acceleration? It is

$$\bar{a}_x = \left[\int_{t_1}^{t_2} \frac{d^2x}{dt^2} dt\right] / \left[\int_{t_1}^{t_2} dt\right] \tag{4.38}$$

which reduces to

$$\bar{a}_x = \frac{v_x(t_2) - v_x(t_1)}{t_2 - t_1}. \tag{4.39}$$

To calculate the average acceleration between times -1 and +1 you need the velocity at +1, which is

$$v_x(1) = 2(1)^3 + 7(1)^2 + \cos(1) - 5, \tag{4.40}$$

so that the average acceleration between times -1 and +1 is

$$\bar{a}_x = \frac{v_x(1) - v_x(-1)}{1 - (-1)}. \tag{4.41}$$

The average velocity is given by

$$\bar{v}_x = \frac{x(1) - x(-1)}{1 - (-1)}. \tag{4.42}$$

This page reserved for your notes.

Chapter 5

Derivatives of Vectors, Circular Motion

5.1 Introduction

In this chapter, we consider derivatives of vectors, circular motion, and ballistic motion.

5.2 Derivatives of Vectors: Cartesian Coordinates

We have introduced vectors and vector notation, and we have reviewed calculus. We now turn to look at a particular example of derivatives, an example which is extremely important for the rest of this course. We now consider derivatives of vectors. In particular, we will consider the vector $\boldsymbol{r}$, the vector from the origin $\mathcal{O}$ of the coordinate system out to some point in space where a particle is located. The particle may be moving, so that $\boldsymbol{r} = \boldsymbol{r}(t)$. Corresponding to the position vector, there is a velocity vector $\boldsymbol{v}(t)$, which gives the velocity of the particle as a function of time, and an acceleration vector $\boldsymbol{a}(t)$, which gives the acceleration of the particle as a function of time. There is no requirement that the particle's acceleration be a constant. In general, $\boldsymbol{v}$ and $\boldsymbol{a}$ depend on time.

The position vector, giving the location of a particle in space relative to the origin of the coordinate system, can be written in Cartesian coordinates as

$$\boldsymbol{r} = x\hat{\boldsymbol{i}} + y\hat{\boldsymbol{j}} + z\hat{\boldsymbol{k}}, \tag{5.1}$$

where (x, y, z) are the coordinates of the particle and the three basis vectors $(\hat{\boldsymbol{i}}, \hat{\boldsymbol{j}}, \hat{\boldsymbol{k}})$ were introduced in Chapter 2. To obtain the velocity, we now take the time derivative of $\boldsymbol{r}$, finding the velocity vector

$$\frac{d\boldsymbol{r}}{dt} = \frac{dx}{dt}\hat{\boldsymbol{i}} + \frac{dy}{dt}\hat{\boldsymbol{j}} + \frac{dz}{dt}\hat{\boldsymbol{k}} + x\frac{d\hat{\boldsymbol{i}}}{dt} + y\frac{d\hat{\boldsymbol{j}}}{dt} + z\frac{d\hat{\boldsymbol{k}}}{dt}. \tag{5.2}$$

In obtaining this result, the rules for the derivatives of a sum and a product were applied repeatedly.

On the right hand side, the first three terms of the equation refers to the certainty that, as the particle moves, its x, y, and/or z coordinates change. On the right hand side, the last three terms of the equation refers to the possibility that at different points in space the unit vectors $\hat{\boldsymbol{i}}$, $\hat{\boldsymbol{j}}$, and $\hat{\boldsymbol{k}}$ do not point in the same directions, so that, as the particle moves, the time derivatives $\frac{d\hat{\boldsymbol{i}}}{dt}$, $\frac{d\hat{\boldsymbol{j}}}{dt}$, or $\frac{d\hat{\boldsymbol{k}}}{dt}$ might be nonzero. For Cartesian coordinates and Newtonian mechanics, this concern is a non-issue. In Newtonian mechanics the three Cartesian unit vectors point in the same three directions at every point in space. The first three terms of the equation are the velocity, while the last three terms all vanish. We can therefore write

$$\frac{d\boldsymbol{r}}{dt} = \frac{dx}{dt}\hat{\boldsymbol{i}} + \frac{dy}{dt}\hat{\boldsymbol{j}} + \frac{dz}{dt}\hat{\boldsymbol{k}}, \tag{5.3}$$

or equivalently

$$\boldsymbol{v} = v_x\hat{\boldsymbol{i}} + v_y\hat{\boldsymbol{j}} + v_z\hat{\boldsymbol{k}}, \tag{5.4}$$

in which $\boldsymbol{v}$ is the velocity vector and v_x, v_y, and v_z are the x, y, and z components of the velocity vector $\boldsymbol{v}$. In particular, we make the identifications

$$\frac{d\boldsymbol{r}}{dt} = \boldsymbol{v}, \tag{5.5}$$

$$\frac{dx}{dt} = v_x, \tag{5.6}$$

$$\frac{dy}{dt} = v_y, \tag{5.7}$$

$$\frac{dz}{dt} = v_z. \tag{5.8}$$

Earlier in the chapter, and in previous chapters, we discussed the velocity, for a system in which motion was confined to motion parallel to the x-axis. The object we there referred to as the velocity was a scalar; we see that the quantity we were talking about was in fact v_x, the x-component of the velocity vector $\boldsymbol{v}$.

A further time derivative brings us to the acceleration vector $\boldsymbol{a}$. We may write

$$\boldsymbol{a} = \frac{d\boldsymbol{v}}{dt} = \frac{d^2\boldsymbol{r}}{dt^2}, \tag{5.9}$$

and also

$$\boldsymbol{a} = a_x\hat{\boldsymbol{i}} + a_y\hat{\boldsymbol{j}} + a_z\hat{\boldsymbol{k}}, \tag{5.10}$$

Here $\boldsymbol{a}$ is the acceleration vector while a_x, a_y, and a_z are its x, y, and z components. On calculating the acceleration vector by taking the derivatives for its components, we find

$$\boldsymbol{a} = \frac{d\boldsymbol{v}}{dt} = \frac{dv_x}{dt}\hat{\boldsymbol{i}} + \frac{dv_y}{dt}\hat{\boldsymbol{j}} + \frac{dv_z}{dt}\hat{\boldsymbol{k}} \tag{5.11}$$

or

$$\frac{d\boldsymbol{v}}{dt} = \frac{d^2x}{dt^2}\hat{\boldsymbol{i}} + \frac{d^2y}{dt^2}\hat{\boldsymbol{j}} + \frac{d^2z}{dt^2}\hat{\boldsymbol{k}}, \tag{5.12}$$

where once again the three basis vectors are independent of time. We then recognize the equivalences

$$\boldsymbol{a} = \frac{d^2\boldsymbol{r}}{dt^2}, \tag{5.13}$$

$$a_x = \frac{d^2x}{dt^2}, \tag{5.14}$$

$$a_y = \frac{d^2y}{dt^2}, \tag{5.15}$$

$$a_z = \frac{d^2z}{dt^2}. \tag{5.16}$$

Finally, in some engineering applications it is important to recognize the *jerk*, a measure of how rapidly the acceleration is changing. The jerk $\boldsymbol{J}$ is the third time derivative of position, namely

$$\boldsymbol{J} = \frac{d^3\boldsymbol{r}}{dt^3}. \tag{5.17}$$

5.3 Circular Motion

We now turn from Cartesian coordinates to circular motion. In circular motion, a particle moves around a central point, remaining always at the same distance from that point. The particle's location can be represented by the circular polar coordinates r and θ. The vertical coordinate z remains the same as it was in Cartesian coordinates.

5.3. CIRCULAR MOTION

The Figure indicates the polar coordinates r and θ and how they are related to Cartesian coordinates x and y. A vector r may be written in terms of r and θ as

$$\boldsymbol{r} = r\cos(\theta)\hat{\boldsymbol{i}} + r\sin(\theta)\hat{\boldsymbol{j}} + z\hat{\boldsymbol{k}}. \tag{5.18}$$

For positions in the (x,y) plane r, θ, and the distance s measured along the circumference of a circle of radius r are related by

$$\theta = \frac{s}{r}. \tag{5.19}$$

as seen in Figure 5.1.

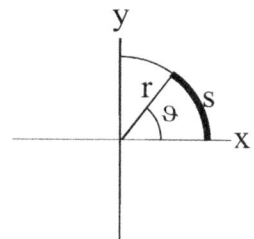

Figure 5.1: Coordinates r and θ, compared with x and y.

Equation 5.19 is the *definition* of θ. This equation also fixes the units used for θ, namely they are not degrees or grads or brads or mils but radians, a full circle being 2π radians.

For motion along the circumference of a circle, r is a constant, so

$$s = r\theta \tag{5.20}$$

and for the speed v_s around a circle we have

$$\frac{ds}{dt} = r\frac{d\theta}{dt}. \tag{5.21}$$

For the angular velocity we customarily introduce the symbol ω

$$\omega = \frac{d\theta}{dt}, \tag{5.22}$$

so that

$$v_s = r\omega \tag{5.23}$$

is the speed of an object moving in a circle at angular velocity ω.

For motion along a circle we may also start with

$$\boldsymbol{r} = r\cos(\theta)\hat{\boldsymbol{i}} + r\sin(\theta)\hat{\boldsymbol{j}} \tag{5.24}$$

and take a time derivative, obtaining

$$\frac{d\boldsymbol{r}}{dt} = \hat{\boldsymbol{i}}r(-\sin(\theta)\frac{d\theta}{dt} + \hat{\boldsymbol{j}}r\cos(\theta)\frac{d\theta}{dt} \tag{5.25}$$

which may be written

$$\boldsymbol{v} = -r\omega\sin(\theta)\hat{\boldsymbol{i}} + r\omega\cos(\theta)\hat{\boldsymbol{j}}. \tag{5.26}$$

The above equation refers only to motion of an object moving in a circle at a fixed distance from a center point. Correspondingly, it is the case that $\boldsymbol{r} \cdot \boldsymbol{v} = 0$. As shown in plane geometry, at any point on a circle the radius vector is perpendicular to the circle's tangent at that point. $\boldsymbol{v}$ is therefore purely tangential, so it is called the *tangential velocity*.

A further derivative with respect to t gives the circular acceleration. There are then two interesting cases. On one hand, the angular velocity ω may be constant, as seen to good approximation in the rotation of the earth, which turns on its axis once every 23 hours and 56 minutes (plus a few seconds). On the other hand, the angular velocity may be a non-constant function of time, i.e., $\omega = \omega(t)$.

For the case that ω is a constant, the acceleration becomes

$$\frac{d\boldsymbol{v}}{dt} = -r\omega\cos(\theta)\frac{d\theta}{dt}\hat{\boldsymbol{i}} - r\omega\sin(\theta)\frac{d\theta}{dt}\hat{\boldsymbol{j}} \tag{5.27}$$

or

$$\boldsymbol{a} = -r\omega^2\cos(\theta)\hat{\boldsymbol{i}} - r\omega^2\sin(\theta)\hat{\boldsymbol{j}}. \tag{5.28}$$

By inspection, $\boldsymbol{a} = -\omega^2\boldsymbol{r}$. That is, for circular motion at constant angular velocity, $\boldsymbol{v}$ and $\boldsymbol{r}$ are perpendicular to each other, while $\boldsymbol{a}$ and $\boldsymbol{r}$ are antiparallel. Also, by calculating $\boldsymbol{a} \cdot \boldsymbol{a}$, one can show that $|\boldsymbol{a}| = \omega^2 r$.

5.4 Ballistic Motion

In many cases, particle motion at constant acceleration is confined to a plane, traditionally labelled as the x- (horizontal) - z- (vertical) plane. The key physics result, on which more in the next chapter, is that in some interesting cases motions in the x and z directions are independent. Movement in one direction then does not create motion in perpendicular directions. A traditional case is *ballistic motion*, the motion of a rock hurled by a ballista (Roman bolt-thrower), in which on ignoring air resistance the acceleration of the particle in the vertical direction is $-g$ while the acceleration in the horizontal direction is 0. For ballistic motion, the motions of the particle in the x and z directions are independent of each other. The motions in the x and z directions share a common time coordinate t, but the position and velocity components in the x direction have no effect on the position and motion in the z direction, and vice versa.

The trajectory of a rock after being thrown is an example of ballistic motion. Often it is the case that the rock is thrown with some speed V while making an angle θ with respect to the horizontal, so that the initial velocity of the rock is

$$\boldsymbol{v} = v\cos(\theta)\hat{\boldsymbol{i}} + v\sin(\theta)\hat{\boldsymbol{k}}. \tag{5.29}$$

We now take a specific example. It's the Presidential Memorial Trajectory Problem. After all, President Garfield gave us a novel mathematics contribution, a new proof of the theorem of Pythagoras. Why should it be surprising that another President, who I shall neglect to name except to observe that he lived before almost all readers were born, gave us a harmless problem in trajectory calculations?

In any event, once upon a time, as covered in period news magazines, there was a President of the United States who would retire to his vast estate not in California and go for a very-high-speed drive on his private roads. Every so often, the driver's side car front window would go down, the Presidential arm would reach out, and another beer can would be dropped at the side of the road. He viewed this as a pleasant and much-needed bit of relaxation.

Now we ask the question. Ignore air resistance, which is admittedly an unrealistic approximation in this case. The driver simply lets go of the can, so it is initially travelling with the velocity of the car. Where does the can land, relative to where the car is, at the moment the can lands?

Solution:

We start with a figure. The car and can are initially each moving at V in the horizontal direction. The car has acceleration $\boldsymbol{a} = \boldsymbol{0}$. The can starts a distance H above the ground and has

$$\frac{d^2z}{dt^2} = -g. \tag{5.30}$$

At the start, the can and car therefore have

$$t = 0, \tag{5.31}$$
$$v_x = V, \tag{5.32}$$
$$v_z = 0, \tag{5.33}$$
$$z = H, \tag{5.34}$$
$$x = 0. \tag{5.35}$$

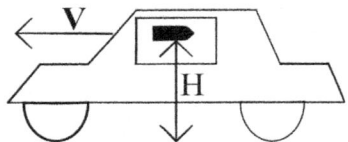

Figure 5.2: The Presidential Memorial Trajectory Problem.

The can hits the ground when it reaches height $z = 0$, which it does at some unknown time T. Where in the horizontal direction are the car and the can at time T?

The equations of motion for the can and for the car, in the horizontal direction, are

$$v_x(t) = v_{0x} + a_x t, \tag{5.36}$$
$$x(t) = x_0 + v_{0x}t + \frac{1}{2}a_x t^2, \tag{5.37}$$

and in the vertical direction are

$$v_z(t) = v_{0z} + a_z t, \tag{5.38}$$
$$z(t) = z_0 + v_{0z}t + \frac{1}{2}a_z t^2. \tag{5.39}$$

5.4. BALLISTIC MOTION

I have not yet substituted for the knowns. The above four equations are the correct starting point. The acceleration has components a_x and a_z. The can and the car move independently from each other, so in generating a solution the constants of integration must be found separately for the can and for the car. Some of the constants of integration may happen to be equal to each other, but that can't be assumed. You have to solve for all of them separately.

We now insert the known values into these equations. I have replaced variables that are equal to zero with their numerical values, but all other quantities are replaced by letters.

For the can in the horizontal direction at time $t = 0$:

$$V = v_{0x} + 0 \cdot 0, \tag{5.40}$$

$$0 = x_0 + v_{0x} \cdot 0 + \frac{1}{2} 0 \cdot 0^2. \tag{5.41}$$

Observe that when I insert values into the original equations *all* I do is to make a character by character replacement with no rearrangements at all. That way I can immediately see if I plugged the right number into the right equation. I then do a check: Do I have at least as many equations as I have unknowns? If I do not, solution is usually impossible. In fact, there are two equations and two unknowns (x_0, v_{0x}), so solution is usually possible. The solutions are $v_{0x} = V$ and $x_0 = 0$.

Usually possible? Why not *always*? Consider the two equations in two unknowns $x + y = 2$ and $x + y = 3$. There are two equations and two unknowns, but no solution is possible.

For the can in the vertical direction at time $t = 0$:

$$0 = v_{0z} - g \cdot 0, \tag{5.42}$$

$$H = z_0 + v_{0z} \cdot 0 - \frac{1}{2} g \cdot 0^2. \tag{5.43}$$

Once again, all I did is a character-by-character replacement, inserting the unknowns. I check: I have two unknowns, but I also have two equations. On solving, I find $v_{0z} = 0$ and $z_0 = H$.

I can now calculate when the can hits the ground, and where along the x-axis the can is located when it hits the ground. At that moment, $z = 0$, $t = T$, and $x = X$. We start with the equation for the vertical position, writing

$$0 = H + 0 \cdot T - \frac{1}{2} g T^2, \tag{5.44}$$

which is a quadratic in the unknown T. The equation has two solutions for T, namely $T = \pm(2H/g)^{1/2}$. In this case, the negative root is not the answer; it has the can reaching the road while the President was still drinking from it. The correct answer is the positive root $T = (2H/g)^{1/2}$.

Inserting this time into the equation for the horizontal motion of the can,

$$x = 0 + V(2H/g)^{1/2} + \frac{1}{2} \cdot 0 \cdot (2H/g), \tag{5.45}$$

which simplifies to $x = (2HV^2/g)^{1/2}$.

Now where is the car at this moment T? We need to start again at the beginning, and find the constants of integration for the motion of the car. Only the car's horizontal motion is of interest here. Beginning with equations 5.36, and noting that at time $t = 0$ the car is at $x = 0$ and $v_x = V$, on substituting we have for the car

$$V = v_{0x} + 0 \cdot 0, \tag{5.46}$$

$$0 = x_0 + v_{0x} \cdot 0 + \frac{1}{2} 0 \cdot 0^2. \tag{5.47}$$

These look much like the equations for the can, but we are in fact finding a new pair of constants of integration, the constants referring to the car. The solutions are $v_{0x} = V$ and $x_0 = 0$. For the position of the car at time $T = (2H/g)^{1/2}$ we have

$$x(T) = 0 + V(2H/g)^{1/2} + \frac{1}{2} \cdot 0 \cdot (2H/g) \tag{5.48}$$

or $x = (2HV^2/g)^{1/2}$.

Behold! At the moment the can strikes the ground, it has travelled exactly as far horizontally as the car has, so it lands immediately outside the driver's side car door. You can test this experimentally, without using a car, by holding an object out to the side, running fast in a straight line, and dropping the object while continuing to run. The object will be observed by others to land at your feet.

5.5 Discussion

In the real world, the statement that the directions of the Cartesian unit vectors for a moving object do not depend on time is incorrect. That's an outcome of Einstein's General Relativity. An object orbiting around a stationary or rotating mass is subject to de Sitter precession or the Lense–Thirring effect, respectively, in which the object's coordinate axes change as time advances. These effects have in this century been seen experimentally. (And now you have had a very small window open on another branch of physics.)

5.6 Worked Problems

1. *Calculus with vectors.* A particle has an acceleration $d^2\boldsymbol{r}/dt^2 = 3\hat{\boldsymbol{i}} - 5\hat{\boldsymbol{j}} + 6t\hat{\boldsymbol{k}}$. At $t = 2$ the particle is stationary and located at $\boldsymbol{r} = 2\hat{\boldsymbol{i}} - 5\hat{\boldsymbol{j}} - 5\hat{\boldsymbol{k}}$. Write the position of the particle explicitly (meaning that you solve for all the constants for which there are solutions) as a function of time (*as a function of time* means that your answer is a function of t; if I plug in for t I get the right answer for my t. Special case issue: If the particle is stationary at all times, then $\boldsymbol{v} = \boldsymbol{0}$ is correct and is a function of time, namely it is $\boldsymbol{v} = \boldsymbol{0}t^1$.).

2. *Ballistic motion.* The Royal Army of Grand Fenwick (hint: literary reference) is testing a new low-power crossbow. On its first trial, the bolt (arrow to non-SCA types) has a speed at launch of 24.2 m/s. It is launched by an archer standing in a trench, so the launch altitude is $z = 0$. The firing field is flat. The bolt returns to earth 44.5 m from the launch location. What was the firing angle (the angle between the initial velocity vector and the horizontal)? What altitude did the bolt reach at the top of its trajectory?

3. The *jerk* $\boldsymbol{J}$ is the third derivative of position against time, i. e.,

$$\boldsymbol{J} = \frac{d^3\boldsymbol{r}}{dt^3}. \tag{5.49}$$

For an object performing uniform circular motion (an object moving in a circle at constant speed), so that its location satisfies

$$\boldsymbol{r} = r\cos(\omega t)\hat{\boldsymbol{i}} + r\sin(\omega t)\hat{\boldsymbol{j}}, \tag{5.50}$$

compute $\boldsymbol{J}$, and find the angles that $\boldsymbol{J}$ forms with $\boldsymbol{r}$, $\boldsymbol{v}$, and $\boldsymbol{a}$.

5.7 Homework

1. According to the discussion following equation 5.26, for motion along a circle at constant speed $\boldsymbol{r} \cdot \boldsymbol{v} = 0$. By direct calculation, prove this claim.

2. Recalling the rotation of the earth, which turns on its axis once every 23 hours and 56 minutes, what is the Earth's rotational velocity in radians per second? Why is the required time 23 hours and 56 minutes, not 24 hours exactly?

3. We presented a calculation of $\boldsymbol{v}$ and $\boldsymbol{a}$ for an object moving in a circle at constant angular velocity. Was the speed of this object constant? Was the velocity of this object a constant? Repeat the calculation in the text, so as to obtain $\boldsymbol{v}$ and $\boldsymbol{a}$, for an object moving in a circle with an angular velocity that is not a constant, i.e., $\omega \equiv \omega(t)$.

5.7. HOMEWORK

4. Calculate $\boldsymbol{a} \cdot \boldsymbol{a}$ for circular motion at constant speed, and from that calculation show that $|\boldsymbol{a}| = \omega^2 r$. For circular motion, does constant speed imply constant velocity? Why or why not?

5. For a particle moving with constant jerk $\boldsymbol{J} = \frac{d^3 \boldsymbol{r}}{dt^3}$, find the particle's position as a function of time. (Hint: you will need at least three constants of integration.)

6. For circular motion at constant speed, calculate $\boldsymbol{v} \cdot \boldsymbol{a}$. What is the numerical value of this scalar product? What does that value say about the angle between $\boldsymbol{v}$ and $\boldsymbol{a}$?

7. In yet another scene from the motion picture, the villain has borrowed a school bus (represented in the Figure as a rectangular block) from its rightful owners and is taking an exit from Route 290 (curved line). The exit is a circle with a radius R of 150 m. The origin of the coordinate system is indicated. Because the State Police are in hot pursuit, the driver takes the exit at a constant speed of 65 mph (27 m/s). At the point indicated in the sketch, find the bus's position, velocity, and acceleration. The angle θ between the perpendicular to I-290 and the radius vector from the center of the circle to the bus is 60 degrees. We are on flat ground. (Hint: Is velocity a scalar or a vector?)

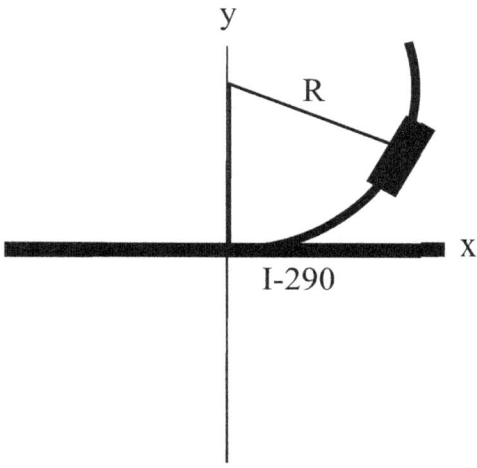

Figure 5.3: The escaping school bus.

8. While I am trying to work in my garden, a mosquito goes into a circular orbit just above my head. It flies at a constant speed. The circular orbit has a radius of 0.25 m. Because the mosquito is already overfed, it manages one complete circle in 20 s. Find the angular velocity of the mosquito relative to the center of the circle. Find the acceleration of the mosquito.

9. A particle has an acceleration $\boldsymbol{a} = 13t\hat{\boldsymbol{i}} + (6t^2 - 3t^5)\hat{\boldsymbol{k}}$. At $t = 3$, the particle is stationary and located at $\boldsymbol{r} = 5\hat{\boldsymbol{i}} - 11\hat{\boldsymbol{j}} + 7\hat{\boldsymbol{k}}$. Write the position of the particle explicitly as a function of time. The modifier *explicitly* means that you have solved for all constants that you can solve for. The phrase *as a function of time* means that your answer is a function of time; if I plug in some value for t I get the correct position at that time.

    Clue #1 Is this a constant acceleration problem?

    Aside #1 I spent a few lines explaining what "explicitly as a function of time" means. This is a standard phrase for specifying what problem is to be solved. You'll all be seeing it and its variations for the rest of your professional lives (And then after you retire, if you have grandchildren or great-grandchildren.)

10. Solve problem algebraically; insert numbers at the end. In the new extreme sport of skateboard-cliff diving, contestants on skateboards are brought up to a speed v (that they choose), and are fired off an inclined ramp into the waters of the Pacific. The ramp is located 20 m above the water; the objective of the contest is to hit the water exactly 20 m out from the edge of the cliff. The launch ramp is inclined at an angle of + 53 degrees from the horizontal. Find the initial speed needed by the skateboarder,

and the time needed from launch to reach the water. If your solution process is correct, you should have found two times. Why? What does the second time mean?

11. A group of engineering students are testing their new super-high-power slingshot by standing at the top edge of a 30m cliff above a lake, firing a ball bearing at an angle of 30 degrees above the vertical at an speed of 25 m/s. How far out from the cliff edge will the ball hit the water? A new, allegedly improved, firing mechanism is now introduced. The ball now lands in the water 100m out from the cliff edge. Assuming the same firitng angle was used, what was the speed of the ball at the moment of launch?

12. An object is located at $t = 0$ at $(x, y, z) = (2, 1, 4)$. Its velocity at this moment is $v(t) = -3\hat{i} + 7\hat{i} + 4\hat{k}$. Its acceleration is a constant $a(t) = -5\hat{i} - 2\hat{i} - 7\hat{k}$. What is $v(t)$ at $t(2)$? Where is the position $r(4)$ at $t = 4$? What is $\hat{r}(4)$?

13. A particle has an acceleration $\frac{d^2r}{dt^2} = 3t\hat{i} + 2\hat{j} + \cos(4t)\hat{k}$. At $t = 0$, the particle has velocity $v = 5\hat{i} - 7\hat{j} + \hat{k}$. At $t = 4$, the particle is located at $r = 3\hat{i} - 7\hat{j} + 4\hat{k}$. For simplicity, do not replace trig functions with their evaluated numerical values. (a) Write the position of the particle explicitly as a function of time. The phrase *explicitly* means you solve for all constants that you can solve for. The phrase *as a function of time* means that your answer is supposed to be a function of t.

    Clue #1 Wake up! Is this a constant acceleration problem?

    Aside #1 I just spent time explaining the phrase "explicitly, as a function of" means. The phrase is commonly used, and has the meaning explained here.

14. Consider a moving particle whose position is r, whose velocity is v, and whose acceleration is a. In the following, M, N, H, J, Q, and R are constants. (i) The x-component of the acceleration is $a_x = Mt^4 + Nt^3$. What are the x-components of the position and the velocity? (ii) The y-component of the position of the particle is $y = H\cos(Jt)$. From the above, what are the y-components of the velocity and the acceleration? (iii) The z-component of the velocity of the particle is $v_z = Q\exp(Rt)$. From the above, what are the z-components of the position and the acceleration? (iv) From the above, what is v? (v) Give an expression for the component of v in the direction $D = \hat{i} + \hat{j} + \hat{k}$.

15. In the opening scene of yet another film, the heroine is making her way into the villain's base by sliding down the inside of a large pipe. Unfortunately for the heroine, the pipe is a cleverly disguised air cannon. The villain fires the cannon, sending the heroine out the end of the pipe at a speed of 14.1 m/s² at an angle of 45 degrees above the horizontal. The end of the pipe is located at the edge of a cliff 20 m above the ocean surface. 40 m out from the edge of the cliff, cutting the water of the ocean in twain, is a steel net, beyond which is confined the villain's pet great white shark. Where does the heroine land? Will the hero need to remind her that great white sharks are an endangered species?

16. We are on the Moon, where there is no air resistance. You are emulating an early astronaut, who brought along a golf club and ball, and took a single successful swing. Assuming the ball starts off on the ground at speed V and an angle from the horizontal that is under your control, at what angle should the ball be launched in order to have a maximum distance out when it returns to the ground?

17. The motion picture crew is back. In the current episode, the heroine is loaded into a cannon and fired from the top of a cliff. The cliff is 8m tall. The air cannon has a muzzle velocity of 10 m/s and fires the heroine at an angle of 30 degrees above the horizontal. The producer and director are on a small boat at the rear of a 20m deep cave under the cliff. Naturally, they are the center of the universe, so they are located at the coordinate origin. The cannon is fired at $t = -5$. (a) Write equations for the position and velocity of the heroine, after the cannon is fired, in the forms

$$x = x_0 + v_0 t + \frac{1}{2}at^2, \qquad (5.51)$$

$$v_x = v_0 + at, \qquad (5.52)$$

and similarly for z and v_z. (b) At what value of t does the heroine hit the water? At what value of t is the heroine at the top of her trajectory?

18. A particle has an acceleration $\frac{d^2\mathbf{r}}{dt^2} = 5t\hat{\mathbf{i}} + t^3\hat{\mathbf{j}}$. At $t = 2$, the particle has velocity $\mathbf{v} = -2\hat{\mathbf{i}} + 4\hat{\mathbf{j}} + \hat{\mathbf{k}}$. At $t = 4$, the particle is located at $\mathbf{r} = 3\hat{\mathbf{i}} - 7\hat{\mathbf{j}} + 4\hat{\mathbf{k}}$.

5.8 Solutions to the Worked Problems

1. The acceleration vector in terms of its components is

$$\frac{d^2\mathbf{r}}{dt^2} = (3, -5, 6t). \tag{5.53}$$

The velocity and position result from two integrals with respect to time, leading to

$$\frac{d\mathbf{r}}{dt} = (3t + v_{0x}, -5t + v_{0y}, 3t^2 + v_{0z}), \tag{5.54}$$

$$\mathbf{r} = (\frac{3}{2}t^2 + v_{0x}t + x_0, -\frac{5}{2}t^2 + v_{0y}t + y_0, t^3 + v_{0z}t + z_0). \tag{5.55}$$

We have the conditions $\mathbf{v}(2) = \mathbf{0}$ and $\mathbf{r}(2) = (2, -5, -5)$, which lead to the vector equalities

$$(3 \cdot 2 + v_{0x}, -5 \cdot 2 + v_{0y}, 3 \cdot 2^2 + v_{0z}) = (0, 0, 0), \tag{5.56}$$

$$(\frac{3}{2}2^2 + v_{0x}2 + x_0, -\frac{5}{2}2^2 + v_{0y}2 + y_0, 2^3 + v_{0z}2 + z_0) = (2, -5, -5). \tag{5.57}$$

The above two vector equations correspond to six scalar equations, the first of which is $3 \cdot 2 + v_{0x} = 0$. Their solutions, written in vector form, are

$$(v_{0x}, v_{0y}, v_{0z}) = (-6, 10, -12), \tag{5.58}$$

$$(x_0, y_0, z_0) = (8, -15, 11). \tag{5.59}$$

which gives for the position as a function of time

$$\mathbf{r} = (\frac{3}{2}t^2 - 6t + 8, -\frac{5}{2}t^2 + 10t - 15, t^3 - 12t + 11). \tag{5.60}$$

2. You should recognize this problem as a variation on the Presidential can drop problem.

The bolt (crossbows fire bolts, not arrows) is subject to gravity, so the horizontal and vertical motions are described by

$$x(t) = x_0 + v_{0x}t, \tag{5.61}$$

$$z(t) = z_o + v_{0z}t - \frac{gt^2}{2}. \tag{5.62}$$

At time $t = 0$, we are at the origin, at which $x = 0$ and $z = 0$. This result lets you solve for x_0 and z_0, namely $x_0 = 0$ and $z_0 = 0$.

For the velocities, the basic equations are

$$v_x(t) = v_{0x}, \tag{5.63}$$

$$v_z(t) = v_{0z} - gt. \tag{5.64}$$

The bolt starts at $V \approx 24.4$ m/s while making an angle θ with respect to the horizontal. The components of the bolt's velocity vector at $t = 0$ are therefore $V_x = V\cos(\theta)$ and $V_z = V\sin(\theta)$. Substitution for v_x and v_z in the above pair of equations reveals $v_{0x} = V\cos(\theta)$ and $v_{0z} = V\sin(\theta)$. Those equalities are true because we start at $t = 0$, which in general is not the case.

At some later time T, the bolt is at the unknown $x(T) = X$ and $z(T) = 0$. What is X? (which requires solving first for T.) We can insert these quantities in the equations for motion at constant acceleration, getting

$$0 = 0 + V\sin(\theta)T - gT^2/2 \tag{5.65}$$

for motion in the z direction, and

$$X = 0 + V\cos(\theta)T \tag{5.66}$$

for motion in the x direction.

The first equation is a simple quadratic in T, whose solutions are $T = 0$ and $T = (2V\sin(\theta))/g$. The first root says correctly that the bolt started at height $z = 0$ and time $t = 0$, while the second root tells us when the bolt fell back to earth. Substituting the second root into the equation for X, we get

$$X = \frac{2V^2\cos(\theta)\sin(\theta)}{g} \tag{5.67}$$

for the bolt's range. In solving this problem, you were not actually told or required to compute T, but calculating T gives a straightforward path for finding X.

We now want to find θ, given that X was measured. Solving the above equation for θ, and using the trig identity $2\sin(\theta)\cos(\theta) = \sin(2\theta)$, we have

$$\sin(2\theta) = \frac{Xg}{V^2} \tag{5.68}$$

and hence

$$\theta = \frac{1}{2}\arcsin\left(\frac{44.5 \cdot 10}{(24.2)^2}\right). \tag{5.69}$$

Readers will note I have used a slightly crude approximation for g.

The trajectory is symmetric around its midpoint. The bolt is therefore at its maximum altitude at time $T/2 = (V\sin(\theta))/g$. Substituting this value of T in the equation for $z(t)$ gives the maximum altitude.

3. We have a definition of Jerk: $\boldsymbol{J} = \frac{d^3\boldsymbol{r}}{dt^3}$. Why are we interested in the jerk, the time derivative of the acceleration? Jerk is a good path to breaking things. We will now calculate $\boldsymbol{J}$ for circular motion at constant speed.

For circular motion we have already shown for the position, velocity, and acceleration that

$$\boldsymbol{r} = r\cos(\omega t)\hat{\boldsymbol{i}} + r\sin(\omega t)\hat{\boldsymbol{j}}, \tag{5.70}$$

$$\boldsymbol{v} = -r\omega\sin(\omega t)\hat{\boldsymbol{i}} + r\omega\cos(\omega t)\hat{\boldsymbol{j}}, \tag{5.71}$$

$$\boldsymbol{a} = -r\omega^2\cos(\omega t)\hat{\boldsymbol{i}} - r\omega^2\sin(\omega t)\hat{\boldsymbol{j}}. \tag{5.72}$$

The jerk is the time derivative of the acceleration, so

$$\boldsymbol{J} = r\omega^3\sin(\omega t)\hat{\boldsymbol{i}} - r\omega^3\cos(\omega t)\hat{\boldsymbol{j}}. \tag{5.73}$$

The magnitudes r, v, a and j are thus r, $r\omega$, $r\omega^2$, and $r\omega^3$, respectively.

If you calculate the scalar products, you find

$$\boldsymbol{r} \cdot \boldsymbol{J} = r^2\omega^3\sin(\omega t)\cos(\omega t) - r^2\omega^3\sin(\omega t)\cos(\omega t) = 0. \tag{5.74}$$

A similar calculation shows $\boldsymbol{a} \cdot \boldsymbol{J} = 0$. $\boldsymbol{J}$ is therefore perpendicular to $\boldsymbol{r}$ and $\boldsymbol{a}$

On the other hand,

$$\hat{\boldsymbol{v}} \cdot \hat{\boldsymbol{J}} = (-r^2\omega^5\sin^2(\omega t) - r^2\omega^5\cos^2(\omega t))/(r^2\omega^5). \tag{5.75}$$

Cancelling the $r^2\omega^5$ terms, and applying a trig identity,

$$\hat{\boldsymbol{v}} \cdot \hat{\boldsymbol{J}} = -1. \tag{5.76}$$

For uniform circular motion, $\boldsymbol{v}$ and $\boldsymbol{J}$ are antiparallel.

Chapter 6

Newton's Laws of Motion

6.1 Introduction

We have now considered units and dimensions, vectors, calculus, and a few simple motion problems generally described as kinematics. We are finally ready to discuss real physics.

Physics is the study of matter and motion, of forces and fields. Classical mechanics is based on the existence of particles, identifiable forces that act on those particles, and the response of the particles to those forces. The responses to those forces are described by Newton's three Laws of Motion. Newton worked in the 17th century, inventing the key mathematical tool, the calculus.

In the late 19th century, it became apparent that Newtonian mechanics did not appear to be entirely adequate. The specific heat of air, the specific heat of a vacuum within a metallic box, electrochemistry, the stability of atoms, the perpetual emission of heat by radium metal, the precession of Mercury's perihelion, and the tetrahedral bond arrangement of carbon in simple organic compounds all seemed beyond the ability of Newtonian mechanics to explain. In this century, quantum mechanics and general relativity explained these phenomena, which arise from processes taking place over very small distances, very short times, very high speeds, or in intense gravitational fields.

However, there are an extremely wide range of phenomena that are adequately described by Newtonian mechanics, because over macroscopic distances and speeds, not near stars, quantum mechanics and general relativity predict Newtonian behavior, at least very nearly.

Interpreting the Laws of Motion require some discussion of *reference frames*; we'll get that in the next chapter.

6.2 Newton's Laws

Newton's Laws of Motion begin with the existence of the quantity *momentum*. The momentum **p** is a vector. *In Cartesian coordinates*, the momentum of a particle is

$$\mathbf{p} = m\mathbf{v}. \tag{6.1}$$

In this equation, m is the mass of the particle and **v** is the particle's velocity. This equation satisfies the Curie principle. On the left-hand-side, **p** is a vector. On the right-hand-side, **v** is a vector, while $m\mathbf{v}$, being the product of a vector and a scalar, is also a vector. This equation equates two vectors, as is allowed by the Curie principle.

To repeat, equation 6.1 for the momentum *is only correct in Cartesian coordinates*. If you want to go to some other coordinate system, you have to use techniques beyond what is covered in this text to determine how to write **p**.

Newton's three Laws of Motion are:
The Second Law, which tells us that

$$\frac{d\mathbf{p}}{dt} = \mathcal{F}. \tag{6.2}$$

in this equation, $\mathcal{F}$ is the total force on the mass m due to all forces acting on it. Forces are vectors and add as vectors.

The First Law (which is actually a corollary of the Second Law) tells us that if there is no force on an object, then the object continues to move in a straight line without changing its speed.

And, finally, there is **The Third Law**, which I will give as *all forces come in action-reaction pairs*. Newton gave a somewhat different phrasing for this law, one which is a bit hard to follow. You'll find that my phrasing and explanation of the Third Law makes things much clearer.

Why should we believe that Newton's Laws of Motion are correct, at least over a large range of times, distances, masses, and velocities? The answer is that they reduce an extremely large number of different natural phenomena to a coherent whole. They provide a very simple description for a large part of natural behavior. They are not complete. At large velocities, the behavior of two parallel lines of clocks moving with respect to each other is not quite what Newton had expected. These behaviors are correctly described by invoking special and general relativity. How large is large? Actually, one deviation from Newton's understanding of time has been measured using a very good clock and a jet airliner. By the time you have an object moving as fast as an earth satellite, the effects are significant for engineering purposes. The satellite-based Global Positioning System would not work correctly unless corrections due to special and general relativity were applied. Newton's Laws of Motion also do not work on small distance scales. If Newton's Laws of Motion were correct on extremely small distance scales, atoms and molecules as we know them could not exist. However, there is a wide range of phenomena, including most of engineering, in which deviations from Newton's Laws of Motion are of minimal practical consequence.

We begin with Newton's Second Law, which tells us that the time rate of change of the momentum is determined by the total force on the object. The components of that equation read

$$F_x = \frac{dp_x}{dt}, \tag{6.3}$$

$$F_y = \frac{dp_y}{dt}, \tag{6.4}$$

$$F_z = \frac{dp_z}{dt}. \tag{6.5}$$

I've use the usual notation for the Cartesian components of the vector $\mathcal{F}$. This second set of equations begs the question of how we know what the forces on an object are. We'll get to that later.

In many problems, the mass of the object being considered is a constant that does not change in time. That's not always true; for example, when a rocket takes off and flies into orbit, it burns fuel as it accelerates. It becomes lighter and lighter as it climbs. If the mass of the object being considered does not change as time passes, we may rewrite Newton's Second Law as

$$F_x = m\frac{dv_x}{dt} = m\frac{d^2x}{dt^2}, \tag{6.6}$$

$$F_y = m\frac{dv_y}{dt} = m\frac{d^2y}{dt^2}, \tag{6.7}$$

$$F_z = m\frac{dv_z}{dt} = m\frac{d^2z}{dt^2}. \tag{6.8}$$

Mathematically, the First Law is simply a corollary of the Second Law. If the total force on an object is zero, then equation 6.8 becomes

$$0 = m\frac{d^2x}{dt^2}, \tag{6.9}$$

$$0 = m\frac{d^2y}{dt^2}, \tag{6.10}$$

$$0 = m\frac{d^2z}{dt^2}. \tag{6.11}$$

The accelerations (the second time derivatives of the position) are all zero. Dividing out the m and integrating

6.2. NEWTON'S LAWS

these three equations with respect to time, we obtain three constants of integration v_{0x}, v_0, and v_{0z}.

$$\frac{dx}{dt} = v_{0x}, \tag{6.12}$$

$$\frac{dy}{dt} = v_{0y}, \tag{6.13}$$

$$\frac{dz}{dt} = v_{0z}. \tag{6.14}$$

According to Newton's First Law, the velocity of an object subject to a net force of zero is a constant.

One might then ask why Newton called these two Laws the Second Law and the First Law, rather than numbering them the other way around. Part of the answer is that Newton did not work in a vacuum. Prior to Newton, there was an extensive intellectual belief as to the nature of motion. Newton rejected a large part of this belief.

Some authors will trace the theory of motion in question back to Aristotle, but Aristotle's idea of change was very different from anything we would recognize as a derivative with respect to time. In defense of Aristotle, he did not know calculus or algebraic geometry. Indeed, the Greeks did have excellent plane and solid geometry, but they were reasonably careful most of the time not to use diagrams in their proofs. Diagrams mislead, because they provide a specific example of a class of objects. It is then very easy to incorporate properties of that specific example into a proof, even though those properties are concomitant and not true for all objects in the class. Aristotle had one modest advantage over the ancient Romans, namely the Greek system for writing large numbers was much better than the Roman system, at least if you wanted to write a large number, say the number of grains of sand in a bucket.

[Aside: Why did the Greeks avoid diagrams? Consider a parallelogram with a line connecting the two opposite corners. The line divides the parallelogram into two similar triangles. That's a standard middle school plane geometry problem. However, that middle school proof actually contains an enormous hole. The hard part of the proof is to prove that the line between the two opposite corners lies someplace inside the parallelogram, so that it actually divides the parallelogram to two triangles, as opposed to lying outside the parallelogram, thus failing to divide the parallelogram into two triangles.]

What was the pre-Newtonian, pre-Galileo belief as to the nature of motion? A substantial body of thinkers believed in the notion of *impetus*. You fired a cannon at a wall. The cannon transferred to the cannonball a quantity of impetus. The cannon ball traveled through the air. As it moved, it exhausted its supply of impetus. When it ran out of impetus, it crashed to the ground. You may think of impetus as functioning like the fuel for a jet airliner. The airliner flies along until it runs out of fuel. It then goes into a glide until it hopefully reaches a runway. (At least one modern western airline has managed to do this. Twice.)

In Newton's time, there were good reasons to believe in the impetus model. You fired a cannon at a besieged fortress, while standing a safe distance behind the cannon. You saw the cannonball sail out, climbing as it went, and at some point the cannonball appeared to drop quite quickly to earth. Part of this sudden drop with less forward motion is an optical illusion. Part of this is that period cannon balls were very poorly made spheres, so that cannon balls had very large amounts of drag due to the air and did indeed slow down at the far end of their trajectories. To make matters more challenging, period muskets and cannon could not be aimed effectively. A standard European manual on siege cannon – you fire cannon balls at a wall until is has a hole in it – estimates that a cannon ball strikes the wall within something like fifteen degrees of the direction in which the cannon was pointed.

Galileo Galilei had done experiments which more or less demonstrated the First Law, the Law of Inertia, which says that an object moving under the influence of no forces will simply keep on moving. He experimented with rolling a smooth sphere down a ramp and then letting it climb back up a second ramp. He observed that if you made this second ramp shallower and shallower the sphere would still travel until it reached the same height as its starting point. (You may recognize this as the law of conservation of energy.) He reasoned that if the second ramp were made horizontal the rolling sphere would never get back to its original height and therefore would keep on going indefinitely. (Actually, there is something called rolling friction that will eventually bring the sphere to a stop.)

The other significance of the First Law is that it says there are coordinate systems in which if there is no force the velocity remains constant. When we discuss reference frames, we will come back to this point.

And now we reach the *Third Law*. My statement of the Third Law is quite different from the one you will have encountered elsewhere, the statement represented as an exact translation of Newton's words. The form you will read here does have exactly the same content. However, the point of having the Third Law is to get across a certain amount of information, not to repeat the not very good English translation, from centuries ago, of Newton's original book, which was, of course, written entirely in Latin, in the form of a plane geometry text.

The Third Law tells us:

All forces come in action-reaction pairs. If we call the two forces of an action-reaction pair $\mathbf{F}_1$ and $\mathbf{F}_2$, their properties are as follows.

1) The forces of an action-reaction pair are always on two different bodies. Two forces on the same body are never an action-reaction pair.

2) The forces in an action-reaction pair are equal in magnitude, so that $F_1 = F_2$. The forces in an action-reaction pair point in opposite directions, so that $\hat{\mathbf{F}}_1 = -\hat{\mathbf{F}}_2$.

3) As a result, $\mathbf{F}_1 = -\mathbf{F}_2$.

4) The forces in an action-reaction pair have the same physical basis. What do I mean "physical basis"? The answer will become clearer after further discussion.

5) The action and reaction forces are simultaneous. By 'simultaneous' I mean that it is incorrect to ask which force is the action force and which force is the reaction force. The two forces always come as a pair. If you call one of the two forces the "action force", then the other of the two forces is the "reaction force", but it does not matter which of the two forces you called the action force. (Aside: Some of you have already heard about special relativity, and the question of what 'simultaneous' means in comparing clocks. I am here using 'simultaneous' in a completely different sense.)

As a modest qualification, if one studies intermolecular forces there are also *three-body forces*, in which three molecules put a force on each other that cannot be split into a trio of pair forces. Three-body forces are actually quite important in properties of liquids, but for this course we will stay with two-body forces.

Critics of Newton propose that Newton's laws are actually content-free. That is, the way we know that there is a force is that there is a response, an acceleration or other forces, and therefore the Second Law is a definition of 'force', not a law of nature. We could imagine how this claim could be true. People have put into orbit around the earth thousands of earth satellites. If the force between each earth satellite and our planet was different from the force on each of the other satellites, so that corresponding to each earth satellite there was a different law of gravity determining its orbit, the laws of gravity would be little more useful than a list of the positions and velocities of each orbit at different times. That's not what happens. Instead, there is one law of gravity, with all satellites being subject to the same law of gravity. There is a single gravitational force that rules them all and in their orbits binds them. By introducing the correct forces, we take a huge numerical description of planetary orbits, and reduce it to a few equations and a relatively small number of constants. That's a strong argument that forces are real, and that we more or less know what they are.

The reality, a third of a millennium after Newton did his work, is that a very small number of forces and a few other bits and pieces describe a huge number of natural phenomena. The traditional list of forces that I learned as an undergraduate, is

- The strong nuclear force.

- The weak nuclear force.

- The electromagnetic force, as explained by Maxwell and Dirac.

- Gravity, as explained by Newton and Einstein

Separate from these is the Pauli exclusion principle, which constrains how electrons can move, but is not a force.

Now an amusing historical anecdote arises. I was at a small student seminar run by the physicist who supervised my bachelor's thesis. He came in with this neat new experiment he had just invented, and told no one else about yet, a device for detecting gravity waves, a definitive test of Einstein's General Relativity theory of gravity. We were the first people to hear about it. Four decades, thousands-of man-years, and a hundred million dollars later, the Large Interferometric Gravitational Observatory worked as hoped, and Rainer Weiss shared in the Nobel Prize for inventing it and making it work.

6.3 Applications of Newton's Laws

We now turn to some simple applications of Newton's Laws. As a simple example, consider me, having mass m, standing on a chair. I'll first show the process, and then explain what the steps were.

What do Newton's Laws tell us? We start by drawing a sketch, and identifying the forces acting on the person. There are two of them, the force of gravity $-mg\hat{\mathbf{k}}$ pulling the person downward, and the *normal force* $\mathbf{N}$ due to the chain pushing him upward. Gravity is always with us. The normal force may be less familiar. The normal force is a contact force; it arises because objects are touching each other. It keeps objects from moving through each other. $\mathbf{N}$ arises from the electron clouds surrounding atoms and molecules. The clouds very much resist moving through each other, so material objects cannot readily interpenetrate. The normal force can be explained in detail in terms of the forces on the previous page, but we will not do that here. $\mathbf{N}$ is called the *normal force* because it points normal (perpendicular) to the surface of the two objects that are touching each other.

Figure 6.1: your author standing on a chair.

Let's consider the person standing on the chair. Like everyone else, the person is subject to a force of gravity $-mg\hat{\mathbf{k}}$. So long as the person stays on the surface of the earth, that gravitational force, the person's weight, stays approximately the same.

The normal force differs from many other forces in that it does not have a fixed value; the strength of the normal force is what you need to keep two objects from moving through each other. In the case at hand, the person is standing on the chair. The normal force points in the direction opposite to the force of gravity and is exactly strong enough that the person does not go through the chair. Quantitatively, the normal force is not stronger than the force of gravity, so the person is never pushed up away from the chair. The normal force is also never weaker than the force of gravity, so the person never floats ghost-like through the chair.

Let's consider these forces as parts of action-reaction pairs. There is a normal force of the chair pushing up on the person. That force has a reaction force, the normal force of the person pushing down on the chair. Those two forces satisfy Newton's Third Law. They are on two different objects, the person and the chair. They are equal in magnitude and opposite in sign; one pushes up, the other pushes down. They are of the same physical nature; they are both contact forces, the normal force.

There is also the gravitational force of the earth on the person. That force pulls the person down. What is the reaction force to the gravitational force of the earth on the person? Why, it's the gravitational force of the person on the earth. Those two forces are on two different objects, the person and the earth. They point in opposite directions, and as we will see later in the course are indeed equal in magnitude. They have the same physical nature; they are both gravitational. The force and reaction force act at the same time, so they are *simultaneous* in the sense of this Chapter.

You might reasonably question my claim that the two gravitational forces are equal in magnitude. After all, the planet Earth is much more massive than I am, so it has a much stronger gravitational field. However, the Earth's gravitational field can only act on my near-hundred kilograms. My gravitational field, while very weak, acts on the chair, the foundations of the building, the core of the earth, and the bright sunlit waves breaking across the windswept Indian Ocean. My gravitational field is acting on the huge mass of planet Earth, and therefore manages to create on the Earth a force equal in magnitude and opposite in direction to the force of the Earth's gravitational field on me.

Are the forces $\mathbf{N}$ and $mg\hat{\mathbf{k}}$ acting on the person an action-reaction pair? No! They are acting on the same body, and have different physical natures, so they cannot possibly be an action-reaction pair.

Minor aside: mg is the magnitude of the gravitational force on the person. g is the constant that converts m into the gravitational force. No matter that it is often done, it is incorrect to call g as used here *the acceleration of gravity* because, so long as the chair does not collapse, the person is not accelerating. He is subject to a force, but he isn't accelerating.

We now take this list of forces and use it to create a *force diagram* for the person. Force diagrams are not a part of Newton's Laws of Motion. You can perfectly well solve all classical mechanical problems without using them. Indeed, you can march all the way through the full series of physics degrees and never hear of them. After all, I did. However, the force diagram is a useful mnemonic tool. It helps you confirm that you have identified all the forces in the problem and which bodies they are acting on.

In a force diagram, you represent the mass of interest by a large dot labeled with the mass of the object of interest. You now indicate on the force diagram all of the forces acting *on* the object of interest, representing each of them as an arrow pointing from the object in some direction, and a label indicating the magnitude of the force. In some cases, you actually don't know which way the force is pointing, so the arrow is inserted without believing that the force necessarily points in that direction. It might be the case that several forces are acting on a mass, all pointing in the same direction. The appropriate response is to show them deviated slightly to one side or the other so that each force is clearly visible. Figure 6.2b gives an example. The general rule is that a force diagram is a qualitative sketch of the problem, not a scale drawing. Finally, you indicate on the force diagram the coordinate axes being used to solve the problem.

Here are force diagrams for the person of interest and for a toy helium balloon.

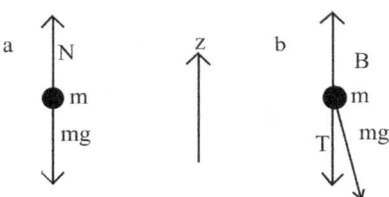

Figure 6.2: Two force diagrams. (a) Force diagram of the man standing on a chair. (b) Force diagram of a balloon on a string, including the buoyant force **B** due to the helium in the balloon, the force of gravity $-mg\hat{\mathbf{k}}$ on the balloon and the tension $-T\hat{\mathbf{k}}$ exerted on the balloon by its string. The two downward forces are exactly parallel to each other, but they are drawn slightly splayed with respect to each other so that they are both clearly visible.

The arrow pointing up labeled with the letter z is the vertical, z, coordinate. The black dot labeled m is the person. The two arrows pointing up and down are labeled N and mg. The labels could be the force components, or the magnitude of the forces. Writing the force of gravity on the force diagram as $-mg$ instead of $+mg$ is of no consequence. When we insert a force into a Second Law equation, the sign given to each term is determined by comparison between the direction of the force and the direction of the coordinate axes.

The force diagram emphatically *does not* include an arrow labeled $m\mathbf{a}$ to represent the mythical 'force of acceleration' or 'force of inertia'. There are no such forces, and therefore they are not shown on the force diagram. The choice of coordinate directions is up to the person solving the problem. Some choices of coordinate directions make a problem easier to solve. Some choices of coordinate directions make a problem more challenging to solve. However, a problem that is soluble with one choice of coordinate directions is equally soluble (the solution may look a bit different) with any other choice of coordinate directions.

We now invoke the Second Law and do substitution of the known forces. Starting with the Second Law, we have

$$m\frac{d^2\mathbf{r}}{dt^2} = \mathcal{F} \tag{6.15}$$

from which the z-component is

$$m\frac{d^2z}{dt^2} = F_z. \tag{6.16}$$

The force F_z can be read off the force diagram. One inserts a list of all forces pointing along the z-axis, and the z-component of any force pointing in some other direction. Each force must be inserted with the correct sign so that it is pointing in the right direction. In the case at hand, one obtains

$$m\frac{d^2z}{dt^2} = N - mg. \tag{6.17}$$

The person is not accelerating up or down, so d^2z/dt^2 is zero. Inserting this result into the Second Law, we obtain

$$N - mg = m \cdot 0, \tag{6.18}$$

leading to the solution $N = mg$. The normal and gravitational forces on the man are of equal magnitude and point in opposite directions, but they are not an action-reaction pair, because they act on the same object and have different physical natures.

A simple lecture demonstration brings this out. For the case of the hypothetical lecture-demonstration, I am standing on the chair. The legs of the chair have been carefully wired with small amounts of plastic

6.3. APPLICATIONS OF NEWTON'S LAWS

explosive. The plastic explosive is now detonated. The legs are reduced to tiny particles that recede in all directions. Of course, if this were physics as seen in children's cartoons, I would be standing on the chair, perfectly safely, until I looked down and noticed that the legs of the chair were no longer present. I would then fall. We are not doing cartoon physics, we are doing real physics. The moment we detonate the legs of the chair, the chair can provide no support, so I would begin to fall groundward.

Figure 6.3 shows the explosion and the force diagram. Note the direction that $+z$ is assigned in the figure, namely **straight down**. There is no physics reason for choosing this direction. The reason I reoriented the coordinate axis is to demonstrate that the signs chosen for different forces are not intrinsic; they are determined by the directions of the coordinate axes. In this case, gravity is pointing in the $+z$ direction, as is my acceleration, while the normal force is $-N\hat{\mathbf{k}}$, so the z-component of the Second Law becomes

$$-N + mg = mg. \qquad (6.19)$$

This equation has as its solution $N = 0$. N and mg are thus shown to be independent.

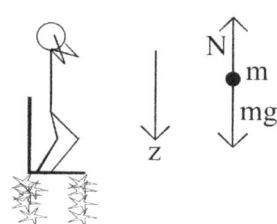

Figure 6.3: The chair legs have exploded! I fall downward at an acceleration g. A force diagram for this circumstance is displayed.

We now turn to a second problem illustrating the Laws of Motion. It's a propeller airplane pulling a glider. What is the acceleration of the glider, given that there is a force **P** on the airplane centered at the propeller? I begin by drawing a sketch of the problem, showing the airplane and glider, and collecting in one place all the information that I appear to need. The figure shows the propeller airplane, labeled '1' with a propeller in front and a tow rope tying it to a glider. The glider is labeled '2'. The masses of the airplane and glider are m_1 and m_2, respectively.

I now set up force diagrams for the airplane and the glider. In the vertical direction, the airplane and the glider are subject to gravitational forces $m_1 g$ and $m_2 g$ pointing down and lift forces L_1 and L_2 pointing up. In addition, the glider is subject to a force due to the tension T in the tow rope, that force pulling the glider in the $+\hat{\mathbf{i}}$ direction. The tension T also acts on the airplane in front, pulling on it in the $-\hat{\mathbf{i}}$ direction.

Finally, there is a force **P** present because the aircraft's propeller is spinning rapidly. The force that acts on the airplane is a force due to the air pushing on the propeller blade. The reaction force to that force on the airplane is a force due to the propeller blade pushing on the air. The force that moves the aircraft is the force on the propeller, this being the force of the air on the propeller. The force that the propeller puts on the air does move the air, but that is not a force on the airplane, so it does not contribute to moving the airplane.

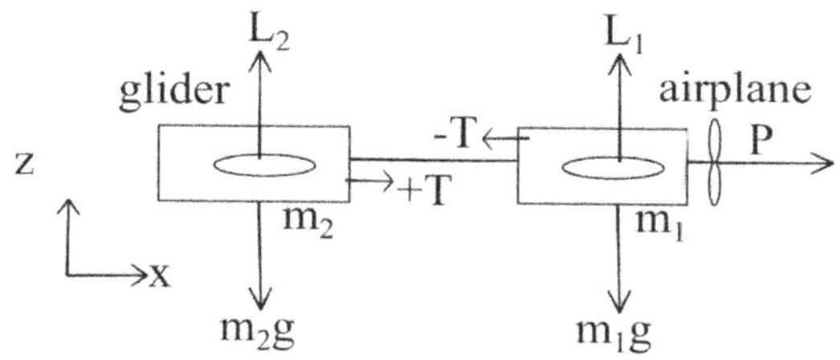

Figure 6.4: This figure is a *sketch* for the glider problem. *It is not a force diagram!* I indicate the airplane with propeller, headed right, the glider being towed behind, and the forces on each of them. The rope exerts a tension force on both vehicles. The sign on T is only a reminder to whoever is solving the problem that the two tension forces do not point in the same direction.

Let us pause to say something about tension. Tension is the force created by a solid object, such as a string or a rope, when it is stretched. On a microscopic level, when the object is stretched, the component atoms and molecule change their shapes, orientations, and distances apart, so that the object tries to pull its two ends in toward the center. If the object is very light, the forces it exerts at its two ends, on whatever is holding it, are very nearly equal in magnitude and opposite in direction. The forces exerted by a rope are always parallel to the rope. In the case here, where the rope is very light relative to the aircraft and the glider, the tension forces on the two ends of the rope are equal in magnitude. We will discuss 'very light' in a future chapter. The tension force never pushes away from the center of the rope, the principle being 'You can't push on a rope.'

We may now draw the force diagrams for the glider and the airplane. In the force diagram, each mass is indicated as a point. It is incorrect, because it leads to deceptive and wrong logic, to replace the point with a little sketch of the airplane. We need one force diagram for the glider and a separate force diagram for the airplane. It is simplest, but not required, that we put the coordinate axes in the same directions for the glider and for the airplane. We could switch one of them around, for example making $+z$ to be in the downward rather than the upward direction for the glider, but this leads to confusion without in this case netting any advantage.

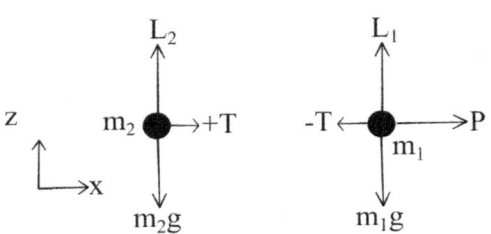

Figure 6.5: The force diagram for the glider problem,

We now write the Second Law for the two aircraft. To do things slightly differently, I will write the Second Law in its vector form. There is no new physics here relative to writing the Second Law in its scalar, component-by-component, form. In some problems, writing the Second Law componentwise makes life easier. In some cases, you are much better off writing the Second Law in its vector form.

From the force diagrams we obtain two force equations, one each for the propeller plane and the glider.

$$-T\hat{\mathbf{i}} + \mathbf{P} + L_1\hat{\mathbf{k}} - m_1 g\hat{\mathbf{k}} = m_1 \frac{d^2\mathbf{r}_1}{dt^2}, \qquad (6.20)$$

$$+T\hat{\mathbf{i}} + L_2\hat{\mathbf{k}} - m_2 g\hat{\mathbf{k}} = m_2 \frac{d^2\mathbf{r}_2}{dt^2}. \qquad (6.21)$$

In writing these equations, there is no reason to suppose that the lift forces on the two aircraft, the masses of the two aircraft, or the position vectors of the two aircraft are the same. To distinguish between them, I have labeled L, m, and r with subscripts, the numbers 1 and 2 referring to the propeller plane and the glider, respectively.

How do we solve these equations? Suppose we specify that the aircraft and glider are in level flight, with a known $\mathbf{P}$, pointing horizontally, that causes the aircraft and glider to accelerate in the horizontal direction. The masses m_1 and m_2 of the airplane and glider are take to be known. In level flight, the vertical components of the two accelerations are zero, so the vertical components of the two Second Law equations are

$$L_1 - m_1 g = m_1 \cdot 0, \qquad (6.22)$$
$$L_2 - m_2 g = m_2 \cdot 0. \qquad (6.23)$$

Here the two vertical accelerations were replaced by their values, these being zero. We have two equations and two unknowns, so we expect to be able to solve for L_1 and L_2.

The horizontal components of these equations are

$$-T + P = m_1 \frac{d^2 x_1}{dt^2}, \qquad (6.24)$$

$$T = m_2 \frac{d^2 x_2}{dt^2}. \qquad (6.25)$$

If we look carefully, we see there is a problem. P, m_1, and m_2 are all known. However, that leaves us with three unknowns, namely T and the two accelerations. We have therefore two equations and three unknowns. How can we solve?

The answer is that there is another equation, a *constraint*. Constraints put limits on the values of different variables. In the case at hand the constraint is that the distance between the airplane and the glider is fixed at some distance k that is determined by the length of the rope. This gives us a new equation

$$x_2 - x_1 = k. \qquad (6.26)$$

Taking the second time derivative of this equation, we find

$$\frac{d^2 x_2}{dt^2} - \frac{d^2 x_1}{dt^2} = 0. \qquad (6.27)$$

Equation 6.27 is a third equation connecting the three unknowns, so we can solve for T and the accelerations. The constraint that the accelerations of the airplane and glider are equal, rather than having, e.g., different values in a fixed ratio, is an accident: This constraint happens to be true in this particular problem. In another problem, there may be a different constraint. Finally, note that we omitted any drag force on the airplane and glider; that neglect is not a good approximation here.

6.4 Discussion

We have now introduced Newton's Three Laws, which tell us
First, the Second Law

$$\mathcal{F} = \frac{d\mathbf{p}}{dt}. \tag{6.28}$$

Then, the First Law, which is simply a corollary of the second: If $\mathcal{F} = \mathbf{0}$, then the momentum $\mathbf{p}$ is a constant.

Finally, the Third Law: All forces come in Action-Reaction pairs. If we call the two forces of an action-reaction pair $\mathbf{F}_1$ and $\mathbf{F}_2$, their properties are as follows.

1) The forces of an action-reaction pair are always on two different bodies. Two forces on the same body are never an action-reaction pair.

2) The forces in an action-reaction pair are equal in magnitude, so that $F_1 = F_2$.

3) The forces in an action-reaction pair point in opposite directions, so that $\hat{\mathbf{F}}_1 = -\hat{\mathbf{F}}_2$. As a result, $\mathbf{F}_1 = -\mathbf{F}_2$.

4) The forces in an action-reaction pair have the same physical basis. What do I mean "physical basis? The answer will become clear after further discussion.

5) The action and reaction forces are simultaneous.

Having said this, there is a simple grammatical trick for identifying the reaction force to a specified force. The specification of a force indicates the symbol for the force, what type of force we are talking about, the source of the force, and the target of the force. Thus, mg is the magnitude of the force of gravity of *the earth* on **the man**. The reaction force to the force of gravity on the man is a force of magnitude mg of **the man** on *the earth*.

To find the reaction force, all we have to do is to exchange the source of the force and the target of the force. Because the action and reaction forces are simultaneous, we could then write: The reaction force to the force of gravity mg by the man on the earth is a force of magnitude mg of *the earth* on **the man**.

6.5 Worked Problems

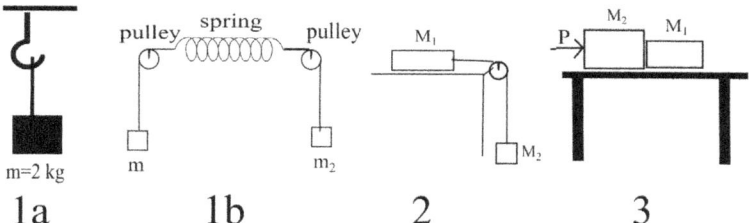

Figure 6.6: Figures for Worked Problems 1, 2, and 3.

1. (a) Figure 6.6, part 1a, shows a 2 kg mass hanging at the end of a rope. The top of the rope is wrapped securely around a hook that is attached to the ceiling. For the 2 kg mass, draw the force diagram. *In complete sentences*, identify each of the forces acting on the mass, including the nature of the force and the object applying it. In complete sentences, for each force in your force diagram, identify the reaction force, the object applying it, the nature of the force, and the object on which the reaction force is acting. (b) Figure 6.6, part 1b, shows two masses suspended by ropes; the ropes at their tops ends each go over a pulley and are connected by a spring. Repeat part (a) of this problem for the left-hand-mass and the left hand rope in the Figure. Hint: Including its own weight, there are four forces acting on the rope. The answer to this problem is quite long but should not be complicated. Hint: The phrase "complete sentence" has an exact meaning.

2. For the masses shown in Figure 6.6, part 2, find the force diagrams for the two masses, and for each force identify the object applying it, the nature of the force, and the object on which the force is acting.

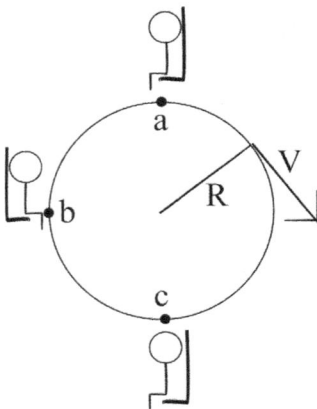

Figure 6.7: The Ferris Wheel problem.

M_1 and M_2 are connected by a rope that goes over a pulley, the pulley being represented in the figure by a circle.

3. For the masses shown in Figure 6.6, part 3, find the force diagrams for masses M_1 and M_2, and for each force identify the object applying it, the nature of the force, and the object on which the force is acting. The masses are resting on a horizontal table surface. [Hint: The force **P** pushing M_1 to the right is "external"; its nature and source cannot be determined from the problem as given.]

4. A stereo speaker is suspended from the ceiling by two wires. The speaker has mass m; the tensions in the two wires are T_1 and T_2. Draw the force diagram for the speaker. In complete sentences, identify each of the forces acting on the speaker. In complete sentences, for each force in your force diagram, identify the reaction force, the object applying it, the nature of the force, and the object on which the reaction force is acting.

5. An interesting example of a circular motion problem that turns out to be best solved by keeping the Second Law in its vector form is provided by that amusement park devisement, the Ferris Wheel. The wheel, radius R, is shown in Figure 6.7. Someone is sitting in a seat as the wheel goes around in a circle at constant speed v. What is the force on the person due to the chair at points a, b, and c? Hints: Two forces are acting on the person, a force **N** due to the chair, whose direction you do not know, and a force $-mg\hat{\mathbf{k}}$ due to gravity.

6.6 Homework

The first five problems all refer to Figure 6.4.

1. For the two aircraft in Figure xx, break the vector form of the Second Law into its components.

2. Solve for the tension in the rope and the acceleration of each aircraft.

3. The forces on the two ends of the rope due to the airplane and the glider are equal in magnitude, opposite in direction, and are both contact forces. Are they an action-reaction pair? Why not?

4. The lift force L_1 is the aerodynamic force of the air on the airplane. What is the reaction force to L_1?

5. For each of the forces acting on the airplane and the glider, identify the reaction force.

6. Calculate the acceleration of a falling body having mass m, using the skew coordinates indicated in Figure 6.8, in which the x and y axes are each at angle θ with respect to the horizontal or vertical, respectively.

7. A bow is used to fire an arrow into an arc in the air. Point a is after the archer has released the arrow, but before the arrow has left the bow. Point b is after the arrow has left the bow. Point c is the highest point reached by the arrow. Points d and e are on the way down. Write the free body diagram for the arrow at each point. What is the acceleration of the arrow at point c?

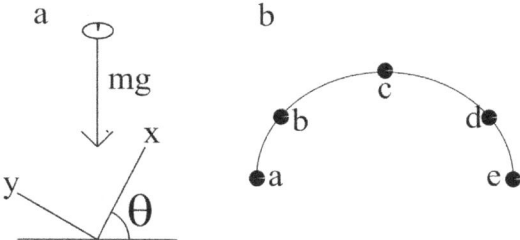

Figure 6.8: Figures for problems 6 and 7.

8. A man and a boy, masses M and m, respectively, both wearing ice skates, stand facing each other on wet, slick ice. The man pushes on the boy with a force **P** of magnitude 300 Newtons. Draw the complete force diagrams for the man and the boy. In complete sentences, specify the **reaction** forces corresponding to each force in your two force diagrams. Is the magnitude of the force with which the boy pushes back on the man smaller, larger, or the same as the force with which the man is pushing on the boy? In specifying a force, give the physical nature of the force, the object on which it is acting, and the object supplying the force.

9. A decorative sign is attached to the ceiling by two wires. The wires are tied to the sign at two different points, and slant away from each other and up toward the ceiling. Draw the complete force diagram for the sign. In complete sentences, specify the forces shown on your force diagram, and identify for each force its corresponding reaction force. In specifying a force, give the physical nature of the force, the object on which it is acting, and the object supplying the force.

10. (a) A bug collides with an automobile windshield. The bug exerts a force **F** on the windshield. The windshield exerts a force **F'** on the bug. What is the quantitative relationship between **F** and **F'**? (b) The Earth exerts a gravitational force $-mg\hat{\mathbf{k}}$ on the car. Correspondingly, the car's wheels exert a force $N\hat{\mathbf{k}}$ on the earth. Are $-mg\hat{\mathbf{k}}$ and $N\hat{\mathbf{k}}$ an action-reaction pair? Why or why not?

11. A mass m hangs from a string. The far end of the string is attached to the ceiling. For the mass, and for the string, write the force diagram, labeling each force. Identify the reaction force to each force. In identifying each force and reaction force, specify the nature of the force, the object exerting the force, and the object on which the force is being exerted. Bonus solution: We have a mass m resting on top of another mass m', which in turn rests on the ground. Find the force diagrams for m and m', and specify the reaction forces to each of the forces in your diagrams.

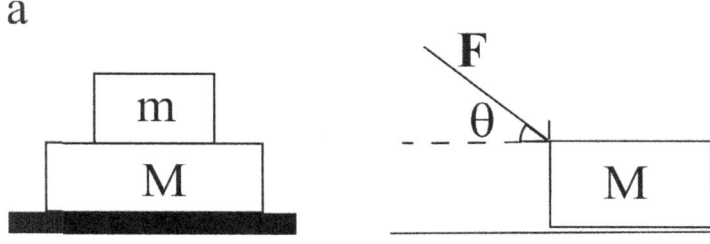

Figure 6.9: Figures for problems (a) 12 and (b) 13.

12. See Figure 6.9. A book having mass m is resting on a table having mass M in my office. The table is resting on the floor, which is part of the planet Earth. Find the force diagrams for the book, the table, and the Earth, and identify each force in a few words. For each force, identify the reaction force, including the nature of the force, the object exerting the force, and the target of the force. *Using the Second Law*, calculate the normal force exerted by the floor on the table.

13. See Figure 6.9. I push a trunk, mass M, along a well-waxed (frictionless) surface by applying to it a constant force **F** as shown. (a) Give the force diagram of the trunk. (b) Compute the normal force of the floor on the trunk. (c) Are (1) the force of the Earth's gravity on the trunk, and (2) the normal force of the floor on the trunk, an action-reaction pair? Why? (d) Obtain $x(t)$ and $v(t)$ of the trunk as functions of time, giving the most general correct forms for the solutions. (You may need to use some unknown constants.)

14. The new and improved rocket car has been given a new engine. The force that the engine applies on the car, over the time of interest, increases with time as $\mathbf{F}(t) = F_0 t^2 \hat{i}$. Here F_0 is a constant. The rocket car with engine has mass m. The rocket car is parked on flat horizontal ground. Find the force diagram. Find the acceleration of the rocket car, the velocity of the rocket car, and the position of the rocket car as functions of time.

15. The perpetrator of a prank gone wrong is hanging onto a rope a small distance out from the side of a building. The rope exerts on him a tension force of 800N straight up. He has a mass of 90 kg. Find his acceleration.

16. For the next scene in the movie, the lead actress is strapped into the rear seat of a high-performance combat aircraft, for whose use a considerable sum of money has been paid by the studio. [Some years ago, in a certain foreign country, you could actually do this.] The aircraft takes off and flies in a vertical circle at a constant speed of 300 m/s. The circle has a radius of 2000 m. At the top and bottom of the circle, and half-way in between, find the force that the aircraft exerts on the actress, who is taken to have a mass of 65 kg.

17. An automobile accelerates in the $+x$ direction in a straight line from a standing start, Draw a complete force diagram for the automobile. For each force on the automobile, identify the source and direction of the force.

18. We have a jogger on a running track. Discuss whether or not the following pairs of forces are an action-reaction pair: (a) The jogger's feet push down on the running track. The surface of the running track pushes up on the jogger's feet. (b) The earth's gravity pulls down on the jogger. The surface of the running track pushes up on the jogger's feet. (c) The jogger's feet push down on the running track. The earth's gravity pulls down on the jogger. (d) (b) The earth's gravity pulls down on the jogger. The jogger's feet push down on the running track.

19. Consider a crate resting on an inclined board. The crate is being pulled up the board by a rope that lies parallel to the board. Draw a complete force diagram for the crate. For each force on the crate, identify the nature of the force, the object exerting the force, and the direction of that force. Identify the reaction forces to the force on the crate. Identify the reaction forces to the forces on the crate, including the object exerting the force, the object on which the force is exerted, the physical nature of the reaction force, and the direction of the reaction force.

20. An experimental aircraft of mass M has a novel engine whose thrust depends on time as $T = T_o(1 - \exp(-\Gamma t))$, where T_o and Γ are constants determined by the details of the engine design. Here t is the time measured from the instant that the engine is started. The airplane is placed on a runway aligned along the x-axis and pointed in the $+\hat{\mathbf{i}}$ direction. The engine is started. This is a taxi test; the airplane does not leave the ground. The airplane may or may not have been at a stop when the engine is started. a) Find the x-component of the velocity of the aircraft as a position of time. b) Find the x-component of the position of the aircraft as a position of time.

6.6. HOMEWORK

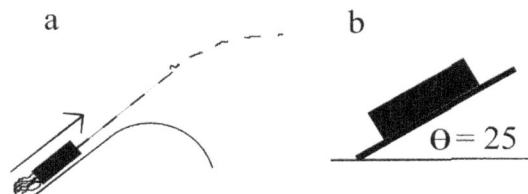

Figure 6.10: Figures for problems (a) 21 and (b) 22.

21. See Figure 6.10. Our rocket-propelled car, as described in lecture, reaches Worcester. In Worcester, it is to go up one of our steeper hills and over the top, where it goes airborne. (Sketch gives trajectory.) At $t = 0$, the car reaches the base of the hill, located at $(x, y, z) = (0, 0, 0)$. The car rolls up the hill at a constant speed (measured parallel to the hill) of 70 m/s. The slope of the hill is 45 degrees. At $t = 3$, the car reaches the top of the hill, located at $x = 150, z = 150$. At this point, the car leaves the ground, and the engine is turned off. The car moves under the influence of gravity until it hits the ground. For credit, you must use my coordinate origins to solve the problem. (a) Give the correct force diagrams for the car before and after the car leaves the ground. Identify each force with a short phrase. (b) For each force in your free body diagram, identify fully the corresponding reaction force. (c) A photographer wishes to take a picture of the car when it is 100m above the ground, and descending. The photographer needs to know the x coordinate and the exact time at which the car will be 100m above the ground. What x and t will give the photograph showing the car at this altitude?

22. See Figure 6.10b. A 200 kg desk is firmly bolted to a horizontal office floor. (i) Give the free body diagram for the desk. Label all forces, and identify each force in a short phrase. (ii) This is California. There is now an earthquake. The earthquake causes the office and desk to have a time dependent displacement which for most of the earthquake may be written

$$\vec{r} = 0.2\cos(7t)\hat{i} + 0.1\sin(7t)\hat{k} - 0.1(1 - \exp(0.8t))\hat{j}. \quad (6.29)$$

Here the x and z motion describe the earthquake oscillations, while the y motion refers to the local continental plate sliding toward the ocean. Find the velocity and acceleration of the desk during the earthquake. (iii) What was the total force on the desk during the earthquake? (iv) After the earthquake, the floor has gained a tilt of 25 degrees as shown in the sketch. Give the free body diagram for the desk. Label and identify in **complete English sentences** each of the forces on the desk. What is the reaction force to the force of the earth's gravity on the desk?

23. We now transport ourselves forwards to the twenty-second century, onto a manned space station in a low orbit around the earth. The station is a cubical box whose base always faces the earth. An astronaut is floating, stationary with respect to the walls of the space station, in the middle of the space station, not in contact with any object. Draw the free body diagram for the astronaut. Is the total force on the astronaut on the astronaut zero, or not? Why? Is the astronaut accelerating or not? Why?

24. A 4-kilogram mass is subject to a force $3\hat{i}+7\hat{k}t$ and a force $5\cos(\omega t)$ in the $+y$ direction. The frequency ω is 5 radians per second. At $t = 0$, the mass is stationary and located at the origin. (i) Find the position $\mathbf{r}$ and the velocity $\mathbf{v}$ of the mass as functions of time. (ii) What is the location of the mass at time $t = 2$?

25. It is a typical winter day in Worcester. The streets are all covered with two inches of polished ice, reducing friction to zero. The rocket-engine-powered car our second lecture has reached Worcester, and is trying to climb the hill on Salisbury Street located just west of Park Avenue. (i) Give the force diagram for the car. (ii) Write component by component the equations of motion $\vec{F} = m\frac{d^2\mathbf{r}}{dt^2}$ for the car. (iii) Find the minimum required thrust $|\mathbf{T}|$ (force applied to the car due to the rocket engine) if the car is to climb the hill at constant speed. (iv) If the car is to climb the hill at a constant road speed of 5 m/s, assuming a 5.0×10^3 kg car and a hill slope of 10 degrees, find as exactly as possible

the thrust of the rocket engine. (v) Find the normal force $\mathbf{N'}$ applied to the road by the car's tires. Assume that the tires carry the full weight of the car.

26. Consider an car accelerating up a steep hill. The wheels and engine are part of the car. (i) Give the force diagram for the driver. For each force, identify the nature of the force, the direction of the force, and the reaction force. I'll give you the first entry: Force mg is the force of the earth's gravity on the driver, pointing straight down. The reaction force is the force of the driver's gravity pulling up on the earth. (ii) Repeat part (i) of this problem, but this time give the force diagram for the car. For each force, identify the nature of the force, the direction of the force, and the reaction force. (iii) The car is now parked in a flat driveway. The forces on the car are the earth's gravitational force $-mg$, pointing straight down, and the normal force N due to the driveway, pointing up. Is N the reaction force to $-mg$? Why or why not? Explain in words, in one or more complete sentences.

27. This question asks for thought and discussion, not equations. Answer in complete sentences. Do not resort to writing equations in place of words. (i) According to Newton's Second Law, the vector sum of all forces acting on a body is equal to the mass of the body times its acceleration. What can we conclude about a body's motion if the vector sum of all forces acting on the body is zero? (ii) Some sources state Newton's third law in the following way: "When two bodies exert mutual forces on one another, the two forces are always equal in magnitude and opposite in direction." If the two forces are always equal in magnitude and opposite in direction, then their total is always zero. Consider your answer to part (i). Assume that Newton's laws of motion are correct. If the total force is always zero and $\mathbf{F} = m\mathbf{a}$, how is it possible for any object to be accelerated by forces due to other bodies?

28. A 37 kg mass hangs from the end of a spring whose equilibrium length is 2.00 m. The mass oscillates up and down over a distance $A = 10$ cm, at a frequency $\omega = 3$ radians/s, so that we may write the position of the mass as
$$z = A\cos(\omega t). \tag{6.30}$$
z is the displacement of the mass from its equilibrium position. What is the force F on the mass? (Hint: F depends on time.)

29. For each of the following, identify the correct answer, and explain why each choice is right or wrong:
(i) An airplane ($m = 100,000$ kg, $v = 340$ m/s) collides with a moth ($m = 0.001$ kg, $v = 1$ m/s). During the collision, the airplane applies a force of magnitude F to the moth; the moth applies a force of magnitude f to the airplane. We may say with certainty that:

(i) $F > f$

(ii) $F = f$

(iii) $F < f$

(iv) Either none of the above are true, or more than one of the above could be true.

(ii) The same airplane ($m = 100,000$ kg, $v = 340$ m/s) now suffers a second and more serious collision, this with a giant flying creature escaped from a 1950's sci-fi thriller. The giant flying creature is large and fast (wingspan = 300 m, $m = 30,000,000$ kg, $v = 2000$ m/s) During the collision, a few pieces of the airplane are knocked free and thrown violently upwards. For a piece having mass m of 1 kg, we may say with certainty that

(i) The force of gravity on the piece is substantially changed during the collision by the acceleration of the collision.

(ii) The force of the planet earth's gravity on the piece is larger in magnitude than the force of the piece's gravitational field on the planet earth.

(iii) During the collision, the force of gravity on the piece is approximately 10 Newtons straight down, so from $F = ma$ the piece accelerates at 10 m/s² straight down.

(iv) The above three statements are all false.

(iii) On a fine winter day, an automobile skids across the library parking lot. The parking lot is covered with sheet ice, so the brakes have no effect. Until the collision, the car is subject to no horizontal force.

6.7. SOLUTIONS TO THE WORKED PROBLEMS

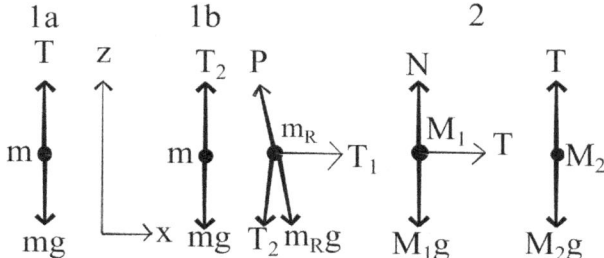

Figure 6.11: Force Diagrams for Worked Problems 1 and 2; labels 1a, 1b, and 2 correspond to those problems and parts..

The automobile then collides with a student on a skateboard. If the magnitude of the force the car exerts on the skater is F, and the magnitude of the force that the skater exerts on the car is f, we may say with certainty that

(i) $F > f$.

(ii) $F < f$.

(iii) Since there is no horizontal friction, the car and the skateboarder have no way to exert a force on each other.

(iv) None of the above statements is true, or more than one of the above statements could be true.

30. (i) In complete sentences or valid equations, state Newton's three laws of motion. (ii) List the properties that all action-reaction force pairs must have.

31. An A. D. 1940 jet aircraft, mass M, is parked at one end of the runway, preparing to take off. At time $t = 0$ the engine is engaged, and the aircraft begins to move down the runway. This is a very early jet airplane, so the engine takes from time $t = 0$ to some much later time T to come to full power. Between times 0 and T the thrust (the force on the airplane due to the engines) may be written

$$\mathbf{F} = \hat{\mathbf{i}} ct. \tag{6.31}$$

c is a constant. $\mathbf{F}$ depends on time. The aircraft remains on the ground until after time T. (i) Beginning with Newton's Laws of Motion, obtain explicit equations for the acceleration, velocity, and position of the aircraft as a function of time. Note that $|\mathbf{F}|$ is increasing with increasing time. (ii) At $t = T/2$, the aircraft's horizontal speed is 30 m/s. What is the aircraft's horizontal speed at $t = T$? (iii) At time $T/2$, the aircraft has travelled a distance $\Delta x = 300$m down the runway. How far down the runway has the aircraft travelled by time T?

6.7 Solutions to the Worked Problems

1. (a) $\mathbf{T}$ is the tension in the rope, a contact force applied to the mass. The reaction force to $\mathbf{T}$ is a contact force $-\mathbf{T}$ from the mass onto the rope. $-mg\hat{\mathbf{k}}$ is the force of the earth's gravity on the mass. The reaction force to $-mg\hat{\mathbf{k}}$ is the force $mg\hat{\mathbf{k}}$ of the mass's gravity on the earth.

(b) Mass m: $-mg\hat{\mathbf{k}}$ is the force of the earth's gravity on the mass. The reaction force to $-mg\hat{\mathbf{k}}$ is the force $mg\hat{\mathbf{k}}$ of the mass's gravity on the earth. $\mathbf{T}$ is the tension in the rope, a contact force applied to the mass. The reaction force to $\mathbf{T}$ is a contact force $-\mathbf{T}$ from the mass onto the rope.

(c) Rope: $\mathbf{P}$ is the contact force of the pulley on the rope. The reaction force to $\mathbf{P}$ is the contact force $-\mathbf{P}$ of the rope on the pulley. How can I be sure that $\mathbf{P}$ points in the direction indicated in the figure? I can't, other than from experience in solving this sort of problem. The actual direction comes out of the algebraic calculation when the problem is solved. The force diagram is a qualitative tool, a mnemonic device to help you be sure that you have not forgotten a force when you set up the second law equations, not a devisement that gives you quantitative solutions.

T_1 is the force due to the spring acting on the rope. The reaction force to T_1 is a contact force $-T_1$ from the rope acting on the spring.

$-m_R g \hat{k}$ is the force of the earth's gravity on the rope. The reaction force to $-m_R g \hat{k}$ is the force $mg\hat{k}$ of the rope's gravity on the earth.

T_2 is the contact force due to the left-hand-mass acting on the rope. Its reaction force is $-T_2$, the force exerted by the rope on left-hand mass. T_2 and $-m_R g \hat{k}$ are drawn splayed apart for clarity; they may well actually be parallel.

2. (a) The force diagram shows the forces N, T, and $M_1 g$ acting on M_1. N is the normal force, a contact force, of the table acting on M_1. T is the tension force, a contact force, of the rope acting on M_1. $M_1 g$ is the gravitational force of the earth acting on M_1.

   (b) The force diagram shows the forces T and $M_2 g$ acting on M_2. T is the tension force, a contact force, of the rope acting on M_2. $M_2 g$ is the gravitational force of the earth acting on M_2.

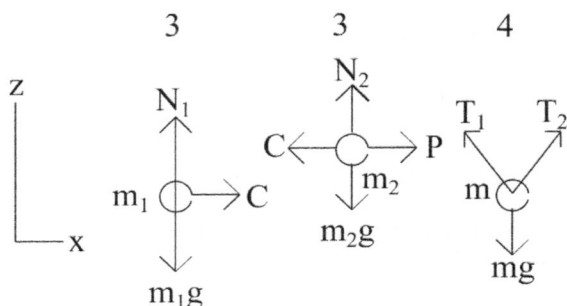

Figure 6.12: Force Diagrams for Worked Problems 3 and 4.

3. (a) The force diagram shows the forces N_1, C, and $M_1 g$ acting on M_1. N_1 is the normal force, a contact force, of the table acting on M_1. C is the contact force of M_2 pushing on M_1. $M_1 g$ is the gravitational force of the earth acting on M_1. $M_1 g$ always points vertically downward.

   (b) The force diagram shows the forces N_2, C, P, and $M_2 g$ acting on M_2. N_2 is the normal force, a contact force, of the table acting on M_2. C is the contact force of M_1 pushing on M_2. $M_2 g$ is the gravitational force of the Earth acting on M_2. P is the external force of the man pushing on M_1.

4. mg is the force of the Earth's gravity acting on the stereo speaker. T_1 and T_2 are the tension forces, contact forces where the wires touch the speaker, acting on the stereo speaker. $-mg$ is the force of the stereo's gravitational field acting on the Earth. $-T_1$ and $-T_2$ are the contact forces, where the wires touch the speaker, of the stereo speaker acting on the wires.

   Note that $-mg$, $-T_1$, and $-T_2$ are not forces applied to the speaker.

5. **The Ferris Wheel**

   We apply the Second Law $\mathbf{F} = m \frac{d^2 \mathbf{r}}{dt^2}$, finding

   $$\mathbf{N} - mg\hat{k} = -\frac{mv^2}{R}\hat{\mathbf{R}}. \tag{6.32}$$

   Solving for $\mathbf{N}$, this equation becomes

   $$\mathbf{N} = mg\hat{k} - \frac{mv^2}{R}\hat{\mathbf{R}}. \tag{6.33}$$

At the three points A, B, and C, $\hat{\mathbf{R}}$ is $+\hat{\mathbf{k}}$, $+\hat{\mathbf{i}}$, and $=-\hat{\mathbf{k}}$, respectively. At point A, the force due to the seat is

$$\mathbf{N} = (mg - \frac{mv^2}{R})\hat{\mathbf{k}}, \tag{6.34}$$

so the force due to the seat is less than would be expected from the rider's weight. At point C, the force due to the seat is

$$\mathbf{N} = (mg + \frac{mv^2}{R})\hat{\mathbf{k}}, \tag{6.35}$$

so the force due to the seat is more than would be expected from the rider's weight. In both cases, $\mathbf{N}$ is parallel or antiparallel to the radius vector $\mathbf{R}$.

Point B is a bit more interesting. At point B,

$$\mathbf{N} = mg\hat{\mathbf{k}} - \frac{mv^2}{R}\hat{\mathbf{i}}. \tag{6.36}$$

This $\mathbf{N}$ has both horizontal and vertical components, and is not parallel or antiparallel to $\mathbf{R}$.

This page reserved for your notes.

Chapter 7

Inertial and Non-Inertial Reference Frames

We now come to the topic of inertial and non-inertial reference frames.

I will actually begin with a seemingly unrelated idea, the *ansatz*. That's originally a word in German, describing where you initially place a tool on starting a piece of work. When applied to mathematics or physics, the notion of an *ansatz* is that you make an initially unsupported assumption, and when you are done with the analysis you see everything hangs together. A slightly different description of *ansatz* is 'that's how it's done'; you carry out the analysis in a certain way, and you get an answer that agrees with theory or experiment.

Newton's Laws are an example of an *ansatz*. You do things in a certain way, and you find that everything hangs together. All satellite orbits are reduced to an equation for the gravitational force and a relatively modest number of numerical constants. Philosophers of science who try to force all of Newtonian mechanics into a middle-school plane geometry model, with a rigid starting point and logical deductions from that point, sometimes become argumentative, because in a certain sense Newtonian Mechanics is an *ansatz*, not a plane geometry set of proofs from a few axioms and definitions.

The *ansatz* is hidden in the Second and First Laws of Motion. To see where we have an *ansatz*, we first introduce the notion of *reference frames*. The reference frame is the structure underlying three-dimensional Cartesian coordinates. The reference frame specifies an origin $\mathcal{O}$ and directions for the x, y, and z coordinate axes. We label each reference frame by the location of its origin and assume that we can take the coordinate axes in all reference frames to remain parallel to each other at all times. We also assume that we have an array of clocks spread across space, all running at the same rate and all synchronized (set to agree with each other), so that if we want to know what time it is at some point in space we can just look at the neighboring clock. These are Newtonian physics assumptions, and are excellent approximations for a very large number of purposes. Einstein's special relativity changes these assumptions.

The location $\mathbf{r}(t)$ of a moving particle is the vector $\mathbf{r}$ from the origin $\mathcal{O}$ to the particle's location at the time t. Now suppose we introduce a second reference frame with a second origin $\mathcal{O}'$. We clearly can do that if we want to. As the new origin is $\mathcal{O}'$, we will call the vector, from the new origin to the particle, $\mathbf{r}'(t)$. I will also introduce the vector that takes us from $\mathcal{O}$ to $\mathcal{O}'$. That's the vector $\mathbf{R}$, given by

$$\mathbf{R} = \mathcal{O}' - \mathcal{O}, \tag{7.1}$$

which gives us

$$\mathbf{r}(t) = \mathbf{R} + \mathbf{r}'(t). \tag{7.2}$$

What does this say about the velocity or the acceleration of the particle, as measured in each reference frame? If we take time derivatives, we get for the velocities in the two frames

$$\frac{d\mathbf{r}(t)}{dt} = \frac{d\mathbf{R}}{dt} + \frac{d\mathbf{r}'(t)}{dt}; \tag{7.3}$$

and for the accelerations in the two reference frames we obtain

$$\frac{d^2\mathbf{r}(t)}{dt^2} = \frac{d^2\mathbf{R}}{dt^2} + \frac{d^2\mathbf{r}'(t)}{dt^2}. \tag{7.4}$$

If the two origins are stationary with respect to each other, then

$$\frac{d\mathbf{R}}{dt} = \mathbf{0}, \tag{7.5}$$

$$\frac{d^2\mathbf{R}(t)}{dt^2} = \mathbf{0}. \tag{7.6}$$

In this case, the velocities and accelerations measured in the two reference frames are equal to each other.

Suppose one reference frame is moving at constant velocity with respect to the other. A familiar example is seen in driving a car on the interstate. Suppose car A is in the right lane, driving due east at 65 miles per hour. That's 65 miles per hour relative to the road surface. We also have car B in the left lane, driving due east at 75 miles per hour relative to the ground. "relative to the ground" defines the ground as a reference frame, the reference frame $\mathcal{O}$. Now I define a second reference frame whose origin $\mathcal{O}'$ is the driver's seat in car A. $\mathcal{O}'$ is moving due east with respect to $\mathcal{O}$ at 65 miles per hour. How fast is car B moving in the $\mathcal{O}'$ frame, i.e., how fast is car B moving with respect to car A? The answer is that car B is passing car A with a relative speed of ten miles per hour, meaning that car B's velocity in the $\mathcal{O}'$ frame is 10 miles per hour due east.

If the two cars are moving at constant velocities, then their relative acceleration $\frac{d^2\mathbf{R}(t)}{dt^2}$ is zero. What does this say for the Second Law as applied to car B? We have for the Second Law

$$m\frac{d^2\mathbf{r}(t)}{dt^2} = \mathcal{F}, \tag{7.7}$$

using the acceleration of car B as measured in the $\mathcal{O}$ frame. That gives us $\mathcal{F}$ for the force on car B, as calculated in the $\mathcal{O}$ frame. Now suppose I rewrite this is terms of the acceleration of car B in the $\mathbf{O}'$ frame, and insert it into the Second Law? I obtain

$$m\left(\frac{d^2\mathbf{r}'}{dt^2} + \frac{d^2\mathbf{R}(t)}{dt^2}\right) = \mathcal{F}. \tag{7.8}$$

However, the relative velocity of the two cars is a constant, so the second time derivative of $\mathbf{R}(t)$ is zero. This result gives us

$$m\frac{d^2\mathbf{r}'}{dt^2} = \mathcal{F}. \tag{7.9}$$

Even though equation 7.9 refers to the acceleration of car B as measured in the $\mathbf{O}'$ reference frame, equation 7.9 is still the calculation of the force on car B in the $\mathbf{O}$ reference frame. No matter whether I use the acceleration of car B as measured in the $\mathcal{O}$ reference frame or in the $\mathcal{O}'$ reference frame, I get the same answer for the force on car B. **Warning**. This result only refers to two reference frames that are moving with respect to each other at a constant relative velocity, so their relative acceleration is zero. Remember that $d\mathbf{R}/dt = \mathbf{0}$ is a constant velocity, so if the two cars are stationary with respect to each other the above result still applies.

However, suppose car A has applied its brakes. In that case, the location of its driver's seat, the location $\mathcal{O}'(t)$ of the origin of the $\mathcal{O}'$ reference frame, is accelerating with respect to the location $\mathcal{O}(t)$ of the origin of the $\mathcal{O}$ reference frame. From equation 7.1, in this case

$$\frac{d^2\mathbf{R}(t)}{dt^2} \neq \mathbf{0}. \tag{7.10}$$

and therefore

$$\mathcal{F} = m\left(\frac{d^2\mathbf{r}'}{dt^2} + \frac{d^2\mathbf{R}(t)}{dt^2}\right). \tag{7.11}$$

Equation 7.8 still gives us the force $\mathcal{F}$ on car B as calculated using car B's acceleration in the $\mathbf{O}$ reference frame. However, if we used car B's acceleration as measured in the $\mathbf{O'}$ reference frame, we would obtain for the force $\mathcal{F}'$ on car B as calculated in the $\mathbf{O'}$ frame

$$\mathcal{F}' = m \frac{d^2 \mathbf{r}'}{dt^2}. \tag{7.12}$$

If you compare the above two equations, you notice that $\mathcal{F} \neq \mathcal{F}'$. The force on car B appears to depend on which reference frame you use to calculate it. How can we have laws of nature such as Newton's Law of Gravity or Coulomb's equation for the force between two bodies, if the force depends on your choice of reference frame? We have now reached the *ansatz* I mentioned at the beginning of the chapter.

The answer is provided by the First and Second Laws. The answer is: We distinguish between *inertial* and *non-inertial* reference frames. Newton's Laws of Motion are correct in inertial reference frames. Frames, plural? Yes. The First and Second Laws say that there is some reference frame $\mathcal{A}$ in which Newton's Laws are correct. In that case, from equations 7.7-7.9, in any reference frame $\mathcal{B}$ moving at constant velocity with respect to reference frame $\mathcal{A}$, the calculated forces are exactly the same as they were in the original reference frame $\mathcal{A}$, so Newton's Laws of Motion are as valid in reference frame $\mathcal{B}$ as they were in reference frame $\mathcal{A}$. This collection of reference frames in which Newton's Laws of Motion are valid, these being a collection of reference frames that are all moving with respect to each other at constant velocities, are the *inertial reference frames*.

However, there are also reference frames that have non-zero accelerations with respect to the inertial reference frames. In these reference frames, if you use the Second Law and the acceleration to calculate the force, you get a different answer for the force. These are the *non-inertial reference frames*. The *non-inertial reference frames* can be partitioned into groups; in each group, the members of the group are moving with respect to each other at constant velocity, but are not accelerating with respect to each other. The First Law does not directly tell you which group of reference frames are the inertial reference frames. In each group, there is some arrangement that causes things to move in straight lines at constant speed. People in other reference frames would say that those things are not moving in straight lines at constant speed, and therefore they have forces on them.

How do we sort out which reference frames are the inertial frames? The now ansatz rears its beautiful head. There is a set of reference frames on which the forces have a simple and attractive set of forms. If we say these are the inertial reference frames, we find a simple set of force laws, and everything hangs together in a consistent manner. It's not 'we use the reference frames and accelerations to find the forces' or 'we use the forces to tell when the total force is zero so the First Law applies'; it's 'we can work this relationship in either direction, and everything is consistent with everything else.'

(Special relativity aside: Newton's assumptions about space and time are not quite correct in the real world. The assumption that coordinate axes can all be made parallel to each other, and that they stay that way as time goes on, is invalid. The assumption that clocks are as described by Newton fails twice. First and more important, imagine I have a line of clocks sitting on the ground next to an extremely long straight railroad track. I am being passed by an extremely long railroad train, with a clock in each car. I and my team of students confirm that my clocks are all synchronized. The conductor and his team aboard the train confirm that all the clocks in the train are synchronized. We then each look at the other person's clocks, correct for speed of signal delays, and discover that the other person is wrong: The other person's clocks are not synchronized. Instead, we each observe that the clocks in a line of moving clocks are out of synchrony with each other. We also both observe that the time interval measured on any one clock, when measured in the other reference frame, necessarily on two clocks as one measures the rate of the one clock by recording when it goes by first one and then the other of the two clocks, is shorter as measured on the one clock than on the two clocks.)

Discussion Aside: Suppose I am in an elevator, standing on a bathroom scale, and find that I am exerting a force mg straight down on the scale, where m is my body mass. It might be the case that I am on Earth in a stationary light-tight elevator cage. It might be I am in outer space, distant from any source of gravity, and the elevator is accelerating straight up with acceleration g. Einstein's derivation of General Relativity begins with the assertion that these two circumstances are exactly the same. From a strictly local measurement made inside the elevator, you cannot tell whether you are in a gravitational field or are inside an accelerating body. From that statement and special relativity, you eventually reach the Einstein Field Equations for gravity.

This space reserved for your notes.

Chapter 8

Applications of Newton's Laws of Motion

8.1 Introduction

In this chapter, we consider using Newton's Laws of Motion to solve problems in mechanics. As I said at the start of the book, everything after the last Chapter is simply an application. We now start on applications. I've already outlined a general approach to applying the Laws of Motion. That solution method is

1. draw a sketch of the system, assigning an algebraic character to each variable, including variables whose values are known.

2. for each mass in the problem, sketch a force diagram. The force diagram for each mass should show the symbol for the mass, an arrow representing each force in the problem, its corresponding algebraic symbol, and a corresponding direction.

3. Choose coordinates for each mass, and indicate those coordinates on the force diagram for that mass.

4. Write the Second Law for each mass.

5. Break the forces and the accelerations into their components, and write the Second Law for each component.

6. Identify any constraints in the problem.

7. Confirm that you are ready to solve. Do you have as many equations, counting constraints, as you do unknowns?

8. Solve for the unknowns.

Each problem begins with the needed sketches and force diagrams.

8.2 The Rocket Car on a Hill

Let us consider an example. We have the rocket car from Chapter 3. Now it has encountered a steep hill and is headed uphill. It has the clever design feature that if it's on a hill the wheels are automatically raised or lowered so that the passenger cabin remains comfortably horizontal. You are asked to calculate the acceleration **a** of the rocket car.

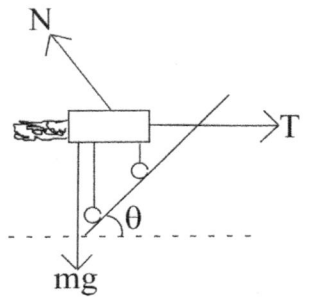

Figure 8.1: The rocket car climbing the hill.

First draw a sketch. The angle between the road surface and a horizontal is θ. What are the forces on the rocket car? The car has a mass m, so correspondingly it is subject to a force of gravity mg in the vertical downward direction. The rocket motor creates a thrust T on the car in the horizontal direction. Finally, there is a normal force N keeping the car from moving through the hill rather than over the hill. The normal force points perpendicular to the road surface. The car is accelerating parallel to the road surface.

The sketch is not a force diagram. In a force diagram, each object is indicated as a point, labelled with the symbol for its mass. Corresponding to each force on the mass, an arrow labeled with the symbol for the force is drawn emanating from the point. For the rocket car, the mass and its forces are shown in Figure 8.2.

Is this diagram a correct force diagram? Yes, it is. A correct force diagram indicates the coordinates, the directions of the x, y, and/or z axes. By choosing one set of coordinates or another, one may make the problem easier or more interesting to solve. However, no matter the orientation of the perpendicular x, y, and z axes, the problem remains soluble. It is not correct to believe that a particular problem can only be solved if you orient the coordinate system in a particular way.

If you listen to fellow students, you might find someone giving you as a force diagram Figure 8.3. Take a few moments to look at the second sketch. Why is it completely wrong? First, the sketch is completely wrong because it indicates the presence of a force ma, to which the wrong names *force of acceleration* or *force of inertia* are sometimes applied. There is no such force. Figure 8.3 is also incorrect because it does not indicate coordinate axes for the problem.

How, then, should you choose to orient your coordinates, before you put them down on the force diagram? All choices are correct, but some choices are better than others. Here there are three useful paths to choosing directions.

On one hand, in some problems there is a direction along which the acceleration has value zero. For example, in the case of the rocket car going up the hill, the rocket car is accelerating in the horizontal and vertical directions, but stays parallel to the road surface. That means there is no acceleration of the rocket car in the direction perpendicular to the road surface. Choosing a coordinate system in which the acceleration, parallel to one of the coordinate axes, is zero means that the

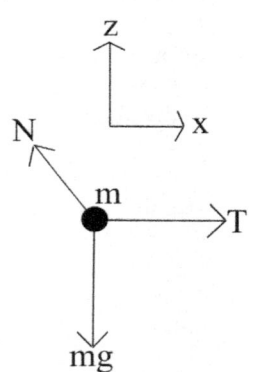

Figure 8.2: A correct force diagram for the rocket car climbing the hill.

corresponding Second Law equation simplifies, namely one has a set of forces on one side, and a zero on the other. If there is only acceleration along a single coordinate axis, then the acceleration along that axis is not only a component of the acceleration, but is also the total acceleration. Contrariwise, if the acceleration is in one direction, which is not parallel to any of the coordinate axes, then the acceleration will need to be split into its components along the three coordinate axes, perhaps with a constraint linking the acceleration components.

On the other hand, one might choose the orientation of the coordinates so that as many forces as possible are parallel to a coordinate axis, meaning that as many forces as possible do not have to be split into distinct components along several coordinate axes. On the third hand, the engineering specifications or problem statement may specify how the coordinate axes are to be oriented.

In this problem, there are two obvious sets of coordinates. In one set of coordinates, the X and Z axes point in the horizontal and vertical directions, so that the thrust and the force of gravity point along those two coordinate axes. In the other plausible set of coordinates, one of the coordinate axes points parallel to the road surface, meaning the acceleration is only nonzero along that coordinate axis, and the other coordinate axis points perpendicular to the road surface, that is, it points parallel to the normal force. You could, of course, point the two coordinate axes that lie in the plane of the paper in any other pair of perpendicular directions, and the problem would continue to be soluble.

For better or worse, we are going to take our two coordinate axes to lie parallel to the road surface and

8.2. THE ROCKET CAR ON A HILL

perpendicular to the road surface, i.e., parallel to the normal vector, respectively. The corresponding force diagram is Figure 8.4.

We now write Newton's Second Law. The law may be written either in the vector form, or in the form of its three scalar components, namely

$$\mathbf{F} = m\frac{d^2\mathbf{r}}{dt^2}, \text{ or} \qquad (8.1)$$

$$F_x = m\frac{d^2x}{dt^2}, \qquad (8.2)$$

$$F_y = m\frac{d^2y}{dt^2}, \text{ and} \qquad (8.3)$$

$$F_z = m\frac{d^2z}{dt^2}. \qquad (8.4)$$

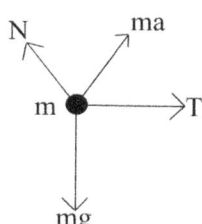

Figure 8.3: Wrong force diagram for the rocket car climbing the hill.

The y direction is perpendicular to the paper. There are no forces in that direction, so for that direction we may write

$$0 = m\frac{d^2y}{dt^2}, \qquad (8.5)$$

which tells us that $\frac{dy}{dt}$ is a constant.

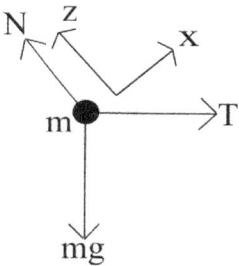

Figure 8.4: A second correct force diagram for the rocket car, with a different pair of coordinate axes.

We now partition the three forces into their components parallel to the coordinate axes, these being the coordinates parallel and perpendicular to the road surface. For the normal force $\mathbf{N}$ the partition is straightforward, namely $\mathbf{N}$ has no component parallel to the road surface. It only has a component perpendicular to the road surface, letting us write $\mathbf{N} = N\hat{\mathbf{k}}$. Corresponding to this analysis is a sketch of the normal force vector and the two coordinates.

For the other two forces, a little more attention to trigonometry is needed. Remember that *the total force always forms the hypotenuse of a right triangle, while the force's components form the legs of a right triangle.* The legs of the right triangle are always shorter than the hypotenuse.

Decomposing the thrust T into its x and z components is shown in figure 8.6. The thrust vector $\mathbf{T}$ is shown as the horizontal, while the components become

$$T_x = T\cos(\theta), \qquad (8.6)$$
$$T_z = -T\sin(\theta). \qquad (8.7)$$

We will come back in a moment to discuss the signs given to these components. Figure 8.7 shows the decomposition of the gravitational force into its components. From the Figure, we can use trigonometry to calculate the length of the two legs, namely the two legs have length $mg\cos(\theta)$ and $mg\sin(\theta)$ as indicated in the figure. Note that in this case the x-component of the force is associated with the sine of the angle, while the z-component of the force is associated with the cosine of the angle.

Figure 8.5: The normal force $\mathbf{N}$ has only one non-zero component.

In the figures, I have marked various components as being positive or negative. The signs of the components are determined by which way the forces point relative to the coordinate axes. The signs of the components are not copied from any sign that the force vector has. As seen in Figure 8.7, the two components of gravity have both been assigned negative signs because the components of gravity point in the $-\hat{\mathbf{i}}$ and $-\hat{\mathbf{k}}$ directions. On the other hand, in Figure 8.6 for decomposing the thrust into its components, the x-component of the thrust is $+T\cos(\theta)$, but the z component is $-T\sin(\theta)$, The z component of the thrust is given a minus sign because the thrust points somewhat in the negative z direction.

Having decomposed the forces into their components, we are now ready to write the Second Law. The relevant equations become

$$T\cos(\theta) - mg\sin(\theta) = m\frac{d^2x}{dt^2}, \quad (8.8)$$

$$-T\sin(\theta) - mg\cos(\theta) + N = m\frac{d^2z}{dt^2}. \quad (8.9)$$

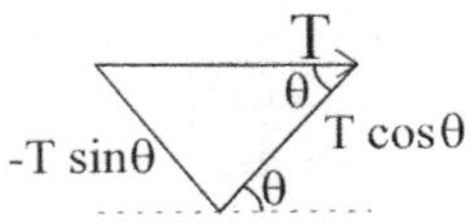

Figure 8.6: Decomposition of the thrust **T** into its x and z components.

However, the rocket car does not fly through the hill, because its velocity in the direction perpendicular to the road surface remains zero. The direction perpendicular to the road surface is the direction of the z-axis, so we can conclude that $\frac{d^2z}{dt^2} = 0$. Substituting zero into the Second Law,

$$-T\sin(\theta) - mg\cos(\theta) + N = m \cdot 0, \quad (8.10)$$

and therefore

$$N = T\sin(\theta) + mg\cos(\theta). \quad (8.11)$$

Can we check this result? What happens if the slope is horizontal? In that case, the thrust is entirely horizontal, so the force of gravity and the normal force should exactly balance each other. If you substitute $\theta = 0$ into equation 8.11, you indeed get $N = mg$.

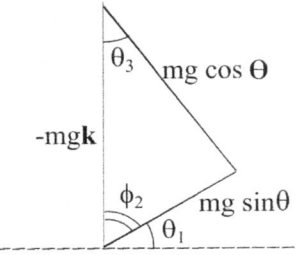

Figure 8.7: Decomposition of the gravitational force $-mg\hat{\mathbf{k}}$ into its x and z components. Subscripts on the angles mark the logical path by which they were deduced, starting with θ_1, the angle given in the problem. Φ_2 is the complement of θ_1 and also of θ_3.

What is the acceleration of the car up the slope? Dividing the first of equations 8.9 by m, one finds

$$\frac{d^2x}{dt^2} = \frac{T}{m}\cos(\theta) - g\sin(\theta). \quad (8.12)$$

Does this equation make sense? What if $T\cos(\theta)/m - g\sin(\theta) < 0$? Then the car's acceleration would be backward, down the hill. The car would slow to a stop and roll back downhill.

We can integrate equation 8.12 with respect to time to get the velocity and position of the rocket car as functions of time. When we do this, we pick up two constants of integration, v_{0x} and x_0, just as we did earlier.

8.3 Coupled Masses

Suppose we have two masses that put forces on each other. How do we apply the Second Law to this situation? We treat a simple example. We are out-of-doors in winter. We have a person, and two large sleighs m_1 and m_2 resting on sleigh tracks. The person applies a force of magnitude P horizontally on mass m_1. What are the accelerations of the two masses? I supply a crude sketch of the situation. The two masses are putting on each other a contact force having magnitude F_{12}. Each mass is subject to a normal force upward, keeping it from falling through the snow, and a gravitational force pulling it downward. Here is a crude sketch of the situation.

The next step is to draw the force diagram for the two masses. I'll choose the coordinate system so that x is in the horizontal direction and z is in the vertical direction, the same directions applying to both masses. 'Same directions' is a choice, not a requirement from natural law.

For each mass, I have chosen a coordinate system, each with an x and a z axis. When I write the equations of motion I will put subscripts on the coordinates. The coordinates associated with mass m_1 are called x_1 and z_1, while the coordinates associated with mass m_2 are called x_2 and z_2. We need separate x and z coordinates for the two masses because the positions, velocities, and accelerations of the two masses are independent of each other; they may have different values.

8.3. COUPLED MASSES

For mass 1, there is a force of gravity $m_1 g$ pulling down and a normal force N_1 pushing up. The man is pushing on m_1, so I show on m_1 a force P pointing horizontally. The two sleighs are putting contact forces on each other, the magnitude of these contact forces being F_{12}. The normal forces keep the two sleighs for moving through each other, so they act to push the two sleighs apart.

Shouldn't there be a force P on mass m_2? After all, when the man pushes, both masses are going to move. The answer is no. The force P is only exerted on mass m_1, so it only appears on the force diagram for this mass. Mass m_2 will move because the action-reaction pair of forces F_{12} and $-F_{12}$ apply a horizontal force on mass m_2.

Let us now set up the Second Law and solve. The general forms are

$$\mathbf{F_1} = m_1 \frac{d^2 \mathbf{r_1}}{dt^2}, \tag{8.13}$$

$$\mathbf{F_2} = m_2 \frac{d^2 \mathbf{r_2}}{dt^2}. \tag{8.14}$$

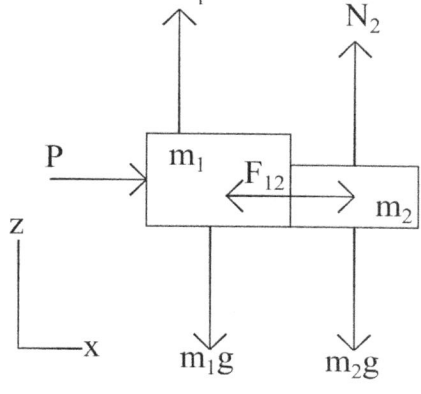

Figure 8.8: Sketch of the two sleighs.

We break these equations into components and replace the general symbols with symbols or numbers appropriate to the problem.

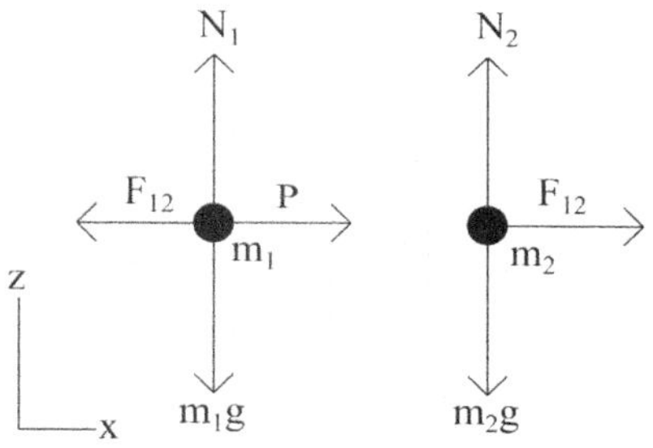

Figure 8.9: Force diagrams for the two sleighs.

For the first sleigh, we have

$$m_1 \frac{d^2 x_1}{dt^2} = -F_{12} + P, \tag{8.15}$$

$$m_1 \frac{d^2 z_1}{dt^2} = N_1 - m_1 g. \tag{8.16}$$

These are the equations of motion for mass m_1. The Equations of Motion are the Second Law, with values inserted for the forces. The signs on the force components were chosen to correspond to the directions of the coordinate axes. Why is F_{12} shown with a minus sign, given that F_{12} is the magnitude of the contact force between the two sleighs. Isn't m_1 pushing on m_2? Answer: Yes, but the forces in the force diagram for m_1 are the forces **on** m_1, not the forces exerted **by** m_1.

For the second sleigh, we have

$$m_2 \frac{d^2 x_2}{dt^2} = F_{12}, \tag{8.17}$$

$$m_2 \frac{d^2 z_2}{dt^2} = N_2 - m_2 g. \tag{8.18}$$

These equations are the equations of motion for m_2. They are almost the same as the equations for m_1. However, m_1 and m_2 each have their own x, y, and z coordinates, so the subscripts on the coordinates have been changed. The two sleighs each have their own mass and normal force, so those subscripts have changed. Finally, the contact forces on masses 1 and 2 are an action-reaction pair. They point in opposite directions, so the sign of F_{12} reverses between the equations of motion for m_1 and m_2.

We also have constraints. The motion is on a horizontal surface, so z_1 and z_2 are both constant. The vertical accelerations $\frac{d^2 z_1}{dt^2}$ and $\frac{d^2 z_2}{dt^2}$ therefore are both equal to zero. From the equations of motion, one then has

$$0 = N_1 - m_1 g, \tag{8.19}$$

$$0 = N_2 - m_2 g, \tag{8.20}$$

leading to $N_1 = m_1 g$ and $N_2 = m_2 g$.

Aside: The two vertical forces on either sleigh are equal in magnitude and opposite in direction. Are they an action-reaction pair? No. They are on the same body. They have different physical natures, namely contact force and gravity. They do not satisfy two of the requirements for a pair of forces to be an action-reaction pair. Therefore, they are not an action-reaction pair.

What about the horizontal constraints? The two sleighs stay in contact, so the distance $x_2 - x_1$ between them remains constant. If $x_2 - x_1$ is a constant, it must be the case that

$$\frac{d^2(x_2 - x_1)}{dt^2} = 0, \tag{8.21}$$

leading to

$$\frac{d^2 x_1}{dt^2} = \frac{d^2 x_2}{dt^2}. \tag{8.22}$$

What, then, is the horizontal acceleration of the two sleighs? We have three unknowns, namely the two accelerations and the contact force. We also have three equations, namely one constraint and one equation of motion in each x coordinate. We have three equations for our three unknowns, so there is a possibility that we can solve for the unknowns. If we add equations 8.16 and 8.18, we find

$$-F_{12} + P + F_{12} = m_1 \frac{d^2 x_1}{dt^2} + m_2 \frac{d^2 x_2}{dt^2}. \tag{8.23}$$

Using the constraint to eliminate one of the accelerations, and solving, we find

$$\frac{d^2 x_1}{dt^2} = \frac{P}{m_1 + m_2} \tag{8.24}$$

for the acceleration of either mass.

The equation of motion for the second mass gives us the magnitude of the contact force. Using equation 8.24 for the acceleration, equation 8.18 becomes

$$F_{12} = m_2 \frac{P}{m_1 + m_2}. \tag{8.25}$$

8.4 The Hanging Mass

The sketch shows the view from the side. There is a table, shown in cross-section from the side. There is a mass m resting on top of the table and at the start being held in place. A cord connects the mass m to a second mass M that is hanging off the side of the table. The top surface of the table has very little friction. When the mass m is released, the weight of the mass M drags m toward the edge of the table. The cord runs over a pulley, shown as a small circle, so that the mass M is hanging well away from the side of the table. What is the acceleration of the mass m?

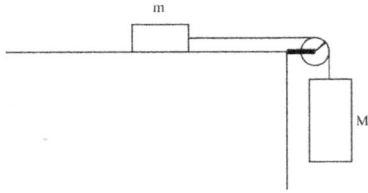

Figure 8.10: Sketch for the hanging mass problem.

To solve this problem, we first need to discuss how a rope works; we also need to discuss the notion of *massless-ness*. The rope has a weight hanging from it, corresponding to which there is a tension T in the rope. The tension T creates forces on the masses to which the rope is attached. The tension force due to the rope always pulls along the line of the rope toward its center. The tension in the rope thus pulls mass m to the right and mass M in the upward direction. Where the rope slides over the pulley, the rope is putting a force on the pulley, but the tension in the rope is approximately the same everywhere in the rope. We will eventually discuss rotational motion, at which point it will become clear that the tension in the rope below the

8.4. THE HANGING MASS

pulley and the tension in the rope to the left of the pulley actually do not have to be equal. As a start, we approximate the two tensions being very nearly equal, and will correct that approximation in a future chapter.

We are making the approximation that the rope is massless. Of course, the rope isn't actually massless. If you put it on a sensitive balance, you can weigh it. That tells you its mass. However, if the mass of the rope is much less than m or M, we can neglect the mass of the rope, calculate the acceleration, and still get an answer that is quite close to being correct.

We now draw the two force diagrams, one for the mass m and one for the mass M. The forces on mass m are a normal force $\mathbf{N}$ due to the table, a force of magnitude mg due to the Earth's gravity, and a force $\mathbf{T}$ due to the tension in the rope. The forces on the mass M are the tension $\mathbf{T}$ in the rope and the weight Mg of mass M due to the earth's gravity. Note that the two tension forces are pointing into different directions. Are the two forces on mass M, namely the tension and the force of gravity, an action-reaction pair? Absolutely not! Those two forces are acting on the same body, and therefore cannot be an action-reaction pair.

We again start with the second law $\mathbf{F} = m d^2 \mathbf{r}/dt^2$ and write the equations of motion for the two masses. For mass m, the equations of motion are

$$m \frac{d^2 z_1}{dt^2} = N - mg, \quad (8.26)$$

$$m \frac{d^2 x_1}{dt^2} = T. \quad (8.27)$$

For mass M, the equations of motion are

$$M \frac{d^2 z_2}{dt^2} = T - Mg, \quad (8.28)$$

$$M \frac{d^2 x_1}{dt^2} = 0. \quad (8.29)$$

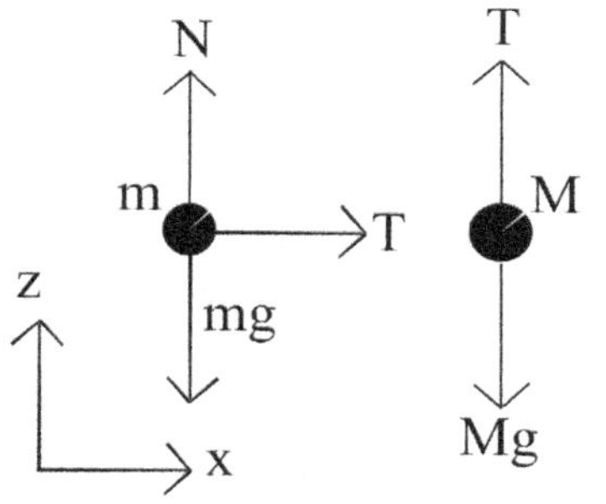

Figure 8.11: Force diagrams for the hanging mass problem.

Mass M has no forces on it in the horizontal direction, so its acceleration in that direction must be zero. Mass m is not being held in place. As a result, m accelerates to the right while M accelerates in a downward direction. Because M is accelerating downward, we can be confident that $T \neq Mg$. If instead we believed for mass M that $T = Mg$, which we do not, then the total force on M in the vertical direction would be zero, and the mass M would not accelerate. It would instead simply float in midair. That's not what happens.

Can we solve? We have three unknowns (two accelerations and the tension), but only two equations. We are not ready to solve. We need a constraint. Here the constraint is that the length of the rope is a constant, so the velocity at which m is moving in the $+x$ direction is equal in magnitude but opposite in sign to the velocity with which M is moving along the z-axis. The constraint may be written

$$\frac{dx_1}{dt} = -\frac{dz_2}{dt}. \quad (8.30)$$

The corresponding two accelerations are therefore equal in magnitude but opposite in sign. In general, if you have several objects connected by a rope, the constraint is the total length of the rope. If you choose the coordinates well, you can write the length of the rope in terms of the coordinates of the objects with perhaps some extra constants being needed.

We are now ready to solve. The simplest approach is to eliminate the tension between two of the equations via subtraction of one of the equations from another, leading to

$$T - T + Mg = m \frac{d^2 x_1}{dt^2} - M \frac{d^2 z_2}{dt^2}. \quad (8.31)$$

If you use the constraint to eliminate the second acceleration, you can show

$$\frac{d^2 x_1}{dt^2} = \frac{Mg}{M+m}. \quad (8.32)$$

Observe that once again I have eliminated an unknown by means of subtraction, not substitution. If the variable being eliminated appears only linearly in the equations, subtraction is generally the cleaner, less error-prone, approach to eliminating the variable.

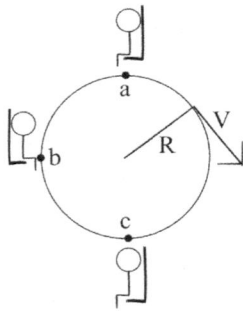

Figure 8.12: The Ferris Wheel.

8.5 Worked Problems

1. While driving along the New York Thruway, I realized that the vehicle now visible over the top of the hill was stopped and had no brake or hazard lights showing. Worse, its passengers were outside the car blocking the other Thruway lane and the emergency lane. I applied my brakes. Assuming that my Toyota with luggage had a mass of 1500 kg and that my acceleration was $-0.5g\hat{\mathbf{i}}$, that the road was smooth and horizontal, and that wind resistance was not significant, calculate the contact force of the road on my automobile.

2. A pair of railroad cars sit on railroad tracks in contact with each other. They are attached by a coupler, so they may pull as well as push on each other. Their masses are m_1 and m_2. The rear cart is pushed from its rear by an oscillating force $F_o(1 + \cos(\omega t))$. Give the force diagram for each of the two carts. Write the Second Law for each of the two carts. What is the acceleration of each cart as a function of time? What is the velocity of each cart as a function of time?

3. Consider a child seated in a Ferris Wheel, as seen in Figure 8.12. Where the child is sitting, the tangential velocity due to wheel rotation is V. For points a, b, and c, give the force diagram for the child, and compute the force that the seat is putting on the child at each point.

4. (a) A rock is spun on a string in a vertical circle. When the rock is headed sideways and up at a 45 degree angle to the horizontal, the string is cut. Its speed at this moment is v. Give the force diagram and the equations of motion for the rock immediately before and immediately after the string was cut.
(b) A rock is thrown through the air; it describes a parabola before returning to the ground. What are the force diagram, the acceleration, and the velocity of the rock when it reaches the top of its trajectory?

5. A retired faculty member constructs a mansion modestly to the east of the San Andreas fault. No sooner has he moved into the mansion than the really big earthquake occurs, and all land east of the San Andreas fault begins its terminal slide to the bottom of the Atlantic ocean. In the dining room (7 m ceiling) is a chandelier of total length (from the ceiling to the bottom) 3 m. Assuming that the building's vibration isolation protected the chandelier from all vibrations of the earthquake, so that the chandelier was stationary when the slide began, and that the acceleration of the building is initially due east and horizontal at 0.1 m/s^2, calculate the deflection of the bottom of the chandelier from the vertical. You may approximate the chandelier as a point mass at the end of a piece of massless string having length 3 m.

8.6 Homework

1. For the rocket car discussed earlier in the book, going up the slope, find its velocity and position as functions of time while it is climbing the slope.

2. For mass m in Figure 8.10a, what are the forces on m? For each force, identify the nature of the force, the object exerting the force, and the object on which the force is exerted. For each force, identify the reaction force: Indicate the nature of the force, the object exerting the force, and the object on which the force is exerted.

3. The text claimed you can derive equation 8.32 from the previous equation. Show that the claim is correct.

4. An old exam problem: In the elsewise undistinguished action film, the 70 kg heroine wearing 40 kg of body armor and other equipment is lying without moving on a steep, extremely slippery roof, hanging on to a 5m rope as shown in Figure 8.13a. The roof makes an angle of $\Theta = 55$ degrees with respect to the horizontal.

    a) Give the force diagram for the heroine. Identify each force in a complete sentence.

    b) In complete sentences, identify the reaction forces for each force shown in part a)

    c) Starting with $\mathbf{F} = m\frac{d^2\mathbf{r}}{dt^2}$, write the equations of motion for the heroine.

    d) What is the tension in the rope?

    e) The heroine is wearing a rocket belt activated by a delay timer, which would allow her to fly to safety. For technical reasons, the belt must be activated *after* she slides off the edge of the roof. What is the shortest time after she lets go of the rope that she can activate the rocket belt? Assume that she needs to slide down the roof through a distance of 5 m and may then fire the belt.

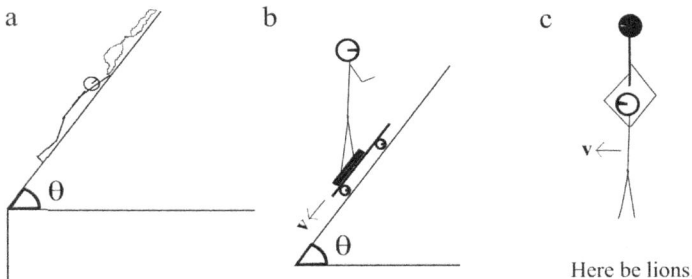

Figure 8.13: (a) Heroine on roof. (b) The intrepid skateboarder. (c) Rope swing above hungry lions.

5. An old exam question:

    a) In complete sentences or valid equations, state Newton's three laws of motion.

    b) List the properties that all action-reaction force pairs must have.

    c) A Pharaoh in ancient Egypt had the fortune to have a chariot pulled by a talking horse. The horse and Pharaoh have both been gifted, slightly ahead of historical period, with a knowledge of Newton's Laws of Motion. The Pharaoh demands that the horse accelerate the chariot into a charge, so that he may ride into battle. The horse patiently explains that it is unable to accelerate the chariot, because: "The force that I can exert on the chariot and the force that the chariot exerts on me are an action-reaction pair. The two forces in question are therefore equal in magnitude and opposite in sign, so they add to zero. The total force is therefore zero, so there will be no acceleration."

    Is the horse's claim right or wrong? For credit, explain why in a short paragraph.

6. We now transport ourselves forwards to the twenty-second century, onto an manned space station in a low orbit around the earth. The station is a cubical box whose base always faces the earth. An astronaut is floating, stationary with respect to the walls of the space station, in the middle of the space station, not in contact with any object. Draw the free body diagram for the astronaut. Is the total force on the astronaut on the astronaut zero, or not? Why? Is the astronaut accelerating or not? Why?

7. A circular pendulum is a mass at the end of a string. Instead of going back and forth, the mass in a circular pendulum rotates in a circle at constant height and angular velocity. The circular pendulum in this problem is hung from the ceiling of the room. It has a 20.0 kg mass suspended at the end of a 3.00m long rope. The mass rotates in a circle with its string making an angle of 15.0 degrees from the vertical. a) Give the force diagram for the circular pendulum. In a few words, identify each force on the pendulum. b) In complete sentences, identify the reaction force to each force on the pendulum, including the physical nature of the force and the object on which the force is applied. c) Write the equations of motion for the pendulum. d) Find the tension in the string. e) Find the speed with which the pendulum moves in a circle. f) The elevator within which the physics building is mounted – this is a very large elevator – is now activated. The physics building accelerates downward toward the Indian Ocean with an acceleration of magnitude 1.5 m s^{-2}. The pendulum is again set spinning, at the same angle θ. Now what is the rotational speed of the pendulum relative to the lecture hall?

8. An intrepid skateboarder (Figure 8.13b) takes a long straight slope that rises thirty degrees above the horizontal while standing on a bathroom scale placed on top of the skateboard. [SAFETY WARNING: Do NOT try this yourself.] The bathroom scale reports the normal force on the scale due to the person standing on it. The person has mass m. What does the bathroom scale report for the person's weight, which is equal to the normal force he puts on the scale.

9. In yet another scene (Figure 8.13c) from our interminable motion picture, the heroine is escaping from a pride of hungry lions by using a conveniently placed rope to swing across a gorge filled with more lions. At the bottom of her swing, the rope is stretched vertically downward. Note sketch. For this moment, give the force diagram for the heroine. For each force in your diagram, identify **in one or more complete sentences** the force, the object on which the force is acting, the physical nature of the force, and the object exerting the force. For each force in your diagram, identify the reaction force. For each reaction force identify **in one or more complete sentences** the force, the object on which the force is acting, the physical nature of the force, and the object exerting the force.

10. Assuming in the immediately previous problem that the length of rope between the pivot point and the heroine's hands is 20 m, at the bottom of the swing the heroine is travelling at 15 m/s, and the heroine with body armor and other equipment masses 70 kg, find the tension in the rope.

11. In another scene of an elsewise undistinguished action film, the 70 kg heroine wearing 50 kg of body armor and other equipment is in an elevator when the elevator lines are cut, and the emergency brakes partially fail. The elevator begins to accelerate toward the basement. a) Give the force diagram for the heroine (including her armor as part of "her"). b) Identify each force in a complete sentence. Include the nature of the force, the source of the force, and the object to which the force is applied. c) In complete sentences, identify the reaction forces for each force shown in part a). d) Starting with $\mathbf{F} = m\frac{d^2\mathbf{r}}{dt^2}$, write the equations of motion for the heroine. e) The pressure sensors in the heroine's boots report that she is applying a normal force of 800 Newtons downward on the floor of the elevator. What is the acceleration of the elevator? f) The top of the elevator and elevator shaft are both open. The heroine is wearing a rocket belt which would allow her to fly to safety. For technical reasons, the belt must be activated *after* the elevator floor is below ground level. The building has multiple subbasements, so this step is actually possible. The elevator was at rest when it started falling from the third floor, 20m off the ground. What is the shortest time after the elevator starts falling that she can activate the rocket belt?

12. In a later scene, the heroine (her plus body armor = 120 kg) is shown crossing a deep chasm using the rocket belt. The director has cunningly arranged to use a real rocket belt, not wires and a crane. Her

agent advises the actress that it's in the contract. The rocket belt generates a thrust of 1500 Newtons. The actress leans forward so as to maintain a constant altitude. a) Give the free body diagram for the actress. b) Write the equations of motion for the actress. c) What is the actress's horizontal acceleration?

13. Resting on a frictionless tabletop are a 2 kg mass, a 1 kg mass in contact with the 2 kg mass, and a mechanical pusher that pushes on the 2 kg mass. The pusher, 2 kg mass, and 1 kg mass lie in a straight line. The pusher generates a force of 5 N. Find the force diagrams for the two masses, the equations of motion (Newton's second law, with "**F**" replaced by the actual forces), the accelerations of the two masses, and the net force (total of all forces) on the 2 kg mass.

14. Your trusting lecturer is persuaded by a group of former students to attempt as his first time on the skateboard the simplest possible skateboard maneuver, namely a reverse double somersault with half twist from the kneeling reverse position. This is perfectly safe, and besides the landing point is nice and soft. Having been loaded onto the skateboard, the board is set into motion by a group of people pushing, who let go and wave as the board goes up the ramp as indicated. (Figure 8.13b, except this time the skateboarder is headed up hill.) For this problem, treat the skateboard and its passenger as a single object of mass m. There is no friction on the launching ramp. (a) give the force diagram for the skateboard and rider. Use my coordinates. (b) Report in a table listing the forces acting on the skateboard, their physical nature, their source, and the object to which they are applied. (c) Report in a table the reaction forces to the forces acting on the skateboard, their physical nature, their source, and the object to which they are applied. (d) What is the acceleration of the skateboard?

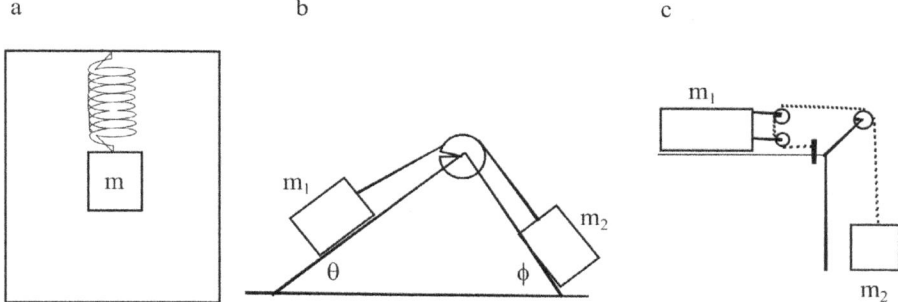

Figure 8.14: (a) The mass in the elevator. (b) A double ramp. (c) a complicated constraint: The dotted line is a rope going back and forth in the air above the table.

15. A 2 kg mass is suspended from a spring in the physics building elevator, which begins at the second floor. (a) The elevator is set into motion and begins accelerating toward the basement. The acceleration has magnitude 0.01 m s^{-2}. Beginning from the force diagram and the Second Law of Motion, calculate the force on the suspended mass due to the spring. (b) The elevator approaches the basement. It is moving downward at 0.1 m/s. It slows to a stop, the magnitude of the acceleration being the same as in part (a). Beginning from the force diagram and the Second Law of Motion, calculate the force on the suspended mass due to the spring. See Figure 8.14a.

16. Two masses m_1 and m_2 lie on opposite slopes of a triangular wedge. The angles from the horizontal are θ and ϕ for m_1 and m_2, respectively. The two masses are connected by a rope slung over a pulley at the apex of the wedge. Find the acceleration of the two masses as they slide up or down the slope. What is the relationship between the two masses if the acceleration is zero? See Figure 8.14b.

17. We have a mass m_1 lying flat on a table, and a mass m_2 hanging from a pulley. They are connected by a rope that goes from a fastener on the table to a pulley attached to mass m_1 back to the second pulley, and then down to the second mass. Find the force diagrams and the constraint. Find the acceleration of the two masses and the tension in the rope in general, and in the special case $m_1 = 20$, $m_2 = 5$ kg. See Figure 8.14c.

18. For the next scene in our interminable motion picture production effort, the heroine is taken to a certain foreign capital, put in the rear seat of a Sukhoi-27, flown to an altitude of 10,000 m, and flown in a horizontal circle of radius 2000 m at a speed of 300 m/s. Assuming her body mass is 65 kg, find the force exerted on her by the seat of the aircraft. To simplify grading, you may assume that the vector from her to the center of the circle, at the moment you are answering the question, is in the $-\hat{\mathbf{j}}$ direction. [Aside: for a not-entirely-modest fee, at one time you could go to Moscow and more-or-less duplicate her experience. Safety warning: If you do this by using your tuition money to pay for the flight, your parents will kill you.]

19. A 10 kg mass is subject to a 20 N force in the $-y$ direction and a 30 N force in the $x - y$ plane at an angle of $+30$ degrees from the $+x$ axis. The sum of all forces on the mass in the z direction is zero and need not be considered further. (a) Find the acceleration of the mass. (b) An additional force $\mathbf{F}''$ is now applied to the mass. The additional force reduces the acceleration of the mass to zero. Find $\mathbf{F}''$.

20. A railroad train consists of four cars, namely an engine and three boxcars, each of the four vehicles having mass m. The train moves down the track at a constant acceleration to the right. Contacts between the rails and the boxcars are (very nearly) frictionless. Draw a force diagram for each of the four cars in the train. For each force, identify the reaction force and the object on which the reaction force acts. The acceleration is caused by a friction force (the 'tractive' force, discussed in a future Chapter) between the wheels of the locomotive and the railroad tracks. For this force causing the acceleration, be sure to identify the object causing the force, the object on which the force is acting, and the direction in which the force points.

21. One of your fellow students has volunteered to help move a friend to a new apartment. He is pushing a steamer trunk up a ramp. The ramp makes an angle θ with respect to the horizontal. Thanks to a clever use of rollers, the ramp is very nearly frictionless. Unfortunately for the student, the trunk is filled with the friend's collection of tropical bird feathers, packed on top of her collection of lead bricks, giving the trunk a mass M. The student pushes parallel to the ramp with a force P. (a) Give the force diagram of the trunk. (b) compute the normal force on the floor of the trunk. (c) Compute the value of P needed so that the steamer trunk moves with a constant speed V.

22. Two carnival ride cars float on a frictionless surface. Their masses are m_1 and m_2. They are connected by extremely light ropes having lengths L_1 and L_2 to a central pivot point. They are set spinning in circles around the central point, with a period T. You are the safety physicist, hired to analyze the carnival ride for potential hazards. Computer the tensions T_1 and T_2 in the two ropes.

23. Two masses, m_1 and m_2, lie in contact with each other on a flat frictionless surface, with $m_2/m_1 = n$. They are subject to gravity. m_1 is subject to a time-dependent external force $F = F_o \exp(-at)$ pointing horizontally toward mass m_2. During the problem, the two masses remain in contact with each other. Find the force diagrams for the two masses, the contact force F_c between the two masses, and the velocity and position of the masses as functions of time. From the available information, you may not be able to solve for every constant of integration.

24. We have two 10 kg masses hanging from strings. The other end of each string is attached to one end or the other of a spring. For the first mass, the first rope, and the spring, (a) write the force diagram, (b) identify in a complete sentence each force in the force diagram, including the nature of the force, the object on which the force is acting, and the object exerting the force. See Figure 8.15a.

25. A right triangular block sits on a frictionless table. Resting on the frictionless hypotenuse of the block (see figure) is a mass M. The hypotenuse makes an angle θ with respect to the horizontal. A rocket motor has been attached to the back (vertical) side of the block. The rocket motor is fired, bringing the block to a constant acceleration $d^2x/dt^2 = a$; the mass M is then released so it is free to slide up or down the block. If the rocket's thrust is properly chosen, there is a value for a such that the mass slides neither up nor down the block. What is a? If a is increased, does the mass slide up or down the side of the block? See Figure 8.15b.

8.6. HOMEWORK

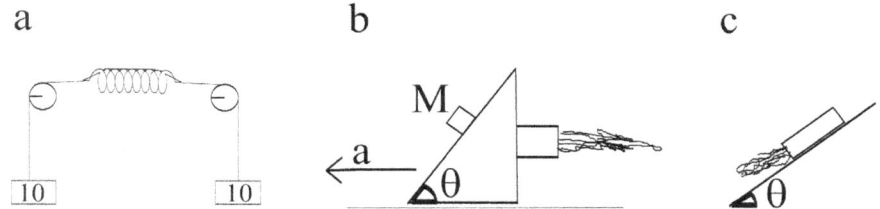

Figure 8.15: (a) Masses and spring. (b) Mass on an accelerating ramp. (c) The rocket car.

26. The rocket car is now operated with the leveler system turned off, so that when the vehicle of mass m goes up a slope at an angle θ relative to the horizontal, the rocket thrust $\mathbf{T}$ is aligned parallel to the road surface, pointing uphill. Using coordinates in which the x axis lies parallel to the road surface and the z axis is perpendicular to the orad surface, calculate the normal force on the rocket car due to the road. Calculate the vehicle's acceleration. See Figure 8.15c.

27. Repeat the calculation of the previous problem, except this time the x axis is horizontal and the z axis is vertical. The point of this problem is to demonstrate that there are no privileged coordinate systems. If you can solve a problem in one coordinate system, you can solve it (not necessarily equally easily) in any other valid coordinate system. See Figure 8.15c. Confirm that the magnitudes of the acceleration as calculated in this and the previous problem are equal to each other.

28. A gymnast of mass m is hanging vertically, bobbing up and down at the end of a long, non-linear spring in the hold of a moving ocean liner. The gymnast's position is given by $\mathbf{r} = 9t\hat{\mathbf{i}} + [3\cos(3\pi t) + 2\sin(5t)]\hat{\mathbf{k}}$.
a) Find the velocity and acceleration of the gymnast. b) Find the z-component of the force that the gymnast exerts on the spring.

29. In one of my future novels (yes, I write novels) a small child is sitting on a swing just east of the San Andreas Fault, when a giant earthquake strikes. The entire United States east of the San Andreas Fault begins its inexorable slide due east into the waiting waters of the Atlantic Ocean. The swing itself is moving in the x direction at 2.0 m/s. The child swings back and forth with respect to the swing center, with a horizontal velocity $0.3\cos(2t)$ and a vertical velocity $0.2\sin(2t)$. The child has a mass of 20 kg.

A) Write the velocity of the child as a vector in $(\mathbf{i}, \mathbf{j}, \mathbf{k})$ notation.

B) Find the total force on the child as a function of time.

C) Find the position of the child as a function of time.

30. A man is standing in an elevator. We have put a detector into his shoes, so we can measure the upwards normal force N of the elevator floor on the man's feet. At one instant in time, the normal force (straight up) is 800 Newtons. The man's mass is 100 kg. Six students discuss the results of the single measurement of N, and what it says about the motion of the elevator. They propose, in order:

Student 1: The elevator must be going up.

Student 2: The elevator must be going down.

Student 3: The elevator may be going up or down, but may not be stationary.

Student 4: The elevator may be going up, coming down, or be stationary. The above paragraph gives no information of any kind about the elevator's motion.

Student 5: Students 1-4 are all wrong; from the above information we can deduce something (maybe not a lot) about the elevator's motion.

Student 6: Students 1-5 have all missed the point. If you are standing on the floor, Newton's third law (action equals reaction) guarantees that $N = mg$; the detector in the shoes must be broken.

Explain whether each student's answer is true or false, and why.

8.7 Solutions to the Worked Problems

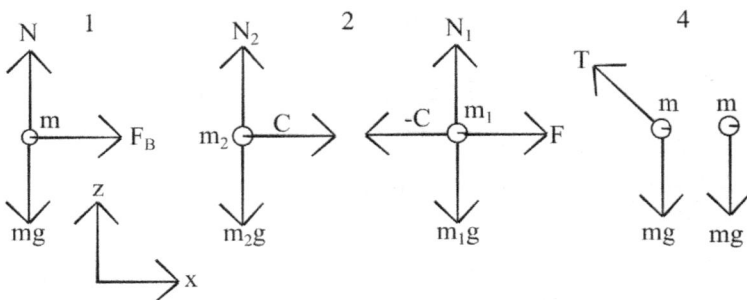

Figure 8.16: Force Diagrams for worked problems 1, 2, and 4. The same x and z axes are used in all three problems.

1. We are trying to calculate forces. We start with drawing a force diagram (Figure 8.16(1)) for my car and labelling the forces. The force diagram shows forces $\mathbf{N}$, $-mg\hat{\mathbf{k}}$, and $\mathbf{F}_B$ acting on my car. $\mathbf{N}$ is the normal force of the road on the car. $-mg\hat{\mathbf{k}}$ is the gravitational force of the earth on the car. $\mathbf{F}_B$ is a force of the *road* on my car. At this point, we do not have a value for $\mathbf{F}_B$; we only know that this force must point in the horizontal direction. The indicated direction for $\mathbf{F}_B$ is therefore arbitrary and may not be the actual direction that $\mathbf{F}_B$ points. That uncertainty is irrelevant; force diagrams are not scale diagrams. The reaction force to $\mathbf{F}_B$ is a force $-\mathbf{F}_B$ that my car's wheels are exerting on the road. That reaction force is not a force *on my car* and therefore does not contribute to slowing my car down.

   The Second Law tells us
   $$\mathbf{F} = \frac{d^2\mathbf{r}}{dt^2}. \tag{8.33}$$

   We substitute into the Second Law the forces. In this problem, we happen to know the acceleration, but do not know all of the forces, so
   $$N\hat{\mathbf{k}} + F_B\hat{\mathbf{i}} - mg\hat{\mathbf{k}} = m(-0.5g\hat{\mathbf{i}}). \tag{8.34}$$

   The contact force of the road on the car is $N\hat{\mathbf{k}} + F_B\hat{\mathbf{i}}$.

   Resolve things into components. In the z direction,
   $$N - mg = 0, \tag{8.35}$$
   so in this case $N = mg$.

   In the x direction,
   $$F_B = -0.5mg. \tag{8.36}$$

   The contact force on the car was then
   $$\mathbf{F}_C = mg\hat{\mathbf{k}} - 0.5mg\hat{\mathbf{i}}, \tag{8.37}$$
   or, as a rough estimate,
   $$\mathbf{F}_C = 1.5 \cdot 10^4\hat{\mathbf{k}} - 7.5 \cdot 10^3\hat{\mathbf{i}} \text{ Newtons}. \tag{8.38}$$

2. We begin by drawing the force diagrams (Figure 8.16(2)). The forces on m_1 are a normal force N_1, the force of gravity $m_1 g$, the external force $\mathbf{F}$, and the contact force $-C$ from m_2. I'm giving the forces as single components. Until coordinate axes are chosen, signs and directions in terms of the coordinates are not known. The forces on m_2 are a normal force N_2, the force of gravity $m_2 g$, and the contact force C from m_1.

The contact forces C and $-C$ are an action-reaction pair, reflecting the two carts jostling against each other as they are shoved along the ground. The force $\mathbf{F}$ does not appear in the force diagram for m_2. Why? The force $\mathbf{F}$ is not being applied to m_2; it is only being applied to m_1.

We choose the horizontal coordinate to be the x-coordinate and the vertical coordinate to be the z-coordinate. For motion in the horizontal plane, the Second Law

$$\mathbf{F}_{\text{total}} = m \frac{d^2 \mathbf{r}}{dt^2} \tag{8.39}$$

gives us

$$-C + F = m_1 \frac{d^2 x_1}{dt^2}, \tag{8.40}$$

$$C = m_2 \frac{d^2 x_2}{dt^2}. \tag{8.41}$$

In these equations, x_1 and x_2 are the x-coordinates of m_1 and m_2, respectively.

We now have two equations, but we also have three unknowns, the unknowns being C, x_1, and x_2, all of which are functions of time. in order to solve, we need a third equation. That equation is provided by a constraint. So long as $\mathbf{F}$ is always in the $+x$ direction, the two carts will stay in contact, so that $x_2 = x_1 + b$, where b is a constant. Taking the time derivative of the constraint, twice, we find

$$\frac{dx_2}{dt} = \frac{dx_1}{dt}, \tag{8.42}$$

$$\frac{d^2 x_2}{dt^2} = \frac{d^2 x_1}{dt^2}. \tag{8.43}$$

Using the constraint to eliminate references to x_2, we have

$$C = m_2 \frac{d^2 x_1}{dt^2}. \tag{8.44}$$

Adding the two Second Law equations, with references to x_2 eliminated, we find

$$(m_1 + m_2) \frac{d^2 x_1}{dt^2} = F - C + C, \tag{8.45}$$

or on substituting for F

$$\frac{d^2 x_1}{dt^2} = \frac{F_o(1 + \cos(\omega t))}{m_1 + m_2}. \tag{8.46}$$

Finally, to obtain the velocity, we integrate $\frac{d^2 x_1}{dt^2}$ with respect to time, giving

$$v_x = \int dt \left[\frac{F_o(1 + \cos(\omega t))}{m_1 + m_2} \right] = v_{ox} + \frac{F_o t}{m_1 + m_2} + \frac{F_o \sin(\omega t)}{\omega(m_1 + m_2)}. \tag{8.47}$$

3. We worked this one in a prior chapter.

4. (a) We supply two force diagrams (Figure 8.16(2)). In the first diagram, before the string is cut, there are two forces acting on the rock, namely the force of gravity $-mg\hat{\mathbf{k}}$ and the tension T in the string. After the string is cut, only one force, the force of gravity, acts on the rock. The equations of motion for the rock, before and after the string was cut, are

$$-T\hat{\mathbf{r}} - mg\hat{\mathbf{k}} = -\frac{mv^2}{r}\hat{\mathbf{r}}, \tag{8.48}$$

$$-mg\hat{\mathbf{k}} = m \frac{d^2 z}{dt^2} \hat{\mathbf{k}}. \tag{8.49}$$

in which $\hat{\mathbf{r}}$ is the unit vector toward the center of the circle in which the rock was initially spinning.

(b) At the top of the rock's trajectory, from the above the rock's acceleration is $-g\hat{\mathbf{k}}$, straight down.

5. We have a chandelier in a mansion sliding into the Atlantic, with acceleration 0.1 m/s^1. There is a hanging chandelier. What is its inclination from the vertical? This is a somewhat different Second Law problem, in which we know the acceleration and the forces, but need to calculate the geometry. We start by constructing a force diagram for the chandelier, which is subject to the force of gravity and to a tension force.

The two forces on the chandelier are the tension in the string and the force of gravity. The string has a length L and is displaced from the vertical by a distance L. We begin with the Second Law $\mathbf{F} = md^2\mathbf{r}/dt^2$. The conditions of the problem give us the accelerations:

$$\frac{d^2x}{dt^2} = a_x \equiv 0.1 \text{m/s}^1, \tag{8.50}$$

$$\frac{d^2z}{dt^2} = 0. \tag{8.51}$$

The above two equations show a useful problem solving technique, not a vast and powerful technique, but an effective technique for reducing error. We have a number, the horizontal acceleration of the house, so I have assigned it a symbol, a_x which will be carried through the problem until the end. Why does this technique reduce error? First, copying a_x is much more reliable than copying, say, 2.913576. It's also less tedious. Second, one path for seeing if the final answer makes sense is to ask how the final answer depends on the input parameters. If a_x increases, you should expect that the deviation x of the string from the vertical will also increase. That's practical to see if a_x is carried through to the end. If your answer claims that x decreases as a_x increases, something is wrong. When you get more practice with looking at equations and asking how an answer depends on its input parameters, it will become easier to judge if your answer is reasonable.

On the other hand, if you solve a problem by inserting numbers at the beginning, which is a bad problem solving technique when it can be avoided, especially if you simplified numerical forms as you went, for example if you replaced a_x/g with a numerical value, it would soon become impossible to tell by inspection how the answer depends on the starting point. Of course, you have already seen a family of problems where numbers needed to be inserted in the equations early on, namely the rocket launch problems, because a purely algebraic solution can become intractable. In those problems, if you want to ask whether, say, the calculated acceleration depends in a reasonable way on the final altitude, you have to do numerical experiments, in which you change a number slightly, repeat the calculation, and see if the change in the final answer makes sense in terms of your change in an input parameter.

We also need to decompose T into its components. We do this with the similar triangle process, in which T is the hypotenuse of a right triangle, with its components T_z and T_x as the two legs of the triangle. This triangle is similar to the geometric triangle that has L as its hypotenuse, x as one leg of the triangle, and from Pythagoras $(L^2 - z^2)^{1/2}$ as the triangle's other leg. The decomposition gives us

$$T_x = T\frac{z}{L}, \tag{8.52}$$

$$T_z = T\frac{(L^2 - z^2)^{1/2}}{L}, \tag{8.53}$$

as the two components. As it happens, the ceiling leads the chandelier, so x is positive, but if the ceiling turned out to be behind the chandelier (it isn't; that's a hypothetical), x would be negative, but T_x would still be correct as to sign.

We now write the Second Law in component form, using the above to substitute for the components and the accelerations.

$$T\frac{x}{L} = ma_x, \tag{8.54}$$

$$T\frac{(L^2 - x^2)^{1/2}}{L} - mg = 0 \tag{8.55}$$

8.7. SOLUTIONS TO THE WORKED PROBLEMS

We have two equations, and the two unknowns T and x. It is not hopeless to advance toward a solution. A reasonable first step is to get rid of the square root by squaring everything.

$$T^2 x^2 = (ma_x L)^2, \tag{8.56}$$
$$T^2(L^2 - x^2) = (mgL)^2. \tag{8.57}$$

We can use the first equation to eliminate $T^2 x^2$ from the second equation, leading to

$$T^2 L^2 - (mgL)^2 + (ma_x L)^2, \tag{8.58}$$

which brings us to a solution for T

$$T = m(g^2 + a_x^2)^{1/2}, \tag{8.59}$$

and therefore a solution for $x = ma_x L/T$:

$$x = \frac{a_x L}{(g^2 + a_x^2)^{1/2}}. \tag{8.60}$$

It is finally appropriate to plug in numbers, which I will do to one significant figure.

$$x = \frac{0.1 \cdot 3}{(10^2 + 0.1^2)^{1/2}} \approx 3 \cdot 10^{-2} \text{ m}. \tag{8.61}$$

This page reserved for your notes.

Chapter 9

Tribology

9.1 Introduction

Tribology is the study of friction, the resistance to motion. Friction is actually a very complicated problem. In this course, all that we will do is to describe a few of its effects. We usefully draw a line between friction involving solid objects, friction involving gases, friction involving simple "Newtonian" liquids, and friction involving viscoelastic liquids. You've actually encountered all of these in your life, though you may not have noticed what you were seeing.

Between two solid objects, we may distinguish between static friction, kinetic friction, rolling friction, and the tractive force that allows wheeled vehicles to propel themselves.

Static friction arises if we have two objects in contact with each other, for example one resting on top and the other below, and we try to push the top object sideways. If you try to slide a piece of furniture across the carpet, and don't push hard enough, the piece of furniture just sits there. That's static friction.

If you do push the piece of furniture hard enough, you may notice a bit of a jerk, and then the piece of furniture slides across the floor. The jerk is the static frictional force releasing. However, there is resistance to sideways motion. That resistance to sideways motion is sliding friction.

If you find a spherical object, for example a child's marble, and set it rolling across the floor, at first the object rolls, but if you wait a while it comes to a stop. That's rolling friction.

Finally, wheeled vehicles such as automobiles and locomotives propel themselves by exerting tractive forces, a type of friction, on the rails or road surface.

Friction is also found in gases and liquids. Suppose you have resting on your desktop a fragment of something very light, say a bit of cat fur. If you pass your hand quickly above the cat fur, not touching it, you will observe that the fur moves some modest distance. You set the air moving. The cat fur is now moving with respect to the air, namely the fur is initially stationary while the air is moving, so the relative motion of the air and the cat fur creates a force on the cat fur and a reaction force on the air. That's friction, termed 'viscosity', in a gas.

You can get the same apparent effect in a liquid. If you have a mug filled with hot milk, sprinkle some cocoa powder on top, and gently stir the liquid with a spoon, not where the cocoa is, you will observe that the liquid moves, setting up a flow pattern, and after a moment the cocoa also moves. That's viscosity in a liquid.

Some liquids are much more complicated than water. They're called viscoelastic liquids, because in addition to showing a resistance to flow, they also behave as though they contain molecule-size springs (the details are complicated). Many of you have actually seen this effect. If you take a tube of thick shampoo, squeeze on the end so some shampoo is flowing out, and then take your hand off the tube so you're no longer putting pressure on it, you may observe that the shampoo near the hole in the tube flows back into the tube. That's viscoelasticity. There exists an extremely wide range of viscoelastic phenomena, of which the shampoo behavior is perhaps the most likely to be familiar.

Viscosity in a gas and viscosity in a simple liquid look much like each other. You apply a force to a fluid, a gas or a liquid, and the fluid moves. You might therefore think that the molecular basis of viscosity in a gas and liquid are much the same as each other. That logic is an example of the logical error *post hoc ergo propter hoc*, ("like consequences imply like causes"), the logical error that if two effects are the same, they must have arisen from the same cause. In fact, the molecular basis of viscosity in a gas is almost completely unlike the molecular basis of viscosity in a Newtonian liquid. We'll discuss this point at the end of the chapter.

Newtonian? Newton? Yes, Isaac Newton, in addition to setting down his three laws of motion and inventing the differential calculus, also worked out the behavior of light on mixing light of different colors, and made serious studies of flowing liquids, finding the result that describes, for example, water flowing down a pipe. (On the other hand, Newton made a heroic effort to reform our understanding of alchemy, arguably becoming one of its finest practitioners, including preparing a large and well-formed Star of Regulus, but admitted that he was unable to transmute lead to gold or create a philosopher's stone. He kept sufficiently quiet about his efforts to improve Christian theology that he was not imprisoned for heresy.)

In this course, we will discuss friction between static objects at length, but have little to say about viscosity and viscoelastic effects in gasses and liquids.

9.2 Kinetic Friction

As we now know, sliding contact between solids was first studied systematically by Leonardo da Vinci (1452–1519). Alas, da Vinci did not publish his work. His notes showing what he had learned were only found recently. Most textbooks, in treating these results, will point you at Amonton's laws. The Amonton in question is Guillaume Amonton (1663 - 1705), who did his work three and a half centuries ago. Amonton limited himself to studying static friction, but he is sometimes credited with studies of kinetic friction.

We begin by considering kinetic friction, the resistance to sliding of two surfaces pressed against each other. As a simple example, consider a box having mass m sliding across the top of a table.

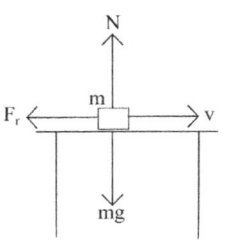

Figure 9.1: Sketch of a box sliding across a table top.

Figure 9.1 shows the box and the table. For the sake of argument, the box is moving to the right, as indicated by the arrow lebelled v. It has acting on it a gravitational force mg, a normal force $\mathbf{N}$, and a kinetic friction force $\mathbf{F}_r$, which tends to slow it down. This sketch is not a force diagram. It includes an arrow indicating the direction of motion of the box. A force diagram for the box is seen in the next figure.

The force diagram has one modest peculiarity, namely we only know the direction that the kinetic friction force points because we know which way the box is moving relative to the table. If the box were moving in the opposite direction, toward the left, the friction force would still oppose motion, meaning the friction force on the box would now point to the right.

The magnitude of the kinetic friction force is given by a simple equation, namely

$$|\mathbf{F}_r| = \mu_k |\mathbf{N}|. \tag{9.1}$$

In this equation, $\mathbf{F}_r$ is the kinetic friction force, μ_k is the coefficient of kinetic friction, and $\mathbf{N}$ is the normal force between the two objects that are sliding with respect to each other.

This equation for the force of kinetic friction has several noteworthy features. First, the equation only gives the magnitude, not the direction, of the force. If the object is moving with velocity $\mathbf{V}$, the direction of the kinetic friction force is determined by the unit vector $\hat{\mathbf{V}}$ for the velocity, namely the kinetic friction force points in the direction $-\hat{\mathbf{V}}$, i.e., the friction force points in the direction opposite the direction in which the sliding object is moving. However, the friction force for sliding friction does not depend on how fast the sliding object is moving, only that it is not stationary.

Second, the friction force is proportional to the magnitude of the normal force. The normal force of the sliding object on the object below it, and the normal force of the object below on the sliding object, are an action-reaction pair, so they are equal in magnitude. Determine the normal force on the sliding object, or the normal force the sliding object is applying, and you have the $|\mathbf{N}|$ needed for equation 9.1. The friction force is determined by the normal force, not by the pressure (the force per unit area). Within broad limits, if you keep constant the normal force between the sliding object and the object adjoining it, but change the contact area between the two objects, the friction force does not change.

9.2. KINETIC FRICTION

For familiar materials, coefficients of sliding friction are usually in the range 0.1 to 1.0.

What is the reaction force to $|\mathbf{F}_r|$? The kinetic friction is the frictional force of the lower object on the moving object, so the reaction force is the friction force that the moving object applies to the lower object. The area of contact does not matter. The two forces are an action-reaction pair, so the reaction force on the lower object points in the direction that the upper object is moving. The reaction force on the lower object may be important in engineering applications. For example, if you have a railroad train headed across a bridge, and the train applies its brakes, a large frictional force acts to slow the train. Corresponding to that frictional force, there is a reaction force, the frictional force that the train applies to the bridge beneath it. If the bridge is not sufficiently heavily built to withstand large forces parallel to its tracks, the bridge is pulled from its foundations, with negative consequences for the train's insurers.

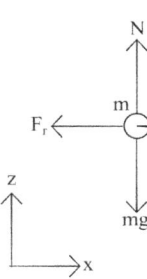

Figure 9.2: Force diagram for a box sliding to the right across a table top.

We first consider the simplest example of sliding motion, a box sliding across the table. The sketch in Figure 9.1 shows the sliding box. It has some initial velocity $\mathbf{V} = V\hat{\mathbf{i}}$. The forces on the box are the force of gravity, the normal force, and the frictional force. There is a corresponding force diagram, seen in Figure 9.2. Is this a legitimate force diagram? Yes, it shows the box, which has a mass m, as a point. The individual forces on the box are shown as arrows, labeled with appropriate algebraic symbols, and pointing in approximately the correct directions. The coordinates are shown to the side of the figure.

Using the force diagram, we can transform the Second Law

$$\mathbf{F} = m\frac{d^2\mathbf{r}}{dt^2} \tag{9.2}$$

into the equations of motion for the box. For vertical motion, we have

$$N - mg = m\frac{d^2z}{dt^2}. \tag{9.3}$$

If the table is flat, horizontal, and stationary, the acceleration in the z-direction is zero, so $N - mg = 0$ and therefore $N = mg$. If the table were on an elevator accelerating up or down, the z-acceleration might not be zero, in which case $N \neq mg$ would be found.

In the horizontal direction, the only force acting is the kinetic friction. If the box is moving in the $+x$ direction, the frictional force points in the $-x$ direction, leading to an equation of motion

$$-\mu_k N = m\frac{d^2x}{dt^2}. \tag{9.4}$$

Substituting mg for N, cancelling the two factors of m, and integrating with respect to time, one obtains

$$v_x(t) = v_{0x} - \mu_k g t, \tag{9.5}$$

$$x(t) = x_0 + v_{0x}t - \frac{1}{2}\mu_k g t^2. \tag{9.6}$$

Suppose that the box has velocity $v_x(t) = V$ at $t = 0$, with $V > 0$. In that case, you can solve for v_{0x}. You should do this. You can confirm that $v_{0x} = V$. Where does the box come to a stop? It comes to a stop at a time T such that $v_x(T) = 0$. At that time,

$$0 = V - \mu_k g T, \tag{9.7}$$

leading to $T = V/\mu_k g$. The stopping point is then

$$x(T) = x_0 + VT - \frac{1}{2}\mu_k g T^2. \tag{9.8}$$

Substituting for T, one obtains for $x(T)$

$$x(T) = x_0 + V \cdot V/\mu_k g - \frac{1}{2}\mu_k g \left(\frac{V}{\mu_k g}\right)^2, \quad (9.9)$$

which simplifies to

$$x(T) = x_0 + \frac{1}{2}\left(\frac{V}{\mu_k g}\right)^2. \quad (9.10)$$

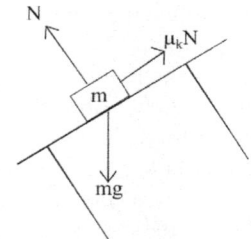

Figure 9.3: Sketch of a box sliding across a tilted table top.

Interesting aside. Suppose $V < 0$. In that case, according to equation 9.7, the stopping time T is also less than zero. That is, the object seems to have come to a stop before it started moving. What has gone wrong here? That's a homework problem.

Suppose that someone had picked up one end of the table, so that the box was on a slope, as seen in Figure 9.3. The box is then given a firm tap to start it sliding. The tabletop makes an angle θ with respect to the ground. Calculate the acceleration, velocity, and position of the box.

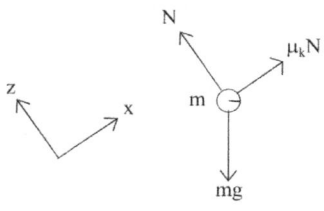

Figure 9.4: Force diagram for a box sliding across a tilted table top.

(Hint: Trick question at one point.) We can readily sketch the table, the box, and the forces and convert these into a force diagram.

There is one complication. Which way should the friction force be pointing? It might be that the box was given a firm tap on its upper side so that it is sliding down the slope, in which case the kinetic friction force should point up the slope. It might be that the box was given a firm tap on its lower end so that it is sliding up the slope, preparing to come to a stop but still moving, in which case the kinetic friction force should point down slope. Finally, suppose that the box is initially sliding up the slope, comes to a stop, and starts sliding down the slope. When the box changes the direction in which it is moving, the kinetic friction force also changes direction.

What was the trick in the question? You might have assumed that you knew which way the box was sliding, in which case you would have assumed that you knew which way the frictional force was pointing. You didn't. Making extra assumptions can readily lead to incorrect problem solutions.

9.3 Static Friction

We now turn to static friction, the frictional force between two objects that are not moving with respect to each other. The static frictional force is created by the other forces in the system. The static frictional force serves to hold an object in place, meaning that force is exactly large enough to cause the total force on an object to be zero. A static frictional force has a maximum, an upper limit on its size, given by

$$|\mathbf{F}_s| \leq \mu_s |N| \quad (9.11)$$

Note that this equation is an inequality, not an equality. It gives an upper bound to the magnitude of the static frictional force, not the value of that force. The force of static friction can have any magnitude, up to the upper bound given by equation 9.11. Static friction acts to hold objects in place; it will not cause an object to accelerate relative to the object with which it is in contact.

As an example of static friction, consider the person in Figure 9.5 pushing on a table. The table declines to move. What are the forces acting on the table? The gravitational force $-mg\hat{\mathbf{k}}$ pulls down on the table. We must have a normal force $\mathbf{N}$ that keeps the table from falling through the floor. The person is exerting a push force $P\hat{\mathbf{i}}$ that unsuccessfully encourages the table to move to the right. A force of static friction $\mathbf{F}_s$ holds the table in place. How large is $\mathbf{F}_s$?

First, equation 9.11 does not give a value for $|\mathbf{F}_s|$. It only supplies an upper limit on how large $\mathbf{F}_s$ can be. To determine the actual value of $\mathbf{F}_s$, we need to solve the corresponding Second Law problem. We first set up the force diagram, with all forces and coordinates marked and labeled, as seen in Figure 9.6.

9.4. COUPLED MASSES WITH FRICTION

There is one constraint: The table is stationary.
We now apply the Second Law

$$\mathcal{F} = m\frac{d^2\mathbf{r}}{dt^2}. \tag{9.12}$$

All forces on the table are parallel to one or the other of the indicated coordinate axes, so it is straightforward to write the Second Law in terms of its components. The table is stationary, so its acceleration is zero. We have

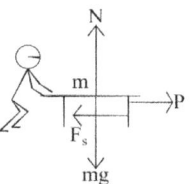

Figure 9.5: Sketch of a person pushing on a stationary table.

$$P - F_s = 0, \tag{9.13}$$
$$N - mg = 0. \tag{9.14}$$

The unknowns are the scalar components N and F_s. We readily solve for them, finding

$$F_s = P, \tag{9.15}$$
$$N = mg. \tag{9.16}$$

Observe that the static frictional force is determined by P, not by equation 9.11. That equation does not tell you how large the static frictional force is. It only tells you the maximum magnitude that the static frictional force can have. At this point, you do need to check that P is not larger than the upper limit on $|F_s|$. Substituting N from equation 9.16 into equation 9.11, one finds

$$|\mathbf{F}_s| \leq \mu_s mg. \tag{9.17}$$

As long as P is smaller in magnitude than $\mu_s mg$, the table does not move. If the person pushes harder, so that $P > \mu_s mg$, the table begins to slide. The frictional force retarding the sliding motion is kinetic friction, as discussed below.

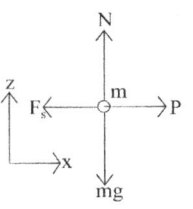

Figure 9.6: Force diagram for a person pushing on a stationary table.

An extremely important effect of static friction is seen in the incorrect operation of a beam balance. A beam balance is used to weigh objects. The simplest beam balance has a pivot point, sometimes described as a *knife-edge*, an arm that rests on the pivot point, and two pans that hang from the two ends of the beam. The arm conveniently has an indicator arranged so that one can tell how far the arm is from being level.

Superficially: One places the object to be weighed on one pan of the balance, and weights of known size into the other pan. When the two pans have equal amounts of weight in them, the balance levels, and the weight of the unknown object has been determined.

Competently: However, static friction plays a critical role in how one uses the balance for accurate measurements. If you adjust weights until the two pans become level, with the balance stationary, static friction between the pan's arm and its pivot point will hold the pans in place, even if the weights on the two pans are not quite equal. To get an accurate measurement, you must have the arm gently swinging up and down, and adjust the weights until the upswing and downswing of the arm are equal.

9.4 Coupled Masses with Friction

Let's consider another friction problem. Once again we have two masses connected by a thin piece of twine.

Mass m_1 is resting on a tabletop, while mass m_2 is hanging over the edge of the table, as seen in Figure 9.7. The two masses have mass m_1 and m_2, respectively. The twine is approximated to be massless. Mass 1 is sliding to the right but is subject to a kinetic friction coefficient μ_k. We'll consider two cases. In the first, μ_k is a constant. In the second case, μ_k depends on the time, namely

$$\mu_k(t) = \mu_0 + A\cos(\omega t), \tag{9.18}$$

where μ_0, A, and μ_k are constants. A is small enough that $\mu_k(t)$ is always positive.

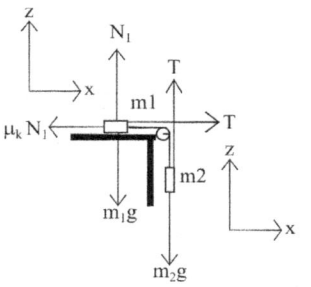

Figure 9.7: Sketch of two masses connected by twine, one mass lying on a table.

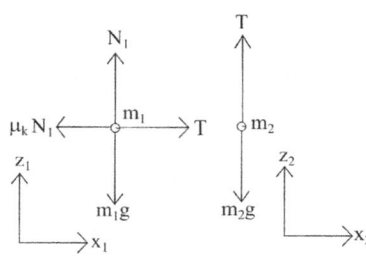

Figure 9.8: Force diagrams for the two masses connected by twine.

The problem objectives are to calculate the acceleration, velocity, and position of the first mass, and the tension T in the twine.

Step one is to draw a sketch and label the known quantities. The known quantities are the two masses and the coefficient of kinetic friction. Aside: What do we mean 'known'? I didn't provide numerical values for m_1, m_2, or T. However, the problem statement specifies these quantities algebraically as symbols, namely that they are m_1, m_2, and μ_k, respectively, so these three quantities are known. We also know the gravitational quantity g.

The next step is to draw the force diagrams.

Each mass is subject to a gravitational force mg. The first mass, but not the second, is subject to a normal force N in the $+z$ direction. The first mass is subject to sliding friction. As it is sliding to the right, the frictional force on the mass is directed to the left. Finally, there is a tension T in the twine. The tension always points toward the center of the twine connecting the two masses. T has the same magnitude everywhere in the twine between the two connection points, so as indicated it points to the right for mass 1 and up for mass 2.

Correct force diagrams always indicate the coordinate systems that you are using. The physics is independent of the choice of coordinate system, so we could choose the x axis to be horizontal for the first mass and vertical for the second mass. We instead take the x-axis to be horizontal for both masses, and the z-axis to be vertical for both masses. Note that the coordinates have been given subscripts corresponding to the particle numbers. The x and z coordinates of the first mass are x_1 and z_1; the x and z coordinates of the second mass are x_2 and z_2.

We invoke the Second Law, written for the special case that the two masses are constant, namely

$$\mathcal{F} = m\frac{d^2\mathbf{r}}{dt^2} \tag{9.19}$$

We now break the forces into their components, and write the Second Law, coordinate axis by coordinate axis, for each of the two particles. We find

$$m_1 \frac{d^2 x_1}{dt^2} = T - \mu_k N, \tag{9.20}$$

$$m_1 \frac{d^2 z_1}{dt^2} = N - m_1 g, \tag{9.21}$$

$$m_2 \frac{d^2 x_2}{dt^2} = 0, \tag{9.22}$$

$$m_2 \frac{d^2 z_2}{dt^2} = T - m_2 g \tag{9.23}$$

The above four equations are the equations of motion for the two masses.

There are also two constraints. First, the table is stationary, so $d^2 z_1/dt^2 = 0$. Second, the speed at which mass 1 slides to the right must equal the speed at which mass 2 descends. (Remember, speed ia a magnitude and always positive.) As a result,

$$\frac{dx_1}{dt} = -\frac{dz_2}{dt}. \tag{9.24}$$

Why is there a minus sign in the equation? The answer is that a positive dx_1/dt for mass 1, meaning mass 1 is moving to the right, means that mass 2 is moving downward, i.e., dz_2/dt must be negative. The minus sign causes the equality to be true.

We now ask if we have as many equations as we have unknowns. x_2 drops out of the problem; mass 2 only moves vertically. The unknowns are then the normal force N, the tension T, and the three remaining

9.4. COUPLED MASSES WITH FRICTION

accelerations, for a total of five unknowns. However, there are also three equations of motion and two constraints, so we have five equations. A solution is possible. What do we find as we try to solve?

First, from the constraints, z_1 is a constant, so the second equation becomes

$$0 = N - m_1 g, \qquad (9.25)$$

which gives the familiar $N = m_1 g$, an equation that is true only because the table is not accelerating in the vertical direction.

Second, the second constraint tells us that

$$\frac{d^2 x_1}{dt^2} = -\frac{d^2 z_2}{dt^2}, \qquad (9.26)$$

allowing us to replace one of these accelerations with the other, giving

$$-m_2 \frac{d^2 x_1}{dt^2} = T - m_2 g. \qquad (9.27)$$

At this point there are two equations in two unknowns. The best way to eliminate an unknown, when it can be done, is by subtraction. We can use subtraction to eliminate T, namely we subtract equation 9.27 from the first of equations 9.20, giving

$$m_1 \frac{d^2 x_1}{dt^2} + m_2 \frac{d^2 x_1}{dt^2} = T - T - \mu_k m_1 g + m_2 g \qquad (9.28)$$

whose solution is

$$\frac{d^2 x_1}{dt^2} = \frac{m_2 g - \mu_k m_1 g}{m_1 + m_2}. \qquad (9.29)$$

A common error is to claim at the start of the solution that $T = m_2 g$, so that the tension in the twine is equal to the gravitational force on mass 2. This claim is completely wrong. $T = m_2 g$ is not a solution of the above equations. If it were true that $T = m_2 g$, which it is not, then in the last of equations 9.20, one would have

$$m_2 \frac{d^2 z_2}{dt^2} = m_2 g - m_2 g = 0, \qquad (9.30)$$

whose right side vanishes, leading to the outcome that mass 2 has vertical acceleration zero and mass 1 does not accelerate horizontally. That's not what happens.

We now integrate equation 9.29 with respect to time to get the velocity and position of mass 1 as functions of time. Following the procedure shown in Chapter 3, the integrals are

$$v_x(t) = v_{0x} + \frac{m_2 g - \mu_k m_1 g}{m_1 + m_2} t, \qquad (9.31)$$

$$x_1(t) = x_0 + v_{0x} t + \frac{1}{2} \frac{m_2 g - \mu_k m_1 g}{m_1 + m_2} t^2. \qquad (9.32)$$

The tension is obtained from equation 9.27 by solving for T and substituting for the acceleration, giving

$$T = m_2 g - m_2 \left(\frac{m_2 g - \mu_k m_1 g}{m_1 + m_2} \right). \qquad (9.33)$$

What if μ is time dependent, as seen in equation 9.18? While tempting, simply replacing μ_k in 9.31 with $\mu_0 + A \cos(\omega t)$ would be incorrect. Equations 9.31 follow from equation 9.29 by integration, not by multiplication with respect to time. To obtain the velocity, the relevant integral is

$$v_x(t) = v_{0x} + \int_{t_i}^{t_f} dt \left(\frac{m_2 g}{m_1 + m_2} - \frac{\mu_0 m_1 g}{m_1 + m_2} + \frac{A \cos(\omega t) m_1 g}{m_1 + m_2} \right), \qquad (9.34)$$

which becomes

$$v_x(t_f) = v_{0x} + \left(\frac{m_2 g - \mu_0 m_1 g}{m_1 + m_2} \right)(t_f - t_i) + \left(\frac{A(\sin(\omega t_f) - \sin(\omega t_i)) m_1 g}{\omega (m_1 + m_2)} \right). \qquad (9.35)$$

Here t_i and t_f are the initial and final times of the acceleration process. A similar process determines $x(t_f)$.

We did not discuss what happens when a mass first starts to slide, i.e., the transition between static and kinetic friction. The frictional force laws discussed here are a somewhat crude approximation that is adequate for many purposes, but do not cope with this situation.

9.5 Rolling Friction and the Tractive Force

We now come to the third sort of friction between solid bodies, rolling friction. Rolling friction is central to the operation of all wheeled vehicles. It also arises in the operation of ball and roller bearings, objects that intermediate between a spinning shaft and its immobile mount. Consider first the simplest case, a wheel of an automobile, truck, or train moving across the ground. The ground supplies a normal force pointing up on the wheel, and the wheel creates a normal force pointing down on the ground. Superficially, the road or rail appears to be flat, and the wheel appears to be nearly circular. However, the normal force from the wheel on the ground creates a deformation in the ground. The reaction normal force of the ground on the wheel deforms the wheel from perfect circularity. An extreme case is described by an underinflated tire moving through soft sand. The wheel leaves a track in the sand; the tire being underinflated deforms visibly. The net result of these processes is that there is a frictional force that opposes the motion of the rolling object. This resistance movement is apparent, for example, in the motion of a billiard ball across a pool table; as the ball rolls, rolling friction causes the ball to come slowly to a stop.

How does rolling friction work? As the wheel passes over a segment of ground, the ground is very slightly indented; when the wheel passes beyond the segment, the ground springs back more or less into position. However, the recovery from being deformed is not perfect. Some of the energy used to deform the ground and wheel is transformed into heat, leading to a resistance to the forward motion of the wheel. The harder the wheel and ground surface are, the less deformation occurs. As a result, for steel wheels on a steel track – the situation encountered in the operation of a railroad train – the rolling friction for a given weight of vehicle is much less than for a rubber tire on concrete or asphalt.

At some level of approximation, the amount of rolling friction is proportional to the normal force between the rolling object and the surface on which it is rolling, together with a proportionality constant C_{rr}. C_{rr} is typically two or three orders of magnitude less than μ_k or μ_s. Rolling friction is rather more complicated than sliding or static friction. For example, there was an extended historical dispute as to whether or not the amount of rolling friction depends on the radius of the rolling object and, if so, how.

In addition to the rolling frictional force, the drive wheels on an automobile or locomotive must supply a *tractive force* to pull the vehicle forward or backward. For an object moving at constant velocity, the moving object supplies a force on the ground opposite to the direction of motion, while the road or rail below supplies a force in the direction of motion. The force in question resembles static or kinetic friction, depending on whether the drive wheels are simply rolling over the ground or spinning, as seen with cars trying to accelerate while on ice. The force arises because the wheel becomes slightly distorted in a direction parallel to the road surface. Traction is actually rather complicated.

9.6 Friction in Fluids

Here we consider two entirely separate phenomena, namely frictional forces – viscosity – in gasses and liquids. In real systems, both phenomena are to some extent present. However, one effect is dominant in dilute gasses, while the other effect is dominant in *simple liquids* such as water. At high frequencies and in *complex fluids* there are additional complications, notably linear and non-linear viscoelastic behavior, in which, for example, a liquid that changes its shape when subject to a force tends to spring back, the way a rubber band springs back if released after being stretched.

A simple demonstration of viscosity is seen whenever sugar is added to hot cocoa, the cocoa cup then being stirred. The spoon exerts a force on the liquid, and parts of the liquid that are not in contact with the spoon are set into motion. The spoon exerts a force on the neighboring molecules of liquid, and that force is transmitted until it reaches the sugar crystals, setting them into motion. Similarly, if we have a swinging pendulum, the pendulum exerts a force on the air, setting the air into motion; the reaction force of the air on the pendulum bob gradually brings the pendulum to a stop.

How does viscosity work? Two completely separate physical activities are present, one being dominant in gasses, the other being dominant in liquids. We will start by considering pictures of atoms and molecules in a gas and in a liquid. Most molecules are not spheres, but for the simplest description we draw them in two dimensions as circles. Real atoms and molecules are not rigid, and cannot move through each other, the way billiard balls are nearly rigid and cannot move through each other, but treating molecules as rigid spheres is a reasonable first approximation at room temperature for atoms and small molecules. Unlike billiard balls,

9.6. FRICTION IN FLUIDS

real atoms and molecules also repel and attract each other when they are not in contact, at least over short distances. These repulsions and attractions are not very important in dilute gasses but are quite important in liquids.

At temperatures above absolute zero, atoms in a gas or liquid are constantly in motion. In a dilute, low-pressure gas, the distance between atoms or small molecules is much larger than their size, so most of the time the atoms in a gas move more or less in straight lines. Occasionally, atoms in a gas collide with each other, bouncing off each other like billiard balls on a pool table. This simple image is the basis of the kinetic theory of gasses. In contrast, in a liquid atoms are permanently in close contact with each other. They do not move in straight lines separated by occasional collisions; instead, they are permanently pushing and pulling on each other. Only over distances much smaller than their atomic diameters do atoms in a liquid move in nearly straight lines. This image is the basis of the theory of simple liquids.

Let us now take these two pictures and see how they lead to descriptions of viscosity. First, how do we measure viscosity? A simple approach is to use two flat, parallel plates that are caused to move with respect to each other. In a standard arrangement, we force one plate to move at a constant speed V, and measure the force that is created by this motion on the other plate. The motion and the force are both in the x-direction, while the separation Y between the plates is in the y direction. In simple cases, the velocity of the fluid varies linearly with the position relative to the two plates, the gradient dv_y/dx in the fluid velocity being V/Y. The force per unit area on the second plate is proportional to the velocity gradient, the proportionality constant being the viscosity η. In more complex cases, one observes, e.g., shear banding, in which dv_y/dx depends strongly on y.

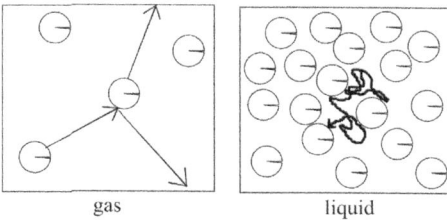

Figure 9.9: Atoms and atomic motions in a gas (arrows) and in a liquid (curved line).

In a dilute gas, when an atom collides with the first plate, on the average it picks up a certain amount of motion in the $+x$ direction. When that atom heads off into the gap between the two plates, and collides with another gas atom, some of that motion in the $+x$ direction is transferred to the second gas atom. This transfer process is repeated many times. Finally a gas atom collides with the second plate. That gas atom is more likely to be moving in the $+x$ than in the $-x$ direction, so on the average the gas atoms push on the second plate in the $+x$ direction. The transfer of motion from one plate to the other occurs because an atom preferentially moving in the $+x$ direction also has some motion in the y direction. The motion in the y-direction serves to carry the motion in the x-direction from one plate to the other. You now have a physical description of how viscosity works in dilute gases.

The quantitative details of gas viscosity are part of the kinetic theory of gases, as first worked out by James Clerk Maxwell. This is the same Maxwell who also worked out Maxwell's equations, which are the classical basis for how electricity and magnetism work. When Maxwell did his calculation he obtained one result that he did not expect. According to his calculation, the viscosity of a gas depends on how warm or cold the gas is, but is independent of the density of the gas. If you stay at constant temperature and raise or lower the pressure, thus raising or lowering the density of the gas, the viscosity of the gas was predicted not to change. Maxwell found this result extremely surprising, so he rushed to his laboratory to make measurements that tested his prediction. His unexpected prediction was correct.

In a liquid, matters are entirely different. In a liquid, atoms are more-or-less but not perfectly close-packed, much like toy marbles in a cloth bag. There is an upper bound to how tightly you can pack marbles (or atoms). In a crystal array, that upper bound has spheres occupying slightly less than three-quarters of the total volume. In a liquid, a random array, the spheres occupy modestly less than two thirds (or sometimes less) of the volume. In both cases, the remainder of the volume is filled by empty space.

Nearby atoms exert forces on each other. In a gas, the distance over which the forces act is much less than the distance between an atom and its nearest neighbors, so the atoms move in nearly straight lines separated by collisions. In a liquid, the distance over which forces act is larger than the typical distance between an atom and its nearest neighbors, so atoms in a liquid are constantly pushing and pulling on each other. How does this lead to viscosity? When the first plate moves, it pushes on the neighboring atoms of the liquid. Those atoms in turn push on atoms further away from the first plate. The force is eventually transmitted from the first plate to the second plate, one atom pushing on the next until the motion of the first plate creates a force on the second plate.

The process that creates viscosity in a dilute gas is also active in a liquid, but in a liquid that process is almost totally unimportant; it creates something like a tenth of a percent of the liquid's viscosity. The contribution of the liquid process to the viscosity of dilute gas is similarly unimportant, because in a gas atomic collisions that bring two atoms close enough together to put a force on each other are rare.

9.7 Discussion

This Chapter discusses a considerable number of topics not found in all texts. I also offer multiple openings to larger areas of physics research. Rolling friction and tractive forces are often skipped over in opening discussions of mechanics. Viscoelastic liquids are examples of *complex fluids*, liquids in which interesting things happen on a range of different time or distance scales; these are a major area of contemporary physics research. The previous few paragraphs describe viscosity in low-pressure gases and in familiar liquids. What happens to the viscosity if we examine a dense, high-pressure gas or a liquid containing molecules of very different sizes, as found in solutions of polymer molecules or micelles? In some cases, part of the study of complex fluids, the viscosity of the small-molecule liquid increases by a great deal. For much more than you wanted to know, see my book *Phenomenology of Polymer Solution Dynamics*.

9.8 Worked Problems

1. Consider a pair of masses m and M, as seen in Figure 9.10a. m lies on a ramp that makes an angle θ with respect to the horizontal. M is hanging in mid air. The two masses are linked by a string that goes over the top of a very light pulley; the tension in the string is T everywhere. The string remains taut for the duration of this problem. Give (a) the force diagrams for m and M, (b) the Second Law equation(s) for each mass, and (c) the constraints. Compute (d) the tension in the string and (e) the acceleration of each mass. Compute the velocity of m as a function of time. If $M \gg m$, do your solutions require that mass m is moving up the ramp? Why or why not? Now solve the problem again, this time with a coefficient of kinetic friction μ_k describing the interaction between m and its ramp.

2. Consider a bag of flour sliding down an extremely long chute having angle θ with respect to the horizontal. The bag has mass M. The flour mill is gradually warming up. The temperature of the slope increases with time, leading to a coefficient of kinetic friction that depends on time as $\mu_k = A + Bt^2$, A and B being constants. Find the position of the bag as a function of position along the slope, using a coordinate system in which one axis lies parallel to the chute, as a function of time. You may assume that B is sufficiently small that the bag of flour does not come to a stop during the time under consideration.

3. Two blocks rest on a horizontal, stationary table, as shown in Figure 9.10b. The 50 kg block rests on top of a 100 kg block. The surface between the 50 and 100 kg blocks has a coefficient of kinetic friction $\mu_k = 0.1$. The surface between the 100 kg block and the table supporting both blocks is frictionless for all practical purposes. A $100\hat{i}$ N force is applied to the upper block. Find the accelerations of the upper and lower blocks.

9.9 Homework

1. For the box on the tilted table, as discussed in the chapter, find the equations of motion and the constraints, using (a) a coordinate system in which the x and z coordinates are horizontal and vertical, and (b) a coordinate system in which the x and z coordinates are parallel and perpendicular to the table surface. Also find the acceleration, velocity, and position as functions of time using each coordinate system. Show that the two accelerations you have calculated are equal in magnitude.

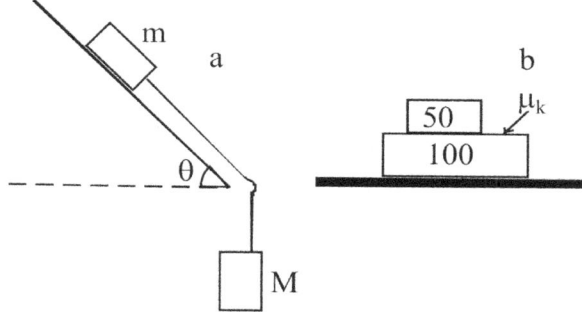

Figure 9.10: Figures for worked problems (a) 1 and (b) (2).

2. For the two coupled masses shown in Figure 9.7, calculate the acceleration, velocity, and position as functions of time, for the case that the mass on the table is sliding to the left.

3. For the two coupled masses shown in Figure 9.7, with the mass on the table sliding to the right, calculate the acceleration of mass m_1 as a function of the distance d of mass m_2 below the table, assuming that the twine has a mass M and total length L.

4. We reach yet another scene in the motion picture, shown in 9.11a. The heroine (70 kg) and her non-simulated body armor (40kg) have landed on the roof of the villain's hideout. The roof makes an angle θ with respect to the horizontal. The villain immediately detonates one of his traps (villainsupply.com, alas, no longer in business as a purveyor of secret island bases), firing the rocket motors under the roof (sketch). The roof (mass 400 kg) and its rider are accelerated perpendicular to its flat surface. The roof has a static friction coefficient of 0.2. Find the minimum thrust (the force on the roof due to the rocket motors) required so that the heroine does not slide down the roof.

5. A mass m_1 lies on an inclined plane that makes an angle θ with respect to the ground. The coefficient of kinetic friction between m_1 and the plane is μ_k. Mass m_1 is connected by a strong wire to a mass m_2 that is hanging off the top end of the ramp. At the moment of interest, mass m_1 is sliding down the ramp at constant speed. Find (a) the force diagrams for the two masses, (b) the second law equations for the two masses, and (c) the tension in the rope. If $m_1 = 8$ kg and $\mu_k = 0.2$ find m_2. What would happen (one-sentence qualitative description) to your calculation if mass a were sliding up the slope?

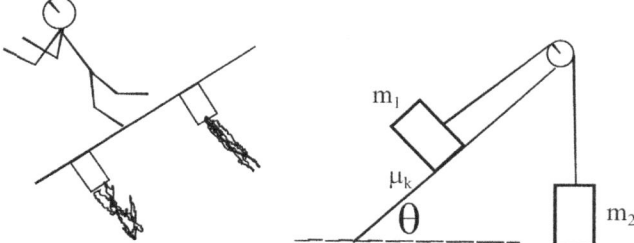

Figure 9.11: Figures for homework problems 4 and 5.

5. A mass m_1 is subject to a horizontal force F. m_1 slides across the top of a second block, mass m_2, which in turn rests on a frictionless surface. The coefficient of kinetic friction between the two blocks is μ. The acceleration of the first block is a known quantity $d^2x/dt^2 = A_0$. Find μ in terms of known quantities. Find an expression for the acceleration of the second block in terms of μ.

6. A mass M rests on a plank, length L, whose coefficient of static friction is μ_s. The plank is initially horizontal. The angle between the plank and the horizontal is gradually increased. When the board

reaches an angle θ with respect to the horizontal, the mass M begins to slide down the board. In terms of the parameters supplied here, including the gravitational constant g, what is the value of μ_s?

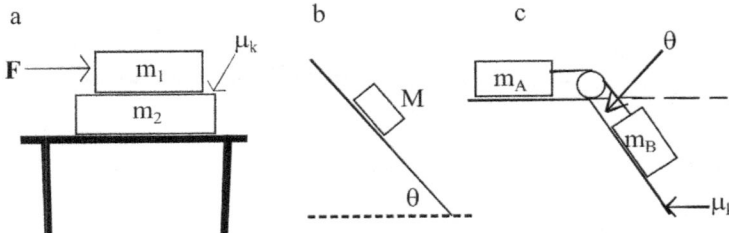

Figure 9.12: Figures for homework problems (a) 6,(b) 7, and (c) 8.

8. The Figure shows two masses linked by a string. Mass B (its mass is m_B) rests on a rough surface that makes an angle of θ with respect to the horizontal. Mass A (its mass is m_A) rests on a flat frictionless surface. The string is stretched over a frictionless pulley. The masses are moving. The surface under mass B has a coefficient of kinetic friction μ. (i) Find the force diagrams for the two masses. (ii) Write the second law for each of the two masses. (iii) Find the constraints relating the accelerations of the two masses. (iv) If the masses are moving to the left, find the acceleration of the first mass and the tension in the string. (v) If you were not told which way the masses were moving, which way (left or right) would the friction force point? Really? Do you know which way that force points, or not?

9. In yet another scene from the interminable motion picture, the villain, mass M, jumps out a window and stands on the steep roof of his secret laboratory. The roof makes an angle θ with respect to the horizontal. Unfortunately for the villain, the roof has been coated with a fine layer of oil; there is no friction. The villain begins his slide down the roof to certain doom.

(i) Give the force diagram for the villain.

(ii) Consider the forces acting on the villain. Which of these forces have reaction forces? Which of these forces do not have reaction forces? For each force that has a reaction force, identify the reaction force: Give the source of the force, the object on which the force is acting, and the nature of the force.

(iii) In terms of the variables given above, how large is the normal force on the villain due to the roof? How does this force change as the villain accelerates down the roof?

10. A child sends a smooth rock sliding across the smooth ice of a frozen pond. The rock slides 15 m in 6 s before coming to rest. Find the coefficient of kinetic friction between the rock and the ice.

11. An aerosan is a propeller-driven motorized sled, used to transport goods and people over snow in areas that are very flat. Starting from rest, for some time after starting the engine the magnitude of the force on an aerosan, mass m, due to the force of the air on its propeller, increases linearly with time, $F_{\text{prop}} = pt$. The aerosan is also subject to kinetic friction with force constant μ_k. For the period that the force is increasing linearly in time, find the displacement of the aerosan as a function of time. Assume that the vehicle starts at $x = 0$ at $t = 0$ and that static friction does not create a delay before the aerosan begins to move.

12. Consider a car being towed up an icy hill. Its wheels are free to rotate. Because the ice is not perfectly polished, there are frictional forces acting on the car. Draw the force diagram for the car. For each force in your free body diagram, in a complete sentence identify the reaction force, its physical nature, and the object on which the reaction force is exerted.

13. Consider a box sliding up a ramp. The coefficient of kinetic friction for the box and ramp is greater than zero. Draw a force diagram for the box. Label each force. For each force identify the nature of the force, e.g., "friction", including specifying the object applying the force to the box. Which of these forces have reaction forces? Which of these forces do not have reaction forces? For each force that has a reaction force, specify the nature of the force.

9.9. HOMEWORK

14. A small bus was proceeding downhill on my street. To the great alarm of the driver, the brakes were floored, the wheels were locked and not turning, and the bus was sliding at constant speed down the icy road. The bus has a mass of 6,000 kg. Across the 115 foot width of my property, the street drops 7 feet; the slope is constant. What was the constant of kinetic friction between the bus and the ice-covered road surface?

15. Two mining carts on opposite sides of a hill are connected by a durable steel cable. The cable is looped over a pulley as shown in Figure 9.13. The carts have masses m_1 and m_2, and the slopes have angles θ_1 and θ_2 as indicated. The axles on the carts have rusted to the cart chassis. The carts slide along the rails with a coefficient of kinetic friction μ_k. I have provided coordinates. Use them. (i) Find the force diagrams for the two carts. (ii) Write the equations of motion for each cart. (iii) Find the acceleration of the first cart. (iv) For each force acting on m_1 identify the reaction force, including the magnitude and direction of the reaction force, the object exerting the reaction force, the object on which the reaction force is exerted, and the nature of the force. A simple table will be adequate for an answer.

16. Two apartment-mates are trying to move a two-piece – pieces chained in the middle – sofa into the next room. Because they are not on speaking terms, roommate number one is trying to pull the sofa straight to the left, to line it up with the left hand door. Roommate number two is trying to pull the sofa straight to the right, to line it up with the right hand door. Each sofa piece has a mass m. The floor is somewhat rough, and has friction coefficients μ_s and μ_k. Roommate number one exerts a force of magnitude F_1 as indicated. Roommate number two exerts a force of magnitude F_2 as indicated. The sofa is currently sliding to the right. (i) Give the force diagrams for each of the sofa pieces. Identify the forces on the left-hand piece of sofa in complete sentences. (ii) In complete sentences, identify the reaction force for each force on the left-hand piece of sofa. (iii) Starting with $\mathbf{F} = m\frac{d^2\mathbf{r}}{dt^2}$, write the equations of motion for the the two sections of sofa. (iv) What is the tension in the chain? (The final answer goes on for a bit. Do not try to simplify.) (iv) What is the acceleration of the sofa? (The final answer goes on for a bit. Do not try to simplify.)

17. A modified tug-of-war [Aside: Once a college varsity sport. Banned: too dangerous.] contest consists of two heavy crates connected by a rope, and two small groups of students pulling the boxes in opposite directions (see sketch). Each crate has a mass of 200 kg. The first student team generates a sustained force of 5000 N on the crates, pulling left. The second student team generates a sustained force of 10,000 N on the crates, pulling right. The ground has a coefficient of static friction of 0.7 and a coefficient of kinetic friction of 0.5. The crates are sliding to the right. (i) Give the correct free body diagrams for each of the two crates. (ii) Write the equations of motion $\mathbf{F} = m\frac{d^2\mathbf{r}}{dt^2}$ for each crate. (iii) Find the acceleration of each crate. (iv) Find the tension in the central rope.

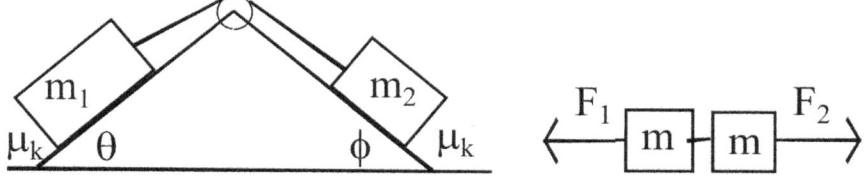

Figure 9.13: Figures for the homework, (a) Problem 15 and (b) Problems 16 and 17.

19. The early-rising victim of this problem stands on a steep, ice-covered slope overlooking a cliff watching the sun rise. The slope makes an angle θ with respect to the horizontal. The victim is initially stationary. However, as the sun rises, the ice becames warmer and warmer, and more and more slippery, until the victim begins to slide down the slope towards the cliff. During the slide, the coefficient of kinetic friction μ_k of the ice follows

$$\mu_k = \mu_o - at^2, \tag{9.36}$$

where μ_o and a are constants, $t = 0$ is the moment at which the slide begins, and a, a positive constant, is sufficiently small that the equation remains applicable until the victim goes off the cliff. Calling the starting point of the slide $s = 0$, where s is the distance measured along the slope, find s as a function of time.

20. It is time for yet another scene from the motion picture. The approach to the villain's headquarters is a steep roof that makes an angle θ with respect to the horizontal. The heroine (mass M including body armor) is crawling up the steep roof to the villain's headquarters when the villain fires the rocket engines under the roof, sending the roof flying throuugh the air. The rocket engines are mounted perpendicular to the roof, giving the roof an acceleration of magnitude a. The roof has a coefficient of static friction μ_s and a coefficient of kinetic friction μ_k. (i) Give the force diagram for the heroine. (ii) Write the equations of motion for the heroine. (iii) The special effects team asks the staff physicist what the coefficient of static friction must be, if the heroine is to be able to hang on to the roof during this process. What value of μ_s is required? Is your value for μ_s a lower or an upper bound? See Figure 9.11 .

9.10 Solutions to the Worked Problems

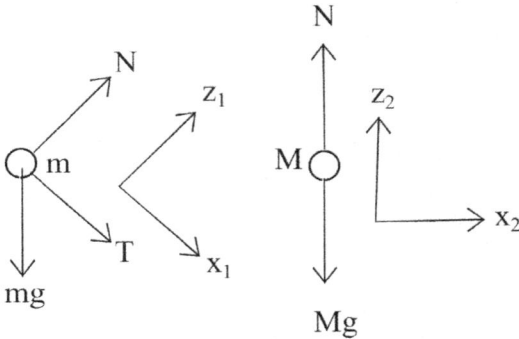

Figure 9.14: Force diagrams for Worked Problem 1

1. (a) We begin by writing the force diagrams for the two masses. Force diagrams are only correct if they include the coordinates to be used for each mass. There is no requirement that the coordinates used for m and for M must be parallel to each other. Instead, a good rule of thumb is to choose coordinate axes that are consonant with any constraints, so that when the Second Law is split into its scalar components some of the components are totally uninteresting.

This problem shows two examples of this rule of thumb. The mass M is hanging from an end of a rope. The forces on it are both vertical, so if we choose one coordinate axis to be vertical and the other two axes horizontal, only the vertical axis is interest. For the components of the forces in the horizontal plane, the x and y components of the forces, the force components and hence the corresponding accelerations are all zero. For the mass m, life is slightly more complicated. The normal force is perpendicular to the plane, the tension is parallel to the plane, but gravity points straight down. No choice of coordinates can make all three forces parallel to a coordinate axis. However, the motion of mass m must lie parallel to the plane, so it is fruitful to make two of the coordinate axes parallel and perpendicular to the plane, as indicated in the figure. The third coordinate axis is perpendicular to the page. There are no forces and hence no accelerations with components perpendicular to the page, so we do not need to consider this coordinate further here.

(b) We now apply the Second Law $\mathbf{F} = M\frac{d^2\mathbf{r}}{dt^2}$. For mass M, only one component of the Second Law's vector form is of interest:

$$T - Mg = M\frac{d^2 z_2}{dt^2}$$

For mass m, we need to resolve the gravitational force into its x and z components. We have done this before. The inclined plane makes an angle θ with respect to the horizontal, while gravity is vertical. In splitting gravity into its x and z components, we recall that the force vector corresponds to the hypotenuse of a right triangle, while the two components are the two legs of that right triangle. Applying a bit of plane geometry, we find that gravity is at an angle θ with respect to the direction of the normal force, so that the gravitational force has a component $-mg\sin(\theta)$ along the x axis and a component $-mg\cos(\theta)$ along the z axis. Two features of these components should be noted. First, the sine and cosine functions have interchanged from their more common positions, so that the sine function is here associated with the x axis. Second, each component has a sign; here both components are negative. That sign is determined by the directions of the coordinate axes with respect to the direction of gravity; those minus signs are not simply copied from the $-mg$ of a common form for gravity. The sign issue is immediately revealed if I make the positive x direction down the slope, because then the x component of gravity becomes $+mg\sin(\theta)$.

For mass m, the Second Law has two interesting components, to whit

$$N - mg\cos(\theta) = m\frac{d^2 z_1}{dt^2}, \tag{9.37}$$

$$T - mg\sin(\theta) = m\frac{d^2 x_1}{dt^2}. \tag{9.38}$$

(c) There is one constraint, namely that the rope connecting the two masses will assuredly stay taut. As a result, the speeds of the two masses must be equal, leading to

$$\frac{dx_1}{dt} = -\frac{dz_2}{dt}$$

and therefore

$$\frac{d^2 x_1}{dt^2} = -\frac{d^2 z_2}{dt^2}.$$

The minus sign appears because if m is moving down the ramp M is moving toward the floor.

(e) We have two equations involving T. If we subtract one from the other we get

$$Mg - mg\sin(\theta) = m\frac{d^2 x_1}{dt^2} - M\frac{d^2 z_2}{dt^2}.$$

If we invoke the constraint to eliminate reference to z_2, we find

$$\frac{d^2 x_1}{dt^2} = \frac{Mg - mg\sin(\theta)}{M + m}$$

and therefore

$$\frac{d^2 z_2}{dt^2} = -\frac{Mg - mg\sin(\theta)}{M + m}.$$

On integrating with respect to time, we obtain for the velocity of the first mass

$$\frac{dx_1}{dt} = \frac{Mg - mg\sin(\theta)}{M + m}t + v_{0x}.$$

If at the start of the problem we set the masses moving, v_{0x} can have any value we want. Even if $M \gg m$, the masses may be moving in either direction, at least initially.

The problem asked us to find the tension and the accelerations, in that order, but here we have found it convenient to solve the problem parts not quite in the order in what they were asked. Finally, what is the tension in the string? From the Second Law for M, and substituting for the acceleration,

$$T = Mg + \frac{M(Mg - mg\sin(\theta))}{M + m},$$

which can be simplified to

$$T = \frac{Mmg}{M + m}(1 + \sin(\theta)).$$

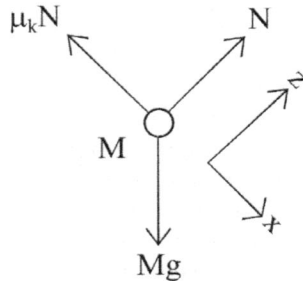

Figure 9.15: Force diagram for Worked Problem

2. We have a mass M sliding down a ramp. The coefficient of kinetic friction is time-dependent. How far does M travel as a function of time? The force diagram should show a force of gravity Mg, a normal force N, and a frictional force $\mu_k N$. Signs depend on how I orient to coordinate axes. I choose to put the x-axis parallel to the ramp pointing downhill and the z-axis perpendicular to the ramp pointing up. The usual similar triangle rationale for a ramp gives us the components of the gravitational force as $+mg\sin(\theta)$ for the gravitational force parallel to the ramp and $-mg\cos(\theta)$ for the gravitational force perpendicular to the ramp pointing down. The signs on the two components are determined by the directions of the force with respect to the two coordinate axes defined in the force diagrams.

Taking the Second Law in terms of its scalar components, in the vertical direction we have

$$N - Mg\cos(\theta) = M\frac{d^2z}{dt^2}. \tag{9.39}$$

The acceleration perpendicular to the ramp is zero, so

$$N = Mg\cos(\theta). \tag{9.40}$$

Along the ramp, the Second Law gives us

$$-\mu_k N + Mg\sin(\theta) = M\frac{d^2x}{dt^2}. \tag{9.41}$$

Substituting for μ_k and N, and solving for the acceleration

$$\frac{d^2x}{dt^2} = g\sin(\theta) - (A + Bt^2)g\cos(\theta).$$

The velocity along the ramp is then

$$\frac{dx}{dt} = gt\sin(\theta) - (At + \frac{Bt^3}{3})g\cos(\theta) + v_{0x}.$$

and the position as a function of time is

$$x(t) = \frac{gt^2\sin(\theta)}{2} - \frac{At^2g}{2}\cos(\theta) + \frac{Bt^4}{12} + v_{0x}t + x_0.$$

B is sufficiently small that the bag does not come to a stop during the problem.

3. We have two blocks m_1 and m_2 resting one on top of the other. The surface between the two blocks has a coefficient of kinetic friction μ_k. The surface under the lower block is frictionless. A horizontal force F_o is applied to the upper block. What are the accelerations of the two blocks?

9.10. SOLUTIONS TO THE WORKED PROBLEMS

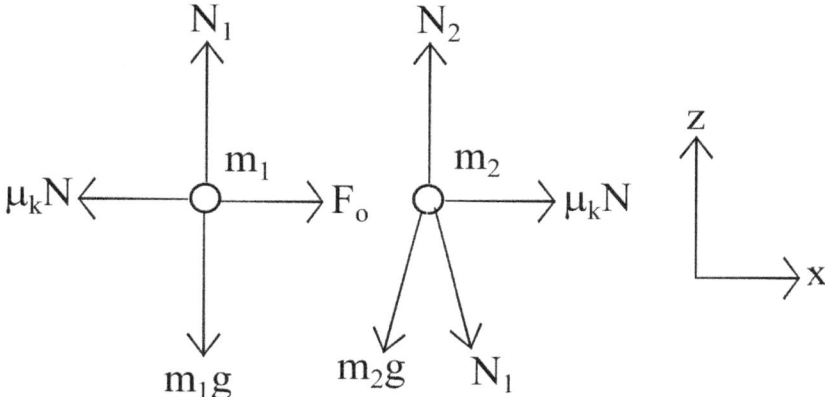

Figure 9.16: Force diagram for Problem Solution 3

We begin with a sketch and then force diagrams. The problem statement supplies numerical values for the algebraic quantities, but there is no use for those numbers until we are almost at the end of the calculation. We write the Second Law in its component form and substitute for the forces. For the upper block we have

$$m_1 \frac{d^2 z_1}{dt^2} = N_1 - m_1 g, \tag{9.42}$$

$$m_1 \frac{d^2 x_1}{dt^2} = F_o - \mu_k N_1, \tag{9.43}$$

while for the lower block we have

$$m_2 \frac{d^2 z_2}{dt^2} = -N_1 - m_2 g + N_2, \tag{9.44}$$

$$m_2 \frac{d^2 x_2}{dt^2} = +\mu_k N_1. \tag{9.45}$$

N_1 is the normal force between the blocks. N_1 for the upper block and N_1 for the lower block are an action-reaction pair, so the sign of N_1 inverts between the two blocks. The vertical acceleration is zero, so for the first block $N_1 = m_1 g$. The accelerations are

$$\frac{d^2 x_1}{dt^2} = \frac{F_o - \mu_k m_1 g}{m_1}, \tag{9.46}$$

$$\frac{d^2 x_2}{dt^2} = \frac{\mu_k m_1 g}{m_2}. \tag{9.47}$$

It is finally appropriate to substitute numbers. To one or so significant figures

$$\frac{d^2 x_1}{dt^2} = \frac{100 - 0.1 \cdot 50 \cdot 10}{50}, \tag{9.48}$$

$$\frac{d^2 x_2}{dt^2} = \frac{0.1 \cdot 50 \cdot 10}{100}. \tag{9.49}$$

This page reserved for your notes.

Chapter 10

Examination 1

First Hour Examination

This is the first hour examination. The examination lasts for 50 minutes. There are three questions, for a total of 100 points, and a bonus problem.

This is a closed book examination. You may not use books, notes, or reference materials during this examination. You may use a pocket calculator to perform numerical calculations.

Begin from first principles, e.g.,

$$\boldsymbol{F} = m\frac{d^2\boldsymbol{r}}{dt^2}. \tag{10.1}$$

Show all work. You are primarily graded on how you reached your answers, though in some cases there is not a lot of work to reach the correct answer. If you do not show your work, you will in general receive little or no credit for your answer.

I [Total of 40 points]

It is time for the trial run of the world's fourth atomic train. Because the last three atomic trains were destroyed in the course of past First Examinations, the crew wisely insisted on being replaced with a robot controller. The train exits the yards with an acceleration 0.1 m/s^2 until it reaches a speed of 10 m/s. It continues at this speed in a straight line in the $+x$ direction until the command is sent to the robot to engage the brakes. At this point, instead of slowing down the train was observed to increase in speed at constant acceleration. When the train reached a speed of 40 m/s, it exited from its tracks and turned into scrap metal. Unfortunately, the instrument recorders on the train mostly failed, but the positions of the train *while it was accelerating* are known at three times. The times and positions are:

$$t = 10, x = 120, \tag{10.2}$$
$$t = 20, x = 230, \tag{10.3}$$
$$t = 40, x = 510. \tag{10.4}$$

A) (25 points) Write the position and velocity of the train in the exact forms

$$x = x_0 + v_0 t + 0.5at^2, \tag{10.5}$$
$$v_x = v_0 + at. \tag{10.6}$$

B) (10 points) At what time did the train begin its final acceleration?

C) (5 points) At what time did the train leave the tracks?

II [Total of 30 points]

In yet another scene of the interminable motion picture, the villain, mass M, flees by jumping out a window and standing on the steep roof of his secret laboratory. The roof makes an angle θ with respect to the horizontal. Unfortunately for the villain, the roof has been coated with a fine layer of oil; there is no friction. The villain begins his slide down the roof to certain doom.

A) (10 points) Give the force diagram for the villain.

B) (10 points) Consider the forces acting on the villain. Which of these forces have reaction forces? Which of these forces do not have reaction forces? For each force that has a reaction force, identify the reaction force: Give the source of the force, the object on which the force is acting, and the nature of the force.

C) (10 points) In terms of the variables given above, how large is the normal force on the villain due to the roof? How does this force change as the villain accelerates down the roof?

III [Total of 30 points]

Two wheeled carts are connected by a heavy rope. One cart, whose mass is m_1, is on level ground. The cart is on rails; friction is negligible. The other cart, whose mass is m_2, has rolled off a cliff and is hanging in mid-air.

A) (10 points) What are the free-body diagrams for the two carts?

B) (10 points) Write the equations of motion (the Second Law) for each cart.

C) (10 points) What is the constraint linking the motion of the two carts?

More Credit (10 points)

For problem III, find the accelerations of the two carts and the tension, if any, in the rope.

*

10.1 Solutions to Examination I

I

A) Our starting points are the equations for motion at constant acceleration

$$x(t) = x_0 + v_0 t + 0.5at^2, \tag{10.7}$$
$$v_x(t) = v_0 + at. \tag{10.8}$$

We have the conditions

$$t = 10, x = 120, \tag{10.9}$$
$$t = 20, x = 230, \tag{10.10}$$
$$t = 40, x = 510. \tag{10.11}$$

which give us three equations, viz.,

$$120 = x_0 + v_0 10 + 0.5a10^2, \tag{10.12}$$
$$230 = x_0 + v_0 20 + 0.5a20^2, \tag{10.13}$$
$$510 = x_0 + v_0 40 + 0.5a40^2. \tag{10.14}$$

where we have substituted for knowns but not reordered, so that corresponding to $v_0 t$ we have $v_0 10$ rather than the orthodox $10v_0$. Note that the values for x replace x on the left side of the equation rather than replacing x_0 on the right side of the equation. x_0 is a constant of integration, not the value of x at some time.

We have three unknowns, but also have three equations, so we can presumably solve. We now solve. We advance by subtraction, if need be multiplying an equation by a constant before subtracting. Subtracting the first equation from the second and the third equations eliminates x_0 from the latter two equations, yielding

$$110 = 10v_0 + 150a, \tag{10.15}$$
$$390 = 30v_0 + 750a. \tag{10.16}$$

Multiplying the first equation by 3 and subtracting, the terms in v_0 cancel, giving us

$$60 = 300a \tag{10.17}$$

so $a = 0.2$. Substituting into equation 10.15a,

$$110 = 10v_0 + 150 \cdot 0.2, \tag{10.18}$$

and therefore $v_0 = 8$. If you had instead substituted a from equation 10.17 into the equation beginning $390 = \ldots$ you would have obtained the same answer for v_0.

Finally, replacing v_0 and a in any of the original equations, we have

$$120 = x_0 + 8 \cdot 10 + 0.5 \cdot 0.2 \cdot 10^2,$$

which leads to $x_0 = 30$.

The position and velocity are therefore given by

$$x(t) = 30 + 8t + 0.2t^2, \tag{10.19}$$
$$v_x(t) = 8 + 0.2t. \tag{10.20}$$

If you had enough time on the exam, you should have substituted 10, 20, and 40 for t in these equations and confirmed that you got the right numbers for $x(t)$ at those times.

B) At the beginning of the acceleration phase, $v(t) = 10$, leading to

$$10 = 8 + 0.2t.$$

so $t = 10$ s.

C) The train derails when $v(t) = 40$, so

$$40 = 8 + 0.2t$$

and $t = 16$ s.

II

A) A force diagram for an object on a ramp appears in Figure 8.2.

B) The reaction force to the normal force N on the villain is the normal force $-N$ exerted by the villain on the roof, these two forces both being perpendicular to the roof.

The reaction force to the force $-mg\hat{\mathbf{k}}$ of the earth's gravity on the villain is the force $+mg\hat{\mathbf{k}}$ of the villain's gravity on the earth.

Each force on the villain has a reaction force someplace else.

The Second Law gives us

$$m\frac{d^2y}{dt^2} = N - mg\cos(\theta).$$

The villain does not accelerate in the y-direction until he slides off the roof, so until then $m\frac{d^2y}{dt^2} = 0$ and $N = mg\cos(\theta)$. N is independent of time until the villain falls off the roof.

III

A) You should recognize the force diagrams, as they appear in Figure 8.11.

B) $\mathbf{F} = m\frac{d^2\mathbf{r}}{dt^2}$ so

$$m_1 \frac{d^2 x_1}{dt^2} = T, \tag{10.21}$$

$$m_1 \frac{d^2 z_1}{dt^2} = N - m_1 g, \tag{10.22}$$

$$m_2 \frac{d^2 x_2}{dt^2} = T - m_2 g, \tag{10.23}$$

$$m_2 \frac{d^2 z_2}{dt^2} = 0. \tag{10.24}$$

C) The constraint is that the rope has a fixed length ℓ, so $x_1 + \ell + z_2 = $ constant. Taking the time derivative twice,

$$\frac{d^2 x_1}{dt^2} = -\frac{d^2 z_2}{dt^2}.$$

Some students who took this exam noted that the tension forces on the two masses had to be equal to each other, which could be said to be a constraint, for which I gave some bonus credit.

To find the accelerations and the tension, apply the constraint to the first of the Second Law equations, giving

$$m_1 \left(-\frac{d^2 z_2}{dt^2}\right) = T.$$

We now have two equations for the acceleration of z_2. Subtracting them, we find

$$(m_1 + m_2)\frac{d^2 z_2}{dt^2} = -m_2 g,$$

10.1. SOLUTIONS TO EXAMINATION I

so the accelerations are

$$\frac{d^2 x_1}{dt^2} = \frac{m_2 g}{(m_1 + m_2)}, \qquad (10.25)$$

$$\frac{d^2 z_2}{dt^2} = -\frac{m_2 g}{(m_1 + m_2)}. \qquad (10.26)$$

The first of these accelerations immediately gives us the tension, namely

$$T = \frac{m_1 m_2 g}{(m_1 + m_2)}. \qquad (10.27)$$

Below reserved for your notes.

This page reserved for your notes.

Chapter 11

Springs

11.1 Introduction

Springs are one of my less favorite topics in freshman mechanics. We will draw springs as coil springs, as shown in Figure 11.1. Springs were originally explained, at first approximation, by Robert Hooke (1635-1703), who wrote *Ut tensio, sic vis*, which on translation from the original Latin tells us "As the extension, so the force." Actually doing calculations with Hooke's Law is challenging. The challenge arises because getting correct answers requires that you be meticulous in assigning signs to algebraic characters. It is straightforward to reduce a discussion of springs to a mis-statement of Hooke's law, a mis-statement that lets you solve some problems and get the correct answer. These are the cooked problems that give you the right answer even if you don't know what you are doing. Having given you these words of warning, let us advance to a correct treatment of springs.

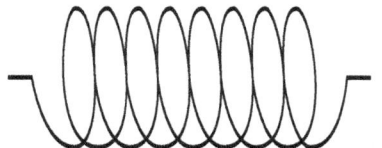

Figure 11.1: Sketch of a coil spring.

We begin with *masslessnessness*, which we have mentioned before. *Massless* refers to the approximation that the mass of an object is negligibly small relative to the masses of the other objects in the system. A mass is negligible if it does not have a significant effect on the calculated answers. That is, a mass is negligible in a particular calculation if you can change the numerical value of the mass by a modest fraction, and the outcome of the calculation does not change by an amount that you care about.

How much of a change is *negligible*? "Negligible" does not tell you how small a mass must be to be negligible. Why are there complications? In part, that depends on how accurate an answer you want. If I am standing out in an open field, and estimate that a tree in the near distance is a hundred yards away, I am probably not accurate to one part in ten. On the other hand, some chapters ago, we mentioned the Large Interferometric Gravitational Observatory, which compares two lengths with an accuracy of one part in 10^{21}.

Aside: However, for a very long time, it was approximately assumed that inaccuracies in physical quantities led linearly (or thereabouts) to inaccuracies in predictions. For example, in some mechanics problems the outcome depends on an input mass as a polynomial in m, so that a 10% change in a mass typically changes the calculated answer by some number like 10%. If I calculate the acceleration of an object falling in vacuum, and my value of g is wrong by 1%, then after some time T my calculation of the velocity of fall will be wrong by something like 1%. The time T required for the object to fall through some distance Z varies as the square root of g. As it turns out, if your error in the value of g is small, your error in calculating T also varies nearly linearly with g. (Surprised? I supply a homework problem that lets you check this claim in two ways, one requiring some knowledge of calculus and the other requiring the use of a pocket calculator.) If I have a mechanical clock that runs slowly, it may lose two minutes a day, but if I wait five days, the clock will be ten minutes slower than it had been.

However, as was only realized recently, some other mechanics problems are exponentially sensitive to input parameters, and have answers that diverge exponentially quickly from each other as input parameters are also changed. In these problems, a 10% change in a mass has an astronomically huge effect. In this course, we mostly stay away from the "some other" problems. Planetary astronomy with multiple planets and billiards played with ideal frictionless billiard balls can be good examples of the "some other" problems.

Let us consider what happens when we apply a set of forces to an object whose mass is taken to be negligibly small. Newton's law of motion states

$$\mathbf{F} = m\frac{d^2\mathbf{r}}{dt^2}. \tag{11.1}$$

If $m \to 0$ while the acceleration remains constant, then the total force on the object must have gone to zero. The condition $\mathbf{F} = 0$ is more clearly understood if we consider a massless object with two forces on it. We have

$$\mathbf{F}_1 + \mathbf{F}_2 = m\frac{d^2\mathbf{r}}{dt^2}. \tag{11.2}$$

In the circumstance that m becomes negligibly small, this equation has as its unique solution

$$\mathbf{F}_1 = -\mathbf{F}_2. \tag{11.3}$$

That is, if a massless object is acted upon by two forces, the two forces must be equal in magnitude and point in opposite directions. Note that the two forces in question *cannot possibly* be an action-reaction pair, because they both act on the same object, while in contrast the two forces of an action-reaction pair always act upon two different objects.

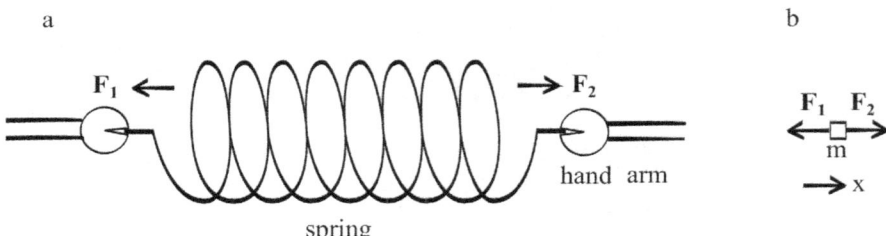

Figure 11.2: (a) Sketch of a coil spring being stretched by someone pulling on it, and (b) force diagram for the spring while being stretched.

11.2 Forces On and By Springs

For the most part, this course will only consider massless springs. We also want to consider a massless spring with two forces on it. You may imagine your hands pulling on the two ends of the spring. The sketch in Figure 11.2 only suggests these. The Figure also shows the corresponding force diagram. In this figure, the two forces that I have sketched are applied *on* the spring *by* the two hands. Because the spring in this figure is massless, the two forces $\mathbf{F}_1$ and $\mathbf{F}_2$ are required by Newton's Laws of Motion to be equal in magnitude and to point in opposite directions.

The two forces $\mathbf{F}_1$ and $\mathbf{F}_2$ are indeed forces, so each of them has a reaction force. The original two forces are applied *on* the spring *by* the two hands, so the reaction forces are necessarily applied *by* the spring *on* the two hands. The reaction forces are appropriately shown in Figure 11.3. Note that in this figure I do not show the spring, so you can see the forces more clearly, but there is indeed a spring, not shown in the figure,

11.3. HOOKE'S LAW SPRINGS

that is supplying the two forces. How large are the forces and reaction forces? To answer that question, we must consider how springs work, as described by Hooke's Law.

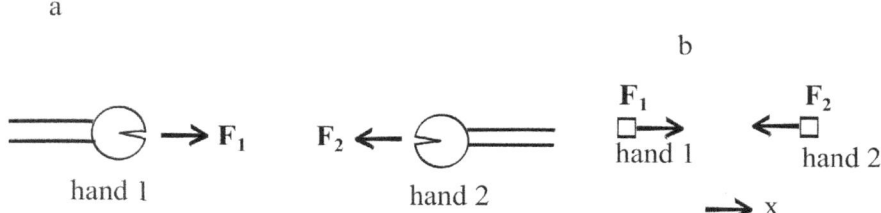

Figure 11.3: (a) Sketch of the forces on the two hands stretching the spring, and (b) a partial force diagram for the two hands. I say *partial* force diagram because the hands are also subject to the force of gravity and to the forces created by the two arms.

11.3 Hooke's Law Springs

The springs that we consider in this course are idealized relative to some springs that you encounter in nature. However, many springs are actually reasonably close to the ideal behavior we treat here. It is generally the case that if you stretch or compress a spring by a tiny amount, it will show the ideal behavior that we are about to discuss. The idealized spring we shall now discuss is known as the Hooke's Law spring or Hookean spring.

The ideal Hookean Spring is massless. It is characterized by two parameters, namely a spring constant k and an unstretched length ℓ_0. When we insert a spring into a mechanics problem, the spring also has an axis along which it lies, and a current length ℓ. The spring is attached at each end to another part of the system. What do we mean *unstretched length*? The notion here is that a spring may be compressed, in which case it becomes shorter, or pulled apart at its ends, in which case it becomes longer. However, if no forces are applied to the spring (meaning, Third Law, that the spring is applying no forces on other objects), the spring comes to a length ℓ_0, this length being the spring's unstretched length. Only when the length of the spring is changed away from ℓ_0 does the spring apply forces to the objects to which its ends are attached.

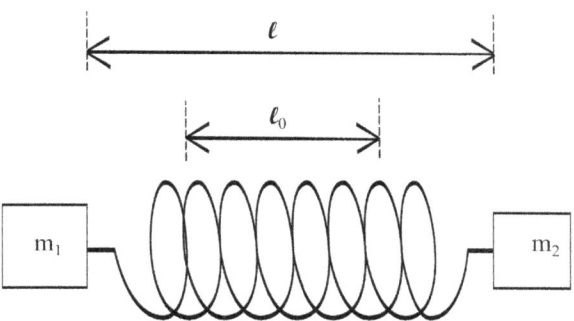

Figure 11.4: Spring, attached to two masses m_1 and m_2, and stretched from its unstretched length ℓ_0 to a stretched length ℓ.

A massless spring with a free end, an end that is not attached to anything else, must have a force **0** applied to its free end, because there is nothing at the free end to apply the force. Because it is massless, the total force on the spring is also **0**, so therefore the force applied to the end of the spring that is not free must also be **0**. A massless spring with a free end can have no mechanical effect on the rest of the system.

The force generated by a spring is a restoring force, which attempts to move the two attachment points until they are separated by the distance ℓ_0. If the spring is stretched, the spring exerts a force at each end, pulling the objects to which it is attached toward its center. If the spring is compressed, the spring exerts a force on each end, pushing the objects to which it is attached away from its center.

We now reach Hooke's Law: The magnitude of the force generated by an ideal Hookean spring is $|k(l-\ell_0)|$. The force generated by an ideal coil spring, as an approximation because the spring is ideal, lies along the axis of the spring. (Real springs can be much more complicated.) The spring generates a force of this magnitude at each end of the spring, so if we take a spring of equilibrium length ℓ_0 attached to masses m_1 and m_2 and stretch it to length $\ell > \ell_0$, as seen in Figure 11.4, the spring applies a force $+k(\ell - \ell_0)$ on mass m_1 and a force $-k(\ell - \ell_0)$ on mass m_2.

A common error is to assume that the force $|k(l - \ell_0)|$ is somehow divided between the two masses. No, the force, magnitude $|k(l - \ell_0)|$, is applied to the mass at each end of the spring. Another common error is to assume that the force only appears at the end whose mass has been moved. For example, suppose we have a wall, a spring attached at one end to the wall and the other end to a mass, as seen in Figure 11.5. Suppose we pull the mass away from the wall. The spring pulls the mass back toward the wall. The spring also tries to pull the wall toward the mass, with a force of the same magnitude. Similarly, if there is an earthquake, and the wall moves back and forth, the spring is stretched and compressed, and a force is created on the mass. A slight variation of the mass on a spring is a physical basis for a seismometer, a device for detecting distant earthquakes. Another variation on the scheme uses a pendulum that swings when the roof shakes. It was Galileo Galilei, who discovered that the period of a swinging pendulum, driven by a distant earthquake, is approximately independent of the size of the swing.

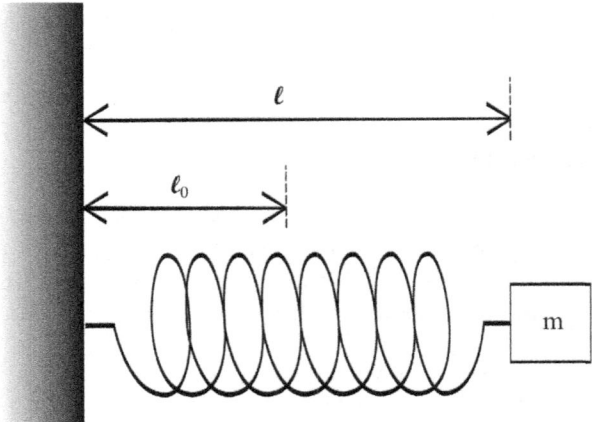

Figure 11.5: A spring attached to a wall and to the mass m, and stretched from its equilibrium length ℓ_0 to a stretched length ℓ.

Return to Figure 11.4. No matter which end of the spring is moved, the force is determined by $\ell - \ell_0$. Moving the left mass to the left, or the right mass to the right, stretches the spring, and creates a net force pulling each mass toward the center of the spring. Moving the right mass to the left, or the left mass to the right, compresses the spring, and creates a net force pushing each mass away from the center of the spring. The pictures here have put the spring along the x-axis. In general, a spring lies along some arbitrary axis in 3-dimensional space. The forces generated by the spring lie along that axis, and in general may need to be resolved into their Cartesian coordinates in order to solve a problem.

Specifying the length of the spring in terms of the coordinates of the spring's ends requires some attention to detail. Consider, Figure 11.6, two masses and a connecting spring. What are the motions of the two masses? Suppose the spring above lies along the x-axis, with the two masses at coordinates x_1 and x_2, respectively, and $x_1 < x_2$. In this case, the forces on the two masses are proportional to $|k(x_2 - x_1 - \ell_0)|$. On the other hand, if $x_1 > x_2$, so that with our indicated coordinate direction mass 1 is to the right of mass x_2, the forces on the two masses are proportional to $|k(x_1 - x_2 - \ell_0)|$.

We now draw the force diagrams for the two masses. The two masses are taken to be resting on a stationary horizontal tabletop; we are only interested in the horizontal motions. From the force diagrams we may immediately write the equations of motion for a spring. The equations of interest are those corresponding to horizontal motion, which are

$$m_1 \frac{d^2 x_1(t)}{dt^2} = k(x_2 - x_1 - \ell_0), \tag{11.4}$$

$$m_2 \frac{d^2 x_2(t)}{dt^2} = -k(x_2 - x_1 - \ell_0). \tag{11.5}$$

11.3. HOOKE'S LAW SPRINGS

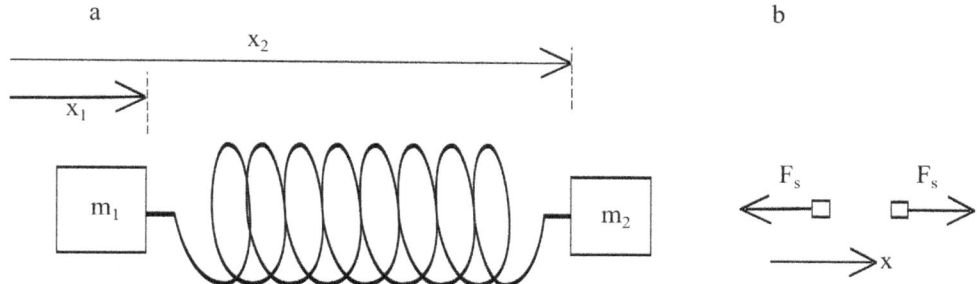

Figure 11.6: a) A spring attached to the masses m_1 and m_2. The coordinates of the two masses are indicated as x_1 and x_2. b) force diagram for the spring attached to two masses, the force being $F_s = k(x_2 - x_1 - \ell_0)$. The sign of F_s for each mass is determined when the equations of motion for the two masses are set down.

Are these equations correct? It's always wise to ask if your equations make qualitative sense. Suppose we start at equilibrium so that $x_2 - x_1 - \ell_0 = 0$. Consider the equation for m_1. If m_2 is displaced to the right, while leaving m_1 in place, the $+x_2$ direction, the force becomes positive, meaning m_1 is dragged toward the right, as expected. Suppose instead that m_1 is moved to the right, while m_2 stays in place. In that case, the force on m_1 becomes negative, meaning m_1 is pushed back toward its equilibrium position, as expected. The signs for the forces on m_1 and m_2 give the expected behaviors. We'll discuss solving these equations in the next Section of this Chapter.

Careful attention is required to ensure that x_2 and x_1 are inserted into the force equation in the correct order. If the force on the masses had been written $\pm k(x_1 - x_2 - \ell_0)$, the force would – incorrectly – fail to vanish when the two masses are separated by their equilibrium distance. If the masses had been labelled oppositely, interchanging m_1 and m_2, the forces would not be the same. After the interchange, the forces would correctly be written $\pm(x_1 - x_2 - \ell_0)$, and a force written $\pm(x_2 - x_1 - \ell_0)$ would fail to vanish with the spring at its equilibrium length.

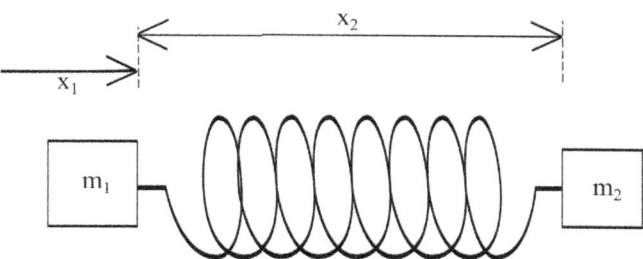

Figure 11.7: a) A spring attached to the masses m_1 and m_2. The coordinates are indicated as x_1 and x_2.

Equal attention is needed on the signs in front of the spring constant k. When the spring is stretched the force on m_1 must be positive, because the force is in the $+\hat{\mathbf{i}}$ direction, while at the same time the force on m_2 must be negative, because the force is in the $-\hat{\mathbf{i}}$ direction. If one were eccentrically to reverse the $+x$ direction – actually, this readily occurs in problems with wedges and ramps – every single sign must be checked again, to ensure that (i) a force of the expected direction appears when each mass is moved in each direction, and (ii) the force vanishes when the spring is at its equilibrium length.

An interesting physics error appears if one defines the first mass to be at x_1, and then uses the location of mass m_1 as the coordinate origin for the coordinate x_2 of the second mass, as seen in Figure 11.7.

It appears as though you could write the force as $k(x_2 - \ell_0)$ with an appropriate sign. If you did that, the equations of motion for the two masses would appear to become

$$m_1 \frac{d^2 x_1(t)}{dt^2} = k(x_2 - \ell_0), \tag{11.6}$$

$$m_2 \frac{d^2 x_2(t)}{dt^2} = -k(x_2 - \ell_0). \tag{11.7}$$

Something is strange here. According to these equations, moving m_1 has no effect on the force on either mass. That's obviously wrong. Why?

The force $k(x_2 - \ell_0)$ is useless for problem-solving because the mass m_1 is free to accelerate. x_2 is then the coordinate of m_2 in a non-inertial (potentially accelerating) reference frame. Newton's Laws of motion are only correct in inertial reference frames. In a non-inertial frame, Newton's law of motion $\mathbf{F} = md^2\mathbf{r}/dt^2$ is incorrect. You can calculate the forces in this frame, but you cannot use that force and Newton's laws to calculate the acceleration of anything.

A further complication arises for hanging masses, such as the one in Figure 11.8. A hanging mass stretches the spring, so that when the mass is stationary it is subject to a force of gravity down and a spring force up. If the mass moves away from this new "equilibrium" position, there is an additional spring restoring force due to the displacement of the mass. In writing $\mathbf{F} = md^2\mathbf{r}/dt^2$ for this problem, if you start with the spring at length ℓ_0 you have the force diagram seen to the left. However, if you start with the mass in its "equilibrium" position, the length of the spring is not ℓ_0 any more. At equilibrium, the force due to the spring exactly cancels gravity. If we call δx the displacement from equilibrium, and write the force $\sim k\delta x$, the complete force diagram has the form seen on the right. Note that gravity does not appear in this force diagram, because it is cancelled by the force created by the spring when the spring is at its equilibrium, greater-than-ℓ_0, length.

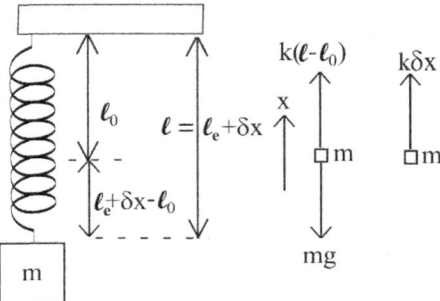

Figure 11.8: a) A hanging spring. The spring has unstretched equilibrium length ℓ_0. With the mass m attached, the spring stretches to rest length ℓ_e. The mass is then displaced from its rest position by an amount δx. b) Two versions of the force diagram for this problem. In the first force diagram, the force due to the spring is the total force it generates due to its being stretched from its unstretched equilibrium length, the part $k(\ell_e - \ell_0)$ of this force being canceled by the force of gravity mg, so that the mass is subject to a net force $k\delta x$. In the second force diagram, the terms $k(\ell_e - \ell_0)$ and mg are taken to have cancelled, so the diagram only shows the remaining part of the force, the term $k\delta x$.

As I said, springs are not trivial, and hanging springs are worse than others.

11.4 Spring Attached to Wall

Having said that, we now consider the more or less simplest example of motion driven by a spring, as shown in Figure 11.5, which shows a spring attached to a massive wall. The spring puts a force on the wall, but because the wall is so massive, the wall's motions are quite entirely negligible. To make life even simpler, the origins of the the mass's coordinates, the point where $x = 0$, is placed at the equilibrium position of the mass. In this case, the force on mass m_2 becomes $-kx$, and the equation of motion for m becomes

$$m\frac{d^2x(t)}{dt^2} = -kx. \tag{11.8}$$

To determine how mass m moves, we need to integrate this equation. That's considerably more complicated than integrating the equation of motion for a falling body, because on the right hand side of the equation x is not a constant; it's a function of time. Because x depends on time, $\int dt\, kx \neq tkx$. How do we solve this equation for $x(t)$? Of course, we could guess an answer, and, if our guess is correct, then on plugging the guess into equation 11.8 we would find that the guess is indeed a solution to the equation.

Here we'll try something more systematical, namely we will guess that the correct solution can be represented as a power series in time, namely

$$x(t) = \sum_{j=0}^{\infty} a_j t^j. \tag{11.9}$$

If your calculus has reached Taylor series, you will recognize that any continuous, uniformly differentiable function of time can be represented by this series. If you are not there yet, you can view this as a guess that turns out to work. If you plug this series for $x(t)$ into equation 11.8, you obtain

$$m \sum_{j=2}^{\infty} a_j j(j-1) t^{j-2} = \sum_{j=0}^{\infty} a_j t^j. \tag{11.10}$$

Two power series are equal to each other if (i) both series are convergent, meaning that their sums converge to a well-defined number, and if (ii) the coefficients of the two series are equal term-by-term. To see if that's the case, we need to rewrite the left-hand-side of the equation as a power series in t^n rather than t^{n-2}. We do this by introducing on the left-hand-side of the equation a new integer variable u, with $u = j - 2$, meaning $j = u + 2$. On the right-hand-side of the equation, we simply replace the summation index j with a summation index u.

With that replacement, we obtain

$$\sum_{u=0}^{\infty} \left(a_{u+2}(u+2)(u+1) + \frac{k}{m} a_u \right) t^u = 0, \tag{11.11}$$

which tells us, for each value of u (except two of them) that

$$a_{u+2}(u+2)(u+1) + \frac{k}{m} a_u = 0, \tag{11.12}$$

or

$$a_{u+2} = -\frac{k}{m} \frac{1}{(u+2)(u+1)} a_u = 0. \tag{11.13}$$

This series gives us the value for each coefficient in terms of the coefficient two lower in the series. It does not give us a_0 or a_1. a_0 and a_1 are not fixed by the original equation of motion; they are independent variables. With some algebra, one finally shows for n even that

$$a_n = (-1)^{n/2} \left(\frac{k}{m} \right)^{n/2} \frac{1}{n!} a_0. \tag{11.14}$$

The solution arising from a_0 is therefore

$$x(t) = a_0 \sum_{j=0}^{\infty} (-1)^j \left(\frac{k}{m} \right)^j \frac{1}{(2j)!} t^{2j}. \tag{11.15}$$

Equation 11.15 is also the power series for $\cos(\omega t)$ with $\omega = (k/m)^{1/2}$. A similar rearrangement for the n-odd terms gives us a power series for $\sin(\omega t)$, so the final answer is

$$x(t) = a_0 \cos(\omega t) + a_1 \sin(\omega t). \tag{11.16}$$

A mass on a spring has a position that oscillates in time at a fixed frequency ω.

11.5 Discussion

The power series approach described above worked here. In general, a truncated power series diverges to $\pm \infty$ at large values of its argument. To avoid this feature, it is sometimes used to examine how the solution behaves in the limit of large values of the argument, choose this asymptotic behavior as an approximate solution, and multiply the approximate solution by a power series that now gives corrections to the approximate solution. Sometimes this approach helps; sometimes it does not.

11.6 Worked Problems

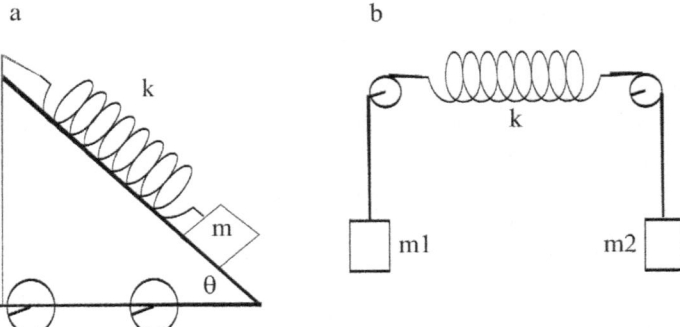

Figure 11.9: Figures a and b are for the worked problems 1 and 2, respectively.

1. See Figure 11.9a. A mass m lies on a ramp. The ramp supports have been cleverly fitted with wheels so that the ramp may be moved to the left or the right. The ramp is frictionless. A spring of spring constant k and unstretched length ℓ is attached to the ramp at its top end and to the mass m at the bottom end. By how much is the spring stretched if the mass m and the ramp are both stationary?

2. See Figure 11.9b. As seen in the figure, two masses m_1 and m_2 are suspended from strings. The strings drape over a pair of pulleys and connect to opposite ends of a nearly massless spring. The spring has spring constant k and unstretched length ℓ. By how much is the spring stretched from its equilibrium length, if the spring's length does not change with time?

11.7 Homework

1. Consider a falling body. Its acceleration is g, straight down, where to one significant figure $g = 10$ m/s^2. Suppose g is slightly different from this approximate value, so that the downward acceleration is actually $g + \epsilon$. ϵ may be positive or negative. Calculate and plot on a graph the time τ required for the object to fall 10,000 m for a series of numerical values of ϵ, keeping *epsilon* small, say $\epsilon < 1$.

2. Consider Problem 1, the downward acceleration of the falling body in this problem now being $g + \epsilon$. Calculate algebraically the time required for the body to fall 10,000 m, leaving g as a symbol rather than a number. You should obtain some function of ϵ. Now do a Taylor series expansion of the solution in powers of ϵ. Truncate the series after the ϵ^1 term, and compare τ with the values obtained in the prior problem. A plot of the fall time against ϵ, showing the solutions to both problems on the same graph, is appropriate.

3. (a) Show that equation 11.15 does indeed follow from equation 11.14. (b) Show that the corresponding sine series does indeed appear for n odd.

4. A spring having force constant k is attached above to the ceiling and below to a 3 kg mass. The equilibrium length of the spring is 2.0 m. After attaching the mass, the spring stretches by 2 cm. What is the numerical value of k?

5. An exam problem, slightly tweaked. A gymnast of mass m is hanging vertically, bobbing up and down at the end of a long, non-linear spring in the hold of a moving ocean liner. The mass of the spring is correctly approximated as zero. A non-linear spring is a spring that does not obey Hooke's law; you cannot compute the force it provides from its extension and the spring constant k, which incidentally I have not given you. The gymnast's position is given by $\mathbf{r} = 9t\hat{\mathbf{i}} + [3\cos(3\pi t) + 2\sin(5t)]\hat{\mathbf{k}}$.

   a) Find the velocity and acceleration of the gymnast.

   b) Find the z-component of the force that the spring places on the roof of the hold.

6. A mass m lies on a ramp. The ramp's angle from the horizontal is θ. The mass is attached on its upslope side to a spring of spring constant k, which has been stretched by an amount X. The constant of kinetic friction of the slope is μ_k. The coordinate x points up hill. What is the acceleration of the mass? Figure 11.9, the ramp being held stationary, describes this problem.

7. We have a spring having spring constant k. Its equilibrium length is ℓ_0. The spring is held between my hands, the x-component of the force on my left hand being F_1 and the x-component of the force on my right hand being F_2. (a) Case 1: The spring is initially at its unstretched length. I move my right hand 5 units to the right while holding my left hand stationary. What are F_1 and F_2? (b) Case 2: The spring is initially at its unstretched length. I move my left hand 2 units to the left while holding my right hand stationary. What are F_1 and F_2? (c) Case 3: The spring is initially at its unstretched length. I move my left hand 2 units to the left and my right hand 5 units to the right. What are F_1 and F_2?

8. Mass m_1 is to the left of mass m_2. The two masses are connected by a spring having a spring constant $k = 100$ N/m. The two masses are 3 kg and 6 kg, respectively. You are being asked for the accelerations at the moment that the two masses are released. (a) The spring is initially at its unstretched length. Mass m_1 is displaced 0.04 m to the right. What are the accelerations of m_1 and m_2? (b) The spring is initially at its unstretched length. Mass m_2 is displaced 0.03 m to the right. What are the accelerations of m_1 and m_2? (c) The spring is initially at its unstretched length. Mass m_1 is displaced 0.06 m to the left and mass m_2 is displaced 0.04 m to the left. What are the accelerations of m_1 and m_2?

9. The ramp and mass from Figure 11.9a are now given a modest horizontal acceleration a. Once matters settle down, the mass M sits at some constant position along the ramp. Find the force diagram and the equations of motion that, if solved, would give the change in the length of the spring. Do not solve.

10. A simple model for a propeller is a very long rod that rotates around a pivot in its center. As seen in the picture, other side of page, the rod rotates in the plane of the page, which is horizontal. The rod rotates counterclockwise around its pivot at angular frequency ω. A mass m is mounted on the rod. m can only move in or out along the rod toward or away from the pivot point, but it has come to rest relative to the rod. A spring having spring constant k and unstretched length ℓ is attached to the mass and to outer end of the rod. (i) (5 points) As the rod rotates, is the spring stretched or compressed, relative to its unstretched length ℓ? (Points are for saying why.) (ii) (5 points) By how much is the spring stretched or compressed?

11. We have two massless springs, spring constants k_1 and k_2, and a mass m. The springs can be attached to the ceiling, to the mass m, or (cf. Figure 11.10b) each other. The position of the lower end-point of spring 2 is measured before and after attaching the mass to the springs. The lengths of the springs are so arranged that in Figure 11.10a the springs are both hanging exactly vertically. For Figures 11.10a and 11.10b, find the force on the mass when the mass has descended by δy from the height where it was attached.

12. Consider two masses resting on a frictionless, level table. The two masses are connected by a nearly massless spring. Draw a force diagram of each mass. For each force in your force diagrams, identify the reaction force to that force, including the object exerting the force, the object on which the reaction force is being exerted, and the physical nature of the force.

11.8 Solutions to the Worked Problems

1. How do we solve this problem? It's basically the same as the rocket car headed up the hill, except that instead of a rocket providing thrust we have a stretched spring providing a Hooke's Law force. A reasonable choice of coordinates puts the x-axis parallel to the ramp surface.

   The x-component of the Second Law equation for the mass is

   $$k(\ell - \ell_s) - Mg\sin(\theta) = M\frac{d^2x}{dt^2},$$

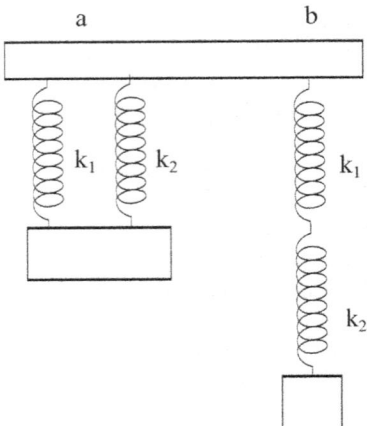

Figure 11.10: Springs having force constants k_1 and k_2 linking a ceiling above to a hanging mass m.

where $(\ell - \ell_s)$ is the amount by which the spring has been stretched. The mass is not moving, so its acceleration is zero, leading to

$$\ell - \ell_s = \frac{Mg\sin(\theta)}{k}.$$

2. We begin with a sketch and force diagrams for the two masses and the spring. Each mass is subject to a tension straight up and a force of gravity straight down. The spring is subject at each end to the tension in the string to which it is attached.

The spring is massless, so it exerts equal and opposite forces at its two ends. Those forces have magnitude

$$|F_s| = |k(\ell - \ell_s)|,$$

where ℓ_s is the length of the unstretched spring and ℓ is the length of the stretched spring. The reaction forces to the forces delivered by the spring are the tensions in the two strings, so both tensions are equal in magnitude to $|F_s|$.

We use the rule that the tension in a string is the same everywhere along a string between points at which it is exerting a force. As a result, the two masses have equal forces T pulling them upwards, leading to

$$T - m_1 g = m_1 \frac{d^2 z_1}{dt^2}, \tag{11.17}$$

$$T - m_2 g = m_2 \frac{d^2 z_2}{dt^2}, \tag{11.18}$$

as the z-components of the Second Law for the two masses.

There is also a constraint. The distance along the strings and spring between the two masses is a constant, so for the two masses

$$\frac{dz_1}{dt} = -\frac{dz_2}{dt}, \tag{11.19}$$

$$\frac{d^2 z_1}{dt^2} = -\frac{d^2 z_2}{dt^2}. \tag{11.20}$$

If we subtract the two component equations from each other and apply the constraint, we get

$$m_1 g - m_2 g = (m_1 + m_2)\frac{d^2 z_2}{dt^2},$$

and therefore
$$\frac{d^2 z_2}{dt^2} = \frac{m_1 - m_2}{m_1 + m_2} g.$$

Inserting this result into the component equation for mass 2 leads to
$$T = m_2 g + \frac{m_2(m_1 - m_2)}{(m_1 + m_2)} g = \frac{2 m_1 m_2 g}{m_1 + m_2}.$$

Finally, for the spring we must have $k(\ell - \ell_s) = T$, so the stretch of the spring is
$$\ell - \ell_s = \frac{1}{k} \frac{2 m_1 m_2}{m_1 + m_2} g.$$

This space reserved for your notes.

This page reserved for your notes.

Chapter 12

Momentum, Conservation, Collisions

We consider the motions of a group of bodies, which leads us to the Law of Conservation of Momentum. We finally take up collisions.

12.1 The Center of Mass

This section considers an *object*, a collection of N smaller masses. We begin by defining the *center of mass*. The smaller masses are labeled with an index i that advances from 1 to N. We also introduce the total mass M, which may be calculated from the masses of the individual smaller masses as

$$M = m_1 + m_2 + \ldots + m_N \tag{12.1}$$

or

$$M = \sum_{i=1}^{N} m_i. \tag{12.2}$$

Each of these smaller masses has a location. To specify location, we first choose an origin for the coordinate system. The origin may be placed in any convenient location, but all choices of origin location are in principle equally valid.

The particle locations are then vectors from the origin to the location of the particles. The location of mass i is assigned the symbol $\mathbf{R}_i$. We now introduce the center of mass vector $\mathbf{R}_{cm}$, often written $\mathbf{R}$. The center of mass vector is the mass-weighted average of the vector locations of the smaller masses.

(We previously discussed weighted averages when we introduced the time-weighted *average velocity* and the time-weighted *average acceleration*. Here we have another weighted average, but this time the weighting factors are the masses m_i of the objects).

We may write this as

$$\mathbf{R}_{cm} = \frac{m_1 \mathbf{R}_1 + m_2 \mathbf{R}_2 + \ldots + m_N \mathbf{R}_N}{\sum_{i=1}^{N} m_i} \tag{12.3}$$

or

$$\mathbf{R}_{cm} = \frac{\sum_{i=1}^{N} m_i \mathbf{R}_i}{M}. \tag{12.4}$$

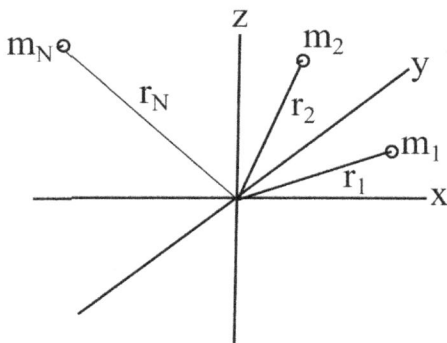

Figure 12.1: Cartesian axes, three masses, and the three location vectors for the three masses.

12.2 Motion of a Group of Bodies

Having defined the location of the center of mass, we can also specify the center-of-mass velocity $\mathbf{V}_{cm}$. The center of mass velocity is the time derivative of the position of the center of mass, namely

$$\mathbf{V}_{cm} = \frac{d\mathbf{R}_{cm}}{dt}. \tag{12.5}$$

Distributing the time derivative over the terms of $\mathbf{R}_{cm}$, one has several equivalent forms, including

$$\mathbf{V}_{cm} = \left(\frac{1}{\sum_{i=1}^{N} m_i}\right) \sum_{i=1}^{N} m_i \frac{d\mathbf{r}_i}{dt} \equiv \frac{\sum_{i=1}^{N} m_i \mathbf{v}_i}{M}. \tag{12.6}$$

That is, just as the center-of-mass location is the mass-weighted average of the locations of the component parts of the object, so also the center-of-mass velocity is the mass-weighted average of the velocities v_i of the component parts of the object.

A further time derivative gives us the center-of-mass acceleration of the object as

$$\mathbf{A}_{cm} = \frac{d\mathbf{V}_{cm}}{dt}. \tag{12.7}$$

If we eliminate $\mathbf{V}_{cm}$ in this equation by applying equation 12.6, we obtain

$$\mathbf{A}_{cm} = \left(\frac{1}{\sum_{i=1}^{N} m_i}\right) \sum_{i=1}^{N} m_i \frac{d^2\mathbf{r}_i}{dt^2} \equiv \frac{\sum_{i=1}^{N} m_i \mathbf{a}_i}{M}, \tag{12.8}$$

where a_i is the acceleration of mass i.

On taking M to the other side of the equation, we can apply the Second Law to replace the accelerations with the forces, obtaining

$$M\mathbf{A}_{cm} = \sum_{i=1}^{N} \mathcal{F}_\mathbf{i}, \tag{12.9}$$

where $\mathcal{F}_\mathbf{i}$ is the force on particle i. We define the total force $\mathcal{F}$ as

$$\mathcal{F} = \sum_{i=1}^{N} \mathcal{F}_\mathbf{i}. \tag{12.10}$$

Divide the forces into two parts, the *internal* forces and the *external* forces. The division into internal and external forces is based on the Third Law, which tells us that all forces come in action-reaction pairs, the two forces in an action-reaction pair being on two separate bodies. We now go down the list of action-reaction pairs in the system. If both bodies in a pair are in the large object, meaning that they are each indexed by i for i in the range $(1, N)$, that action-reaction pair corresponds to an internal force. If one of the bodies in the pair is not part of the object, then the action-reaction pair corresponds to an external force.

The sum of all the internal forces on an object is zero. Why? Every internal force in the object is half of an action-reaction pair, the other half of the pair being equal in magnitude and opposite in direction to first force of the pair. In calculating the total force on the object, the two forces in each internal force pair are added together. Pair by pair they add to zero. The internal force on an object therefore sums to zero. This result is a mathematical justification for the well-known observation that you cannot fly by pulling yourself up by your shoelaces.

Because the sum of the internal forces vanishes, the total force on an object is the sum of the external forces, leading to

$$M\frac{d^2\mathbf{R}_{cm}}{dt^2} = \mathcal{F}, \tag{12.11}$$

in which $\mathcal{F}$ is the total of the external forces on the object.

12.3. COLLISIONS

We now turn to a special case, in which the external force vanishes. In this case, $\mathcal{F} = \mathbf{0}$, leading to

$$M \frac{d^2 \mathbf{R}_{cm}}{dt^2} = \mathbf{0}. \tag{12.12}$$

Integrating this equation with respect to time, one obtains

$$M \frac{d\mathbf{R}_{cm}}{dt} = \mathbf{P}_0. \tag{12.13}$$

$\mathbf{P}_0$ is a constant of integration; it is the *momentum* (plural: *momenta* of the object. The momentum of an object is a constant if the sum of the external forces on the object vanishes. This result is the Law of Conservation of momentum; it applies in the circumstance that the external forces sum to zero. are This result is Newton's First Law of Motion, as earlier identified by Galileo Galilei.

12.3 Collisions

We now reach the notion of a *collision*. The notion is that one or more objects come together. Then something complicated happens, something that takes very little time, but we don't look to see what the complications are. The complications happen in a *black box*. Finally one or more objects, perhaps not the same as the original objects, emerge and go on their way. Immediately before the collision, the objects have masses and velocity vectors. The collision takes very little time, so any external forces have little time to act. Immediately after the collision, the objects have masses and velocity vectors. In problems, some of the masses and/or velocity vectors in a collision are usually the unknowns for which a solution is needed.

During a collision, external forces can have almost no effect, because the collision does not last very long. The initial total momentum $\mathbf{p}_i$ – the total momentum just before the collision – and the final total momentum $\mathbf{p}_f$ – the total momentum just after the collision – must be equal. $\mathbf{p}_i = \mathbf{p}_f$ is the *Law of Conservation of Momentum*. Momentum conservation is only exact if no external forces are being applied. In the approximation that the collision does not last very long, the phrase 'not last very long' means that the external forces do not have enough time to change the momentum appreciably.

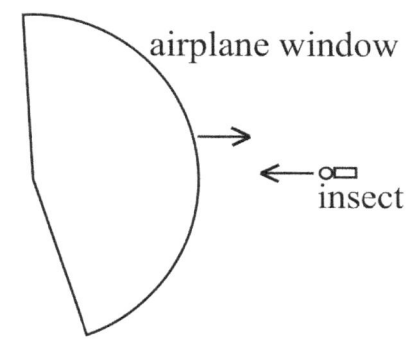

Figure 12.2: Airplane window about to collide with insect.

Let's consider a few collisions. Here we have an airplane window and a fly, about to collide with each other. After the collision, the fly is a mark on the airplane's windshield. What is the airplane's final momentum? The details of the collision are a bit complicated, but are sufficiently short-lived that external forces have approximately no effect on the motion of either object. After the collision, there is one object, whose mass is the sum of the two original masses. The airplane has $m_1 = 2 \cdot 10^5$ kg and $\mathbf{v}_1 = 300\hat{\mathbf{i}}$ m/s. The fly has $m_2 = 3 \cdot 10^{-4}$ kg and $\mathbf{v}_2 = -0.1\hat{\mathbf{i}}$ m/s. The total momentum before the collision is $\mathbf{p}_i = 2 \cdot 10^5 \cdot 300\hat{\mathbf{i}} + 3 \cdot 10^{-4} \cdot (-0.1)\hat{\mathbf{i}}$ kg m/s. From momentum conservation, the final momentum $\mathbf{p}_f$ must equal the initial momentum $\mathbf{p}_f$, while the initial mass and the final mass of the system in this case are equal to each other, so $\mathbf{p}_f$ must equal the initial momentum $2 \cdot 10^5 \cdot 300\hat{\mathbf{i}} + 3 \cdot 10^{-4} \cdot (-0.1)\hat{\mathbf{i}}$.

Let us consider another collision. This time the airplane ($m_1 = 2 \cdot 10^5$ kg and $\mathbf{v} = 300\hat{\mathbf{i}}$ m/s) has an unscheduled interaction with an escapee from one of those giant monster movies. The escapee is large and fast ($m_2 = 3 \cdot 10^6$ kg and $\mathbf{v}_2 = -1000\hat{\mathbf{i}}$ m/s). The airplane flies down the escapee's gullet and more or less instantaneously finds itself travelling at the same velocity as the escapee. Immediately after the collision, what is the escapee's velocity?

We start with the Law of Conservation of Momentum.

$$\mathbf{p}_i = \mathbf{p}_f \tag{12.14}$$

which in this case may be written

$$m_1 \mathbf{v}_1 + m_2 \mathbf{v}_2 = (m_1 + m_2) \mathbf{v}_f. \tag{12.15}$$

Numerical substitution as a first step gives us

$$2 \cdot 10^5 \cdot 300\hat{\mathbf{i}} + 3 \cdot 10^6 \cdot (-)1000\hat{\mathbf{i}} = (2 \cdot 10^5 + 3 \cdot 10^6)\mathbf{v}_f \quad (12.16)$$

To find the final velocity, solve this equation for the final velocity $\mathbf{v}_f$.

Pause: Is solution possible? After all, $\mathbf{v}_f$ is a vector. It has three components, so we need three equations to solve. However, equation 12.16 is a vector equation. A vector equation is a symbol for a set of three scalar equations, so equation 12.16 in fact counts as three equations, which are enough to solve.

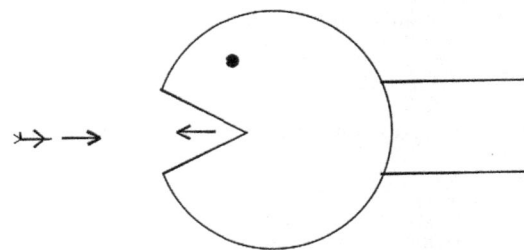

Figure 12.3: Creature about to eat airplane.

We now consider that staple, the vehicle collision. These are not real collisions; no automobiles were injured in writing this problem. We first have a dispute over right-of-way between an automobile and an oncoming truck.

The car initially masses $2 \cdot 10^3$ kg and is traveling at $\mathbf{v} = 75\hat{\mathbf{i}}$ m/s. The truck initially masses $4 \cdot 10^4$ kg and is travelling at $\mathbf{v}_2 = -3\hat{\mathbf{i}}$ m/s. The car imbeds itself in the grill of the truck. No parts of either vehicle are lost. Immediately after the collision, what is the velocity of the car and truck, which are now moving together as a single unit? We may write for the solution

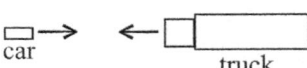

Figure 12.4: The car-truck right-of-way dispute.

$$\mathbf{p}_i = \mathbf{p}_f, \text{ so} \quad (12.17)$$
$$2 \cdot 10^3 \cdot 75\hat{\mathbf{i}} + 4 \cdot 10^4 \cdot (-3)\hat{\mathbf{i}} = 4.2 \cdot 10^4 \mathbf{v}_f, \text{ so} \quad (12.18)$$
$$\mathbf{v}_f = \frac{5}{7}\hat{\mathbf{i}} \text{ m/s} \quad (12.19)$$

In writing the solution, we always start with the fundamental equations being used, not with a hash of numbers. Numerical substitution for algebraic quantities should not happen before the problem is even set up.

Consider another problem: Two automobiles on black ice try to pass through an intersection at the same time. Neither driver even considers slowing down. Instead, the two cars collide, lock bumpers and sail off roughly as indicated in Figure 12.5, leaving no parts behind as they exit the collision zone. The first car masses $3 \cdot 10^3$ kg and is traveling at $\mathbf{v}_1 = 10\hat{\mathbf{i}}$ m/s. The other car masses $4 \cdot 10^3$ kg and is travelling at $\mathbf{v}_2 = 20\hat{\mathbf{j}}$ m/s. What is the final velocity here?

The important new feature of this problem is that it requires that you remember that momentum is a vector, not a scalar, so it adds as a vector, not as a scalar. In solving, we apply the special case circumstance that the final mass is simply the sum of the initial masses, while after the collision the final velocity is the same for all components. A solution reads

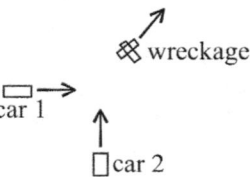

Figure 12.5: The car-car right-of-way dispute.

$$\mathbf{p}_i = \mathbf{p}_f, \quad (12.20)$$
$$m_1 \mathbf{v}_1 + m_2 \mathbf{v}_2 = (m_1 + m_2)\mathbf{v}_f, \quad (12.21)$$
$$3 \cdot 10^3 \cdot 10\hat{\mathbf{i}} + 4 \cdot 10^3 \cdot 20\hat{\mathbf{j}} = 7 \cdot 10^3 \mathbf{v}_f, \quad (12.22)$$
$$\mathbf{v}_f = \frac{30}{7}\hat{\mathbf{i}} + \frac{70}{7}\hat{\mathbf{j}} \text{ m/s}. \quad (12.23)$$

Finally, a collision problem that does not immediately look like a collision problem, but that's what it is. We have two masses, m_1 and m_2, separated by a compressed spring that is not attached to either mass. Initially, they are tied together. The tie is cut. The masses spring apart, moving parallel to the spring axis, but in opposite directions, leaving the spring behind. What can be said about their final momenta?

We apply momentum conservation $\mathbf{p}_i = \mathbf{p}_f$. Initially the two masses are stationary. At the end, they have velocities $\mathbf{v}_{1f}$ and $\mathbf{v}_{2f}$, respectively. From momentum conservation, we then have

$$\mathbf{p}_i = \mathbf{p}_f, \tag{12.24}$$
$$\mathbf{0} = m_1 \mathbf{v}_{1f} + m_2 \mathbf{v}_{2f}, \tag{12.25}$$
$$m_1 \mathbf{v}_{1f} = -m_2 \mathbf{v}_{2f}. \tag{12.26}$$

The two final momenta in this case are equal in magnitude and opposite in direction.

12.4 Discussion

This Chapter introduces the important notion of the *conservation law*. A quantity is said to be *conserved* if it cannot be changed during a process. For example, momentum is conserved during a collision; it must be the same immediately before and immediately after any collision. Conservation laws are a powerful aid to finding problem solutions, though in general conservation laws by themselves will not take you to the answer you want. To solve real problems, you usually still need detailed calculations.

12.5 Worked Problems

1. A particle of mass 37 kg, at rest at time 0, is subject to an external force $3\hat{\mathbf{i}} + 4t\hat{\mathbf{j}} - 3\cos(2t)\hat{\mathbf{k}}$. The particle is floating free in outer space, and is subject to no other forces. What is its momentum at the later time t?

2. Four masses are located at $(0,0,0)$, $(L,0,0)$, $(0,L,0)$, and (L,L,L). The masses have mass m, m, $2m$, and $2m$, respectively. Find the center of mass of the four masses.

3. We have a one-dimensional problem, in which the center of mass only has an interesting x coordinate. You may assume that the y and z coordinates are both equal to zero everywhere along the rod: A thin rod made of foamed plastic has ends at $x = L$ and $x = 2L$, and a density per unit length that varies with position as $\rho = A + Bx^2$. In SI units ρ would have units kg/m and dimensions m/ℓ. Find the center of mass of the rod.

4. A 3000 kg automobile is traveling up George Street–which as may be inferred from its name is one of the more important thoroughfares in our metropolis–at 30 m/s. It encounters a second 2000 kg car exiting a driveway perpendicular to the main road, traveling at 20 m/s. The two vehicles adhere and proceed as a single unit. What is the final velocity of the unit? George Street lies in the $y-z$ plane and makes an angle of 15 degrees upward with respect to the positive y axis, the y axis being horizontal. Take the second vehicle to be proceeding in the $+x$ direction.

5. This is a slight variation on an old exam problem. It requires slightly more attention than it was given by many people taking the examination. A block of wood of mass M lying on a horizontal frictionless table is struck from the rear by three arrows, the arrows having masses m, $2m$, and $3m$, respectively. The block was initially stationary. The arrows each had speed v relative to the ground in the $+x$ direction. The arrows all stuck into the block. (a) What was the final velocity of the block and arrows? (b) What were the velocities of the block after being struck by the first arrow, the second arrow, and finally the third arrow? Show that your answers in parts (a) and (b) agree, or explain why they do not.

12.6 Homework

1. A futuristic non-conventional baseball bat of length L has a mass per unit length ρ that depends on position as $\rho = \lambda_0(1 + x^3/L^3)$. Here λ_0 is a constant having units mass/length. Find the center of mass of the baseball bat.

2. We have a thin rod having length L. Its density varies with length as $\lambda = \lambda_o(1 + x^2/L^2)$. Here x is the position along the rod. The two ends of the rod are at $x = 0$ and $x = L$. In SI units λ_o would have units kg/m and dimensions m/ℓ. Find the center of mass of the rod.

3. Four masses lie in a square in the x,z-plane. Their coordinates are $(0,0,0)$, $(1,0,0)$, $(1,0,1)$, and $(0,0,1)$, respectively. The masses of these four masses are 1, 2, 3, and 4, respectively. Find their center of mass.

4. Two asteroids sail through the remote and trackless waste of far outer space. They are so far from any other objects that the external forces on them are practically zero. At one instant in time their momenta are $\mathbf{p}_1 = 3\hat{\mathbf{i}} - 2\hat{\mathbf{j}} + \hat{\mathbf{k}}$ and $\mathbf{p}_2 = -\hat{\mathbf{i}} + 5\hat{\mathbf{j}}$. They collide and bounce off each other. Somewhat later the momentum of the first asteroid is $\mathbf{p}_1 = 8\hat{\mathbf{i}} - 6\hat{\mathbf{j}} - 5\hat{\mathbf{k}}$. What must the other asteroid's momentum be?

5. We have a very large block of wood, mass M, on a frictionless surface. All motion is along the x-axis, so you can stay with the x component of the velocity. (a) We hurl into the block three darts, each having mass m and velocity $\mathbf{v} = v\hat{\mathbf{i}}$. The darts stick to the block of wood. What is the velocity of the block and attached darts immediately after the darts strike the block? (b) We now repeat the experiment, except that we hurl the darts at the block one at a time, and measure the velocity of the block after each dart strikes it. (i) What is the velocity of the block after the first dart strikes it? (ii) What is the velocity of the block after the second block strikes it? (iii) What is now the velocity of the block after the third dart strikes it? Are the answers to (a) and (iii) consistent? Why or why not?

6. a) Four masses lie in a row along the x axis, their masses being 1, 2, 3, and 4 kg, in order of increasing x. The first two masses are 0.5 m apart; the second and third masses are 0.5 m apart; the third and fourth masses are 0.8 m apart. Where is the center of mass of the four masses? (b) Masses of mass 1, 2, 3, and 4kg are at $(0,0,0)$, $(0.5,0,0)$, $(0.5,0.8,0)$, and $(0.3,0.8,-0.5)$ m, respectively. Where is the center of mass of the system?

7. It is an icy day in Worcester, and two automobiles slide into the Park Avenue/Salisbury street intersection at the same time. Unfortunately, they are going in two more-or-less perpendicular directions and collide. The first vehicle had a mass of 1500 kg, and velocity $3\hat{\mathbf{i}} + 1\hat{\mathbf{j}}$ m/s, while the second vehicle had a mass of 2000 kg and a velocity $1\hat{\mathbf{i}} - 2\hat{\mathbf{j}}$ m/s. No one is injured; no substantial parts of either car are left behind as the two vehicles stick together and slide out of the intersection. How fast are the two cars going? What is their final velocity?

8. In yet another scene from our movie, the villain, sitting at sea level in his volcanic island hideaway, fires a giant cannon at the good guy's ship. Unfortunately for the villain, he uses a cheap homebuilt atomic artillery shell of mass M, rather than a quality sabotage-resistant atomic artillery shell from Villain Supply, LLC(1). The heroine has sabotaged the artillery shell, so that at the top of its trajectory there is an internal explosion. The top of the trajectory is a distance L away from the villain. The explosion sends a section of the shell of mass $M/4$ horizontally backwards; it finally lands at the villain's feet(2). The remainder of the shell, mass $3M/4$, continues on its way until it hits the ocean some distance from the villain. To add insult to injury, the heroine has sent a coded message to the captain of the villain's yacht, ordering him to move his ship to a new location, so that his ship will be sunk by the remainder of the shell. How far from the villain should the villain's yacht be anchored, if the $3M/4$ section of the shell is to strike the yacht? Remark: Instead of calculating out every bit of the various trajectories, a less detailed approach will get you to the right answer.

(1) No, I did not make them up, but they are no longer on the internet.

(2) This is the section with the atomic warhead, set to detonate when it returns to sea level.

12.7. PROBLEM SOLUTIONS

9. Another true story. One fine day I was inbound on Route 2 at the then legal speed limit of 55 mph (25 m/s). On glancing in the rear view mirror, I noted an idiot in the high speed lane doing 80 mph (36 m/s). In the time it took me to look down at my speedometer and check my rear view mirror, the idiot had passed me, noted he was about to miss his exit ramp, and did a right angle turn directly in front of me. In the real world, at the moment I first saw him in front of me, we were sufficiently close that over the low, sloping hood of my car I could only see the top half of his rear door panel. His passenger, staring at me out of their open convertible, appeared to have realized for the first time that I had not noticed them and had not slowed down at all. We missed.

    (i) In an alternative time line, my 1200 kg car and his 800 kg car collided, fused together, and headed in some direction. Assuming that at the moment of collision I was headed in the +y direction, and the idiot was initially headed still at 80 mph in the +x direction, find our velocity after the collision and before deceleration set in. (ii) Assuming that we moved in a straight line after the collision, and that I had to my right a 10 m perpendicular distance of road and shoulder before reaching the grass, how far would we have traveled before we left the paved road area? (iii) Assuming that our effective coefficient of kinetic friction was approximately 0.2 (Remember, we are not moving in the direction that any wheels point), how far would we have travelled before coming to a stop?

10. In yet another scene from the long-lasting motion picture epic, the director arranges a scene in which two spaceships crash into each other. The pyrotechnics budget has been exhausted. The Executive Producer's new daughter-in-law is allegedly a physics major. She is paid a staggeringly huge salary for "scientific accuracy", meaning that you must supply her with the correct answers for the following two scenarios. Before the collision, the first spaceship has a mass of 1×10^{11} kg, and is proceeding in the $\hat{\mathbf{j}} + \hat{\mathbf{k}}$ direction at 300 m/s relative to the camera. The second spaceship has a mass of 3×10^{12} kg, and is proceeding in the $-\hat{\mathbf{i}} - 2\hat{\mathbf{j}} - 2\hat{\mathbf{k}}$ direction at 50 m/s relative to the camera. Hint: Are those unit vectors for the directions? (i) Scenario 1 The two spaceships bounce off each other. After the collision, the 1×10^{11} kg spaceship has a velocity of $100(-\hat{\mathbf{i}} - \hat{\mathbf{j}} - \hat{\mathbf{k}})$. What is the velocity of spaceship 2? (ii) Scenario 2 The two spaceships lock bumpers and sail off as a single lump of wreckage. After the collision, what is the velocity of the lump of wreckage?

12.7 Problem Solutions

1. We advance from the Second Law

$$\mathbf{F} = m\frac{d^2\mathbf{r}}{dt^2}. \tag{12.27}$$

On integrating this equation with respect to time, we see

$$\int \mathbf{F}\,dt = \int m\frac{d^2\mathbf{r}}{dt^2}\,dt. \tag{12.28}$$

But $\mathbf{F} = 3\hat{\mathbf{i}} + 4t\hat{\mathbf{j}} + 3\cos(2t)\hat{\mathbf{k}}$, so on doing the integrals

$$3t\hat{\mathbf{i}} + 2t^2\hat{\mathbf{j}} - \frac{3}{2}\sin(2t)\hat{\mathbf{k}}\big|_0^t = m\frac{d\mathbf{r}(t)}{dt} - m\frac{d\mathbf{r}(0)}{dt}. \tag{12.29}$$

On the left-hand-side the $t = 0$ value is zero, and on the right hand side the mass started at rest, so $\frac{d\mathbf{r}(0)}{dt} = 0$, leading to

$$\mathbf{p}(t) = 3t\hat{\mathbf{i}} + 2t^2\hat{\mathbf{j}} - \frac{3}{2}\sin(2t)\hat{\mathbf{k}}\big|_0^t. \tag{12.30}$$

2. We have four masses. The center of mass is

$$\mathbf{R} = \frac{\sum_i m_i \mathbf{r}_i}{\sum_i m_i}. \tag{12.31}$$

In this case
$$\mathbf{R} = \frac{m(0\hat{\mathbf{i}} + 0\hat{\mathbf{j}} + 0\hat{\mathbf{k}}) + 2m(\ell\hat{\mathbf{j}}) + m(\ell\hat{\mathbf{i}}) + 2m(\ell\hat{\mathbf{i}} + \ell\hat{\mathbf{j}} + \ell\hat{\mathbf{k}})}{m + 2m + m + 2m}, \qquad (12.32)$$

which reduces to
$$\mathbf{R} = \frac{1}{2}\ell\hat{\mathbf{i}} + \frac{2}{3}\ell\hat{\mathbf{j}} + \frac{1}{3}\ell\hat{\mathbf{k}}. \qquad (12.33)$$

3. A rod has ends at L and $2L$ and a density per unit length $\rho = a + Bx^2$. Where is its center of mass? We have to replace the summation definition of the center of mass with an integral. For the x coordinate of the center of mass
$$x_{\text{CM}} = \frac{\int_L^{2L} dx\, x(A + Bx^2)}{\int_L^{2L} dx(A + Bx^2)}. \qquad (12.34)$$

The integral in the denominator is the total mass of the rod, obtained by integrating the mass per unit length of the rod over the length of the rod. In evaluating the integrals, the various quantities are not zero at either of the limits of integration.

The integrals give
$$x_{\text{CM}} = \frac{(\frac{1}{2}Ax^2 + \frac{1}{4}Bx^4)|_{x=L}^{2L}}{(Ax + \frac{1}{3}Bx^3)|_{x=L}^{2L}}. \qquad (12.35)$$

which leads to
$$x_{\text{CM}} = \frac{(\frac{A}{2}((2L)^2 - L^2) + \frac{B}{4}((2L)^4 - L^4)}{(A(2L - L) + \frac{B}{3}((2L)^3 - L^3))}. \qquad (12.36)$$

4. We have a collision. In a collision, momentum is always conserved. The total momentum of the two automobiles before the collision is
$$\mathbf{p} = \mathbf{p}_1 + \mathbf{p}_2. \qquad (12.37)$$

Because we are using Cartesian coordinates, we can write $\mathbf{p} = m\mathbf{v}$, leading to
$$v_{\text{final}} = \frac{\mathbf{p}}{m}. \qquad (12.38)$$

In this case,
$$\mathbf{p}_1 = 2000 \cdot 20\hat{\mathbf{i}}, \qquad (12.39)$$
$$\mathbf{p}_2 = 3000 \cdot 30\cos(15°)\hat{\mathbf{j}} + 3000 \cdot 30\sin(15°)\hat{\mathbf{k}}, \qquad (12.40)$$

so that
$$\mathbf{p}_{\text{final}} = 2000 \cdot 20\hat{\mathbf{i}} + 3000 \cdot 30\cos(15°)\hat{\mathbf{j}} + 3000 \cdot 30\sin(15°)\hat{\mathbf{k}}, \qquad (12.41)$$

while the final velocity is
$$\mathbf{v}_{\text{final}} = 0.4 \cdot 20\hat{\mathbf{i}} + 0.6 \cdot 30\cos(15°)\hat{\mathbf{j}} + 0.6 \cdot 30\sin(15°)\hat{\mathbf{k}}. \qquad (12.42)$$

5. This is a collision problem. In collisions, momentum is always conserved.

(a) The starting point is momentum conservation $\mathbf{p}_i = \mathbf{p}_f$. Applying this result to the three arrows striking the block of wood
$$m\mathbf{v} + m\mathbf{v} + m\mathbf{v} + M\mathbf{0} = (3m + M)\mathbf{v}_f, \qquad (12.43)$$

which tells us that $\mathbf{v}_f$ is
$$\mathbf{v}_f = \frac{3m\mathbf{v}}{3m + M}. \qquad (12.44)$$

(b) Do this one impact at a time. First arrow. Momentum conservation gives
$$m\mathbf{v} + M\mathbf{0} = (m + M)\mathbf{v}_{1f}, \qquad (12.45)$$

12.7. PROBLEM SOLUTIONS

so the velocity after the first arrow strikes is

$$\mathbf{v}_{1f} = \frac{m\mathbf{v}}{m + M}. \tag{12.46}$$

Now the second arrow strikes. Momentum conservation gives us

$$m\mathbf{v} + (M + m)\frac{m\mathbf{v}}{m + M} = (2m + M)\mathbf{v}_{2f}, \tag{12.47}$$

so the velocity after the second arrow strikes is

$$\mathbf{v}_{2f} = \frac{2m\mathbf{v}}{2m + M}. \tag{12.48}$$

Finally the third arrow strikes. Momentum conservation gives us

$$m\mathbf{v} + (M + 2m)\frac{2m\mathbf{v}}{2m + M} = (3m + M)\mathbf{v}_f, \tag{12.49}$$

so the velocity after the third arrow strikes is

$$\mathbf{v}_f = \frac{3m\mathbf{v}}{m + M}. \tag{12.50}$$

Outcomes (a) and (b) agree, as they should.

This space reserved for your notes.

This page reserved for your notes.

Chapter 13

Work, Kinetic Energy, and the Work-Energy Theorem

We now reach a discussion of work, energy, and the Work-Energy Theorem. The Work-Energy Theorem gives us a new way to solve mechanics problems. The theorem is derived by using the Second Law, so in principle the Work-Energy Theorem tells us nothing that we did not already know. However, by invoking the Work-Energy Theorem, we get to skip over all the steps hidden in its derivation. Furthermore, the Work-Energy Theorem allows us to solve some problems that can't readily be solved with the Second Law, because the problem statement is incomplete. We pay a price for being able to solve incomplete problems; we only get incomplete answers.

The Work-Energy Theorem leads us to energy conservation, a conservation law. As noted in the last Chapter, conservation laws provide a powerful tool for analyzing problems in physics, though often they only provide part of a needed answer. We will reach energy conservation in the next Chapter.

13.1 Work

We begin with the definition of work.

$$W = \int d\boldsymbol{s} \cdot \boldsymbol{F}(s). \tag{13.1}$$

Here $\boldsymbol{F}(s)$ is the force on the body at location s. The displayed mathematical function is a path integral. We are integrating along a contour that may or may not be a straight line. The variable s measures the distance along the path from the starting point toward the conclusion. For each differential segment of the contour, the differential vector element $d\boldsymbol{s}$ runs parallel to the contour and has length ds. We dot this differential vector $d\boldsymbol{s}$ into the force $\boldsymbol{F}(s)$ on the object at the same point on the contour and sum them.

We could equally well write this definition of the work as

$$W = \int ds\, \hat{\boldsymbol{s}} \cdot \boldsymbol{F}(s). \tag{13.2}$$

In this equation, s measures the distance along the path. $\hat{\boldsymbol{s}}$ is the tangent to the path at point s.

From this definition of work, we reach a series of topics, including the work-energy theorem, potential energy, energy conservation, power, effects of friction and non-conservative forces, and stable and unstable equilibria.

There is then a math question. How do we evaluate the needed path integrals? In the examples here, we restrict ourselves to a few special cases, notably:

(a) The angle between $\boldsymbol{s}$ and $\boldsymbol{F}$, measured by $\hat{\boldsymbol{s}} \cdot \hat{\boldsymbol{F}}$, is the same everywhere along the path.

(b) The contour is a straight line, so conventional integration techniques are adequate.

(c) $\hat{\boldsymbol{s}} \perp \hat{\boldsymbol{F}}$; the force is always perpendicular to the contour.

As a minor word of warning, some of you will have encountered the false equation $W = Fd$, read as 'work equals force times distance'. Work is an integral, not a product, though in a few special cases – notably if the force is the same everywhere along a straight line path, and points along the path – the integral simplifies into a product.

13.2 The Work-Energy Theorem

This section derives the work-energy theorem. We begin with the Force and velocity vectors, which can be written in terms of their components as

$$\boldsymbol{F} = F_x\hat{\boldsymbol{i}} + F_y\hat{\boldsymbol{j}} + F_z\hat{\boldsymbol{k}} \tag{13.3}$$

and

$$\boldsymbol{v} = v_x\hat{\boldsymbol{i}} + v_y\hat{\boldsymbol{j}} + v_z\hat{\boldsymbol{k}}. \tag{13.4}$$

In these equations, F_x, F_y, and F_z are the three scalar components of the total force $\boldsymbol{F}$. The three scalar components of the velocity $\boldsymbol{v}$ are v_x, v_y, and v_z.

The scalar product of these two vectors is

$$\boldsymbol{F} \cdot \boldsymbol{v} = F_x v_x + F_y v_y + F_z v_z. \tag{13.5}$$

We now invoke the Second Law for an object whose mass does not change as time goes on. The Second Law is a vector equation. A representative one of its scalar components is

$$F_x = m\frac{dv_x}{dt}. \tag{13.6}$$

We then take the right hand side of equation 13.5, set it equal to itself, and on one side of the equation replace F_x with mdv_x/dt, and similarly for F_y and F_z. The result is

$$F_x v_x + F_y v_y + F_z v_z = mv_x\frac{dv_x}{dt} + mv_y\frac{dv_y}{dt} + mv_z\frac{dv_z}{dt}. \tag{13.7}$$

Now we do a time integral. We take t from an initial time τ to a final time T, giving

$$\int_\tau^T (F_x v_x + F_y v_y + F_z v_z)\, dt = \int_\tau^T \left(mv_x\frac{dv_x}{dt}\right) dt + \int_\tau^T \left(mv_y\frac{dv_y}{dt}\right) dt + \int_\tau^T \left(mv_z\frac{dv_z}{dt}\right) dt. \tag{13.8}$$

It is simplest to analyze the two sides of the equation separately. We start the right-hand-side. From the product rule for derivatives, we have

$$\frac{dv_x^2}{dt} = 2v_x\frac{dv_x}{dt}. \tag{13.9}$$

The right-hand-side of the above equation is the quantity being integrated in equation 13.8. Each of the integrals on the right-hand-side of equation 13.8 is therefore an integral over an exact differential, leading to a replacement

$$\int_\tau^T \left(mv_x\frac{dv_x}{dt}\right) dt \Rightarrow \left.\frac{mv_x^2(t)}{2}\right|_\tau^T \Rightarrow \frac{mv_x^2(T)}{2} - \frac{mv_x^2(\tau)}{2}. \tag{13.10}$$

In this equation, nothing depends on which coordinate we are calling x. The x is just a label. Therefore, in equation 13.10 we can replace the label x with the label y or the label z, and the equations continue to be valid.

On the left-hand-side of the equation, we will do a change in the variable of integration, replacing the $\int dt$ with an $\int dx$. The replacement gives

$$\int_\tau^T F_x \frac{dx}{dt}\, dt \Rightarrow \int_{x(\tau)}^{x(T)} F_x \frac{dx}{dt} \frac{dx}{\left|\left(\frac{dx}{dt}\right)\right|}. \tag{13.11}$$

13.2. THE WORK-ENERGY THEOREM

In making this replacement, the original limits of integration τ and T are replaced by the values of the new variable at the same limits, these limits being $x(\tau)$ and $x(T)$, respectively.

After the replacement, we find in the denominator a term

$$\left|\left(\frac{dx}{dt}\right)\right|. \tag{13.12}$$

This term is the Jacobian of the transformation. The Jacobian compensates for the possibility that the differentials dt and dx do not cover exactly equal distances. You have almost certainly seen a Jacobian, though it may not have been called that, in the transformation from Cartesian to circular polar coordinates, namely

$$\int dx\,dy \Rightarrow \int \frac{d\theta dr}{(1/r)}. \tag{13.13}$$

The denominator term $1/r$ is a Jacobian; it corrects for the dependence of the area $d\theta dr$ element on the distance of the element from the origin.

On the left-hand-side of equation 13.11, the two factors of dx/dt cancel. We recognize that $x(\tau)$ and $x(T)$ are the initial and final values of x, namely x_i and x_f. The same notational change replaces $y(\tau)$ and $y(T)$ with y_i and y_f, and replaces $z(\tau)$ and $z(T)$ with z_i and z_f. The left-hand-side of equation 13.8 becomes

$$\int_{x_i}^{x_f} F_x dx + \int_{y_i}^{y_f} F_y dy + \int_{z_i}^{z_f} F_z dz. \tag{13.14}$$

Let's put together all of these results. We will replace

$$v_x(\tau) \Rightarrow v_{xi}, \tag{13.15}$$
$$v_y(\tau) \Rightarrow v_{yi}, \tag{13.16}$$
$$v_z(\tau) \Rightarrow v_{zi}, \tag{13.17}$$
$$v_x(T) \Rightarrow v_{xf}, \tag{13.18}$$
$$v_y(T) \Rightarrow v_{yf}, \tag{13.19}$$
$$v_z(T) \Rightarrow v_{zf}, \tag{13.20}$$

and then have

$$\int_{x_i}^{x_f} F_x dx + \int_{y_i}^{y_f} F_y dy + \int_{z_i}^{z_f} F_z dz = \frac{1}{2}m(v_{xf}^2 + v_{yf}^2 + v_{zf}^2 - v_{xi}^2 - v_{yi}^2 - v_{zi}^2). \tag{13.21}$$

However, recalling the magnitude of a vector in terms of its components, we may replace $v_{xf}^2 + v_{yf}^2 + v_{zf}^2$ with v_f^2 and $-v_{xi}^2 - v_{yi}^2 - v_{zi}^2$ with $-v_i^2$. A definition is in order. Here K, defined by

$$K = \frac{1}{2}mv^2, \tag{13.22}$$

is the kinetic energy of the mass under consideration.

Finally, applying the definition of the scalar product in terms of its components

$$\int (F_x dx + F_y dy + F_z dz) = \int \boldsymbol{F} \cdot d\boldsymbol{s}. \tag{13.23}$$

The integrals are taken along a path reaching from start to finish.

We finally obtain

$$\int_{\boldsymbol{R}_i}^{\boldsymbol{R}_f} \boldsymbol{F} \cdot d\boldsymbol{s} = \frac{1}{2}mv_f^2 - \frac{1}{2}mv_i^2 \equiv K_f - K_i \tag{13.24}$$

This result is the Work-Energy Theorem. $\boldsymbol{F}$ is the force on the body. Correspondingly, $\int_{\boldsymbol{R}_i}^{\boldsymbol{R}_f} \boldsymbol{F} \cdot d\boldsymbol{s}$ is the work done on the body by outside forces.

13.3 Power

Having introduced work, it is appropriate to introduce the *power* P, the rate at which work is done. Power and work are related by

$$P = \frac{dW}{dt} \tag{13.25}$$

To obtain dW/dt, we need to invoke the rule for taking the derivative of an integral I:

$$I = \int_a^b f(x)\,dx. \tag{13.26}$$

We now take the derivative d/dt of this integral. Note that I is a definite integral, so it is not a function of the variable of integration x. We have integrated over x, which eliminates the x-dependence. x is just a label; if you replace the character x with a different character, the value of the integral does not change. The definite integral eliminates all dependence of I on x. The derivative is then

$$\frac{dI}{dt} = \int_a^b \frac{df(x)}{dt}\,dx + f(b)\frac{db}{dt} - f(a)\frac{da}{dt}. \tag{13.27}$$

A simple picture, Figure 13.1, explains the terms on the right-hand-side of this equation.

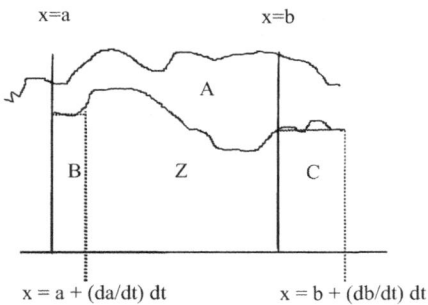

Figure 13.1: Time derivative of the integral I.

The original integral is the area Z under the lower solid curve and between the two solid vertical lines, the vertical lines being the limits of integration. The upper solid line shows $f(x)$ after t is increased to $t+dt$. If $f(x)$ depends explicitly on t, then when we change $t \to t + dt$ the integrand $f(x)$ changes, perhaps over the full range of x, leading to the new area labelled A in the figure. It may also be the case that changing t changes the lower bound of integration from a to $a+(da/dt)dt$, which moves the left-hand limit of integration of the area. If $da/dt\,dt > 0$ the boundary moves to the right, over to the left-hand dashed line, thus reducing the area under the curve by $f(a)(da/dt)\,dt$, the decrease being the area B. It may also be the case that changing t changes the upper bound of integration from b to $b + (db/dt)\,dt$, thereby moving the right-hand limit of integration of the area. If $db/dt\,dt > 0$, the boundary of the curve moves to the right, over to the right-hand dashed line, which increases the area under the curve, the increase being the area C. Richard Feynman is said to have claimed that he had seen this result in introductory calculus, and that remembering it many years later was the basis for his Nobel Prize.

From the above, we can write

$$\int_\tau^T \boldsymbol{F} \cdot \boldsymbol{v}\,dt = K(T) - K(\tau) \tag{13.28}$$

where $K(T)$, also written K_f, is the kinetic energy at time T. We now take the derivative of this integral with respect to the ending time T. The starting time τ is not affected by changing T; the time τ has already happened. The value of the integrand $\boldsymbol{F} \cdot \boldsymbol{v}$ is not affected if T is changed; changing T only changes one of the limits of integration. This time derivative of the definition of work is

$$\boldsymbol{F}(T) \cdot \boldsymbol{v}(T) \frac{dT}{dT} = \frac{dK(T)}{dT}. \tag{13.29}$$

Two of the three general terms in the time derivative of the integral vanish and are not shown explicitly. The power P is the time rate-of-change of the energy – here, the kinetic energy – of the system, leaving as the final expression for the power

$$\frac{dK}{dt} \equiv P = \boldsymbol{F} \cdot \boldsymbol{v}. \tag{13.30}$$

13.4 Examples of the Work-Energy Theorem

We now present some examples of calculations that use the Work-Energy Theorem.

Consider a wooden block sliding across a table top. The block is sliding to the left. It is subject to a force of kinetic friction having magnitude $\mu_k N$. How does its kinetic energy change with time? Given an initial velocity v, what is its final velocity V, so long as the block has not come to a stop first? What is the displacement L of the block at the point that the block comes to a stop? We start with a sketch and a force diagram

To calculate the frictional force, we need to determine the normal force on the block. We can obtain N by writing the Second Law for the vertical direction. We get

$$\boldsymbol{F} = m\frac{d^2\boldsymbol{R}}{dt^2}, \quad (13.31)$$

$$N - mg = 0, \quad (13.32)$$

$$N = mg. \quad (13.33)$$

We now have N. The work-energy theorem tells us

$$\int_{\boldsymbol{R}_i}^{\boldsymbol{R}_f} \boldsymbol{F} \cdot d\boldsymbol{s} = \frac{1}{2}mv_f^2 - \frac{1}{2}mv_i^2. \quad (13.34)$$

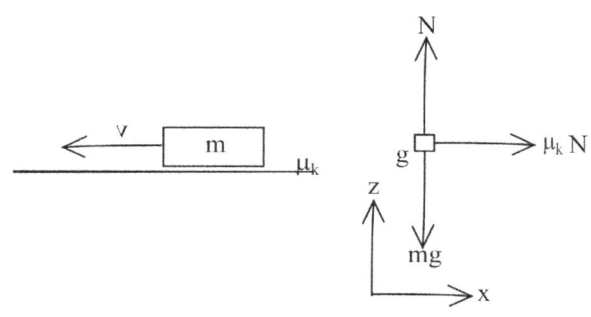

Figure 13.2: A block sliding on a table and its force diagram.

The normal force and the force of gravity act perpendicular to the direction in which the block is moving. The scalar product of two perpendicular vectors, in this problem either the normal force or the gravitational force and the displacement $d\boldsymbol{S}$, is zero. These two components of the total force therefore contribute nothing to $\boldsymbol{F} \cdot d\boldsymbol{s}$.

The remaining force is $\mu_k mg \hat{\boldsymbol{i}}$. Motion is along the x axis, so $d\boldsymbol{s} = \hat{\boldsymbol{i}} dx$. The initial and final positions along the x-axis are x_i and x_f. Substituting all of these quantities in equation 13.34, we have

$$\int_{x_i}^{x_f} \mu_k mg \hat{\boldsymbol{i}} \cdot \hat{\boldsymbol{i}} dx = \frac{1}{2}mV^2 - \frac{1}{2}mv^2. \quad (13.35)$$

On evaluating the integral

$$\mu_k mg(x_f - x_i) = \frac{1}{2}mV^2 - \frac{1}{2}mv^2, \quad (13.36)$$

so the final kinetic energy is

$$\frac{1}{2}mV^2 = \frac{1}{2}mv^2 + \mu_k mg(x_f - x_i). \quad (13.37)$$

In this equation, the final kinetic energy is equation to the initial kinetic energy plus the work done *on* the system by the external frictional force. The work done on the system by friction is in this case negative, because the block is sliding to the left so that $x_f - x_i$ is negative.

The final velocity, after isolating V^2 and taking the square root, is

$$V = \left(\frac{2}{m}(\frac{1}{2}mv^2 + \mu_k mg(x_f - x_i))\right)^{1/2}. \quad (13.38)$$

In writing this answer, you could have used a square root sign instead of the 1/2 power, but the overbar of a radical (the square root sign) is clumsy when it has to be extended over a complicated quantity.

Look carefully at that expression. You should intuitively expect that as the block slides farther and farther the final velocity V becomes smaller and smaller. Is that what is happening? The displayed sign on the $\mu_k mg$ term is positive. However, the block was sliding to the left, so $x_f - x_i$ is negative. V indeed decreases as the block slides along the table.

Now calculate $L = x_f - x_i$ for the point at which the block comes to a stop. Do it yourself, then we'll check your answer.

The check: Is your value for L positive or negative?

Let's do the calculation. We have $V = 0$ when the block stops, so

$$\mu_k mgL = 0 - \frac{1}{2}mv^2 \tag{13.39}$$

and

$$L = -\frac{1}{2}\frac{v^2}{\mu_k g}. \tag{13.40}$$

which is indeed negative.

Let's consider another two problems. They both treat the work done by gravity. We first consider an object staying close to the surface of the Earth, on which the gravitational force $-mg\hat{k}$ is nearly independent of altitude. We throw an object into the air. What is the work done on the object by gravity? Our starting point is the definition of work

$$W = \int \boldsymbol{F} \cdot d\boldsymbol{s}. \tag{13.41}$$

The displacement may be in any combination of directions, but the force is parallel to the z-axis. What result does this lead to?

$$W = \int -mg\hat{k} \cdot (dx\hat{i} + dy\hat{j} + dz\hat{k}). \tag{13.42}$$

The scalar product of $\hat{k}$ with $d\boldsymbol{s}$ will project out the z-component of $d\boldsymbol{s}$. We are in the special case that mg is a constant, not a function of x or y, so we can write

$$W = \int -mg\hat{k} \cdot dx\hat{i} - mg\hat{k} \cdot dy\hat{j} - mg\hat{k} \cdot dz\hat{k}. \tag{13.43}$$

In the integral, the first two scalar products vanish. The object may displace in the horizontal plane via dx or dy, but the unit vectors $\hat{i}$ and $\hat{j}$ are perpendicular to $-mg\hat{k}$. They therefore contribute zero to the scalar products in equation 13.43, so they contribute zero to the work.

The work done by gravity is therefore

$$W = \int_{z_i}^{z_f} -mg\hat{k} \cdot \hat{k} dz \tag{13.44}$$

or

$$W = -mg(z_f - z_i). \tag{13.45}$$

Inserting this result into the work-energy theorem, we have

$$-mg(z_f - z_i) = K_f - K_i. \tag{13.46}$$

Suppose we start on the ground at a height z_i above sea level and throw the mass m vertically into the air. How high is it when it reaches the top of its trajectory? How fast is it going when it reaches the ground again? How fast is it moving when it reaches half of its maximum altitude? The mass has an initial speed v in the vertical direction. The vertical force of gravity has no effect on the mass's horizontal velocity, so those terms drop out of the problem.

At the top of the trajectory, $v_f = 0$, so the work-energy theorem gives us

$$-mg(z_f - z_i) = \frac{1}{2}m0^2 - \frac{1}{2}mv^2, \tag{13.47}$$

whose solution is

$$z_f - z_i = \frac{v^2}{2g}. \tag{13.48}$$

13.4. EXAMPLES OF THE WORK-ENERGY THEOREM

When the mass returns to ground level, $z_f = z_i$, leading to

$$0 = \frac{1}{2}mv_f^2 - \frac{1}{2}mv^2, \tag{13.49}$$

whose solution for the speed is $v_f = v$. v^2 only gives us the magnitude of the velocity, the speed, which is always a positive number, and not the velocity (a vector) or some component of the velocity (a signed number).

Finally, the half-maximum altitude is $z_f - z_i = \frac{v^2}{4g}$. At that altitude, the speed v_f can be calculated using the work-energy theorem as

$$-mg\frac{v^2}{4g} = \frac{1}{2}mv_f^2 - \frac{1}{2}mv^2. \tag{13.50}$$

To solve this equation, the factors of m cancel. On the left-hand-side, the factors of g cancel. Taking the initial kinetic energy to the other side of the equation

$$\frac{1}{2}v^2 - \frac{v^2}{4} = \frac{1}{2}v_f^2. \tag{13.51}$$

so $v_f^2 = v^2/2$ and $v_f = v/\sqrt{2}$. It would indeed be possible to calculate v_f using the Second Law. First one would calculate the time and thence the height at which the mass reached its maximum altitude, then the time needed to reach the half-maximum altitude, and then finally v_f at the half-maximum altitude. Those calculations of times are not needed here.

We now consider an object thrown far into the sky. The force of gravity $\boldsymbol{F}_g$ on an object due to the Earth, over a large range of altitudes, is given by

$$\boldsymbol{F}_g = -\frac{GMm}{r^2}\hat{\boldsymbol{k}}. \tag{13.52}$$

In this equation, r is the distance from the object to the center of the earth, m is the mass of the object, G is the gravitational constant, M is the mass of the earth, and $\hat{\boldsymbol{k}}$ points vertically upward. We return to gravity in a later chapter.

What is the work done on the object by gravity? The only force is gravity, which points vertically downward. We now calculate the work done by gravity. Our starting point is again the definition of work

$$W = \int_{z_i}^{z_f} \boldsymbol{F}_g \cdot d\boldsymbol{s}. \tag{13.53}$$

Substituting for the specifics of this problem,

$$W = \int_{z_i}^{z_f} -\frac{GMm}{r^2}\hat{\boldsymbol{k}} \cdot \hat{\boldsymbol{k}} dr. \tag{13.54}$$

The vector scalar product gives unity. The integral gives

$$W = \frac{GMm}{r}\Big|_{z_i}^{z_f} \tag{13.55}$$

or

$$W = \frac{GMm}{z_f} - \frac{GMm}{z_i}. \tag{13.56}$$

Suppose we throw an object, mass m, straight up with some initial velocity V and hence some initial kinetic energy $\frac{1}{2}mV^2$. As it climbs, its kinetic energy decreases. How much does it decrease? The work-energy theorem tells us, namely

$$\frac{GMm}{z_f} - \frac{GMm}{z_i} = K_f - \frac{1}{2}mV^2. \tag{13.57}$$

This equation has an interesting property. Suppose you throw the object very, very hard, and have made arrangements to cancel air resistance. As it climbs, the object goes more and more slowly. However, if you throw it really hard, the object might keep on going forever, so that $z_f = \infty$. At this point, $GMm/z_f = 0$ and

$$K_f = \frac{1}{2}mV^2 - \frac{GMm}{z_i}. \tag{13.58}$$

As the object climbs from here to infinity, it loses kinetic energy, but there is a maximum of kinetic energy it can possibly lose. So long as it starts with a kinetic energy $\frac{1}{2}mV^2$ that is larger than $\frac{GMm}{z_i}$, the object can keep climbing forever.

The velocity an object needs in order to keep climbing forever is independent of its mass. We show this by writing K_f in terms of m and V, namely

$$\frac{1}{2}mv^2 = \frac{1}{2}mV^2 - \frac{GMm}{z_i}. \tag{13.59}$$

In this equation every term is linear in m. If we divide out m and simplify, we find

$$v^2 = V^2 - \frac{GM}{2z_i}. \tag{13.60}$$

If you launch something into space, and it is going fast enough, it will never come down. The required velocity v is called the *escape velocity*. Space probes sent to other planets are given speeds larger than this. It is possible to launch a space probe that is travels so fast that it eventually escapes the Sun's gravitational field and continues indefinitely into the infinite and starry void. The United States has in fact launched five such probes, three of which are still operating, even though they are billions of kilometers from Earth. Engineering aside: The Earth both rotates on its axis and orbits around the Sun. If you launch a space probe, it starts with the Earth's rotational and orbital velocities, so the velocity you need in order to reach escape velocity from the Sun depends on the direction in which the probe is launched.

We finally consider a problem in which the work done by a spring, and the work done by friction, both come into play. We have a mass sliding on a tabletop. The tabletop is frictionless until the mass reaches the spring. When the mass encounters the spring, it is subject to a friction force, and begins to compress the spring, creating a spring force on the mass. The mass is brought to a stop. How far has it compressed the spring at this point? The spring is assumed to be massless and to be attached to a wall at its far end. We start with a sketch and force diagram for the problem.

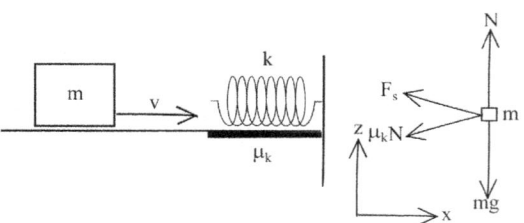

Figure 13.3: A block sliding on a table, a spring with force constant k and frictional surface μ_k, and the block's force diagram.

In this diagram, $\boldsymbol{F}_s$ is the spring force. If we call the x-coordinates of the left and right ends of the spring x_L and x_R, respectively, the magnitude of the spring force is $|F_s| = |k(x_R - x_L - \ell_0)|$. For the spring, equilibrium occurs when neither end of the spring is applying any force to whatever it is attached to. Call the equilibrium length of the spring ℓ_0, and the equilibrium positions of the left and right ends of the spring x_{L0} and x_{R0}, so that at equilibrium $|F_s| = |k(x_{R0} - x_{L0} - \ell_0)| = 0$. The spring is in contact with a wall, so x_R never changes. Call $x_L = x_{L0} + x$. In that case, $|F_s| = |kx|$. For the mass compressing the spring, as drawn, a positive displacement of the mass leads to a force in the $-x$ direction, so we may write

$$\boldsymbol{F}_s = -kx\hat{\boldsymbol{i}}. \tag{13.61}$$

To solve the problem, we also need to apply the Second Law for motion in the z-direction, which gives us

$$N - mg = m\frac{d^2z}{dt^2}, \tag{13.62}$$

which gives the needed result that here $N = mg$.

13.5. WORKED PROBLEMS

We now apply the work-energy theorem. For the mass, $v_i = V$ and $v_f = 0$. The spring starts at $x = 0$ and brings the mass to a stop at $x = X$. As always, I start with the basic equation. We then have

$$K_f - K_i = \int_0^X dx\, (-kx - \mu_k mg) \text{ or} \tag{13.63}$$

$$\frac{1}{2}m0^2 - \frac{1}{2}mv^2 = \left(-\frac{1}{2}kx^2 - \mu_k mgx\right)\Big|_0^X. \tag{13.64}$$

A bit of algebraic rearrangement gives us the final equation for X. We evaluate x as indicated, and take the two terms in X to the left-hand-side of the equation, namely

$$\frac{1}{2}kX^2 + \mu_k mgX - \frac{1}{2}mV^2 = 0. \tag{13.65}$$

The solution to this quadratic is

$$x = \frac{-\mu_k mg \pm \left[(\mu_k mg)^2 + kmV^2\right]^{1/2}}{k}. \tag{13.66}$$

Which of the roots of this equation is correct? Clearly x has to be positive. The quantity in the square root is positive and larger than $\mu_k mg$, so the square-root term is real, and large enough to make x positive if the positive root is selected.

The spring exerts a force $-kx$ on the mass, and therefore a force $+kx$ on the wall. How much work does the spring do on the wall? The correct answer is "none" because the wall does not move appreciably when the force is applied to it and therefore $\int \boldsymbol{F} \cdot d\boldsymbol{s}$ for the wall vanishes. The Third Law guarantees that the massless spring puts equal and opposite forces on the wall and the mass; it does not guarantee that the spring does equal and opposite amounts of work on the two walls.

Finally, a case in which the scalar product matters: Pulling a sled across frictionless wet ice. What work is done by the pulling force $\boldsymbol{P}$? First, the sketch and force diagram, including the pull force $\boldsymbol{P}$ directed at an angle θ above the horizontal.

We start with the work-energy theorem.

$$W = \int_{x_i}^{x_f} \boldsymbol{F} \cdot d\boldsymbol{s}. \tag{13.67}$$

Substituting for the force and the displacement vector,

$$W = \int_{x_i}^{x_f} \boldsymbol{P} \cdot \hat{\boldsymbol{i}}\, ds. \tag{13.68}$$

However, $\boldsymbol{P} \cdot d\hat{\boldsymbol{i}} = P\cos(\theta)$, leading to

$$W = P\cos(\theta)(x_f - x_i). \tag{13.69}$$

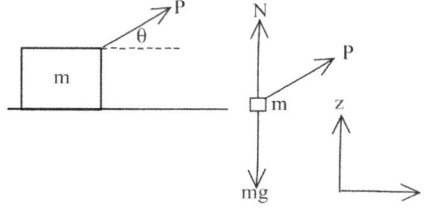

Figure 13.4: Sketch and force diagram for a sled being pulled across wet frictionless ice by a pull force, magnitude P.

When the force is applied at an angle to the direction of displacement, only the component of the force parallel to the displacement contributes to the work done on the object. The normal force $\boldsymbol{N}$ and the force of gravity $-mg\hat{\boldsymbol{k}}$ both act in directions perpendicular to the displacement $\hat{\boldsymbol{i}}\,ds$, so they do no work on the object. Note that if $P \neq 0$ the sled has a non-zero acceleration.

13.5 Worked Problems

1. A particle of mass 98 kg is subject to an external force $(3 + 2x)\hat{\boldsymbol{i}}$ as it moves from $x = 15$ to $x = 10$ along the x-axis. What is the change in the kinetic energy of the particle due to this move? Could the initial (at $x = 15$) kinetic energy of the mass have been zero?

2. A 10 kg box starts at the top of a frictionless inclined plane and slides down it. The incline makes a 50 degree angle with respect to the horizontal. Identify each force acting on the box, and calculate the work done on the block by that force after the box has slid 3 m along the plane. What is the total work done on the box? If the box began its slide from rest, how fast is it moving after it has moved 3 m along the plane. If the box began its slide with an initial down slope speed of 6 m/s, how fast is it moving after it has moved 3 m along the plane?

3. A mass M rests on a horizontal table. It is initially stationary at $x = 0$. It is touching an initially uncompressed spring, fixed at its far (left) end. An outside force P pushes leftward on the mass, compressing the spring very slowly by a distance ℓ. At the end of the compression step, M is again stationary. During the compression step, compute the work done on M by the spring, by the outside force, and by the total force. What is the change in the kinetic energy of M during the compression step? The outside force is then removed. The mass M is pushed to the right by the spring, starting at $x = -\ell$ and ending at $x = 0$. What is the total work done on M while it is being pushed to the right? What is the kinetic energy of M when it reaches $x = 0$? Hint: If the net force on M is zero, does that mean that M's velocity is zero or that M's acceleration is zero? [Graders should check carefully for multiple, canceling, sign errors.]

4. A 10 kg box starts at the top of an inclined plane and slides down it. The incline makes a 50 degree angle with respect to the horizontal. The coefficient of kinetic friction is $\mu_k = 0.3$, Identify each force acting on the box, and calculate the work done by that force after the box has slid 3 m along the plane. What it the total work done on the box? If the box began its slide from rest, how fast is it moving after it has moved 3m along the plane? If the box began its slide with an initial down slope speed of 6 m/s, how fast is it moving after it has moved 3 m along the plane?

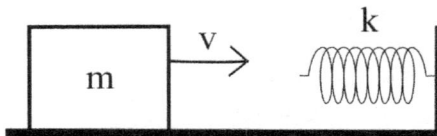

Figure 13.5: Figure for Homework Problem 2.

13.6 Homework

1. Several students are moving from one apartment to another. As part of the process, they are helping a fellow student push one of her 100 kg crates of books up a ramp from the sidewalk into a truck. The ramp is 2m long and is inclined at an angle of 30° with respect to the horizontal. The coefficient of sliding friction μ_k of the crate is 0.5. The pushers started the crate moving before they hit the ramp, so the speed v of the crate was constant during the entire process. Find the work done on the crate during the move up the planks by (a) gravity, (b) friction, (c) the normal force from the planks, (d) the people pushing the crate. The people are exerting a constant horizontal force on the crate. For simplicity, take $g = 10$ in SI units. Demonstrate that the total work done by all forces sums to zero to reasonable approximation.

2. A block of ice slides in the $+x$ direction across a horizontal frictionless surface until it collides with a bumper and spring. The collision compresses the spring. The spring is **not** a Hooke's-Law spring. The spring under compression gives a restoring force $F_x = -kx - ax^2$. The bumper end of the spring is initially at $x = 0$. The spring is compressed until the bumper end reaches $x = A$. Calculate the work done by the spring on the block of ice during the compression.

3. A 100 kg sofa is at rest on a corridor floor. The floor has a friction coefficient of 0.2. (a) The sofa is now pushed along the corridor with a horizontal force of 600 N. The sofa slides parallel to the force through a distance of 10 m. What is the sofa's final kinetic energy? (b) The same sofa begins again

at rest. This time the people pushing it are leaning over, so that they push with 600 N directed at an angle 15 degrees below the horizontal. The sofa again slides through 10 m horizontally. What is the sofa's final kinetic energy?

4. A flat loop of heavy steel chain having length L is resting on the ground. The chain has a mass density μ (mass/length). The chain is picked up by one end and hoisted into the air until the bottom end of the chain is just resting on the ground. At this point, the chain is again stationary. Calculate the total work needed to hoist the chain. Hint: The further the loop is raised, the more of its weight must be canceled by the hoisting force.

5. This time, the sofa of Problem 3, plus some extra boxes so that the total mass of the sofa is now m, is pushed up a ramp. The length of the ramp is L. Fortunately for the pushers, the ramp is frictionless. The ramp makes an angle θ with respect to the ground. Take the $+x$ axis to point downhill parallel to the ramp. Calculate (a) the work done on the sofa by the normal force and (b) the work done on the sofa by gravity. (c) By this time, the volunteer pushers are getting really tired, so they push the sofa up the ramp very slowly, with the sofa coming to a stop at the top of the ramp. Calculate from the force applied by the pushers to the sofa what work they did on the sofa. (d) Calculate the total work done on the sofa as it moves up the ramp, and show that is very nearly zero. Is your result consistent with the work-energy theorem? If you get your answer to work by inserting extra minus signs here or there, you deserve no credit.

6. After the spring and friction of Figure 13.3 bring the mass to a stop, does the mass remain at rest?

7. Assuming the spring of Figure 13.3 rebounds after the mass compresses it, how fast is the mass moving just as the spring releases it?

8. "The solution to this quadratic is equation 13.66" Prove that my solution is correct. Show your work.

9. You are driving down a nearly deserted highway at speed V, come over the top of the hill, and discover that there is a truck blocking the road a distance R in front of you. Your choices are (1) slam on the brakes and hope you stop in distance R, and (2) put the steering wheel over and hope to make a 90 degree turn (a circle of radius R) before you ram the truck. (You will end up safely in a flat field if you do this.) Assume you have modern safety devices so that your car maintains rolling contact with the road at all times. You can solve the problem by calculating the force you need to perform either of these choices. (Aside: Real cars are optimized against collisions from the front rather than the side, so this solution process should not be used as a guide to emergency driving. Also, a head-on collision with inadequate breaking will be at some speed $< V$, while a side collision from an inadequate circle will be at speed V.)

10. A large, flat 1000 kg rock moves virtually without friction over the ice surface of a skating rink. The rock is attached to a nonlinear spring that over the range of interest delivers on the rock a force

$$F_y(y) = -2y - 3y^2 - 4y^3 \qquad (13.70)$$

The rock is initially at $y = 5$. The rock is moving in the $+y$ direction. At $y = +8$ the rock is brought to a stop by the spring. (i) How fast was the rock moving when it reached y= +6? (ii) How fast was the rock moving initially, when it was at $y = +5$?

11. A nonlinear spring develops a restoring force $-kx^3$, where x is the amount the spring has been stretched from its unstretched length. A mass m is attached to the spring, which is hanging vertically. The mass is released from rest while the spring is unstretched. How far does the mass drop before the spring brings the mass to a stop?

12. Summer has returned. A student is moving his belongings from his apartment back to his parents' house. The student has a box of school-books which has a mass of 10 kg; the books will be stored on a loft in his parents' garage. The loft is 5 meters off the ground. To reach the loft, the student will use a 10 meter long roller-ramp; because of the rollers, assume that the ramp is frictionless.

As it is the end of the day, the student becomes so tired that he pushes the box up the ramp with a force that decreases the further up the ramp the box goes. This pushing force is parallel to the ramp and can be modeled as $\mathbf{F}_p = 200 * (1 - x/10)\hat{i}$, where $\hat{i}$ is parallel to the ramp and x is the distance up the ramp relative to ground. When the box was at the bottom of the ramp, at $x = 0$, the box was stationary. (i) Calculate the work done by gravity on the box. (ii) Calculate the work done on the box by the student pushing on it. (iii) Calculate the net work performed on the box. (iv) From the Work-Energy Theorem, calculate the velocity of the box when it reaches the top of the ramp.

13. A box of mass m lies on a horizontal plane with a coefficient of kinetic friction μ_k. This mass is attached to a spring that has a spring constant k which, at its other end, is attached to a wall. When the spring is at its equilibrium length, the mass is located at position $x = 0$. Next, the mass/spring system is stretched so that the mass is moved to position $x = X$. At time $t = 0$, the mass is released; at the moment of release, it is stationary and at $x = X$. At time T, the mass returns to its equilibrium position $x = 0$. (i) Calculate the work done on the mass by friction. (ii) Calculate the work done on the mass by the spring. (iii) Calculate the net work done on the mass. (iv) Using the Work-Energy Theorem, find the velocity of the mass, after it is released, in terms of X, μ_k, k, g, and m. You may use your answers to i, ii, and iii to find the answer to this part of the question.

14. A large crate filled with lead bricks and a collection of tropical feathers sits at the top of a flat, straight boat launching ramp. The crate is prevented from sliding into the ocean by a rope attached to a conveniently placed anchor. The ramp makes an angle θ with respect to the water. The feet of untold generations of sea birds have irregularly worn the ramp, so that the coefficient of kinetic friction of the ramp depends on position as

$$\mu_k = \mu_0 + B\cos(kx). \tag{13.71}$$

x is measured along the ramp from the box to the water. The ramp has length L. $|B| < \mu_0$.

A passer-by cuts the rope, sending the box sliding down the ramp into the water. (i) Find the speed of the sliding crate as a function of position along the ramp. (ii) How fast was the crate moving when it hit the water?

15. Two carts, masses m and M ride on rails on opposite sides of a pair of angled ramps. The cable is looped over a pulley as shown in Chapter 7, Figure 8.14. The ramps are frictionless. The ramp incline angles are θ and ϕ, respectively. When m and M are released from rest, M slides downhill until it encounters a bumper–a massless metal plate with a spring behind it. The spring constant of the spring is k. (i) By what distance parallel to the ramp is the spring compressed when the mass M is brought to a stop? (ii) At the moment that the spring reaches its maximum compression and brings the mass M to a stop, what is the acceleration of mass M?

16. A 5 kg box starts at the top of a ramp and slides down it. The incline makes a 60 degree angle with respect to the horizontal. The coefficient of kinetic friction is $\mu_k = 0.2$. Identify each force acting on the box, and calculate the work done by that force after the box has slid 4 m along the plane. What it the total work done on the box by all forces acting on it? If the box began its slide from rest, how fast is it moving after it has moved 4 m along the plane? If the box began its slide with an initial down slope speed of 6 m/s, how fast is it moving after it has moved 4 m along the plane?

17. Two mining carts on opposite sides of a hill are connected by a durable steel cable. The cable is looped over a pulley as shown in Chapter 7, Figure 8.14. The carts have masses m_1 and m_2, and the slopes have angles θ and ϕ, as indicated. The axles on the carts have rusted to the cart chassis. The carts slide along the rails with a coefficient of kinetic friction μ_k. The coefficient of kinetic friction depends on the cart position as $\mu_k = 0.05 + 0.03x^2$, with x being measured along the ramp. (a) Find the work done on the first cart by the friction as the cart slides from $x = 1$ to $x = 2$m. The incline angle is $\theta_1 = 20$ degrees; the mine cart has a mass of 1.0×10^4 kg. (b) If you apply the work-energy theorem, does your answer to part (a) give the change in the kinetic energy of the cart as it slides from $x = 1$ to $x = 2$m? Explain your answer **in complete, grammatical sentences**.

18. A ride at a Florida amusement park consists of a spring-loaded cannon that propels a swimmer down a straight water-slide into a pool of water. The swimmer starts 10m from the water, and slides towards

the water at an angle 30 degrees from the horizontal. The spring has a spring constant of 250 N/m, and is compressed by 3 meters. A safety device stops the spring after 3 m of slide; after the 3 m, the swimmer is released from the spring. The slide is water covered, and has a coefficient of static friction of 0.3, and a coefficient of kinetic friction of 0.1. (i) Calculate the work done on the swimmer by the spring. (ii) Set up a correct integral for calculating the work done on the swimmer by friction. Perform the integral and compute the work done on the swimmer by friction. (iii) The swimmer is initially stationary. From the above calculations, etc., compute the swimmers (a) kinetic energy and (b) speed when she hits the water.

19. Real World Problem: The real* volksraketenwagon (not its actual name) recently attained a steady sustained speed of 700 mph (for sake of argument, 1100 kph) over an open track. The vehicle is propelled by two jet engines, each generating 100,000 horsepower (75 MW) at full power. Assuming that the two engines were each at 80% of full power [the design top speed is supersonic] during this drive, and assuming that the power supplied by the engines is dissipated by the friction on the car, compute the frictional force on the rocket car.

*You didn't think I was making them up, did you?

13.7 Solutions to the Worked Problems

1. The force $(3 + 2x)\hat{i}$ is parallel to the x-axis. The motion is similarly parallel to the x-axis. Effectively, this is a one-dimensional problem. We want to find the change in the kinetic energy. In identifying problem-solving approaches for motion problems, we have seen two general methods. If the problem somehow relates positions, velocities, accelerations, and times, direct use of the Second Law has been seen to be effective. If the problem interrelates positions, velocities, and energies, the work-energy theorem is often the appropriate path. The question here corresponds to the latter sort of problem, so we begin with the work-energy theorem, written in one dimension as

$$K_f - K_i = \int_{x_i}^{x_f} dx F(x). \tag{13.72}$$

For this problem, the actual force gives

$$K_f - K_i = \int_{15}^{10} dx(3 + 2x) = 3x + \frac{2x^2}{2}\Big|_{x=15}^{10}, \tag{13.73}$$

so that

$$K_f - K_i = 3 \cdot 10 + \frac{2 \cdot 10^2}{2} - 3 \cdot 15 - \frac{2 \cdot 15^2}{2}, \tag{13.74}$$

and therefore

$$K_f - K_i = 130 - 270 = -140 \ J. \tag{13.75}$$

J is the unit of energy, the Joule.

From this calculation K_f is 140 Joules less than K_i. The kinetic energy can never be negative, because m and v^2 are always positive numbers, so K_i must have been at least 140 Joules.

2. We have a mass m sliding down a frictionless inclined plane that makes an angle $\theta = 50$ degrees with respect to the horizontal. How fast is the block moving, given several initial conditions, after it has slid $X = 3$ meters down the ramp? This problem involves positions and velocities, so it appears effective to attack it using the work-energy theorem.

You should have begun by drawing a sketch and a force diagram, including coordinates. The normal force is perpendicular to the plane, so the component N_x of the normal force parallel to the plane is zero. The work done by the normal force, given that the motion is parallel to the plane, is

$$W = \int_0^X N_x \, dx = \int_0^X 0 \, dx = 0. \tag{13.76}$$

The normal force acts on the mass, but does no work on it.

What about the work done by gravity? The component of gravity parallel to the plane is $mg\sin(\theta)$, so the work done by gravity is

$$W = \int_0^X mg\sin(\theta)\,dx. \tag{13.77}$$

We have reached a point where we have to be careful that everything we have done is self-consistent. We have chosen coordinates such that the mass starts at $x = 0$ and advances to $x = +3$, not to $x = -3$. The positive x-direction is down the ramp. That's also the direction that gravity pulls the mass, so the component of the gravitational force parallel to the ramp must also be positive. The work done by gravity is therefore

$$W = mgX\sin(\theta), \tag{13.78}$$

so the work-energy theorem gives us

$$K_f - K_i = W_g. \tag{13.79}$$

Replacing the kinetic energy with the velocity, we have

$$\frac{1}{2}mv_f^2 = \frac{1}{2}mv_i^2 + mgX\sin(\theta). \tag{13.80}$$

We solve for v_f by multiplying through by $2/m$ and taking the square root of both sides of the equation, leading to

$$v_f = (v_i^2 + 2gX\sin(\theta))^{1/2}. \tag{13.81}$$

This problem illustrates the great virtue of solving the problem symbolically until the end, and only plugging in numbers after the problem is solved symbolically. Because we have a symbolic solution, all we need to do is to insert a series of different numbers into the final formula. If we had plugged in all numbers at the beginning and then solved, we would have had to do the entire calculation several times. With all numbers plugged in except v_i, the solution is

$$v_f = (v_i^2 + 2 \cdot 10 \cdot \sin(50))^{1/2}, \tag{13.82}$$

and all we need to do is insert $v_i = 0$ or $v_i = 6$ to obtain the final velocity v_f.

3. The mass moves very slowly from $x = 0$ to $x = -\ell$, $x = 0$ being the position at which the spring exerts no force on the mass. Because the motion of the mass is very slow, it must be the case that the push force P and the spring force F_s are very nearly equal in magnitude, but opposite in direction, at all times. The x-component of the spring force is

$$F_s = -kx, \tag{13.83}$$

so the x component of the push force is

$$P = +kx. \tag{13.84}$$

The work done by each of the two forces is

$$W_s = \int_0^{-\ell} (-kx)dx = \frac{-kx^2}{2}\Big|_{x=0}^{-\ell} = \frac{-k\ell^2}{2}, \text{ and} \tag{13.85}$$

$$W_p = \int_0^{-\ell} (kx)dx = \frac{kx^2}{2}\Big|_{x=0}^{-\ell} = \frac{k\ell^2}{2}. \tag{13.86}$$

The total work done by the two forces is then

$$W_{\text{total}} = \int_0^{-\ell} (F_s + P)dx = \int_0^{-\ell} (-kx + kx)dx = 0. \tag{13.87}$$

13.7. SOLUTIONS TO THE WORKED PROBLEMS

If we calculate the total of the works done separately by each of the two forces, we get the same answer as if we calculate the work done by the total of the two forces. For the compression step, we have $K_f - K_i = 0$.

The push force P is now removed. The spring expands through $x = 0$. How fast is the mass moving at this point? We again apply the work-energy theorem $K_f - K_i = W$ and get

$$K_f - K_i = W_s = \int_{-\ell}^{0} dx(-kx) = -\frac{kx^2}{2}\Big|_{x=-\ell}^{0}. \tag{13.88}$$

Recalling that the mass is initially stationary, $K_i = 0$, so

$$K_f = \frac{-k0^2}{2} - \left(-\frac{k(-\ell)^2}{2}\right) = \frac{k\ell^2}{2}. \tag{13.89}$$

There is a point where multiple sign errors could cancel, namely if one wrote the force as $+k\ell$ and the integration as being taken from 0 to ℓ rather than $-\ell$ to 0, you would get an answer that made sense (the final kinetic energy would be positive) but would still be totally wrong, because the process for reaching it would be wrong. Graders know to look carefully at all the signs in the calculation, because with great regularity people will work through the problem, get an answer that should have found incorrectly to be $-\frac{k\ell^2}{2}$, and report the final answer as $+\frac{k\ell^2}{2}$ so the sign would be right, even though that sign was inconsistent with their calculations. That's not an acceptable path. *Praxis* involves getting the right answer for the right reason.

4. Once again we have a box sliding down a ramp, the same ramp seen as the second of these worked problems. This time the box experiences friction as it slides down the ramp, the coefficient of kinetic friction μ_k being $\mu_k = 0.3$. We need to choose coordinates. For a ramp, putting the two coordinates parallel to the ramp and perpendicular to the ramp will often work out well. With that choice of coordinate axes, the components of the gravitational force are $mg\sin(\theta)$ parallel to the ramp and $mg\cos(\theta)$ perpendicular to the ramp. The signs on these quantities are determined by the direction we assign to the coordinate axes. The normal force has a component N perpendicular to the ramp and a component 0 parallel to the ramp.

As the box slides down the ramp, the work done on it by the normal force is

$$W_N = \int_0^X dx \cdot 0 = 0. \tag{13.90}$$

The work done on the box by gravity is

$$W_g = \int_0^X dx\, mg\sin(\theta) = mg\sin(\theta)(X - 0) = mgX\sin(\theta). \tag{13.91}$$

To determine the friction, we need to know the normal force between the box and the ramp. The ramp applies a normal force having magnitude N on the box, and the box applies a force of the same magnitude but the opposite direction on the ramp. We can obtain a value of the normal force from the Second Law for the force components perpendicular to the ramp, namely

$$m\frac{d^2z}{dt^2} = N - mg\cos(\theta). \tag{13.92}$$

The acceleration perpendicular to the ramp is zero, giving $N = mg\cos(\theta)$. Therefore, for the component of the friction force F_f pointing up the ramp

$$F_s = \mu_k mg\cos(\theta). \tag{13.93}$$

The work done by friction is therefore

$$W_f = \int_0^X dx(-\mu_k mg\cos(\theta)) = -\mu_k mgX\cos(\theta). \tag{13.94}$$

The total work done by all forces when the box slides through a distance L is therefore

$$W = mgL\sin(\theta) - \mu_k mgL\cos(\theta). \tag{13.95}$$

The kinetic energy as a function of the distance slid is given by the work-energy theorem

$$K_f - K_i = W. \tag{13.96}$$

Following the steps of Worked Problem 2, above,

$$v_f = (v_i^2 + 2gL\sin(\theta) - \mu_k gL\cos(\theta))^{1/2}. \tag{13.97}$$

As before, solving symbolically first, when we must solve for several values of v_i, saves us a great deal of work. With numbers substituted for everything except v_i, we have

$$v_f = (v_i^2 + 2\cdot 10\cdot 3\sin(50) - 0.3\cdot 10\cdot 3\cos(50))^{1/2}. \tag{13.98}$$

the angles being given in degrees. We may now solve for v_f for any value of v_i, simply by plugging in a number.

This space reserved for your notes.

Chapter 14

Energy

14.1 Introduction

In this Chapter we discuss energy, in particular the total energy of a system. Our discussion follows from the Work-Energy Theorem.

$$K_f - K_i = \int_{\text{start}}^{\text{finish}} \boldsymbol{F}(s) \cdot d\boldsymbol{s}. \tag{14.1}$$

There are a lot of different forces that could be dropped into this equation as part of $\boldsymbol{F}(s)$. However, we may divide all those forces into three categories, namely *orthogonal*, *non-conservative*, and *conservative* forces.

Orthogonal forces are forces that always act in a direction perpendicular to the particle's direction of displacement. In the last chapter, we gave two examples of forces that were, in the problem under consideration, orthogonal forces. We had a sled sliding in the horizontal direction. The gravitational force $-mg\hat{\boldsymbol{k}}$ and the normal force $\boldsymbol{N}$ both point in the vertical direction, perpendicular to $d\boldsymbol{s}$. The scalar products of $-mg\hat{\boldsymbol{k}}$ and $\boldsymbol{N}$ with $d\boldsymbol{s}$ are therefore each the scalar product of two perpendicular vectors, which evaluates to zero. The projection of these two forces into the $\hat{\boldsymbol{s}}$ direction vanishes. These orthogonal forces therefore do no work on the object of interest.

Another force, one that is always perpendicular to the direction of motion, is the magnetic force $\boldsymbol{F}_B$. The magnetic force is determined by the object's electrical charge q and velocity $\boldsymbol{v}$, and by the local value $\boldsymbol{B}$ of the magnetic field vector, namely

$$\boldsymbol{F}_B = q\boldsymbol{v} \times \boldsymbol{B}. \tag{14.2}$$

Here × denotes the vector product. An important property of the vector product of two vectors is that the product vector, in this case $\boldsymbol{F}_B$, is perpendicular to the two vectors that combined to form it. in particular, $\boldsymbol{F}_B \perp \boldsymbol{v}$, guaranteeing that the scalar product of $\boldsymbol{F}_B$ and $\boldsymbol{v}$ is zero. However, $\boldsymbol{v}dt = d\boldsymbol{s}$, so the scalar product of $\boldsymbol{F}_B$ and $d\boldsymbol{s}$ also vanishes. As a result, magnetism can do no work on a moving charge; it can change the direction of motion but not the speed of a moving charge.

We now draw a distinction between conservative and non-conservative forces. Suppose an object moves from point A to point B. A conservative force is a force that has the property that the work that the force does on an object is independent of the path the object takes in moving from A to B. If an object is moved from point A to point B, a conservative force does the same amount of work no matter which path is followed between the two points. On the other hand, if the same object is moved from point A to point B while being acted on by a non-conservative force, the amount of work done on the object by the non-conservative force depends on which path the object followed.

Two examples may make this distinction a bit clearer. A simple example of a conservative force is gravity. The work done by a gravitational field g is

$$W = \int_{\text{start}}^{\text{finish}} mg \, dz. \tag{14.3}$$

We compare two paths. The starting point is the top of a local mountain and its ski trail.

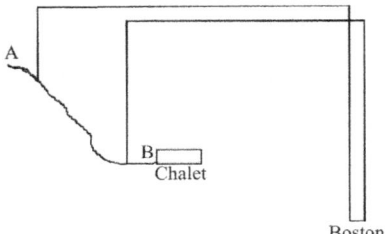

Figure 14.1: Two paths down to the chalet. Ragged line: on skis. Straight lines: With helicopter medivac.

The ragged line shows a skier, who starts at A, the top of the trail, and glides down to the chalet, point B. The straight lines are your author, who was persuaded to strap on skis*. There are then (1) several feet along the trail, (2) the helicopter medivac to the Boston hospital, (3) elevator rides to surgical theaters, and finally (4) the helicopter back to B and the chalet to recover my car. The two paths are very different in length, but have the same total change in altitude. For these two paths the total work $-mg\Delta z$ done by gravity is the same, no matter which path is followed. Gravity is a conservative force.

*Your author actually did not do this, even though it would have been **For Science**!

On the other hand, let us consider two joggers on a California beach. They both run from A to B. One runs on the level sand above the high-tide line. The other runs in waist-deep water through the breaking waves. The resistive force, and the work done by the very complicated frictional forces here, are very different for the runner on dry ground and the runner pushing through the waves. The work done against frictional drag is very strongly path-dependent, because the frictional drag depends on which path was taken.

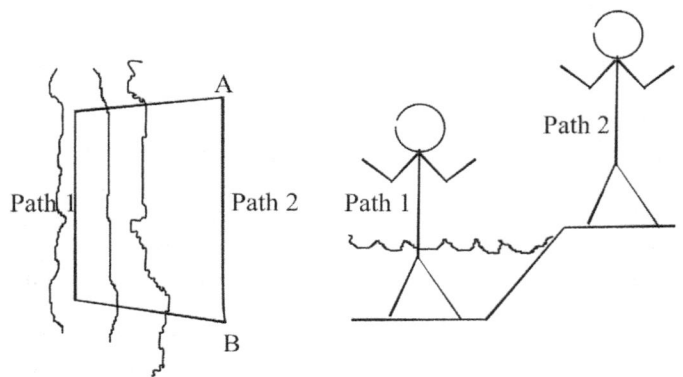

Figure 14.2: Two paths along the beach from A to B. Path 1 goes through the water. Path 2 stays on land.

We may also consider the work done on a runner who runs a closed loop, from A to B and back again to A. There are two obvious forces acting, gravity and friction. For the runner on the beach, half of the loop is on level ground. On the other half of the loop, a runner descends into the surf, runs, and then climbs back out of the ocean again. The work done by gravity is

$$W = -\int_{z_i}^{z_f} mg\, dz. \qquad (14.4)$$

In general $W = mg(z_f - z_i)$. If $z_i = z_f$, then $W = 0$. As with the skiing event, if the runner moves in a closed loop, returning to the starting point, then the work done by gravity is zero. Gravity is therefore a conservative force.

The frictional force always opposes the direction of motion, so on each half of the loop the work on the runner due to friction is negative. Over this closed loop, the frictional work does not vanish, so once again friction is a non-conservative force. A minor but interesting question arises. If friction is always doing negative work on the runner, pulling kinetic energy out of the runner, how can the runner be advancing at constant speed? The answer is that the runner is also subject to a tractive force on his feet, due to the sand pushing on his feet in the forward direction, doing positive work and cancelling the negative work due to friction. The tractive force arises because the runner's feet dig into the sand, giving a non-zero force on the runner in the forward direction.

Let us go back to gravity. If we define the change in altitude Δz to be $\Delta z = z_f - z_i$, then the work due to gravity is $W = -mg\Delta z$. We now propose that we can create a function $U(\boldsymbol{r})$, the *potential energy*. For each conservative force, there is a corresponding potential energy. U is a function of position, so that whenever we return to the same position $\boldsymbol{r}$, U returns to the same value $U(\boldsymbol{r})$. (A simple example of a function that only depends on position is altitude. If you walk in a closed loop, then, no matter your path, when you return to your starting point your change in altitude is zero.)

If we call the potential energy at the starting point U_i and the potential energy at the end point U_f, then the change in potential energy between the starting point and the end point is $\Delta U = U_f - U_i$. Furthermore, U is defined so that the change in U between two points, due to some force, is the negative of the work done

14.2. STABILITY

by that force when we move between the two points. This statement may be written mathematically as

$$\Delta U = -W. \tag{14.5}$$

In this equation, W is the force due to a single conservative force, and U is the potential energy corresponding to that force. In order for this equation to make sense, the force must be conservative, so that the work done in moving from a point A to a point B is the same no matter what path is followed between the two points. If the work done in moving from point A to point B were to depend on which path was used, then ΔU for the movement would depend on the path used, and correspondingly U could not be a simple function of position.

Equation 14.5 only tells us the change in the potential energy between two points. It does not give us an absolute value for the potential energy at a single point. In solving a problem, you are free to choose a single point $\boldsymbol{R}$ at which $U(\boldsymbol{R}) = 0$. There is one bit of caution. In an entire problem, you can only choose one point $\boldsymbol{R}$ at which $U(\boldsymbol{R}) = 0$. All potential energies must then be measured with respect to the potential energy at the same point.

If the path is a closed loop, then ΔU must be zero, no matter if the path is large or infinitesimally small. This requirement corresponds to a statement about the underlying force $\boldsymbol{F}$ that is doing the work. If you have had vector calculus, you will recognize this requirement says $\boldsymbol{\nabla} \times \boldsymbol{F} = \boldsymbol{0}$ for a conservative force.

We now apply this definition of the change in the potential energy to the Work-Energy Theorem. We generalize the discussion by accepting that the moving object may be subject to several conservative forces at the same time, in which case the Work-Energy Theorem applies to the total force due to all those conservative forces. The work-energy theorem tells us that

$$K_f - K_i = W. \tag{14.6}$$

However, $W = -\Delta U$ or $W = -(U_f - U_i)$. Replacing W with U in the Work-Energy Theorem, we find

$$K_f - K_i = -(U_f - U_i). \tag{14.7}$$

We then rewrite this equation by switching terms between its two sides. The final result is

$$K_i + U_i = K_f + U_f. \tag{14.8}$$

From the Work-Energy Theorem, for conservative forces the quantity $K + U$ does not change as time goes on. We say that the quantity $K + U$ is *conserved*. Correspondingly, the Work-Energy Theorem tells us that equation 14.8 represents a *conservation law*.

We define the total energy E of the system to be

$$E = K + U. \tag{14.9}$$

Substituting this definition into equation 14.8, we have

$$E_i = E_f. \tag{14.10}$$

The above equation is *The Law of Conservation of Energy*. So long as only conservative forces are acting, this equation tells us that the total energy of the system does not depend on time. The total energy must be calculated by including all of the conservative forces in the system. This equation is only true if all forces in the system are conservative. If there is friction, the friction does work; that work changes $K_f - K_i$ in accord with the Work-Energy Theorem.

14.2 Stability

The potential energy lets us discuss the mechanical stability of an initially motionless system. By stability we refer to the response of the system if it is ever-so-gently jostled. Three examples clarify how stability works. We start with a table, a hemispherical steel mixing bowl, and a toy marble. If the bowl is set in its normal position, with the marble at the center, and jostle the marble, the marble will roll a short distance uphill and then roll back to its starting location. That's *stable equilibrium*. If we set the bowl on the table upside down, so it is a steel dome, and balance the marble at the top, the slightest disturbance will cause the marble to roll faster and faster downhill, away from its starting position. That's *unstable equilibrium*. If we set the marble on the table top, and displace it slightly from its starting position, it will just sit there. That's *neutral equilibrium*.

The gravitational potential energy is $U = mgz$, so the potential energy of the marble at each position is its height above the floor. Stable equilibrium then corresponds to a potential energy surface that is concave upward, meaning in one dimension that

$$\frac{d^2 U(x)}{dx^2} > 0. \tag{14.11}$$

Unstable equilibrium then corresponds to a potential energy surface that is concave downward, meaning in one dimension that

$$\frac{d^2 U(x)}{dx^2} < 0. \tag{14.12}$$

Finally, unstable equilibrium corresponds to a potential energy surface that is flat, meaning in one dimension that

$$\frac{d^2 U(x)}{dx^2} = 0. \tag{14.13}$$

Thee equations can each be interpreted as indicating which way the force will point, relative to the starting location, after a small displacement, namely toward the rest position, away from the rest position, and in no direction.

14.3 Gravitational Potential Energy of an Extended Body

We begin by considering the potential energy due to a uniform gravitational force $\boldsymbol{F_g} = -mg\hat{\boldsymbol{k}}$. Our discussion of the force of gravity on an object, the work done by gravity, and the gravitational potential energy has consistently treated the object as having a well-defined height z_i. This description is entirely appropriate if the object is a single mass. If the object is not a point, if it has a mass distribution that is spread over some region of space, the point-mass description is not appropriate. What height should be assigned to an object that is not a simple point?

How do we advance? First, we might describe an extended object as a collection of point masses. The individual point masses could be the atoms that comprise the extended object. The gravitational potential energy of the extended object is then the sum of the gravitational potential energies of the point masses, namely

$$U = \sum_{i=1}^{N} g m_i z_i. \tag{14.14}$$

In this equation, i is an index labelling the point masses, the sum is over all N point masses, m_i and z_i are the mass and height of mass point i, and g is the gravitational acceleration, assumed to be the same for each mass.

The vertical component of the center of mass vector of an extended object is

$$Z_{\text{CM}} = \frac{1}{M} \sum_{i=1}^{N} m_i z_i, \tag{14.15}$$

in which M is the total mass of the extended object.

If we multiply the right-hand-side of equation 14.14 by unity in the form M/M, this equation can be written

$$U = Mg \left(\frac{1}{M} \sum_{i=1}^{N} m_i z_i \right). \tag{14.16}$$

The term in parenthesis is the same as the right-hand-side of equation 14.15; it is the z coordinate of the extended object's center of mass. On combining the two above equations, one obtains

$$U = MgZ_{\text{CM}}. \tag{14.17}$$

In a uniform gravitational field, the gravitational potential energy of an extended object is the same as the gravitational potential of a point mass, if the extended object and the point mass have the same total mass M, and if the point mass is located at height Z_{CM}.

The above demonstration assumes that g is a constant, so that it could be factored outside the sum on point masses. However, the real gravitational force is not a constant. Gravity becomes weaker as you increase the distance from the center of the earth. If the gravitational acceleration is not a simple constant, the above calculation is not wrong, but it does not describe the situation at hand. A homework problem asks you to calculate what happens if the gravitational acceleration depends linearly on z.

14.4 I Threw a Rock Into the Air

As a first example, consider throwing a rock vertically into the air. The rock's height is a function of time; $z = z(t)$. The rock starts out with vertical velocity V. As it climbs, it slows down. It reaches a maximum altitude H, and then begins to fall back to earth. What is H? How fast is the rock travelling when it reaches height $z = H/2$? How fast is the rock travelling when it returns to the ground?

How might we solve this problem? In this simple case, we could invoke the Second Law and calculate velocity and position as functions of time. We will instead use energy conservation. The solution starts by writing down conservation of energy

$$E_i = E_f. \tag{14.18}$$

We also need the general form for the total energy of the rock, namely

$$E = \frac{1}{2}mv^2 + mgz. \tag{14.19}$$

In this equation, m and v are the rock's mass and velocity, while z is its height. We'll call the initial velocity and height v_i and z_i. The final velocity and height are v_f and z_f. Combining these two equations, we get as the general form for the rock's travels that

$$\frac{1}{2}mv_i^2 + mgz_i = \frac{1}{2}mv_f^2 + mgz_f. \tag{14.20}$$

How far up does the rock travel? At the start, $v_i = V$ and $z_i = 0$. At the top of the trajectory, $z_f = H$ and $v_f = 0$. Substituting these into equation 14.20, we have

$$\frac{1}{2}mV^2 + mg0 = \frac{1}{2}m0^2 + mgH. \tag{14.21}$$

In writing this equation, all I did was to replace general variables with their values, e.g., $z_f = H$ and $v_f = 0$. A number of terms, e.g., $mg0$, will evaluate to zero. I deliberately wrote them out explicitly so that a check will reveal why they are zero. The next step is to simplify the equation, taking advantage of the zeroes to find

$$\frac{1}{2}mV^2 = mgH. \tag{14.22}$$

We are now at the edge of inferring *a very dangerous error*. It looks as though the Law of Conservation of Energy can be written in the form $K = U$. That's completely incorrect; equation 14.22 simply happens to be true in one part of this problem.

We can solve this equation for H. We get

$$H = \frac{V^2}{2g}. \tag{14.23}$$

Let's look at two cases in which the $K = U$ equation reveals itself to be obviously wrong. First, how fast is the rock travelling when it is at height $z_f = H/2$? We again write out the various steps in the solution process. These are

$$E_i = E_f, \text{ the basic equation,} \tag{14.24}$$

$$E = \frac{1}{2}mv^2 + mgz, \text{ the general form for the energy,} \tag{14.25}$$

$$\frac{1}{2}mv_i^2 + mgz_i = \frac{1}{2}mv_f^2 + mgz_f, \text{ all constants separated,} \tag{14.26}$$

$$\frac{1}{2}mV^2 + mg0 = \frac{1}{2}mv_f^2 + mgH/2, \text{ variables replaced with constants, and finally} \tag{14.27}$$

$$\frac{1}{2}mV^2 = \frac{1}{2}mv_f^2 + mg\frac{V^2}{4g}. \tag{14.28}$$

$$\tag{14.29}$$

The last line certainly does not have the form $K = U$. Dividing out the factor $m/2$, cancelling the factors of g in the final term, and solving for v_f, one finds

$$v_f = \frac{V}{\sqrt{2}}. \tag{14.30}$$

Finally, how fast is the rock going when it hits the ground again? As before, $z_i = 0$ and $v_i = V$. This time we have $z_f = 0$. Steps that are the same as those above lead to

$$\frac{1}{2}mV^2 + mg0 = \frac{1}{2}mv_f^2 + mg0 \tag{14.31}$$

I have been scrupulous about showing the terms with a zero in them, so that when I go back to check my work, I can immediately tell why I set a particular term equal to zero. The solution here is

$$v_f = V \tag{14.32}$$

Remember that v_f is being extracted from the kinetic energy, where it is squared, so we have obtained the magnitude of the final velocity but not its sign.

14.5 The Radical Roller Coaster

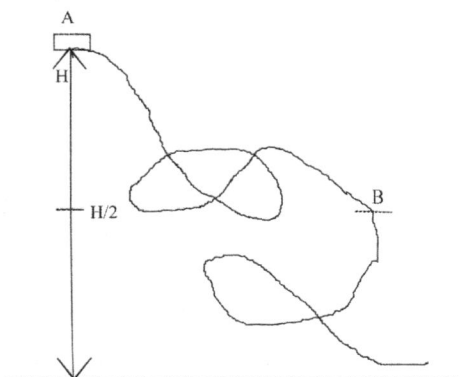

Figure 14.3: A roller coaster – The Crazy Comet.

Let's consider a second problem. It's a frictionless roller coaster, the Crazy Comet, as seen in Figure 14.3. The roller coaster car starts at point A at a height H off the ground. At that point it is almost but not quite stationary. How fast is it going when it reaches point B, a distance $H/2$ off the ground? Solving this problem using the Second Law would be challenging. For starters, you would need to know the exact path that the car follows, my sketch surely being inadequate. Then you would almost certainly need to do numerical integration of the forces and accelerations. On the other hand, solving this problem using energy conservation is entirely straightforward.

At the start, $v_i \approx 0$ and $z_i = H$. At the finish, $z_f = H/2$. What is v_f? In this problem, the only potential energy is due to gravity, so for the total energy we may write

$$E = \frac{1}{2}mv^2 + mgz. \tag{14.33}$$

14.6. TWO MASSES CONNECTED BY A WIRE

Let's now solve the problem. We are using energy conservation, so the starting point is

$$E_i = E_f. \tag{14.34}$$

However, we know what the system's total energy is, so we can write

$$\frac{1}{2}mv_i^2 + mgz_i = \frac{1}{2}mv_f^2 + mgz_f. \tag{14.35}$$

v_i, z_i, and z_f are all known quantities, provided in the problem statement. Substituting without making any simplifications, we get

$$\frac{1}{2}m0^2 + mgH = \frac{1}{2}mv_f^2 + mgH/2. \tag{14.36}$$

The solution here is

$$v_f = (gH)^{1/2}. \tag{14.37}$$

14.6 Two Masses Connected by a Wire

We now consider two masses connected by a thin wire. One mass, m_1, is lying on a tabletop. The other mass, m_2, is hanging off the edge of the table, suspended by the thin wire. The masses are initially stationary. They are then released. Mass 1 slides to the right while mass 2 falls vertically. While the masses are sliding, the wire is under a tension T. How fast are the masses moving when m_2 has fallen through a distance H?

We'll first consider the solution when there is no friction. What terms appear in the energy of the system? Each mass has a kinetic energy. The first mass moves horizontally. The force of gravity and the normal force act perpendicular to the direction in which mass 1 is moving, so they do no work on mass 1; there is no corresponding potential energy. The potential energy of mass 1 does not change when mass 2 falls. Mass 2 has its kinetic energy, and also a potential energy m_2gh.

We now apply energy conservation, so $E_i = E_f$.

The total energy of the system is

$$E = \frac{1}{2}m_1v_1^2 + \frac{1}{2}m_2v_2^2 + m_2gh. \tag{14.38}$$

There is one constraint operating in the system. The two masses are connected by a thin wire that is under tension. The masses must therefore move with the same speed v, in the $+x$ and $-z$ directions, respectively, so $v_1^2 = v_2^2 \equiv v^2$.

At the start, $v = 0$ and $h = Z$. At the end, $v = v_f$ and $h = Z - H$. Energy conservation then gives us

$$\frac{1}{2}m_1 0^2 + \frac{1}{2}m_2 0^2 + m_2gZ = \frac{1}{2}m_1v^2 + \frac{1}{2}m_2v^2 + m_2g(Z-H). \tag{14.39}$$

Some of these terms are zero. Others cancel. After modest simplification we reach

$$m_2gH = \frac{1}{2}(m_1 + m_2)v^2. \tag{14.40}$$

and finally

$$v = \left(\frac{m_2gH}{2(m_1 + m_2)}\right)^{1/2}. \tag{14.41}$$

The speed v increases in proportion to the square root of the distance H through which m_2 has fallen. The speed of descent also depends on the ratio $m_2/(m_1 + m_2)$ of the two masses. If $m_2 \gg m_1$, so that a very heavy descending mass m_2 is coupled to a very light mass on the tabletop, $m_2/(m_1 + m_2) \approx 1$. m_1 then has more or less no effect on the fall of m_2. On the other hand, if $m_2 \ll m_1$, then $m_2/(m_1 + m_2) \approx 0$. m_2 applies a very small force to m_1, so the acceleration of m_1 is very nearly zero.

It might be tempting to write conservation of energy for the second mass, only. In that case, one might be tempted to write

$$E = \frac{1}{2}m_2v_2^2 + m_2gh. \tag{14.42}$$

This equation clearly includes the kinetic energy of mass m_2 and the gravitational potential energy of that mass. However, equation 14.42 is *simply wrong* for m_2. If it were right, m_2 would drop in free fall, accelerating at one gravity. That's not what happens here. What is wrong? We missed a term. m_2 is attached to the wire that leads back to m_1. The wire is under tension. When m_2 moves vertically, work is done on m_2 by the tension in the wire. The amount of work done by the wire on m_2 is

$$W = \int_Z^{Z-H} T dz. \tag{14.43}$$

where I've taken the $+\hat{\boldsymbol{k}}$ direction to be vertically upward. The work done on m_2, as found by doing the integral, is $-TH$.

Why didn't we see this term earlier? Also, as a practical matter, how can we evaluate this expression, given that we don't know what T is? The answer is that the tension in the rope is also doing work on mass m_1. The tension in the two ends of the wire is the same, so that work is

$$W = \int_0^H T dx. \tag{14.44}$$

If you evaluate that integral, you will find that the work done on m_1 is $+TH$, so the total work done by the tension, acting on the two masses, is $-TH + TH = 0$. Equation 14.38 for the energy of the system is therefore correct, because the total work done by the tension in the wire is zero.

There's a general result hiding in there. The wire connecting the two masses is an internal force. The total work done on an object by an internal force of that object is always zero.

14.7 Potential Energy of a Spring

We consider a mass m attached to a spring. The force constant of the spring is k. The equilibrium length of the spring is ℓ. The spring is approximated as being massless. We will then compress or stretch the spring. The far end of the spring is attached to a very heavy object, a wall, which essentially does not move when the length of the spring is changed. All forces in the problem only have a single non-zero component, the x-component.

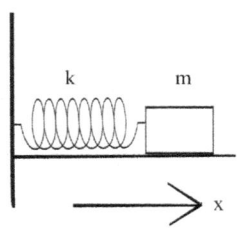

Figure 14.4: A mass on a spring.

In this sketch, $x = 0$ is the location of the wall before we compress the spring, so the mass starts out at $x = \ell$. We now move m to the left through a distance a, $a > 0$, so m is displaced from $x = \ell$ to $x = \ell - a$. The change in the length of the spring is $-a$. The x-component of the force of the spring on the mass is now $+ka$. The corresponding force of the mass on the spring, which is the reaction force, is $-ka$. The x-component of the force of the spring on the mass may also be written $-k(x - \ell)$.

The spring must also be applying a force on the wall, and vice versa. The force of wall on the spring is $+ka$. The force of the spring on the wall, the corresponding reaction force, is $-ka$. The spring is massless, so the total force on the spring must vanish. Indeed, the forces on the spring are $+ka$ due to the mass and $-ka$ due to the wall, which these do indeed total to zero.

The work done in changing the length of the spring follows immediately. Work is defined to be

$$W = \int_{\text{start}}^{\text{finish}} \boldsymbol{F} \cdot d\boldsymbol{s}. \tag{14.45}$$

The work done on the mass in compressing the spring is

$$W = \int_\ell^{\ell-a} -k(x-\ell)\hat{\boldsymbol{i}} \cdot dx\hat{\boldsymbol{i}}, \tag{14.46}$$

so

$$W = \left(-\frac{kx^2}{2} + k\ell x\right)\Big|_\ell^{\ell-a} \tag{14.47}$$

14.8. DISCUSSION

which evaluates to
$$W = -\frac{1}{2}ka^2. \tag{14.48}$$
Recall that $\Delta U = -W$. It is sensible to choose $U = 0$ to be at $a = 0$. We then write for the potential energy of a spring
$$U = \frac{1}{2}ka^2. \tag{14.49}$$

14.8 Discussion

Energy methods follow as a logical deduction from the Second Law, so in principle they only tell you things that you already knew about particle motions. While Second Law methods give relationships between positions, velocities, and times, energy conservation only gives relationships between positions and velocities. However, the solution processes in energy methods are often much simpler than the processes needed to integrate the Second Law to determine positions as functions of time.

In some cases, energy methods give information that could not readily be obtained from the Second Law, because the mechanical details of the system are not given in enough detail. We saw this with the roller coaster problem. To obtain the velocity of the frictionless roller coaster car by integrating the Second Law, you would need to know the exact shape of the track, and then do a complicated, probably numerical, integration. To obtain the velocity of the frictionless roller coaster car by using energy methods, you only need to know the height and velocity of the car at one point on the track, and the height of the car at the point of interest. The solution process then requires only simple algebra.

It is worthwhile to repeat the warning about the insidious $K = U$ error, which, alas, some readers learned in previous physics courses. In the absence of friction, conservation of energy tells us $E_i = E_f$; the initial energy of the system is equal to the final energy of the system. It does not tell us that the kinetic energy at one time is equal to the potential energy at some other time.

14.9 Worked Problems

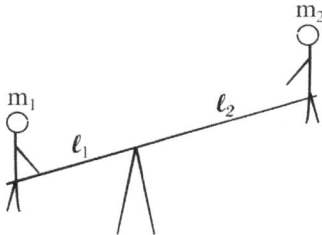

Figure 14.5: Figure for Problem 2.

1. In the space between two flat parallel plates, the electric field vector $\mathbf{E}$ is uniform and perpendicular to the plates. An electric field exerts an electrical force $\mathbf{F}_e = q\mathbf{E}$ on an electrical charge q. In the electrical system as sketched, the electrical field is $\mathbf{E} = 250\mathbf{j}$ in SI units. The plates are a distance L apart. Take the plates to be at $y = 0$ and $y = L$. A tiny, neutrally buoyant helium balloon bearing a charge $q = 1 \cdot 10^{-3}$ in SI units is released (perhaps not from rest) at point $(0.5, 0.6, -.8)$; it moves to point $(1.0, 1.5, 2.2)$. Both points are between the two plates. As a result of this move: What is the change in the potential energy of the balloon? What work is done on the balloon by the electrical field?

2. Two children, masses m_1 and m_2, sit on opposite sides of a teeter-totter at distances ℓ_1 and ℓ_2, respectively, from the center. See Figure 14.5. Find their total gravitational potential energy as a function of the angle θ between the teeter-totter arm and the horizontal. Find the values of ℓ_1 and ℓ_2 (There may be more than one pair of these, though finding the second pair requires thinking outside the box.) for which the gravitational potential energy is independent of θ.

3. A non-linear spring produces a force along the x-axis $F_x = -3x - 7x^3$, where one end of the spring cannot move and x is the location of the other end of the spring. Find the potential energy of the spring when it is stretched or compressed. Take $U = 0$ at $x = 0$.

4. Repeat Chapter 13, worked problem 1, but this time solve using energy conservation.

5. We have a mass m on a ramp that makes an angle θ with respect to the horizontal. The mass slides down the ramp until it encounters an initially unstretched spring. It continues to slide, compressing the spring, until it comes to a stop. For consistency of notation, the mass started a vertical distance h above the top end of the spring. The mass stops when the spring has been compressed by a distance Δ. Find Δ. Why do you find two possible solutions to your calculation, assuming you did it right? [Hint: The place where the mass comes to a stop ($v = 0$) and the point where its acceleration goes to zero ($a = 0$) *are not the same*.] Why?

6. A mass m sits on top of a frictionless sphere having radius R. A vagrant breeze perturbs the mass, which begins to slide down the slope. It moves faster and faster, until at some point the contact force between the mass and the sphere goes to zero and the mass's trajectory takes it away from the sphere. The position of the mass is defined by an angle θ measured with respect to the vertical through the center of the sphere. At what value of θ does the mass cease to move along the surface of the sphere?

7. A somewhat exotic automobile, when the accelerator is applied, approaches its maximum speed exponentially, so that its speed is given by

$$v = v_0(1 - \exp(-\Gamma t)) \qquad (14.50)$$

where v_0 and Γ are constants and t is the time since the accelerator was first applied. The automobile has mass m. Calculate the power being supplied as the car accelerates toward v_0. You may treat the problem as one-dimensional. Hint: The power depends on t.

8. The electrostatic potential energy between two charges in a sodium chloride solution (at a very low approximation, two protein molecules in dilute solution) may be written as a *screened* Coulomb potential

$$U(R) = \frac{kq_1q_2 \exp(-\kappa R)}{R} \qquad (14.51)$$

where k is a remarkably complicated constant, q_1 and q_2 are the electrical charges on the two protein molecules, κ is a constant called the "Debye (inverse) length", and R is the distance between the two protein molecules. What is the magnitude of the electrical force between the two molecules?

14.10 Homework

1. In a small number of crytals, a central atom is held in place by a force that along some axis varies as the cube of its displacement, i.e., $\mathbf{F} = x^3 \hat{i}$ for displacements along the x-axis. If the atom is moved to a distance $x = A$ from the center, released, and allowed to move to a distance $x = B$ from the center, by how much does the kinetic energy of the atom change as it moves from A to B?

2. Due to an unfortunate accident involving our school's neutrino accelerator, our class has accidentally been transported into a parallel universe. All of the laws of nature are exactly the same here and there, except that the force of gravity of a point mass is now

$$\mathbf{F} = -\frac{G'Mm}{r^3}\hat{\mathbf{k}}. \qquad (14.52)$$

Here $\mathbf{F}$ points straight down, $\hat{\mathbf{k}}$ points straight up, G' is a local constant that replaces our gravitational constant G, and M, m, and r are the masses of the two particles and the distance between them, respectively. In the parallel universe, what is the work done on a object by gravity in moving the object from altitude A to altitude B? If $B > A$, so we are moving the object up, is the work done on the object being moved positive or negative?

3. In the section on types of force, I noted that a magnetic field "...can change the direction of motion but not the speed of a moving charge" because the magnetic force on a charged object always acts perpendicular to the direction of motion of the charge. I might instead have said that a magnetic field "...can change the direction of motion but not the velocity of a moving charge." Is this alternative statement correct or incorrect? Explain your reasoning.

4. If we stay above the surface of the earth, but move away from the earth's center, the earth's gravitational field becomes weaker and weaker. The gravitational acceleration g becomes smaller and smaller. Over modest distances, the gravitational acceleration can therefore be written

$$g(z) = g_0 + \alpha z. \tag{14.53}$$

g_0 is a constant. The constant α gives the linear slope dg/dz of $g(z)$. Calculate the gravitational potential energy of an extended object in the gravitational field described by $g(z)$. (You should end up with one sum or integral that you can't actually do, because you don't know the exact shape of the extended object.)

5. Two mountain climbers are connected by a long rope. The climber having mass m_2 is hanging over the edge of a fortunately short cliff. The climber on mass m_1 is lying on a horizontal, ice-coated, frictionless ledge. The climber on the ledge is also hanging on to a second rope, attached to an immoveable tree. Unfortunately, the second rope is made of a new material that in cold weather becomes elastic; it turns into a spring having spring constant k. The unstretched length of the spring is L. The two ends of the spring are at $x = -L$ and $x = X_A$, where X_A is the location of the climber on the ledge. If the climber were at $x = 0$ the spring would be unstretched. (i) Find the force diagrams for the two climbers. (ii) Solve for the acceleration of the two climbers and the tension in the rope connecting them. (iii) Is the acceleration of the climbers a constant (that is, is it independent of time)? Why or why not (one sentence is plenty as an answer here)? (iv) The two climbers start out stationary, with climber m_1 at X_A. Somewhat later, climber m_1 has been dragged to point X_B, where she is about to go over the edge. How fast is she moving when she reaches X_B? Hint: Note Part iii of this question.

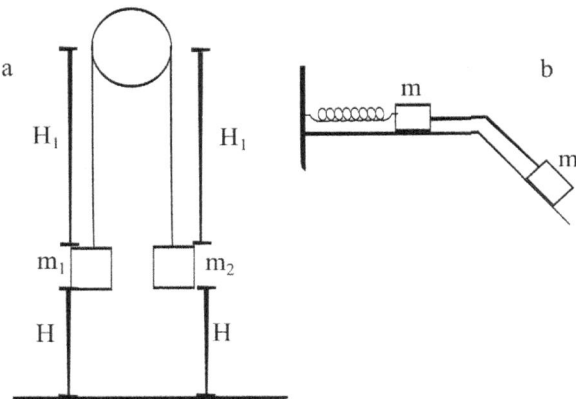

Figure 14.6: Figures for Problems (a) 6 and (b) 7.

6. Two weights (Figure 14.6a) having masses m_1 and m_2 are connected by a light rope stretched over a nearly massless pulley. The weights both begin a distance H off the ground and a distance $H_1 > H$ from the pulley. The weights were chosen so that $m_1 > m_2$. The weights are released from rest; m_1 falls to the ground and pulls m_2 up. (i) Using Newton's Second Law, find the acceleration of m_1 and its velocity at the moment it strikes the ground. (ii) Using energy conservation, find the velocity of m_1 at the moment it strikes the ground. (iii) Using the work-energy theorem, find the velocity of m_1 at the moment it strikes the ground.

7. Two equal masses m (Figure 14.6b) are tied together by a massless rope. One mass rests on a frictionless horizontal surface. The other mass rests on a frictionless ramp that makes an angle α with respect to the horizontal. The top mass is also coupled to a massless spring whose spring constant is k. The spring lies in a horizontal plane; its far end is anchored to the surface and does not move. At the start, the two masses are stationary, while the spring is unstretched. For credit, you must solve using conservation of energy, not the second law or the work-energy theorem. Using the conservation of energy approach: (i) Find the speed v at which each mass is moving, as a function of the distance x that the two masses have moved from their starting positions. (ii) Find the largest distance that the two masses slide before coming to a stop again. The horizontal surface is long enough that the two masses come to a stop before the upper mass reaches the ramp. (iii) Find the distance D that the two masses have moved, at the point at which the speed of the two masses is largest.

8. It is time for yet another scene from the motion picture. This time, the protagonist's twin brother (mass of actor: 70 kg; mass of body armor: 50 kg) is supposed to do a swing across a deep chasm. There is a bridge that partially crosses the chasm. There is a rope, centered directly above the middle of the chasm, fastened to a girder a height h above the bridge surface, making an angle θ with respect to the vertical. The script calls for the actor to step off the bridge (initial speed ≈ 0), swing across to the far side, and step onto the bridge surface at the far side. There is just enough of a breeze blowing to cancel problems with air friction. (i) What is the speed of the actor at the bottom of his swing? (ii) What would be the speed of the actor if he reached the further bridge surface? (iii) Assuming $\theta = 30$ degrees, and that h is 10 m above the bridge surface, find the speed of the actor at the bottom of the swing. (iv) The director has decided that it would be far more artistic if the rope broke at the bottom of the swing. He orders that a shear pin be mounted in the rope. When the tension becomes large enough, the shear pin will break, sending the actor into the water below. For the general case covered by parts (a) and (b) of the problem, what is the tension in the rope at the bottom of the swing? (v) The people installing the shear pin choose a pin that will break when the tension is 10% less than the calculated tension at the bottom of the swing. For the numerical case specified by part (c) of the problem, at what angle does this shear pin break?

9. In yet another scene from our interminable motion picture, the heroine's sidekick attempts to pursue the villain by grabbing a handy rope, length L, and swinging across the gap between two buildings. Unfortunately for the sidekick, at the moment that the rope is vertical it encounters (at a distance $L - \ell$ from its pivot point) an immovable horizontal flagpole. His swing now uses the flagpole as a pivot point. Taking the height of the flagpole off the ground as height zero, what is the highest point of the sidekick's swing?

10. "...which evaluates to ..." equation 14.48. Really? Do the math leading from equation 14.46 to equation 14.48, and see if my final result is correct. In working through the text, you should always do the calculational steps for yourself.

11. A horizontal spring, spring constant k, holds in place a mass m. The spring is initially unstretched. The mass m is free to slide in the horizontal plane. There is no friction in the system. A constant force $F_x \hat{\mathbf{i}}$ is now applied to the mass, tending to compress the spring. Using the work-energy theorem, calculate where the mass come to a stop (the maximum compression of the spring). Hint: Zero net force means zero acceleration, not zero velocity.

12. We have a pendulum, a string of length L with a mass m at its bottom. The mass is released from its start angle θ and swings downward. At the bottom of its swing, the string encounters a nail driven into the wall at a distance x below the original pivot point of the pendulum. The part of the string above the nail comes to a stop. Because the string is nearly massless, the collision of string and nail has no effect on the kinetic energy of the system. The mass continues on its way with no change in its kinetic energy. What is the angle θ_2 at which it comes to a stop on the far side of its swing?

13. Two masses m and M are on opposite sides of a double ramp. They are connected by a nearly massless rope stretched over a nearly massless pulley. The ramp is frictionless. The two sides of the ramp each make an angle θ with respect to the ground. Using conservation of energy, find the speed of each

of the two masses after they have each slid a distance L along their ramp. **This is an extremely important and difficult problem. The standard error, often made by most of the class, is to apply conservation of energy to one of the two masses, writing, e.g., $E = 0.5mv^2 + mgh$. That equation is wrong because the mass m is subject to another external force, namely the tension in the rope. The energy of m changes with time as the rope does work on m and M and transfers energy between them. However, the two masses and the rope put together can be treated using energy conservation.**

14. Masses m_1 and m_2 hang from opposite ends of a rod. The rod is pivoted someplace along its length. The distances from the hangers to the pivot points are ℓ_1 and ℓ_2, respectively. The rod makes an angle θ with respect to the horizontal. (a) Find the total potential energy of the system. Assume that the mass of the rod can be neglected. (b) Find using calculus the values of θ for which U is a maximum or a minimum. Show that your solution finds a neutral equilibrium ($dU/d\theta = 0$) if the masses and lengths are correctly related, and find that relationship.

15. A distinguished Boston Real Estate magnate decides to prove that Boston can outdo Las Vegas by using the wall of the Prudential Tower (height Z) to mount a roller coaster. The roller coaster vehicles are dropped from the top of the tower, fall nearly vertically to the ground, perform a loop-the-loop, and enter Tower Plaza at speed V. What is V? The vehicle then enters a water pool at point A, giving friction with an effective coefficient of kinetic friction μ_k, and comes to a stop at point B. How far is it from A to B?

16. Three masses are connected by ropes and pulleys. Mass m_1 lies on a ramp making an angle θ with respect to the horizontal. Mass m_2 hangs freely in space. Mass m_3 rests on a horizontal frictionless surface. The three masses are not in a state of static equilibrium. When released from rest, they will accelerate in one direction or the other. The mass of the rope is sufficiently small that it can be neglected. **Using energy conservation**, find the speeds of masses m_1, m_2, and m_3 after they have each moved a distance ℓ. The masses were released from rest. The rope between masses 1 and 3 is now replaced with a massless spring. Before releasing the masses, the spring is stretched in such a way that its length does not change significantly with time as the masses accelerate. **Using energy conservation**, find the speeds of masses m_1, m_2, and m_3 after they have each moved a distance ℓ. The masses were released from rest. Also, find the force diagrams and the equations of motion (Second Law) for the masses. Do not solve the equations of motion.

17. A pendulum on a string of length L is initially at an angle θ from the vertical. The pendulum is attached to a wall behind the nail. The bob has a mass m. The bob is released from rest. (i) What is the speed of the bob when it reaches a height H above the bottom of its swing? (ii) What is the speed of the bob when it reaches the bottom of its swing? (iii) A nail has been placed in the wall a distance l directly above the bottom of the swing. The string encounters the nail. The string below the nail continues its swing. The speed of the bob is not changed by the collision. At what angle ϕ does the upwards swing of the bob come to a stop? (iv) What is the tension T in the string immediately before the string strikes the nail?

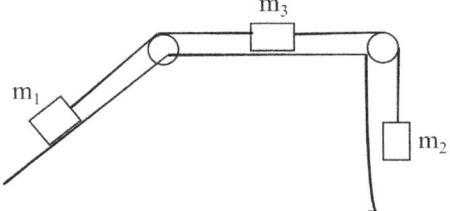

Figure 14.7: The three masses for problem 16

14.11 Solutions to the Worked Problems

1. The force on a charged object between the plates is $\boldsymbol{F} = qE\hat{\boldsymbol{j}}$, with $E = 250$ in SI units. The change in the potential energy is

$$\Delta U = -\int dx F_x - \int dy F_y - \int dz F_z, \tag{14.54}$$

where we have taken advantage of the fact that $\boldsymbol{F}$ is a constant, so that as we take a step in the x direction the values of y and z do not matter, and similarly for the y and z steps. F_x and F_z both vanish, leaving

$$\Delta U = -\int_0^y dy q E_y = -qE_y y. \tag{14.55}$$

In the case at hand,

$$U_f - U_i = -0.9 \cdot 240 \cdot 1 \cdot 10^{-3} J. \tag{14.56}$$

The work done on the charge in moving the charge over the indicated distance is

$$W = -\Delta U = 0.9 \cdot 240 \cdot 1 \cdot 10^{-3} J. \tag{14.57}$$

How did we know that the units of the answer are Joules? We used SI units throughout. U and W both have the dimensions of energy. Therefore, they must have the SI units of energy, which are Joules.

2. We need to start with a little plane geometry. A teeter-totter is a long straight beam with a seat at each end, resting on a perpendicular pivot rod that keeps the teeter-totter from falling to the ground. Each child is some distance ℓ out from the pivot point of the teeter-totter. The teeter-totter itself is a straight beam, which makes an angle θ with respect to the horizontal. Each child is therefore at a vertical distance $\ell \sin(\theta)$ above or below level. For small vertical distances, $U = mgz$. The total potential energy of the two children, who have masses m_1 and m_2, respectively, and sit at distances ℓ_1 and ℓ_2 from the pivot point, must be

$$U = -m_1 g \ell_1 \sin(\theta) + m_2 g \ell_2 \sin(\theta) \tag{14.58}$$

or

$$U = g \sin(\theta)(m_2 \ell_2 - m_1 \ell_1). \tag{14.59}$$

U is independent of θ if

$$m_2 \ell_2 - m_1 \ell_1 = 0, \tag{14.60}$$

i.e., if

$$\frac{m_1}{m_2} = \frac{\ell_2}{\ell_1}. \tag{14.61}$$

If you are thoughtful, you will see the other solution:

If $\ell_1 = \ell_2 = 0$, so the children are sitting on the pivot rod and not the teeter-totter beam, they can have any value for their masses.

3. We apply the definition of the potential energy for a one-dimensional problem

$$\Delta U = -\int dx F(x), \tag{14.62}$$

where here the non-zero component of the force is

$$F_x = -3x - 7x^3. \tag{14.63}$$

For the change in the potential energy from $x = 0$ to $x = X$,

$$\Delta U = \int_0^X (3x + 7x^3) dx, \tag{14.64}$$

14.11. SOLUTIONS TO THE WORKED PROBLEMS

so

$$\Delta U = \frac{3X^2}{2} + \frac{7X^4}{4}. \tag{14.65}$$

This expression agrees that $U = 0$ at $X = 0$, so we may write

$$U(X) = \frac{3X^2}{2} + \frac{7X^4}{4}. \tag{14.66}$$

If we had instead chosen $U = 0$ at $x = A$, the potential energy would instead be written

$$U(X) = \frac{3X^2}{2} + \frac{7X^4}{4} - \frac{3A^2}{2} - \frac{7A^4}{4}. \tag{14.67}$$

4. We repeat our solution of the old problem using the energy conservation approach. In this approach, since there is no friction, the initial and final energies of the system must be equal, i.e.,

$$E_i = E_f, \tag{14.68}$$

where the energy of the system is the sum of the kinetic and potential energies

$$E = K + U. \tag{14.69}$$

The change in the potential energy is

$$U_f - U_i = -\int_{x_i}^{x_f} (3 + 2x)dx, \tag{14.70}$$

which gives us

$$U_f - U_i = -(3x + \frac{2x^2}{2})|_{x_i}^{x_f}. \tag{14.71}$$

Alternatively, we could say that the potential energy at a point x is

$$U(x) = -3x - \frac{2x^2}{2} + U_o, \tag{14.72}$$

where U_o is the potential energy at $x = 0$ and $U(x)$ is calculated from the change in the potential energy between the reference point and the point x.

The law of conservation of energy may then be written as

$$K_i + U_i = K_f + U_f, \tag{14.73}$$

or in this case

$$K_i - 3x_i - \frac{2x_i^2}{2} + U_o = K_f - 3x_f - \frac{2x_f^2}{2} + U_o, \tag{14.74}$$

which may be rearranged as

$$K_f - K_i = 3x_f + \frac{2x_f^2}{2} - U_o - 3x_i - \frac{2x_i^2}{2} + U_o. \tag{14.75}$$

The U_o terms cancel.

We had $x_i = 15$ and $x_f = 10$. Substituting those numbers into the above leads to

$$K_f - K_i = -140J, \tag{14.76}$$

so K falls by 140 J during this process. However, K_f cannot be less than zero, so K must be ≥ 140 J.

5. We have a mass M on a ramp having length L. The ramp makes an angle θ with the horizontal. The mass collides with a massless spring having unstretched length ℓ and spring constant k. The mass sticks to the spring. Where does the mass stop?

We can do this with energy conservation: $E_i = E_f$. The energy of the system is

$$E = Mgz + \frac{1}{2}Mv^2 + \frac{1}{2}kx^2. \tag{14.77}$$

What information do we have? Initially, the mass is stationary ($v = 0$) at the top of the ramp ($z = L\sin(\theta)$), while the spring is unstretched ($x = 0$) At the end, the mass is again stationary ($v = 0$), the stretch of the spring is $x = X$, and the height of the mass off the ground is $z = \ell\sin(\theta) + X\sin(\theta)$. X may be positive or negative. If we substitute all these quantities into the equation for the total energy of the system, we obtain

$$MgL\sin(\theta) + \frac{1}{2}M0^2 + \frac{1}{2}k0^2 = Mg(\ell\sin(\theta) + X\sin(\theta)) + \frac{1}{2}M0^2 + \frac{1}{2}kX^2 \tag{14.78}$$

where I have displayed every term, including the terms that turn out to be zero, so that there is absolutely no doubt as to what substitutions were made.

Simplifying, the conservation of energy equation becomes

$$\frac{1}{2}kX^2 + XMg\sin(\theta) - MgL\sin(\theta) + Mg\ell\sin(\theta) = 0. \tag{14.79}$$

The simplified equation is a quadratic in X. Its solution is

$$X = \frac{-Mg\sin(\theta) \pm ((Mg\sin(\theta))^2 - 4 \cdot \frac{1}{2}k(\ell - L)Mg\sin(\theta))^{1/2}}{\frac{1}{2} \cdot 2k}. \tag{14.80}$$

We started with a quadratic, so we should have two answers. What do the two answers mean? There's a certain tendency to say that if there are two answers, one of them must be unphysical, but that's not the correct response here. The mass has collided with the spring, is now stuck to it, and therefore the spring is alternately compressed and stretched. The mass will oscillate back and forth between the lower and upper points where the spring and gravity bring the mass to a stop, these two points being the two roots of the quadratic.

6. The block is sliding down the sphere. It is subject to two forces, namely a gravitational force $mg\hat{\boldsymbol{k}}$ and a normal force $\boldsymbol{N}$ perpendicular to the surface of the sphere. The block is performing circular motion, so to stay on the surface of the sphere the block must be accelerating toward the center of the sphere with radial acceleration v^2/R. As the block descends, its speed and hence its inward radial acceleration increase. The only inward force is the radial component of the force of gravity, which goes to zero when the block is half way from the top to the bottom of the sphere.

We can create a force diagram for the descending block. θ is the angle between the vertical and the radius vector from the center of the sphere to the block. From that, we can write the Second Law for the radial motion of the block as

$$N - mg\cos(\theta) = -\frac{mv^2}{R}. \tag{14.81}$$

How fast is the block moving? The normal force acts perpendicular to the direction of motion, so it does no work on the block. Only gravity does work on the block, so we can calculate the speed of the block from conservation of energy $E_i = E_f$. The energy of the block is $E = mgz + \frac{1}{2}mv^2$. At the top, on release, $v = 0$ and $z = 2R$. At the moment the block departs from the sphere, $v = V$ and $z = R + R\cos(\theta)$. We may then write from conservation of energy

$$mg \cdot 2R + \frac{1}{2}m0^2 = mg(R + R\cos(\theta)) + \frac{1}{2}mV^2, \tag{14.82}$$

14.11. SOLUTIONS TO THE WORKED PROBLEMS

which leads to
$$V^2 = 2gR - 2gR\cos(\theta). \tag{14.83}$$

Just as the block lifts off from the sphere, the gravitational force component has reached its limit in creating the acceleration, so the normal force goes to zero, meaning

$$-mg\cos(\theta) = -\frac{m(2gR - 2gR\cos(\theta))}{R} \tag{14.84}$$

or
$$2gR = 3gR\cos(\theta), \tag{14.85}$$

meaning $\cos(\theta) = 2/3$ and
$$\theta = \arccos(2/3). \tag{14.86}$$

7. The power is
$$P = \boldsymbol{F} \cdot \boldsymbol{v}. \tag{14.87}$$

From the Second Law, the force is
$$\boldsymbol{F} = m\frac{d\boldsymbol{v}}{dt}. \tag{14.88}$$

We have the velocity as $v = v_o(1 - \exp(-\Gamma t))$, which leads in one dimension to
$$\frac{dv}{dt} = v_o \Gamma \exp(-\Gamma t), \tag{14.89}$$

and therefore to a power
$$P = mv_o\Gamma\exp(-\Gamma t)v_o(1 - \exp(-\Gamma t)), \tag{14.90}$$

which simplifies to
$$P = m\Gamma v_o^2 \exp(-\Gamma t)(1 - \exp(-\Gamma t)). \tag{14.91}$$

8. The potential energy of two small charged bodies in a salt solution is
$$U(r) = \frac{kq_1 q_2 \exp(-\kappa r)}{r}. \tag{14.92}$$

The magnitude of the force between the bodies is
$$F = |\frac{\partial U}{\partial r}| = |-\frac{kq_1 q_2 \exp(-\kappa r)}{r^2} + kq_1 q_2(-\kappa)\exp(-\kappa r)r|, \tag{14.93}$$

or
$$F = |\frac{kq_1 q_2 (1 + \kappa r)\exp(-\kappa r)}{r^2}|. \tag{14.94}$$

This space reserved for your notes.

This page reserved for your notes.

Chapter 15

Energy Conservation, Collisions, and Friction

15.1 Introduction

In this Chapter we consider the effect of collisions on energy conservation. Collisions are not that common in nature, but some of them are very important. For example, the power output of the sun, without which the earth would be a frozen ball covered with ice, arises from collisions between hydrogen atoms and the traces of carbon, nitrogen, and oxygen in the core of the sun. Physicists are fond of collisions, because we use them to study the forces between the particles contained in every atom. The first two interstellar spacecraft, Voyagers 1 and 2, were launched from Florida decades ago, but attained the speed needed to escape from the solar system by means of collisions – well, close passes – near Jupiter.

There is a simple ideon, a simple Platonic ideal, of a collision. There are also more complicated collisions. We have two inbound particles. They approach each other, something happens, and then the two objects head out in two new directions. "Something happens" describes the two objects passing close to each other. However, in the collision ideon, we treat the "something happens" as a black box. We do not look inside the box. Once the objects collide and then emerge from the black box, they are assumed to have stopped applying forces on each other. For example, if we bounce two automobiles off each other, there is a potential energy and frictional forces while they are bouncing, but once they have bounced, they stop putting forces on each other. The interactions between the colliding cars go to zero once the collision is over. The collision ideon requires that the collision time is very short, so that if external forces are acting on the objects, then during the collision the external forces do not have time to have much effect on the objects in the collision.

If two colliding objects have a joint potential energy, a pair potential, that potential is approximated as going to zero once the collision has ended. If the particles are charged, or if gravitational forces are significant during the collision, there is a complication. The electrostatic and gravitational potential energies fall off with distance very slowly, namely inversely in the first power of the distance, so the interactions only disappear asymptotically at large distances.

In all collisions, momentum is conserved. The total momentum immediately before a collision and the total momentum immediately after a collision are always equal to each other. If the objects in a collision are also being acted upon by external forces, those external forces cause the system's momentum to change with time, but during the collision the change in the total momentum due to the external forces is taken to be negligible. 'Negligible' is an approximation. With care, you can move beyond the approximation.

In addition to momentum, we can also talk about energy. The mechanical energy of colliding objects is the total of $\frac{1}{2}mv^2$ for all the objects in the collision. There is also the true total energy of the system which includes not only the mechanical energy but also heat, chemical energy, and other sorts of energy. The true total energy of the system, including all colliding objects and their interactions, is always conserved. However, sometimes it is difficult to determine a system's total energy. If two cars collide and parts become bent, the parts also become warmer. Calculating how much warmer the parts become can become an extremely challenging calculation.

We divide collisions into two general classes, elastic and inelastic. In an elastic collision, the mechanical energy of the system is not changed by the collision. We say that the mechanical energy is conserved in an elastic collision. In an inelastic collision, the mechanical energy of the system changes during the collision process. We then say that in an inelastic collision the mechanical energy of the system is not conserved. The total energy of a complete system is always conserved. In an inelastic collision, the mechanical energy of the system changes. Either some of the mechanical energy goes someplace, for example by being changed into heat, or some mechanical energy comes from someplace.

Many processes can be described as collisions. For example, if we have balls on a billiard table, bouncing back and forth, the collisions are almost instantaneous. Between the collisions, the balls just travel in straight lines at almost constant speed. I say 'almost', because the billiard balls are subject to rolling friction, which gradually slows them down. In Maxwell's description of low pressure gases, this being the kinetic theory of gasses, gas atoms spend most of their time travelling in straight lines, interrupted every so often by near-instantaneous collisions with other gas atoms. During the collisions, the gas atoms exchange momentum and mechanical energy. In the simple kinetic theory model, the collisions between gas atoms are taken to be entirely elastic; the total mechanical energy of colliding gas atoms does not change during the collision. On the other hand, in liquids every molecule is exerting forces more or less constantly one its neighbors, so molecules in a liquid do not travel in straight lines with occasional interruptions due to collisions. Maxwell's kinetic theory is inapplicable to liquids; an entirely different alternative treatment of molecular motion in liquids is required.

15.2 Examples of Collisions

We now treat a few simple examples of collisions. These examples all involve motion in a single dimension. Space is three-dimensional, but during these exemplary collisions the only motion that occurs is along the x-axis.

Consider first two masses sliding toward each other. Their front ends are both coated with superglue. When they reach each other, they stick, and afterwards move as a single object. For simplicity, the two objects each have mass m. Their velocities are $+V\hat{\mathbf{i}}$ and $-V\hat{\mathbf{i}}$, respectively. What is the velocity of the two objects after the collision? What happened to the mechanical energy of the objects as a result of the collision?

How do we solve this? First, momentum is always conserved. The initial momentum of the system $\mathbf{p}_i$ must always equal the final momentum $\mathbf{p}_f$. The total momentum, before or after the collision, can be obtained by summing the momenta of the separate moving objects. We therefore can write

$$\mathbf{p}_i = \mathbf{p}_f, \tag{15.1}$$
$$mV\hat{\mathbf{i}} - mV\hat{\mathbf{i}} = (m+m)\mathbf{v}_f. \tag{15.2}$$

With minimal additional calculation, $\mathbf{v}_f = \mathbf{0}$. After the collision, the fusion of the two masses is stationary.

What happened to the kinetic energy? Before the collision, the kinetic energy was

$$K = \frac{1}{2}mV^2 + \frac{1}{2}mV^2. \tag{15.3}$$

Kinetic energy is a scalar, not a vector. K is always a positive number or zero. The kinetic energies of the two masses are both positive numbers and simply add, giving

$$K = mV^2 \tag{15.4}$$

for the total kinetic energy of the two masses before the collision.

After the collision, the two masses are stationary, so $K_f = 0$.

Where did the energy go? After all, the total energy of the system is conserved. However, the mechanical energy of the system has gone from mV^2 to zero. To good approximation, the mechanical energy of the two masses was converted entirely into heat. In the conversion process, when the two blocks collide, they compress each other's surface, setting up pressure (sound) waves inside the two masses. The sound waves

15.2. EXAMPLES OF COLLISIONS

bounce back and forth inside the solids, breaking up into sound waves of various frequencies and wave lengths, until all the atoms in each solid are in irregular motion. This irregular motion is heat.

Suppose we instead insist that the collision is elastic, so that $K_i = K_f$. We'll consider the general case in which the two masses are m_1 and m_2. All motion is along the x axis, so we only need to consider the x-components of the velocities, namely v_{1x} and v_{2x} before the collision and V_x after the collision. What does that tell us about the collision?

First, momentum is conserved, so as above

$$m_1 v_{1x} + m_2 v_{2x} = (m_1 + m_2) V_x. \tag{15.5}$$

If the collision is elastic, then the amount of mechanical energy is not changed by the collision process, so

$$\frac{1}{2} m_1 v_{1x}^2 + \frac{1}{2} m_2 v_{2x}^2 = \frac{1}{2}(m_1 + m_2) V_x^2. \tag{15.6}$$

We now have two equations, but we have three unknowns, v_{1x}, v_{2x}, and V_x. We can therefore use the equations to eliminate two of the unknowns, meaning that when we are done two of the three unknowns will be written in terms of the third unknown. We want to be learn something about what combinations of initial velocities v_{1x} and v_{2x} lead to an elastic collision, so we start by eliminating V_x. To see this more clearly, solve the equation for momentum conservation for V_x, getting

$$V_x = \frac{m_1 v_{1x} + m_2 v_{2x}}{m_1 + m_2}. \tag{15.7}$$

Use this result to eliminate V_x from the equation for energy conservation, giving

$$\frac{1}{2} m_1 v_{1x}^2 + \frac{1}{2} m_2 v_{2x}^2 = \frac{1}{2}(m_1 + m_2) \left(\frac{m_1 v_{1x} + m_2 v_{2x}}{m_1 + m_2} \right)^2. \tag{15.8}$$

On the right-hand-side of the equation, one factor of $m_1 + m_2$ cancels above and below. The common factor $1/2$ can be multiplied out. Simplify by taking the other $m_1 + m_2$ from the denominator on the right to the numerator on the left, giving

$$m_1(m_1 + m_2) v_{1x}^2 + m_2(m_1 + m_2) v_{2x}^2 = m_1^2 v_{1x}^2 + 2 m_1 m_2 v_{1x} v_{2x} + m_2^2 v_{2x}^2. \tag{15.9}$$

If we eliminate the terms common to the left and right hand sides of this equation, and from the remaining terms divide out the common factor $m_1 m_2$, we reach

$$v_{1x}^2 - 2 v_{1x} v_{2x} + v_{2x}^2 = 0. \tag{15.10}$$

or

$$v_{1x} = v_{2x}. \tag{15.11}$$

The only way we can satisfy conservation of momentum and conservation of energy at the same time, and have two objects fused into one after the collision, is to have the two objects moving with the same velocity before the collision. The only way two objects having the same velocity can be said to have collided and stuck together after the collision point is if they were stuck to each other before the collision.

Our result is true, but mostly interesting in a negative sense. What we just showed is that if we have a collision, and two separate objects fused into one, the collision must not have been elastic. There's only one other choice: If the collision was not elastic, it must have been inelastic.

The above picture of a process can also be run backwards in time. We can imagine starting with a single object having mass $m_1 + m_2$ moving at V_x. At some moment, it splits into two objects, m_1 and m_2, moving at v_{1x} and v_{2x}, respectively. The momentum and energy conservation equations would be exactly the same as in the above calculation, except the before and after sides of the two starting equations would be reversed. We would come to the same conclusion: If one object actually splits into two, the two going off in different directions, the process must be inelastic. Energy was not conserved during the process.

The above calculation also illustrates a general rule on solving a set of equations. Once you use an equation in the solution process, to eliminate some variable, you aren't in general entitled to turn around and use the same equation to eliminate a second variable. We had three variables, but only two equations, so we could not solve for all three variables. All we could so was to solve for two of the variables (V_x, v_{2x}) in terms of the third variable (v_{1x}), though that solution turned out to be enough to give an interesting result, namely that if after a collision two objects fuse into one the collision must have been inelastic. If the process had been elastic, masses m_1 and m_2 must have been fused at all times.

Let's consider a second collision problem, involving two masses that collide with each other, but this time after the collision they move away separately. Once again we'll confine motion to a single dimension. Keeping things simple, the two masses are m_1 and m_2 with $m_1 = m_2$. The first object has some velocity $v_{1x}\hat{\mathbf{i}}$, while the second object has initial velocity $\mathbf{0}$. A billiard ball rolling across a pool table until it collides with a second billiard ball would be an example of this situation. The collision is so arranged that after the collision the two balls only move parallel to the x-axis.

We again have two equations, corresponding to conservation of energy and conservation of the x component of the momentum. However, this time we have two unknowns, the x-components v_{1f} and v_{2f} of the velocities of the two particles. We reasonably expect to be able to find a solution for each of the unknowns in terms of the given variables.

Momentum conservation gives us for the x-components of the momentum vectors

$$mv_{1x} = mv_{1f} + mv_{2f}. \tag{15.12}$$

Before the collision, mass 2 is stationary, so its momentum is $\mathbf{0}$, so the corresponding momentum term $m_2\mathbf{0}$ is in principle present but is not displayed in the above equation.

For conservation of energy, we have

$$\frac{1}{2}mv_{1x}^2 = \frac{1}{2}mv_{1f}^2 + \frac{1}{2}mv_{2f}^2. \tag{15.13}$$

Reminder: In this problem, the potential energy is zero at all times. The above terms represent the total kinetic energy of the particles before and after the collision. We only see the x-components of the velocities, because the y and z components of the velocities are all zero, so they make no contribution to the kinetic energy. Equation 15.13 might appear to refer to a mythical "x-component of the kinetic energy", but that's an illusion. The kinetic energy is a scalar. It does not have x, y, or z components.

The momentum equation lets us solve for v_{2f}, namely

$$v_{2f} = v_{1x} - v_{1f}. \tag{15.14}$$

Substituting this result in equation 15.13, we have

$$\frac{1}{2}mv_{1x}^2 = \frac{1}{2}mv_{1f}^2 + \frac{1}{2}m(v_{1x} - v_{1f})^2. \tag{15.15}$$

Cancelling the common factor $m/2$, we have

$$v_{1x}^2 = v_{1f}^2 + v_{1x}^2 - 2v_{1x}v_{1f} + v_{1f}^2 \tag{15.16}$$

or

$$2v_{1f}^2 - 2v_{1x}v_{1f} = 0. \tag{15.17}$$

This equation is a quadratic in v_{1f}. It has two solutions. We show the solutions, and what they imply for the velocity of the other particle.

$$v_{1f} = 0 \Rightarrow v_{2f} = v_{1x}, \tag{15.18}$$
$$v_{1f} = v_{1x} \Rightarrow v_{2f} = 0. \tag{15.19}$$

The first solution shows that the first mass simply comes to a stop, and the second mass leaves with the original velocity of the first mass. This is the outcome typically desired in billiards: One ball hits another. The first ball stops. The second ball heads off across the table. But what does the second solution even mean, since it says that the second mass simply remains stationary? This is alas the outcome when your author attempts to play pool, namely the first ball completely misses its intended target and keeps on going, and the intended target remains stationary on the table.

Let's consider a further example of the rule on comparing the number of unknowns with the number of equations that we have available to solve for them. We consider a collision between two objects, in which the motion occurs in more than one dimension. We choose a coordinate system in which particle 2 is initially stationary ($\mathbf{v}_{2i} = 0$), particle 1 strikes it, and afterward the two particles depart at velocities $\mathbf{v}_{1f}$ and $\mathbf{v}_{2f}$, making angles θ and ϕ with respect to the original direction of particle 1. The appears to be a two-dimensional collision, but in fact the description is entirely valid for a three-dimensional collision between two objects. Why? First, the two vectors $\mathbf{v}_{1f}$ and $\mathbf{v}_{2f}$ define a plane, because any two vectors define a plane. Second, the motion of particle 1 before the collision is obliged to lie in this plane. Why? Call this plane the $x - y$ plane, and the axis perpendicular to this plane the z-axis. By construction, the velocity vectors of particles 1 and 2 after the collision lie in the $x - y$ plane, so neither of them has a non-zero momentum component in the z direction. The total momentum component in the z-direction, after the collision, must therefore be zero. From momentum conservation, before the collision, the total momentum parallel to the z axis must also have been zero. Before the collision, particle 2 was stationary, so its momentum components were all zero. From momentum conservation, there is then no way before the collision for particle 1 to have had a non-zero momentum component in the z direction, so the velocity of particle 1 before the collision must have been confined to the $x - y$ plane, the plane of the paper. The problem is indeed two-dimensional.

The masses and velocities before the collision of particles 1 and 2 are given in the problem. After the collision, the motions of the two particles are described by four variables, for example their final speeds v_{1f} and v_{2f} and the final angles θ and ϕ. Corresponding to those variables, we have only three equations, namely conservation of energy and momentum conservation for the x and y components of the momentum. (The z component of the momentum is also conserved, but we have already shown that the z component of the momentum is zero for both particles, both before and after the collision, so that equation corresponds to variables for which we have already obtained a solution.) Because we have four unknowns, but only three equations, we can in principle solve for any three of the unknowns in terms of the fourth, but we cannot solve for the all four unknowns.

15.3 Discussion

We've now considered elastic and inelastic collisions. Collisions can be of scientific or practical interest. They also provide good examples of cases in which conservation laws play a central role. Elastic and inelastic collisions differ as to which conservation laws are applicable.

In the absence of external forces, momentum is always conserved. That statement is equally true in elastic and in inelastic collisions. The collision ideon presumes that the duration of a collision is extremely short. If an external force is acting on colliding particles, during the collision the external force does not have time to change the total momentum of the system, at least not appreciably. The effect of external forces on the trajectories of colliding particles, after a collision, may be calculated by assuming that the collision takes precisely no time.

How do we tell if a collision was elastic or inelastic? The important feature of an inelastic collision is that it changes the kinetic energy of a system. If we have two particles going into a collision, but only one coming out, or one particle going in but two coming out, we can be certain that the collision was inelastic. If two particles have an elastic collision with each other, they must emerge from the collision as two separate bodies going in two separate directions. However, a collision between two bodies, in which the two bodies emerge going in different directions, may be either elastic or inelastic. The core question is whether any process changes the mechanical energy of the system, for example by turning kinetic energy into heat.

If there is a collision, the true total energy of a system is not changed. Some of the system's energy may change from mechanical energy to heat or to potential energy (for example if a moving mass collides with a spring and compresses it), but the true total energy of the system is not changed by a collision process.

15.4 Worked Problems

1. In yet another scene from the motion picture, the director stages a car crash, complete with smoke, loud noises, acrobatics, and high explosives hidden at the impact point. Entering stage left we have the villain's 3000 kg convertible SUV traveling at $30\hat{i} + 40\hat{j}$ m/s and the heroine's 1000 kg convertible sports car traveling at $50\hat{j}$ m/s. The quoted masses of the vehicles include the masses of the people in the cars. The two vehicles run into each other. A half-dozen of the villain's henchmen (mass, 700 kg) jump from his car to the heroine's car; the explosives are then detonated. At scene close, the sports car with seven people on board is exiting the collision point at $100\hat{j}$ m/s, while the villain's vehicle departs at $23\hat{i}$ m/s. What was the total work done on the two vehicles and their contents during the scene, which you cannot approximate as a collision?

2. Consider the totally elastic collision between two billiard balls. To simplify the math, all motion is parallel to the x-axis. The balls both have mass m. The initial velocities of the two balls are $v_{1i}\hat{i}$ and 0; the final velocities are $v_{1f}\hat{i}$ and $v_{2f}\hat{i}$. For a totally elastic collision, compute the two final velocities. (Aside: there are actually two answers. Explain.)

15.5 Homework

1. It is a typical Worcester ice storm, and a truck (mass 20,000 kg, speed of 5m/s) is cautiously going North on Park Avenue (North is the $+y$ direction). At the same time, a sports car (mass 2000 kg) is being driven East (the $+x$ direction) on Salisbury street. To avoid driving in a dangerous ice storm any longer than necessary, the driver of the sports car is going as fast as possible, namely 70 m/s. Inevitably, there is a collision in the intersection, leaving the sports car imbedded in the rear body of the truck. Calculate the velocity of the truck and car, which are moving together, immediately after the collision. You may neglect any mass associated with parts, e.g. tires, that head off in their own directions rather than staying with the truck and car. If Salisbury pond is 100 m away along the direct path, and friction is negligible, and the drop to the pond is negligible, how long will it take them to hit the water?

2. One of the more-thoroughly-forgotten episodes in the development of motor vehicles was the spring-cannon-propelled wagon. In the original and totally unsuccessful design, a wagon of mass M has loaded on board a solid marble cannon ball of mass m. The cannon ball is pushed against a spring of spring constant k that has been compressed through a distance x. To move the cart, the spring is released, sending the cannon ball flying out of the rear of the wagon (1). (a) Find the speed of the wagon and of the cannon ball the moment after the cannon ball has sailed off the rear of the wagon. (b) Find the kinetic energy of the wagon and the cannonball immediately after the cannonball has been sent flying.

   (1) An epic fail. This propulsion technique does, however, discourage tailgating.

3. It is the year 1801. You have been tasked by President Jefferson with measuring the speed of a musket bullet. You have the minor difficulty that photography, stop watches, and electronics have not yet been invented. Your solution is the ballistic pendulum. The musket is locked to a table and fired at close range at a large block of timber. The block of timber, mass M, hangs from the ceiling on fine threads. The block of timber swallows the musket ball (mass m) and recoils, eventually rising to a height that can be measured directly. (TRICK: fine cotton threads that break if the block goes through them.) Calculate the original speed of the musket ball if the block of wood rises to a height H.

4. Two cream pies of equal mass m and equal initial speeds v slide across a frictionless surface. They are not moving parallel to each other. Instead, they collide. They stick to each other and slide off along the x-axis toward infinity. (a) If their initial speeds are three times their final speed, find the angle between either of their initial velocities and their final velocity. (b) Initially the two pies had a total kinetic energy that you can calculate. What fraction of the initial kinetic energy of the two pies was turned into heat by the collision? (c) As the two cream pies approach the collision point, their velocity vectors each make an angle with respect to the x axis. Why must these two angles be equal to each other?

15.5. HOMEWORK

5. Late one evening, a pair of volksraketenwagons are cruising through Worcester, one coming north on Park Avenue and the other east on Salisbury Street. Both are steered (loosely speaking) by good Massachusetts drivers who do not believe in traffic lights. Each sees the other, but assumes that he is seeing a reflection, since each believes he has the only volksraketenwagon in Massachusetts. Each vehicle masses 4000 kg and has a speed of 30 m/s. The two vehicles both proceed at constant speed into the intersection. The vehicles collide. Fortunately, volksraketenwagons have external air bags, which deploy between the vehicle and the objects they are about to hit. The two vehicles bounce off each other and (after some steering) continue on their way without significant damage. The first vehicle leaves the collision point at 28 m/s in a direction 75 degrees west of north. The other vehicle leaves the intersection at some other direction and speed. (i) Compute the speed and direction of the second volksraketenwagon immediately after the collision. (ii) Was this collision elastic or inelastic? Compute the change in the total kinetic energy of the two cars as a result of the change? ["zero" is a change, so I did not just tell you whether or not the collision was elastic.]

6. A true story. It is 1947. A good friend of mine, unfortunately no longer with us, is driving his freshly purchased used car (mass, with friend and fuel, 1500 kg) due west at 13 m/s, someplace in Boston. Entering an intersection, he observed a car running a stop sign, entering toward him at an angle of 60 degrees (the 2 o'clock direction) relative to his direction. The other car had a mass of 2000 kg and was travelling at 10 m/s. Fortunately, my friend's military training immediately took over. Unfortunately, his training was as pilot of an Army Air Force B-24 heavy bomber, and as a collision avoidance scheme "pull back on the wheel and go to emergency power" works better with aircraft than with automobiles. However, he was travelling at 15 m/s at the moment of collision. The bumpers locked, and the two cars skidded as a unit across the street, fortunately not hitting anyone. (a) Was this collision elastic or inelastic? (b) Find the kinetic energy of the two automobiles before and after the collision. (c) Find the direction and speed that the two automobiles were travelling after the collision.

7. It is mid-winter, and Worcester is in the grip of an ice storm. Two motorists, eager to enhance traffic safety by spending as little time as possible on the road, are driving at maximum attainable speeds along Park Avenue (north, the $+\hat{i}$ direction) and Salisbury Street (east, the $-\hat{j}$ direction). Their vehicles have equal masses 2000 kg. The vehicles collide and stick together. (i) After the collision, the velocity of the wreckage, all 4000 kg of it, was $v = 20\hat{i} - 25\hat{j}$ m/s. What were the initial velocities of the two vehicles? (ii) What was the change in the kinetic energy of the two vehicles due to the collision? (iii) Soon thereafter, a third driver notices that the light has changed. Her vehicle has a mass of 3000 kg. Fortunately, she is driving a volksraketenwagon (the rocket car from an early lecture), and fires up the rocket motor. She is headed south on Park Avenue on horizontal, frictionless pavement. The velocity of her car is $\mathbf{V} = -(0 + 2t + 0.5t^2)\hat{i}$ m/s. As a function of time, what power is being supplied to the car by the rocket engine, assuming that no other horizontal forces are acting on the car?

8. It is a typical winter day in Worcester, and the interesting drivers are out in full force. A 2000 kg car proceeding due east on Salisbury street at 5 m/s enters the intersection at the same time as a 3000 kg SUV. The SUV was traveling due north at 10 m/s.

(i) After the collision, the two vehicles have stuck to each other. Assuming no significant mass loss, calculate the velocity (speed and direction) of the stuck-together vehicles immediately after the collision.

(ii) A second pair of vehicles, identical with the first pair, enter the intersection. This pair of vehicles is being filmed for our interminable motion picture, and various spring bumpers, explosive charges, etc. as needed have been mounted on the two automobiles, as may or may not have been needed, so as to ensure that the two vehicles bounce off each other as required by the script. After the collision, the first car is proceeding due north, and the second car is proceeding due east. How fast are they each moving? Was this collision elastic or inelastic?

9. A rock of mass M slides across wet, frictionless ice with velocity $\mathbf{V} = V\hat{i}$. It strikes a second rock, mass m, that sits on the ice, not moving. The second rock slides off in a direction that makes an angle θ with respect to the x-axis. Calculate its momentum $\mathbf{p}$ in terms of M, m, V, and θ. Calculate the momentum $\mathbf{P}$ of the first rock, after the collision, in terms of M, m, V, and θ. If you did not know θ, but did know M, m, and $\mathbf{V}$, could you calculate $\mathbf{p}$ and $\mathbf{P}$?

10. You are again a consultant specializing in collisions and other disasters involving high speed vehicles. Once again the corner of Park Avenue and Salisbury Street has endured a high-speed collision. While immediately before collision the client's car was going due east (the $+\hat{j}$ direction) at 10 m/s, witnesses do not agree as to which way or how fast the other vehicle was going before the collision, including witnesses who maintain the other vehicle had entered the intersection via a neighboring lawn. Fortunately, both vehicles had the same mass, 1000 kg. (i) After the collision, the velocity of the wreckage, all 2000 kg of it, was $v = 10\hat{i} + 15\hat{j}$ m/s. What was the initial velocity of the other vehicle? (ii) What was the change in the kinetic energy of the two vehicles due to the collision? (iii) Soon after the first collision two more vehicles entered the intersection. Their masses were m and $2m$. They were travelling north and south, at speeds $2v$ and 2, respectively, but in the same lane, so this was a head on collision. Their bumpers were made of a new experimental plastic, so their collision was elastic. The cars simply bounced off each other, staying in the same lane. Each exited the intersection, the m car now going south and the $2m$ car now going north. What were the velocities of the two vehicles after the collision?

15.6 Problem Solutions

1. We start with the work-energy theorem $W = K_f - K_i$ with $K = \frac{1}{2}mv^2$. For once, there is little to do other than insert numbers.

$$W = \frac{1}{2}(2300(23^2) + 1700(100^2) - 3000(50^2) - 1000(50^2)),$$

which leads to $W = 4.11$ MJ.

2. We consider two billiard balls making a dead center collision. The first ball initially has speed v_{1i}; the other ball is stationary. At the end, the velocities of the two balls are v_{1f} and v_{2f} where, since this is a one-dimensional problem, the velocities are entirely described by their indicated x-components. The collision is elastic, so energy is conserved. We can write the equations for conservation of momentum and conservation of energy as

$$mv_{1i} = mv_{1f} + mv_{2f}, \qquad (15.20)$$

$$\frac{1}{2}mv_{1i}^2 = \frac{1}{2}mv_{2i}^2 + \frac{1}{2}mv_{2f}^2. \qquad (15.21)$$

The first of these equations leads to

$$v_{1i} = v_{1f} + v_{2f}.$$

Using this equation to eliminate v_{1i} from the conservation of energy equation, we obtain

$$\frac{1}{2}m(v_{1f} + v_{2f})^2 = \frac{1}{2}mv_{1f}^2 + \frac{1}{2}mv_{2f}^2.$$

On expanding and cancelling terms, we get

$$v_{1f}v_{2f} = 0,$$

which means that either $v_{1f} = 0$ or $v_{2f} = 0$.

If $v_{1f} = 0$, then from conservation of momentum $v_{1i} = v_{2f}$. The first billiard ball strikes the second, and is brought to a stop, and the second billiard ball goes on its way with the velocity originally had by the first billiard ball.

On the other hand, If $v_{2f} = 0$, then from conservation of momentum $v_{1i} = v_{1f}$. The first billiard ball completely misses the second and keeps on rolling.

Chapter 16

Examination 2

Second Hour Examination

This is the Second Hour Examination. There are three questions, each worth some number of points, for a total of 100 points, and an extra credit question worth ten points. You are urged to read all of the questions before answering any of them.

This s a closed book examination. You may not use notes, reference books, crib sheets, electronic devices other than pocket calculators being used for the purpose of numerical calculation, or other material memory aids in your work. You may not consult with other persons during the examination, except to ask the proctor what the examination questions mean.

Begin from first principles, e.g. $\boldsymbol{F} = d\boldsymbol{p}/dt$. Show all work. Physics is a matter of praxis, not faith. You are graded on how you reach your answers, not whether or not they happen to be right. Do not invoke complex memorized formulae for, e.g., the moment of inertia of an icosahedron. If the final answer is numerical, it should be presented in a valid form.

I. [Total of 30 points] A chalkboard eraser slides along an unpolished wooden surface. The coefficient of kinetic friction depends on the local surface roughness, which is a function of position, so that

$$\mu_k(x) = \mu_0 + \mu_1 \exp(x/\ell). \tag{16.1}$$

Here μ_0, μ_1, and ℓ are constants. The eraser has an initial speed v_1 and initial position $x = 0$.

(A) [10 points] Find the kinetic energy of the eraser after it has slid a distance L along the surface. Assume that the kinetic energy of the eraser is still positive after this slide.

(B) [10 points] After completing the slide, the eraser has a speed v_2. It now slides down a tilted ramp. The ramp is frictionless, and has a total vertical drop of H. Using energy conservation, find the final speed v_2 of the eraser.

(C) Repeat the calculation of part (B), this time using the work-energy theorem to find the final speed v_2 of the eraser.

(D) Do your calculations in parts (B) and (C) give the same answer? Should they? Why?

II. [Total of 30 points] It is a typical day on Route I-190). A first automobile, mass 3000 kg, is going due north in the right hand lane with a speed of 30 m/s. A second automobile, mass 2000 kg, is coming down an entrance ramp (direction 30 degrees west of north) at a speed of 40 m/s, to merge onto I-190 from the right. The first driver has decided to advance in a straight line at constant speed. The second driver has decided to advance in a straight line at constant speed until he is in the left hand lane. Unfortunately, the two cars reach the same point in the right hand lane at the same time and collide, locking bumpers and cars doors in such a mess that afterwards they proceed as a single fused object.

(A) [10 points] What is the velocity $\boldsymbol{V}$ of the fused object after the collision? Ignore any loss of mass arising from small parts falling off the two cars.

(B) [10 points] What is the fractional loss, relative to the initial total kinetic energy of the two cars, of the kinetic energy of the two cars due to the collision?

(C) [10 points] Consider a repeat of the collision with the same drivers having the same intents, and a second pair of cars. This time, however, the two cars collide, spin around each other, and go their separate ways, except that this time, after the collision, the first car is headed due west at half of its original speed. Immediately after the repeat collision has taken place, what is the velocity v of the center of mass of the two cars?

III. [Total of 40 points] Two masses are connected by a rope and a pair of pulleys, as shown in the sketch. The blocks have masses m_1 and m_2; the coefficient of kinetic friction of the ramp under m_1 is μ_k.

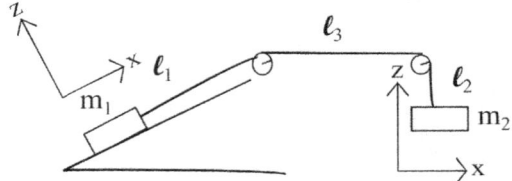

Figure 16.1: Two masses connected by a rope and a pair of pulleys.

(A) [8 points] Give the force diagrams of the two masses. Use my coordinates.
(B) [7 points] Write the equations of motion $\boldsymbol{F} = m\frac{d^2 \boldsymbol{r}}{dt^2}$, component by component, for the two masses.
(C) [5 points] Identify and specify the constraints, if any, in the problem.
(D) [20 points] Calculate the acceleration of the first mass.

Extra Credit

Why are there two correct values for the acceleration in question III? Or is there really only one value of the acceleration, so that the first sentence of this question is a red herring?

16.1 Examination 2 Solutions

I. (A) We start with the work-energy theorem

$$E_i + W = E_f, \tag{16.2}$$

so in this case

$$\frac{1}{2}mv_i^2 + \int_0^L dx \; - (\mu_0 + \mu_1 \exp(x/\ell))mg = E_f. \tag{16.3}$$

The integrals are standard forms. They lead to

$$E_f = \frac{1}{2}mv_i^2 - \mu_0 mgx\big|_{x=0}^L - \frac{\mu_1 mg}{1/\ell}\exp(x/\ell)\big|_{x=0}^L \tag{16.4}$$

or

$$E_f = \frac{1}{2}mv_i^2 - \mu_0 mgL - \mu_1 mg\ell \exp(L/\ell) + \mu_1 mg\ell). \tag{16.5}$$

There are two points in this calculation which require alertness. First, the friction force is a function of position, so in applying the work energy theorem the integral $\int F_x(x)dx$ must actually be performed as an integral, not simply replaced with a multiplication by x. Second, in evaluating the resulting integral, the integral does not vanish at its lower bound $x = 0$.

(B) Conservation of energy tells us $E_i = E_f$. In this case

$$\frac{1}{2}mv_i^2 + mgh_i = \frac{1}{2}mv_f^2 + mgh_f. \tag{16.6}$$

We have a change in height $h_i - h_f = H$, so

$$\frac{1}{2}mv_i^2 + mgh_i - mgh_f = \frac{1}{2}mv_f^2. \tag{16.7}$$

Simplifying,

$$v_i^2 + 2gH = v_f^2, \tag{16.8}$$

and therefore

$$v_f = (v_i^2 + 2gH)^{1/2}. \tag{16.9}$$

(C) The work-energy theorem gives us

$$K_f - K_i = \int_0^L dx F_x. \tag{16.10}$$

From plane geometry, $L = H/sin(\theta)$. The component of the gravitational force parallel to the ramp is $mg\sin(\theta)$. That component points parallel, not antiparallel, to the direction of motion, so

$$K_f - K_i = \int_0^{H/\sin(\theta)} mg\sin(\theta)dx = mgH, \tag{16.11}$$

and therefore

$$\frac{1}{2}mv_f^2 - \frac{1}{2}mv_i^2 = mgH, \tag{16.12}$$

or

$$v_f = (v_i^2 + 2gH)^{1/2}. \tag{16.13}$$

(D) The two results must agree, if the calculations are done correctly, because energy conservation is a corollary of the work-energy theorem.

II. This problem was a homework problem.

III. We have a mass m on a ramp that makes an angle θ with respect to the horizontal. The mass slides down the ramp until it encounters an initially unstretched spring. It continues to slide, compressing the spring, until it comes to a stop. For consistency of notation, the mass started a vertical distance h above the top end of the spring. The mass stops when the spring has been compressed by a distance Δ. Find Δ. Why do you find two possible solutions to your calculation, assuming you did it right? [Hint: The place where the mass comes to a stop ($v = 0$) and the point where its acceleration goes to zero ($a = 0$) *are not the same.*] Why?

The Second Law equations are

$$m_1 \frac{d^2 z_1}{dt^2} = N - m_1 g \cos(\theta) \tag{16.14}$$

$$m_1 \frac{d^2 x_1}{dt^2} = \pm \mu_k N - m_1 g \sin(\theta) + 2T \tag{16.15}$$

$$m_2 \frac{d^2 z_2}{dt^2} = T - m_2 g \tag{16.16}$$

$$m_2 \frac{d^2 x_2}{dt^2} = 0 \tag{16.17}$$

The last of these might have been omitted. In the second of these, there is one point where careful attention is needed. The friction force is written $\pm \mu_k N$, with uncertainty of sign, because kinetic friction opposes the direction of motion, and we do not know which way the masses are moving.

There are two constraints. In terms of the three lengths indicated in the sketch, $\frac{d\ell_1}{dt} = \frac{d\ell_2}{dt}$ and $\ell_1 + \ell_2 + \ell_3 =$ constant. In terms of the coordinates

$$\frac{d^2 x_1}{dt^2} = \frac{d^2 \ell_1}{dt^2} = \frac{d^2 \ell_2}{dt^2}, \tag{16.18}$$

which leads to

$$2 \frac{d^2 x_1}{dt^2} = -\frac{d^2 z_2}{dt^2}. \tag{16.19}$$

Combining equations

$$-4 m_2 \frac{d^2 x_1}{dt^2} = 2T - 2 m_2 g. \tag{16.20}$$

A subtraction takes us to

$$(-4 m_2 - m_1) \frac{d^2 x_1}{dt^2} = \pm \mu_k N + m_1 g \sin(\theta) - 2 m_2 g, \tag{16.21}$$

and thus

$$(-4 m_2 - m_1) \frac{d^2 x_1}{dt^2} = \frac{\pm \mu_k N - m_1 g \sin(\theta) + 2 m_2 g}{(4 m_2 + m_1)}. \tag{16.22}$$

Extra Credit The friction force is written $\pm \mu_k N$, with uncertainty of sign, because kinetic friction opposes the direction of motion, and we do not know which way the masses are moving.

Chapter 17

Descriptions of Rotation

17.1 Rotation

This chapter treats the kinematics of rotation. Kinematics refers to the mathematical description of particle motions, where particles are and how they move as time goes on. Kinematic descriptions do not reach to Newton's laws of Motion. Kinematic descriptions of particle motion do not refer to the forces acting on particles or the masses of those particles; they refer only to where particles are at different times. A representative kinematic description of motion, this for motion at constant linear acceleration, is $\boldsymbol{v}(t) = \boldsymbol{v_0} + \boldsymbol{a}t$. We choose to emphasize kinematic descriptions that correspond to solutions to Newton's Laws of Motion, but the kinematic descriptions of motion would remain as descriptions even if Newton's Laws were incorrect.

When we include particle masses and forces between particles in a discussion of rotation, we reach a description of angular momentum, angular momentum conservation, and torque. Those topics will be considered in future chapters. However, to understand these issues, you first need to understand descriptions of rotation, so that is where we begin.

17.2 Vector Products

Rotations and angular momenta are naturally describer in terms of the *vector product*, also termed the *cross product*. The vector product is the product of two vectors that yields another vector. You may contrast the outcome of the vector product, a vector, with the other two vector products, the scalar product $\boldsymbol{A} \cdot \boldsymbol{B}$ and the outer or tensor product $\boldsymbol{A} \otimes \boldsymbol{B}$. The scalar product yields a simple number, a scalar. For the vectors we are considering, the tensor product is a 3×3 matrix. The cross product of vectors $\boldsymbol{A}$ and $\boldsymbol{B}$ is written $\boldsymbol{A} \times \boldsymbol{B}$, the symbol $\times$ denoting the cross product. The product is called the vector product because the quantity $\boldsymbol{A} \times \boldsymbol{B}$ is itself a vector.

The vector product ignores associated scalars, so if $\boldsymbol{A}$ and $\boldsymbol{B}$ are vectors and a and b are scalars,

$$a\boldsymbol{A} \times b\boldsymbol{B} = (ab)(\boldsymbol{A} \times \boldsymbol{B}). \tag{17.1}$$

In particular, this equation remains correct if $\boldsymbol{A}$ and $\boldsymbol{B}$ are two of the basis vectors, so that

$$a\hat{\boldsymbol{i}} \times b\hat{\boldsymbol{j}} = (ab)(\hat{\boldsymbol{i}} \times \hat{\boldsymbol{j}}). \tag{17.2}$$

The vector product is also distributive, so that

$$\boldsymbol{A} \times (\boldsymbol{B} + \boldsymbol{C}) = \boldsymbol{A} \times \boldsymbol{B} + \boldsymbol{A} \times \boldsymbol{C}. \tag{17.3}$$

The magnitude of the cross product vector is

$$\mid \boldsymbol{A} \times \boldsymbol{B} \mid = AB \sin \theta. \tag{17.4}$$

Here θ is the angle between $\boldsymbol{A}$ and $\boldsymbol{B}$. This final result actually follows from the other properties given here for the vector cross product; it is not an independent result.

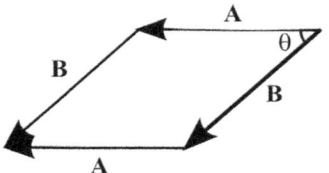

Figure 17.1: Corresponding to any two vectors, here **A** and **B**, there is a single parallelogram.

The next two figures give a geometric interpretation for the magnitude of the vector product. In the figures, the original two vectors are shown as being joined tail to tail. Because a vector has a magnitude and a direction, but not a location, two vectors can always be shown as being attached tail-to-tail. Having drawn the two vectors joined in this way, the two vectors can then always be represented by three points, one point for the head of each vector, and one point for the location of the two tails. Three points define a plane, so to discuss the two vectors in a cross product we can always draw them as lying in the plane of the paper. The orientation of the paper relative to the x, y, and z axes depends on the two vectors, but it is always possible to draw two vectors as lying in a common plane. However, two vectors also define a parallelogram, as shown in Figure 17.1.

What is the area of this parallelogram? That's revealed by Figure 17.2.

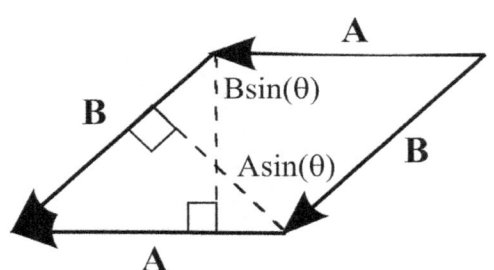

Figure 17.2: The parallelogram of the previous figure. The two dashed lines are the perpendiculars from each side to the opposite vertex. The lines, labeled, have lengths $A\sin(\theta)$ and $B\sin(\theta)$.

The area of a parallelogram is the length of its base times its height, the height being the component of one side that is perpendicular to the other side of the parallelogram. The two construction lines show the component of **A** that is perpendicular to **B** and the component of **B** that is perpendicular to **A**. The two construction lines have lengths $A\sin(\theta)$ and $B\sin(\theta)$, namely they are the two representations of the height of the parallelogram. The area of the parallelogram is therefore $A(B\sin(\theta))$ or equivalently $(A\sin(\theta))B$. The area of the parallelogram formed by **A** and **B** therefore is the magnitude $|\,\boldsymbol{A}\times\boldsymbol{B}\,|$ of their cross product.

The vector product of **A** and **B** is a vector. It must point in some direction. The only unique direction for the vector to point is perpendicular to the plane formed by **A** and **B**. If you think of **A** and **B** as defining a plane, $\boldsymbol{A}\times\boldsymbol{B}$ is a vector perpendicular to that plane. The cross-product vector may therefore also be thought of as a *directed area vector*, the vector that describes a local flat area. In order to describe an area, in which vectors may point in a range of different directions in a plane, with a vector, we use the vector perpendicular to the area, because a flat area has a unique (up to sign) perpendicular vector.

But which perpendicular? A plane has two of them pointing in opposite directions. If you actually have two vectors whose direction you know, and want to know which way their cross product points, you can use the right hand rule. Put your right hand out flat, with the palm up and all fingers together. Now point your thumb out to the side at 90 degrees to the other four fingers. Finally bend your outer three fingers (middle, ring, little) until they are pointing perpendicular to your palm. If the cross product is $\boldsymbol{A}\times\boldsymbol{B}=\boldsymbol{C}$, have your thumb point in the **A** direction, and your index finger point in the **B** direction. Your outer three fingers now point the direction of the **C** vector.

It would be equally possible to find the direction of **C** using your left hand – the left hand rule. If you do this with the same two vectors **A** and **B**, you find that the left-hand-rule cross product vector and the right-hand-rule cross product vector point in opposite directions. We will always use the right-hand rule, but so long as you always use the same hand, there will be no effect on any physical quantity considered in this course. This lack of a physical effect is described as *parity conservation*, a significant topic in *high energy physics*.

The vector product is entirely defined by its effect on basis vectors. The vector product of the basis vectors $\hat{i}$, $\hat{j}$, and $\hat{k}$ and

$$\hat{i} \times \hat{i} = \mathbf{0}, \tag{17.5}$$
$$\hat{j} \times \hat{j} = \mathbf{0}, \tag{17.6}$$
$$\hat{k} \times \hat{k} = \mathbf{0}, \tag{17.7}$$
$$\hat{i} \times \hat{j} = \hat{k}, \tag{17.8}$$
$$\hat{j} \times \hat{k} = \hat{i}, \tag{17.9}$$
$$\hat{k} \times \hat{i} = \hat{j}, \tag{17.10}$$
$$\hat{j} \times \hat{i} = -\hat{k}, \tag{17.11}$$
$$\hat{k} \times \hat{j} = -\hat{i}, \tag{17.12}$$
$$\hat{i} \times \hat{k} = -\hat{j}. \tag{17.13}$$
$$\tag{17.14}$$

This series of equations looks rather complicated, but it actually is composed of three simple parts and a mnemonic device.

First, the cross product of any vector with itself is zero. That's true for the basis vectors, and it is true for every other vector.

Second, the cross product of any two basis vectors is the third basis vector.

Third, the vector product is *anti-commutative*, so that $\mathbf{A} \times \mathbf{B} = -\mathbf{B} \times \mathbf{A}$, where $\mathbf{A}$ and $\mathbf{B}$ are unit vectors. Anti-commutative? Years ago, you were probably taught that addition and multiplication are commutative, so that $4 + 3 = 3 + 4$ and $3 \times 4 = 4 \times 3$. You might have wondered why mathematicians bother to define 'commutative' as a property, if all compositions of two quantities are commutative. The answer is that there are products that are *not* commutative, and we have reached one of them.

Now the mnemonic, which I shall credit to my classical mechanics professor, the late Kenneth Johnson (1931-1999), a mnemonic that I have never seen elsewhere:

$$+\overrightarrow{ijkijk}. \tag{17.15}$$

The arrow indicates the 'plus' direction. How does the mnemonic work? If you read the letters from left to right, the plus direction, the mnemonic generates the positive cross-products. Thus ijk corresponds to $\hat{i} \times \hat{j} = +\hat{k}$ If you read the letters from right to left, the minus direction, the mnemonic generates the negative cross-products. For example, the sequence jik (reading in the mnemonic from right to left) corresponds to the negative product $\hat{j} \times \hat{i} = -\hat{k}$. From the above, one can readily show that $\mathbf{A} \times \mathbf{B} = -\mathbf{B} \times \mathbf{A}$ for any pair of vectors $\mathbf{A}$ and $\mathbf{B}$.

We've actually made a considerable number of different claims about vector cross products. In particular, the rules for multiplication completely define what the cross product of two vectors is. The area of the parallelogram must therefore be and is a logical conclusion from the other statements about vector multiplication, not something additional.

17.3 Circular Motion

We now advance to objects moving in circles. The figure shows a circle drawn in the x-y plane with the coordinates r and θ marked. The distance demarked by θ but measured as a linear distance measured along the perimeter of the circle is s. We define the value of θ to be

$$\theta = \frac{s}{r}. \tag{17.16}$$

If s goes all the way around the circle, the usual rule for the circumference of a circle gives us $\theta = 2\pi$. Our definition of the value of the angle θ therefore chooses which unit we are using to measure θ, namely θ is being measured not in degrees or in grads but in *radians*, a full circle being an angle of 2π radians.

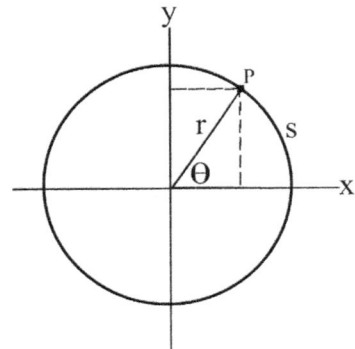

Figure 17.3: Cartesian (dashed lines) and circular polar (radial line) coordinates for a point P.

Suppose θ is a function $\theta(t)$ of time, but r is constant. A particle subject to these two results moves along the circumference of a circle, but does not change its distance from the center of the circle, so that its motion along the circumference is

$$s = r\theta(t). \qquad (17.17)$$

A variety of problems fit this description. For example, we might choose

$$\frac{d^2\theta(t)}{dt^2} = \alpha. \qquad (17.18)$$

Here α is the angular acceleration. If we take angular acceleration to be a constant, independent of time, we can integrate equation 17.18 with respect to time. (We can also integrate this equation if α depends on time; here we are considering an example.) Actually, we've already done this in Chapter 3, when we discussed motion at constant linear acceleration. Equation 17.18 for motion at constant angular acceleration is mathematically exactly the same as equation 3.12 for motion at constant linear acceleration, except for the symbols used. Here the time-dependent variable is called $\theta(t)$; there the time-dependent variable was called $x(t)$. Here the constant value of the second time derivative of $\theta(t)$ is called α; there the time-dependent value of the second time derivative of $x(t)$ was called a. These changes in the symbols have no effect on the mathematics. If we integrate equation 17.18 once or twice with respect to time, we get

$$\frac{d\theta(t)}{dt} = \omega_0 + \alpha t \qquad (17.19)$$

and

$$\theta(t) = \theta_0 + \omega_0 t + \frac{1}{2}\alpha t^2. \qquad (17.20)$$

As we started with the second derivative $d^2\theta/dt^2$, we end up with two constants of integration, ω_0 and θ_0. Except for the change in symbols, the integrals are precisely the same as those seen in Chapter 3 leading to equations 3.21, so they are left as a homework exercise. By convention, we define $\omega(t)$ by

$$\omega(t) = \frac{d\theta(t)}{dt}, \qquad (17.21)$$

where $\omega(t)$ is the angular velocity and $\theta(t)$ is the angular position.

There's an important general result hiding here. We are calculating quantities such as $\theta(t)$. However, the mathematical processes don't care about the physical meaning of the symbols. From the standpoint of the mathematical processes, we have a quantity whose second derivative (it happens to be a time derivative, but that does not matter for finding the math solution) is a constant. From that statement, it follows that the quantity follows equation 17.20. That equation is a quadratic in time, with two constants of integration. The physics supplies the meaning of the symbols, t being time and θ being an angular position, but so long as the starting point is "the second derivative of this function is a constant", the mathematical solution of the problem is a quadratic in the derivative variable.

We can describe the speed and acceleration of the object moving in a circle at constant angular acceleration. We have $s = r\theta$. From this we have

$$\frac{ds}{dt} = r\frac{d\theta}{dt} \equiv r\omega(t) \qquad (17.22)$$

and

$$\frac{d^2s}{dt^2} = r\frac{d^2\theta}{dt^2} \equiv r\alpha. \qquad (17.23)$$

17.4. CIRCULAR MOTION IN AN ARBITRARY PLANE

We can also use a bit of trigonometry to get the x and y coordinates of this moving object.

$$x(t) = r\cos(\theta) \equiv r\cos(\theta_0 + \omega_0 t + \frac{1}{2}\alpha t^2), \tag{17.24}$$

$$y(t) = r\sin(\theta) \equiv r\sin(\theta_0 + \omega_0 t + \frac{1}{2}\alpha t^2). \tag{17.25}$$

That's well and good, but what do we do if we want to tilt the plane in which there is circular motion? That's where vector products come in.

17.4 Circular Motion in an Arbitrary Plane

How do we describe rotation in an arbitrary plane? Let's start by trying to describe rotation using vectors. Why does it make sense to use vectors? Vectors in and of themselves exist independent of their coordinate representations. If we move from the x-y plane at $z = 0$ to an arbitrary plane and center, we can view ourselves as have changed all of the positions in space and all the directions of the basis vectors. However, the vector $\boldsymbol{r}$ from the center of the circle to the point of interest, and the velocity vector $\boldsymbol{v}$ of the point moving around the circle, relative to the circle, have not changed. We have in effect only changed the representation of the vectors.

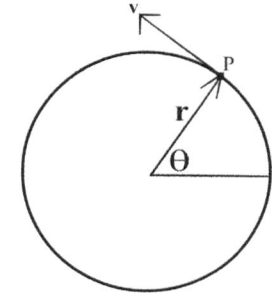

Figure 17.4: Circular motion of a point P in the plane, showing the point's position vector $\boldsymbol{r}$ and velocity vector $\boldsymbol{v}$.

The figure shows a circle in a plane, with an object moving in a circle. The radius vector $\boldsymbol{r}$ is the vector from the origin to the location of the object. The object is moving in a circle, so at each point its velocity vector $\boldsymbol{v}$ is the tangent to the circle. From plane geometry, $\boldsymbol{v}$ is therefore perpendicular to $\boldsymbol{r}$.

We may usefully introduce two new unit vectors, $\hat{\boldsymbol{r}}$ and $\hat{\boldsymbol{\theta}}$. These two unit vectors lie in the plane of the circle, and are perpendicular to each other. At each point on the circle, $\hat{\boldsymbol{r}}$ points parallel to the radius vector. Because $\hat{\boldsymbol{\theta}}$ is perpendicular to $\hat{\boldsymbol{r}}$, it must lie parallel to the circle's tangent line. These two unit vectors have a property that makes them unlike the unit vectors $\hat{\boldsymbol{i}}, \hat{\boldsymbol{j}}$, and $\hat{\boldsymbol{k}}$ that we discussed previously. No matter where we are in the (x, y, z) coordinate system, the unit vectors $\hat{\boldsymbol{i}}, \hat{\boldsymbol{j}}$, and $\hat{\boldsymbol{k}}$ point in the same three directions. $\hat{\boldsymbol{r}}$ and $\hat{\boldsymbol{\theta}}$ are not like this. Depending on where the object is on the circle, the unit vectors $\hat{\boldsymbol{r}}$ and $\hat{\boldsymbol{\theta}}$ may point in completely different directions in space.

However, because $\hat{\boldsymbol{\theta}}$ is always parallel to the tangent vector, and $|\boldsymbol{r}|$ is a constant so that $\frac{dr}{dt} = 0$, we may write

$$\frac{d\boldsymbol{r}}{dt} = 0\hat{\boldsymbol{r}} + r\omega\hat{\boldsymbol{\theta}} \tag{17.26}$$

[Aside: Buried in the statement that $\hat{\boldsymbol{i}}, \hat{\boldsymbol{j}}$, and $\hat{\boldsymbol{k}}$ point everywhere in the same directions is a physics assertion, namely that if you take a vector and displace it to a different point in space, it can without difficulty be displaced parallel to itself. The actual physical universe is not quite so simple. The actual behavior of a vector subject to an infinitesimal displacement refers to the curvature of space, which leads us to Einstein's

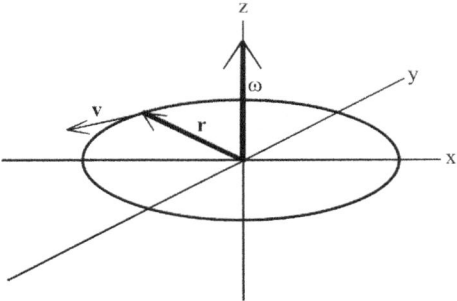

Figure 17.5: A point rotating in the plane, showing the point's position vector $\boldsymbol{r}$ and velocity vector $\boldsymbol{v}$ and the angular velocity vector $\boldsymbol{\omega}$.

General Relativity. In Newtonian mechanics, space is *flat*, meaning we can displace vectors and they stay parallel to their original positions. In Einstein's treatment of gravity in General Relativity, space is not flat, so displacing a vector placed near a mass can cause it to change direction. This change of direction on displacement is much more subtle than the behavior of $\hat{r}$ and $\hat{\theta}$. In Newtonian mechanics, if I pick up a circle and its moving object, when I set them down someplace else $\hat{r}$ and $\hat{\theta}$ are still pointing in the same directions as before. In General Relativity life is more complicated.]

We now ask how we can represent rotation as a vector. The answer is that we use a cross-product. We begin by introducing an angular velocity vector $\boldsymbol{\omega}$. The magnitude of $\boldsymbol{\omega}$ is chosen to be the angular velocity of rotation of the object of interest, so that

$$|\boldsymbol{\omega}| = |\frac{d\theta}{dt}| \qquad (17.27)$$

When $\boldsymbol{r}$ and $\boldsymbol{v}$ are in the x-y plane, where $z = 0$, $\boldsymbol{\omega}$ is taken to be perpendicular to the plane formed by $\boldsymbol{r}$ and $\boldsymbol{v}$. Unlike standard vectors, $\boldsymbol{\omega}$ has a fixed location; it must (perhaps after being extended) pass through the center of the circle.

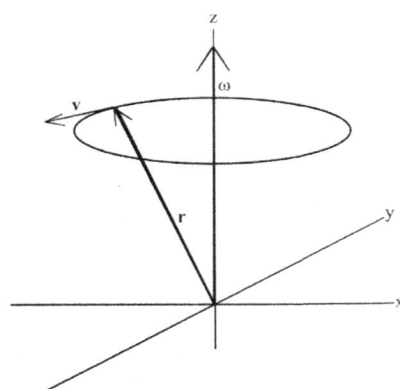

Figure 17.6: A point rotating in a plane parallel to the x-y plane, showing the point's position vector $\boldsymbol{r}$ and velocity vector $\boldsymbol{v}$ and the angular velocity vector $\boldsymbol{\omega}$.

Begin with $\boldsymbol{r}$ and $\boldsymbol{v}$, which between them define the plane in which rotation is occurring, to be in the x-y plane. In that case, $\boldsymbol{\omega}$, being perpendicular to this plane, must be parallel to the z-axis, so that we can write

$$\boldsymbol{\omega} = \omega \hat{\boldsymbol{k}}. \qquad (17.28)$$

We can now use $\boldsymbol{R}$ and $\boldsymbol{\omega}$ to compute $\boldsymbol{v}$. The obvious form is

$$\boldsymbol{v} = \boldsymbol{\omega} \times \boldsymbol{r}. \qquad (17.29)$$

Because $\boldsymbol{\omega}$ and $\boldsymbol{r}$ are perpendicular to each other, the vector calculated in this way has the correct magnitude ωr. This $\boldsymbol{v}$ lies in the plane perpendicular to $\boldsymbol{\omega}$ and is perpendicular to $\boldsymbol{r}$, just as $\boldsymbol{v}$ is. There is one minor uncertainty. The vector cross product is not commutative. Should the correct form be $\boldsymbol{\omega} \times \boldsymbol{r}$ or should it be $\boldsymbol{r} \times \boldsymbol{\omega}$? The answer to that is seen in Figure 17.5, in which $\boldsymbol{\omega}$ is in the $+\hat{\boldsymbol{k}}$ direction and $\boldsymbol{r}$ is as indicated. If θ is increasing as time goes on, meaning $\frac{d\theta}{dt} > 0$, $\boldsymbol{v}$ must also be as indicated, this being the direction that results from $\boldsymbol{v} = \boldsymbol{\omega} \times \boldsymbol{r}$.

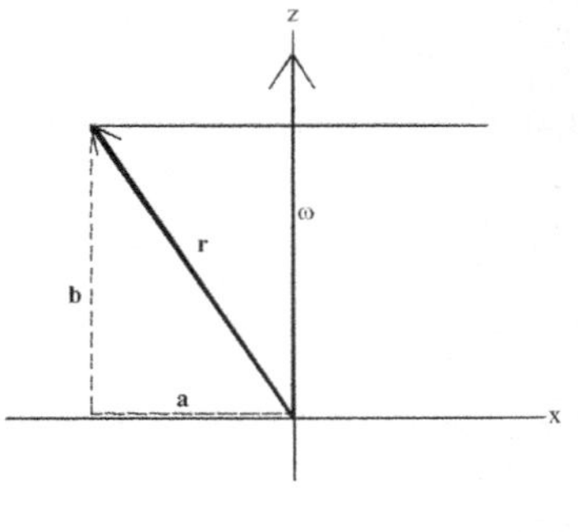

The above discussion refers to objects rotating in the plane that contains the origin, so that

Figure 17.7: The same rotating point as in Figure 17.6, but now we have rotated the projection so that the vectors $\boldsymbol{r}$ and $\boldsymbol{\omega}$ are in the plane of the paper and at this moment in the point's rotation the point's velocity is directly out of the page toward the reader. The vectors $\boldsymbol{a}$ and $\boldsymbol{b}$ are the components of $\boldsymbol{r}$ that are perpendicular or parallel, respectively, to the vector $\boldsymbol{\omega}$.

the $\boldsymbol{r}$ vector is perpendicular to the $\boldsymbol{\omega}$ vector. What if the rotating object is are not in the x-y plane, but is instead well above or below it? Figure 17.6 shows a perspective drawing of just such an object. The object is supposed to be rotating in a circle around the $\boldsymbol{\omega}$ axis. Is $\boldsymbol{v} = \boldsymbol{\omega} \times \boldsymbol{r}$ still applicable? Matters are made clearer in the Figure 17.7, in which the $\boldsymbol{\omega}$ and $\boldsymbol{r}$ vectors are in the plane of the paper.

First, $\boldsymbol{\omega} \times \boldsymbol{r}$ is perpendicular to the plane of the paper, meaning it lies in a plane perpendicular to $\boldsymbol{\omega}$. The $\boldsymbol{r}$ vector has been split into two parts, namely a part $\boldsymbol{a}$ perpendicular to $\boldsymbol{\omega}$ and a part $\boldsymbol{b}$ that is parallel to $\boldsymbol{\omega}$. Because the cross product is distributive, we may write

$$\boldsymbol{\omega} \times \boldsymbol{r} = \boldsymbol{\omega} \times \boldsymbol{a} + \boldsymbol{\omega} \times \boldsymbol{b}. \qquad (17.30)$$

However, $\boldsymbol{\omega}$ and $\boldsymbol{b}$ are parallel, so their cross product is zero, giving

$$\boldsymbol{\omega} \times \boldsymbol{r} = \boldsymbol{\omega} \times \boldsymbol{a}. \tag{17.31}$$

For the magnitude of this product

$$|\boldsymbol{\omega} \times \boldsymbol{a}| = \omega r \sin(\theta). \tag{17.32}$$

However, from the construction of Figure 17.7 we see that $r \sin(\theta)$ is a, the perpendicular distance from the line of the ω vector to the object. a is the radius of the circle in which the object is performing circular motion. From the right hand rule, $\boldsymbol{\omega} \times \boldsymbol{a}$ is a vector pointing directly out of the page, exactly the direction being taken if the object is performing circular motion in a plane perpendicular to the ω axis, the plane being displaced above the origin by a distance b. Equation 17.29 is a vector equation; it does not depend on the orientation of the coordinate axes with respect to $\boldsymbol{\omega}$.

What good is ω? Consider a drill, placed at an angle to the coordinate axes. Suppose you would like to know the velocity at which the surface of the drill bit is moving at some point $\boldsymbol{r}$, given that the rotation speed and the axis along which the drill points are known. $\boldsymbol{\omega} \times \boldsymbol{r}$ gives the answer. Of course, you would need to remember to convert, for example, rotations per minute to radians per second. Also, while $\boldsymbol{\omega}$ can be calculated as $\omega \hat{\boldsymbol{\omega}}$, one would need to remember to check that one's $\hat{\boldsymbol{\omega}}$ is in fact the unit vector.

17.5 Discussion

[Aside: What about quantities not considered in this course? In an early chapter we mentioned the basic forces of nature, including in particular the weak nuclear force that is involved in particle decays that produce neutrinos. It was long believed that all forces in nature satisfied parity conservation, but in 1956 American physicist Chien-Shiang Wu and her collaborators at the National Bureau of Standards demonstrated that parity is not conserved in the weak interaction. It was later realized that a 1928 experiment by R. T. Cox, G. C. McIlwraith, and B. Kurrelmeyer had actually seen the effect, but the theoretical background needed to identify the experiment's significance had not yet been developed, so the 1928 work remained lost until after Wu had done her work.].

17.6 Worked Problems

1. It is once again time to boot up a slightly older computer. The flash boiler is started at $t = 0$. At $t = 30$ seconds, the flash boiler has worked up a sufficient head of steam, and a valve is thrown, supplying steam to the engine that spins the 2048 bit drum memory. The drum, initially at rest, accelerates at 15 rad/s^2 up to its maximum 120 radians/second. Assuming that the drum started at $\theta = 0$, calculate the angular position and velocity of the drum at $t = 70$ seconds. For credit, do all calculations using my origin for the time coordinates.

2. A pin is stuck into a rigid rod. The pin is not perpendicular to the rod. At a particular moment in time, the rod lies along the line $-\hat{\boldsymbol{i}} - \hat{\boldsymbol{j}} - \hat{\boldsymbol{k}}$ and has angular velocity $\boldsymbol{\omega} = -2(\hat{\boldsymbol{i}} + \hat{\boldsymbol{j}} + \hat{\boldsymbol{k}})$. (Yes, the angular velocity is parallel to the rod, which is rotating around its long axis.) The pin intersects the rod at the origin, and at time 0 has its head at $3\hat{\boldsymbol{i}} + \hat{\boldsymbol{j}}$. (a) What is the angular rotation rate ω of the rod? (b) What is the velocity of the head of the pin? Hint: The standard error is to fail to notice that $\boldsymbol{\omega}$ is *not* written in the form $\omega \hat{\boldsymbol{\omega}}$. Look carefully and you will find why.

17.7 Homework

1. Consider the relation $\boldsymbol{A} \times \boldsymbol{B} = \boldsymbol{C}$. Write the components of the vector $\boldsymbol{C}$ in terms of the components a_x, a_y, a_z, b_x, b_y, and b_z of $\boldsymbol{A}$ and $\boldsymbol{B}$.

2. Show that $\boldsymbol{A} \times \boldsymbol{B} = -\boldsymbol{B} \times \boldsymbol{A}$ for any pair of vectors. (Hint: Expand vectors into their components and basis vectors.)

3. Show that the cross product vector $A \times B$ is perpendicular to the two vectors A and B. Remark: If two vectors are perpendicular, what is their scalar product?

4. Starting with the component form for the vector A, show $A \times A = 0$.

5. Prove that $\frac{d(A \times B)}{dt} = \frac{dA}{dt} \times B + A \times \frac{dB}{dt}$. Remark: This is not the product rule you studied in calculus. That rule refers to the scalar multiplication of scalars. A sound approach expands the two vectors in terms of all their scalar components.

6. Derive equations 17.19 and 17.20 from equation 17.18. Procedure: Write out the steps for integrating motion at constant linear acceleration $d^2x/dt^2 = a$. Now go through and cross out, for example, the acceleration a, and replace it with α. Verify that nothing changes except the spelling of the symbols.

7. Consider a point mass performing circular motion at constant angular velocity. Prove that the mass's position and velocity vectors are perpendicular to each other. Prove that the mass's velocity and acceleration vectors are perpendicular to each other.

8. A pin is stuck into a rigid rod. The pin is not perpendicular to the rod. At a particular moment in time, the rod lies along the line $2\hat{i} - \hat{j} - \hat{k}$ and has angular velocity $\boldsymbol{\omega} = 3(2\hat{i} - \hat{j} - \hat{k})$. The pin lies in the x-z plane. The pin intersects the rod at the origin, and has its head at $\mathbf{r} = 2\hat{i} + \hat{k}$. (a) What is the velocity of the head of the pin at the moment of interest? (b) What is the angular rotation rate (a scalar) of the rod?

9. It is time to boot up a low-power-consumption ecologically compliant waste-cereal-recyling computer. Instead of an electric motor the computer uses a cage of well-trained hamsters. The hamsters respond—eventually—to a bell. The bell is sounded at $t = 5$. At $t = 30$ seconds, the hamsters get onto their rotating wheel and start running. A train of gears connects the wheel to the computer's drum memory. The drum, initially at rest, accelerates at 15 rad/sec^2 up to a maximum of 90 radians/second. Assuming that the drum was originally at $\theta = 20$, calculate the angular position and angular velocity of the drum at $t = 35$ seconds. For credit, do all calculations using my origin for the time coordinates.

10. We have a long wooden stick that is rotating at constant angular velocity ω. The stick passes through the origin. The stick points in the direction $\hat{i} + \hat{j} + \hat{k}$. The angular velocity of the stick is $\boldsymbol{\omega} = 3(\hat{i} + \hat{j} + \hat{k})$. A nail has been driven into the stick, very nearly passing though the origin. (As the stick rotates, the end of the nail may point in different directions, but the point of the nail continues to be at the origin.) At the time of interest, the other end of the nail – the head – is at $\hat{i} + \hat{j}$. (a) Find the velocity at the time of interest of the head of the nail. (b) Find the angular rotation rate ω of the stick (Hints: a scalar, not a vector. Not equal to three.) (c) Find the unit vector $\hat{\omega}$.

11. It is necessary to reboot a (slightly older) steam-powered computer. The flash boiler is re-ignited at $t = 0$. Because the steam in the system is still warm the angular acceleration of the drum memory begins at $t = 0$. The drum may or may not have spun down to a stop before the flash boiler was re-ignited. Starting at zero at $t = 0$, the angular acceleration increases exponentially with time as $\frac{d^2\theta}{dt^2} = A\exp(Bt)$, where A and B are constants. Calculate the angular velocity and the angular position of the drum as functions of time.

12. Consider the vectors $\mathbf{A} = A_x\hat{\mathbf{i}} + A_y\hat{\mathbf{j}} + A_z\hat{\mathbf{k}}$ and $\mathbf{B} = B_x\hat{\mathbf{i}} + B_y\hat{\mathbf{j}} + B_z\hat{\mathbf{k}}$. $(\mathbf{A} \times \mathbf{B})$ is not zero. By direct calculation, show that $(\mathbf{A} \times \mathbf{B}) \cdot \mathbf{A} = 0$. What does your calculation prove about the angles between $\mathbf{A}$, $\mathbf{B}$, and $\mathbf{A} \times \mathbf{B}$?

17.8 Solutions to the Worked Problems

1. Here we have a problem for which only the terminal numerics are central. We are discussing rotation at constant angular acceleration, for which

$$\theta(t) = \theta_o + \omega_o t + \frac{1}{2}\alpha t^2, \tag{17.33}$$

$$\omega(t) = \omega_o + \alpha t. \tag{17.34}$$

17.8. SOLUTIONS TO THE WORKED PROBLEMS

We would like to know the angular position of the drum memory at a specified future time. Our first step is to determine θ_o, ω_o, and α. For the acceleration phase, $\alpha = 15$. At time $t = 30$, $\omega(30) = 0$ and $\theta(30) = 0$, meaning

$$0 = \omega_o + 15 \cdot 30, \tag{17.35}$$
$$\omega_o = -450, \tag{17.36}$$

and

$$0 = \theta_o - 450(30) + \frac{1}{2} \cdot 15 \cdot 30^2, \text{ so} \tag{17.37}$$
$$\theta_o = 6750. \tag{17.38}$$

During the acceleration phase

$$\theta(t) = 6750 + -450t + \frac{1}{2}15t^2, \tag{17.39}$$
$$\omega(t) = -450 + 15t. \tag{17.40}$$

The acceleration phase continues until $\omega(t) = 120$, which occurs at

$$120 = -450 + 15t, \tag{17.41}$$
$$t = 38. \tag{17.42}$$

At $t = 38$, $\theta(t)$ becomes

$$\theta(t) = 6750 - 450 \cdot 38 + \frac{1}{2}15 \cdot 38^2, \tag{17.43}$$
$$\theta(t) = 480. \tag{17.44}$$

For the next 32 seconds, $\omega(t) = 120$, so

$$\Delta\theta = \omega \Delta t, \tag{17.45}$$
$$\Delta\theta = 120 \cdot 32 = 3840. \tag{17.46}$$

So at $t = 70$, $\theta(70) = 4320$ radians.

2. The angular rotation vector of the rod is $\boldsymbol{\omega} = -2(\hat{\boldsymbol{i}} + \hat{\boldsymbol{j}} + \hat{\boldsymbol{k}})$.

   (a) The rotation rate is $|\boldsymbol{\omega}| = (2^2 + 2^2 + 2^2)^{1/2} = 2\sqrt{3}$ radians/second.

   (b) $\boldsymbol{v} = \boldsymbol{\omega} \times \boldsymbol{r}$. $\boldsymbol{\omega}$ is given above. $\boldsymbol{r} = 3\hat{\boldsymbol{i}} + \hat{\boldsymbol{j}}$, so taking the product and using the fact that the cross-product is distributive,

$$\boldsymbol{v} = -2(\hat{\boldsymbol{i}} + \hat{\boldsymbol{j}} + \hat{\boldsymbol{k}}) \times (3\hat{\boldsymbol{i}} + \hat{\boldsymbol{j}}), \tag{17.47}$$
$$\boldsymbol{v} = -2\hat{\boldsymbol{i}} \times 3\hat{\boldsymbol{i}} - 2\hat{\boldsymbol{i}} \times \hat{\boldsymbol{j}} - 2\hat{\boldsymbol{j}} \times 3\hat{\boldsymbol{i}} - 2\hat{\boldsymbol{j}} \times \hat{\boldsymbol{j}} - 2\hat{\boldsymbol{k}} \times 3\hat{\boldsymbol{i}} - 2\hat{\boldsymbol{k}} \times \hat{\boldsymbol{j}}. \tag{17.48}$$

On carrying out the multiplications term by term,

$$\boldsymbol{v} = 2\hat{\boldsymbol{i}} - 6\hat{\boldsymbol{j}} + 4\hat{\boldsymbol{k}}.$$

This page reserved for your notes.

Chapter 18

Motion in Cylindrical Polar Coordinates

18.1 Introduction

This Chapter considers a description of motion in cylindrical polar coordinates. We have previously discussed arbitrary motions in Cartesian coordinates. We separately discussed circular motion. This Chapter discusses arbitrary motions, as described by using cylindrical polar (r, θ, z) coordinates.

We begin by recalling the relationship between Cartesian coordinates and cylindrical polar coordinates. The first figure shows a vector from the origin to a point, with the Cartesian x-y axes and the cylindrical polar coordinates of a point marked. We recall that

$$x = r\cos(\theta), \tag{18.1}$$
$$y = r\sin(\theta), \tag{18.2}$$
$$z = z. \tag{18.3}$$

In these equations, the left-hand-sides are the three Cartesian coordinates. The right-hand-sides of the equations are the values of those coordinates in terms of the coordinate values in cylindrical polar coordinates. A position vector in Cartesian coordinates can then be written as

$$\boldsymbol{r} = \hat{\boldsymbol{i}} r \cos(\theta) + \hat{\boldsymbol{j}} r \sin(\theta) + \hat{\boldsymbol{k}} z. \tag{18.4}$$

If r is a constant, meaning we have circular or helical motion,

$$\frac{d\boldsymbol{r}}{dt} = -\hat{\boldsymbol{i}} r \frac{d\theta}{dt} \sin(\theta) + \hat{\boldsymbol{j}} r \frac{d\theta}{dt} \cos(\theta) + \hat{\boldsymbol{k}} \frac{dz}{dt}. \tag{18.5}$$

Helical motion? A circular spring gives an example of a helix. In helical motion, an object moves in a circle around a central axis (the motion may or may not be at constant angular velocity $\frac{d\theta}{dt}$) and at the same time moves in the $\hat{\boldsymbol{k}}$ direction. For a simple helix, $\frac{d\theta}{dt}$ and $\frac{dz}{dt}$ are both non-zero constants.

We now consider representing motion using cylindrical polar coordinates, in which the basis vectors of the coordinate system are $\hat{\boldsymbol{r}}$, $\hat{\boldsymbol{\theta}}$, and $\hat{\boldsymbol{k}}$. To do this, we break a position vector $\boldsymbol{R}$ into its $\hat{\boldsymbol{k}}$ component and its component perpendicular to $\hat{\boldsymbol{k}}$, as shown in the following perspective drawing. The position vector is defined to be the vector from the origin to the location of interest. The vector labelled $\boldsymbol{r}$ is the projection of the $\boldsymbol{R}$ vector into the plane of the page. The projection $\boldsymbol{r}$ has the same $\hat{\boldsymbol{r}}$ and $\hat{\boldsymbol{\theta}}$ components that the $\boldsymbol{R}$ vector has, but its $\hat{\boldsymbol{k}}$ component is zero.

Examples of the $\hat{\boldsymbol{r}}$ and $\hat{\boldsymbol{\theta}}$ basis vectors are shown in the next figure. In the figure, the $\hat{\boldsymbol{k}}$ vector is perpendicular to the page. The basis vector $\hat{\boldsymbol{r}}$ is always parallel to $\boldsymbol{r}$. The basis vector $\hat{\boldsymbol{\theta}}$ is always perpendicular to $\hat{\boldsymbol{r}}$. At different points in the figure, the $\hat{\boldsymbol{r}}$ and $\hat{\boldsymbol{\theta}}$ vectors each point in different directions. As a result, the components of the velocity vector, for a particle moving in a straight line at constant non-zero velocity in the x-y plane, depend on the position of the particle. For the same reason, if we take a vector and translate it horizontally, the vector itself does not change, but its $\hat{\boldsymbol{r}}$ and $\hat{\boldsymbol{\theta}}$ components do change as the vector is translated.

Chapter 18. Motion in Cylindrical Polar Coordinates

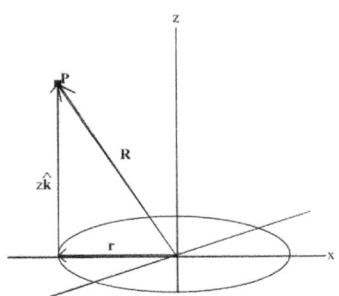

Figure 18.1: Vector $\boldsymbol{R}$ to a point P and its decomposition into a vertical vector $z\hat{\boldsymbol{k}}$ and a horizontal vector $\boldsymbol{r}$. The vectors $\boldsymbol{R}$ and $\boldsymbol{r}$ have their tails at the origin O.

Consider again the position vector $\boldsymbol{r}$ of a particle. By construction, $\boldsymbol{r}$ can be written

$$\boldsymbol{R} = r\hat{\boldsymbol{r}} + z\hat{\boldsymbol{k}}. \quad (18.6)$$

The $\hat{\boldsymbol{\theta}}$ component of $\boldsymbol{R}$ is always zero. It is entirely possible to construct vectors whose $\hat{\boldsymbol{\theta}}$ components are non-zero ($\hat{\boldsymbol{\theta}}$ itself is an example), but these vectors have the common feature that they do not begin at the origin.

The unit vectors $\hat{\boldsymbol{r}}$ and $\hat{\boldsymbol{\theta}}$ are readily written in terms of the (r, θ, z) components of the vector $\boldsymbol{R}$, namely

$$\hat{\boldsymbol{r}} = \hat{\boldsymbol{i}}\cos(\theta) + \hat{\boldsymbol{j}}\sin(\theta) \quad (18.7)$$

and

$$\hat{\boldsymbol{\theta}} = -\hat{\boldsymbol{i}}\sin(\theta) + \hat{\boldsymbol{j}}\cos(\theta) \quad (18.8)$$

These two basis vectors have a zero component in the $\hat{\boldsymbol{k}}$ direction.

We now consider a subtle feature of the Cartesian coordinate system. Suppose we have a vector $\boldsymbol{R}$, which can be written in Cartesian coordinates as $x\hat{\boldsymbol{i}} + y\hat{\boldsymbol{j}} + z\hat{\boldsymbol{k}}$. Its full time derivative is

$$\frac{d\boldsymbol{R}}{dt} = \frac{dx}{dt}\hat{\boldsymbol{i}} + \frac{dy}{dt}\hat{\boldsymbol{j}} + \frac{dz}{dt}\hat{\boldsymbol{k}} + x\frac{d\hat{\boldsymbol{i}}}{dt} + y\frac{d\hat{\boldsymbol{j}}}{dt} + z\frac{d\hat{\boldsymbol{k}}}{dt}. \quad (18.9)$$

The second line of this equation arises automatically from applying the product rule $d(ab)/dt = (da/dt)b + a(db/dt)$. However, the basis vectors $\hat{\boldsymbol{i}}$, $\hat{\boldsymbol{j}}$, and $\hat{\boldsymbol{k}}$ are the same everywhere, so, in particular, for an observer moving with the particle, the local directions of the Cartesian basis vectors are independent of time. The final three terms of equation 18.9 are therefore each equal to zero.

However, for cylindrical polar coordinates,

$$\hat{\boldsymbol{r}} = \hat{\boldsymbol{r}}(x, y, z), \quad (18.10)$$

$$\hat{\boldsymbol{\theta}} = \hat{\boldsymbol{\theta}}(x, y, z), \quad (18.11)$$

$$\hat{\boldsymbol{k}} = \hat{\boldsymbol{k}}. \quad (18.12)$$

If we write

$$\boldsymbol{R} = r\hat{\boldsymbol{r}} + z\hat{\boldsymbol{k}}, \quad (18.13)$$

then the particle's velocity in spherical polar coordinates becomes

$$\frac{d\boldsymbol{R}}{dt} = \frac{dr}{dt}\hat{\boldsymbol{r}} + r\frac{d\hat{\boldsymbol{r}}}{dt} + \frac{dz}{dt}\hat{\boldsymbol{k}}. \quad (18.14)$$

On the right-hand-side of this equation, the displayed terms are all in general non-zero. The formal term $zd\hat{\boldsymbol{k}}/dt$ is not displayed because it is zero.

What does

$$\frac{d\hat{\boldsymbol{r}}}{dt}$$

mean? The symbols arises because, as the particle moves, $\hat{\boldsymbol{r}}(t)$ is always evaluated at the current position of the particle, and $\hat{\boldsymbol{r}}(t + \delta t) - \hat{\boldsymbol{r}}(t) \neq 0$.

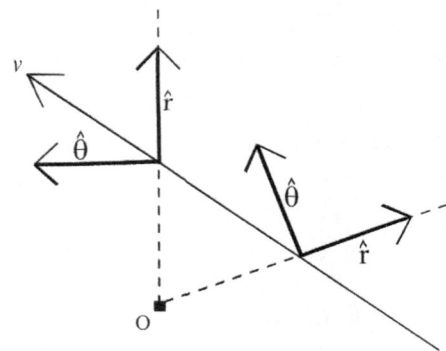

Figure 18.2: Velocity vector $\boldsymbol{v}$ and trajectory of a particle moving in a straight line, showing the directions of its $\hat{\boldsymbol{r}}$ and $\hat{\boldsymbol{\theta}}$ components at two points along its path. The velocity vector is a constant; its vector components in (r, θ, z) coordinates are not.

Said differently, $(\hat{\boldsymbol{i}}, \hat{\boldsymbol{j}}, \hat{\boldsymbol{k}})$ and $(\hat{\boldsymbol{r}}, \hat{\boldsymbol{\theta}}, \hat{\boldsymbol{z}})$ are vector fields. What is a vector field? A vector field is vector quantity whose value is determined by position. An example of a vector field is a weather map showing the wind speed and direction at each of a series of cities. The wind velocity is a vector. However, its value depends on where you are, so the wind velocity may be 5 mph from the east where the reader is

18.1. INTRODUCTION

located, and 200 mph from the west on top of Mount Washington. The position matters. If the velocities are interchanged, so that the wind velocity at the location of the reader becomes 200 mph from the west, there may be a change in the reader's circumstances.

It happens that the basis vectors $(\hat{i},\hat{j},\hat{k})$ are the same at all positions r. In contrast, the basis vectors $(\hat{r},\hat{\theta},\hat{z})$ change as r is changed. If we consistently evaluate $(\hat{r},\hat{\theta},\hat{z})$ at the currently location of the particle, and the particle is moving, then $(\hat{r},\hat{\theta},\hat{z})$, being non-trivial functions of r, are implicitly functions of the time, as seen in Figure 18.2.

Consider a particle acted on by no external forces. As time evolves, the particle translates along the momentum vector p. This vector and its Cartesian components, e.g., $p_x = p \cdot \hat{i}$ do not change with time , because the momentum is conserved if there are no external forces acting, and the Cartesian basis vectors are the same everywhere. However, if $p = a\hat{r} + b\hat{\theta} + c\hat{k}$, p is still constant, because it is conserved, but a and b depend on time because $\hat{r}$ and $\hat{\theta}$ change as the particle moves.

We now advance to a more systematic mathematical description. First, note that if $\hat{u}$ is a unit vector, then the infinitesimal change $\delta \hat{u}$ in $\hat{u}$ during some infinitesimal time step must be perpendicular to $\hat{u}$, i.e. $\hat{u} \cdot \delta \hat{u} = 0$. If this condition were not satisfied, them the change $\delta \hat{u}$ in $\hat{u}$ would have the effect of changing the magnitude of $\hat{u}$, which is not permitted.

Starting at the beginning:

We have a vector R, with $R = (x,y,z)$. For clarity, we will leave the vector R lying in the x-y plane. A vertical component $z\hat{k}$ of R remains unperturbed by the calculational steps here, so it can simply be inserted at the end of the calculation without disturbing the validity of the arguments here.

We now write the vectors $\hat{r}$ and $\hat{\theta}$ in circular polar coordinates. The character θ risks being overworked, so we will here call the polar angle ϕ rather than θ. Why are we using ϕ and not θ as the circular-polar angle for the direction of $\hat{r}$? θ is already in use as the $\hat{\theta}$ component of R. Using the same symbol to mean two different things can lead to confusion and error, so we introduce an extra symbol.

We write

$$\hat{r} = \hat{i}\cos(\phi) + \hat{j}\sin(\phi), \tag{18.15}$$

$$\hat{\theta} = -\hat{i}\sin(\phi) + \hat{j}\cos(\phi). \tag{18.16}$$

In cylindrical polar coordinates we may write for a particle position $R \equiv r\hat{r}$. The velocity is then the time derivative of this quantity, so

$$\frac{dR}{dt} = \frac{dr}{dt}\hat{r} + r\frac{d\hat{r}}{dt}. \tag{18.17}$$

In this equation, the term on the left is the particle's velocity $\frac{dR}{dt}$, the term $\frac{dr}{dt}\hat{r}$ is the contribution to the velocity due to the particle changing its distance from the origin, and the term $r\frac{d\hat{r}}{dt}$ corresponds to the basis vector changing which way it points in space as the particle moves. How are we to calculate this last term? Beginning with eq. 18.15, one finds

$$\frac{d\hat{r}}{dt} = -\frac{d\phi}{dt}\hat{i}\sin(\phi) + \frac{d\phi}{dt}\hat{j}\cos(\phi) \tag{18.18}$$

or

$$\frac{d\hat{r}}{dt} = \frac{d\phi}{dt}\hat{\theta}. \tag{18.19}$$

By the same approach

$$\frac{d\hat{\theta}}{dt} = -\frac{d\phi}{dt}\hat{i}\cos(\phi) - \frac{d\phi}{dt}\hat{j}\sin(\phi) \tag{18.20}$$

or

$$\frac{d\hat{\theta}}{dt} = -\frac{d\phi}{dt}\hat{r}. \tag{18.21}$$

You may confirm for yourself that the claim $d\hat{u}/dt \cdot \hat{u} = 0$ is correct for both of these derivatives.

We now calculate the velocity and the acceleration in cylindrical polar coordinates. We'll write it out starting at the very beginning. That's an important, sound practice for doing calculations. Start at the very beginning, and write out every step. Again, the start is

$$R = r\hat{r}. \tag{18.22}$$

For each step we actually write out in words what we are doing algebraically. That's also a sound practice, focussing your mind on what you are supposed to be doing. We now take the time derivative, finding

$$\frac{d\boldsymbol{R}}{dt} = \frac{dr}{dt}\hat{\boldsymbol{r}} + r\frac{d\hat{\boldsymbol{r}}}{dt}. \tag{18.23}$$

In this equation, the left hand side is the velocity, while the rightmost time derivative was found in equation 18.19. Substituting for these, we have

$$\boldsymbol{V} = \frac{dr}{dt}\hat{\boldsymbol{r}} + r\frac{d\theta}{dt}\hat{\boldsymbol{\theta}}. \tag{18.24}$$

We have now reached a form for the velocity of a particle in circular polar coordinates. The derivative $\frac{d\theta}{dt}$ is the time rate of change of the polar angle θ, which in our discussion of circular motion we denoted ω. If the particle is moving on a circle, a curve of constant radius, then the radial velocity dr/dt is zero, in which case *for circular motion only*, we recover the earlier form

$$\boldsymbol{V} = r\omega\hat{\boldsymbol{\theta}}. \tag{18.25}$$

Why did we make this slight diversion in our calculation? We've analyzed motion in a circle before. We therefore took the general expression given in equation 18.24 and subjected it to the constraints that make it a description of circular motion. We obtained the answer we have previously. This step is a check, not a proof. We have advanced from a special case that we solved first to a more general case that we solved later; we insist that the more general case agrees with the special case when there should be agreement. If there is not agreement, we need to understand why. Did we make a mistake someplace? Is the special case (here, $r = $ constant) not quite the special case we thought it was?

We now push ahead to compute the acceleration. The first step is to take another time derivative, being careful to take the derivative of the general-case equation 18.24. The derivative is

$$\frac{d\boldsymbol{V}}{dt} = \frac{d^2 r}{dt^2}\hat{\boldsymbol{r}} + \frac{dr}{dt}\frac{d\hat{\boldsymbol{r}}}{dt} + \frac{dr}{dt}\frac{d\phi}{dt}\hat{\boldsymbol{\theta}} + r\frac{d^2\phi}{dt^2}\hat{\boldsymbol{\theta}} + r\frac{d\phi}{dt}\frac{d\hat{\boldsymbol{\theta}}}{dt}. \tag{18.26}$$

The right hand side of this equation was obtained by applying the derivative chain rule, writing down the terms in the order that they appear, and making no simplifications. The time derivatives of $\hat{\boldsymbol{r}}$ and $\hat{\boldsymbol{\theta}}$ are given by equations 18.19 and 18.21, respectively.

Applying these two equations, we find

$$\frac{d^2 \boldsymbol{R}}{dt^2} = \frac{d^2 r}{dt^2}\hat{\boldsymbol{r}} + \frac{dr}{dt}\frac{d\phi}{dt}\hat{\boldsymbol{\theta}} + \frac{dr}{dt}\frac{d\phi}{dt}\hat{\boldsymbol{\theta}} + r\frac{d^2\phi}{dt^2}\hat{\boldsymbol{\theta}} - r\left(\frac{d\phi}{dt}\right)^2 \hat{\boldsymbol{r}}. \tag{18.27}$$

Once again, all we did here was to make the substitutions, without trying to simplify anything. However, there are common factors. The most important of these are the unit vectors $\hat{\boldsymbol{r}}$ and $\hat{\boldsymbol{\theta}}$, which sort out these terms into the two directions in which they point

The final result is

$$\frac{d^2 \boldsymbol{R}}{dt^2} = \hat{\boldsymbol{r}}\left(\frac{d^2 r}{dt^2} - r\left(\frac{d\phi}{dt}\right)^2\right) + \hat{\boldsymbol{\theta}}\left(r\frac{d^2\phi}{dt^2} + 2\frac{dr}{dt}\frac{d\phi}{dt}\right). \tag{18.28}$$

There are four terms here. Two are the simple acceleration terms $\frac{d^2 r}{dt^2}$ and $\frac{d^2 \phi}{dt^2}$. The term $-r\left(\frac{d\phi}{dt}\right)^2$ is the centrifugal acceleration. An object moving in a circle at constant angular velocity $d\theta/dt$ is not moving in a straight line; it is accelerating. To be precise, it is accelerating toward the center of the circle, its acceleration being $-r\left(\frac{d\phi}{dt}\right)^2$.

The final term is the coriolis acceleration $2\frac{dr}{dt}\frac{d\theta}{dt}$. A mental image help to clarify the basic of this term. Suppose you are standing on a very large horizontal disk that rotates around its vertical axis at constant angular velocity $\frac{d\phi}{dt}$. Suppose you are walk outward, away from the center, at speed $\frac{dr}{dt}$. You are moving farther from the center so your r increases. However, the tangential velocity of the disc at each r is $v_t = r\frac{d\phi}{dt}$. As your r increases, so does your v_t, so as you walk outward from the center of the disc, you accelerate in the $\hat{\boldsymbol{\theta}}$ direction.

18.2 Worked Problems

1. The *beanstalk* is a hypothetical spaceship launching device composed of an asteroid and a long (e.g., 50,000 km) rope that is tied to the earth's equator at one end and the asteroid at the other. The asteroid is not in orbit around the earth; it is being held in place by the rope, like a stone being whirled around at the end of a long string. The rope is under a large tension T. An elevator starts at time zero, climbing up the rope from the earth's surface. In circular polar coordinates, the position of the elevator follows $r = r_0 + 0.5 a_r t^2$. Its angular velocity is a constant $\dot{\theta} = \omega$, ω being the angular rotation rate of the earth in radians/second. In polar coordinates, find the elevator's acceleration as a function of time.

2. In yet another scene from our motion picture, the heroine has been trapped in a steel cage on a giant merry-go-round. The villain is spinning the merry-go-round faster and faster, so that $\omega = \alpha t$. Also, the cage is gradually being winched from the center to the outside edge of the merry-go-round, where dire consequences will ensue; the radial position of the heroine depends on the as $r = r_0 + v_r t$. Compute the heroine's velocity and acceleration in circular polar coordinates.

18.3 Homework

1. A beanstalk is a previously mythical spaceship launching system made possible by recently advances in materials science. The launcher consists of an anchor at the equator, 40,000 or so miles of extremely strong cable going straight up from the anchor, and a large asteroid at the other end of the cable. The asteroid is not in orbit; it is held in place like a stone tied to the end of a string and whirled in circles around one's head. The launcher is a motor driven car that ascends the cable to an altitude of ca. 23,000 miles, at which point anything released from the car will be in orbit around the earth. Consider a car climbing the rope at constant speed v_o. The earth has rotational angular velocity ω. Find, in circular polar coordinates centered on the center of the earth, with the z axis passing through the south and north poles, the acceleration of the car.

2. In the Hollywood movie version of the beanstalk, the elevator cage is enormous, larger than an airship hangar. Its floor has coefficient of kinetic friction μ_k. Instead of climbing at constant speed, the elevator car has a vertical acceleration a. The heroine is strapped to a rocket sled, total mass m, inside the elevator cage. The rocket sled has a horizontal acceleration that is well approximated as $\frac{d^2 \phi}{dt^2} = bt$. Find the acceleration of the rocket sled relative to the inertial frame in circular polar coordinates.

3. The villain has returned with yet another giant merry-go-round. This time, he has chained the heroine's sidekick to the merry-go-round at a distance R from its center. The merry-go-round is horizontal. The rotation rate of the merry-go-round increases as time goes on, with $\omega = 3t$. Also, the villain has mounted a large set of rockets under the merry-go-round, so that it has vertical acceleration $2\hat{k}$. In Cartesian coordinates, find the position of the sidekick as a function of time. Find the velocity of the sidekick as a function of time. Reminder: Vectors.

4. This is another old exam problem. An otherwise rational Faculty member travels to a well-known foreign country where for a moderate sum of money you actually can go for a ride in the Navigator's seat of a very-high-performance fighter aircraft. Have paid for the flight, she and her pilot take the aircraft to an altitude of 10,000m, at which point the aircraft does a nose-up and flies in a series of vertical circles with a radius of R m. The aircraft has a high thrust-to-weight ratio; it maintains a constant speed V during the circle. Referring to Figure 18.3, for points a), b), c), and d) draw the force diagram for the Faculty member, who is in contact with the aircraft because she is firmly belted to a high-backed chair. At each point, compute the normal force, a vector, on the pilot. Aside: This is a slight variation on an old exam question. Some people made erroneous assumptions as to the direction of the normal force on the Faculty member due to the chair, leading them to the interesting conclusion that the force of gravity had been turned off. Remember, $\boldsymbol{N}$ is a vector.

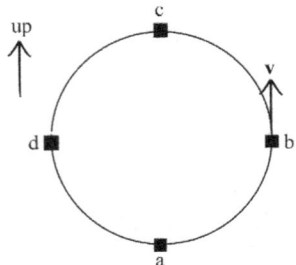

Figure 18.3:

5. A mass moving in a spiral has a position (r, ϕ, z) in polar coordinates. The individual coordinates depend on time as $r = a\phi$, $\phi = \alpha t^2/2$, and $z = 0$. Here a and α are constants. Sketch or graph the spiral for a few turns. (a) Find the acceleration of the mass in circular polar coordinates. (b) Find the angle $\theta > 0$ at which the radial component of the acceleration goes to zero. (c) Find a time $t > 0$ at which the radial and angular components of the acceleration are equal.

6. Confirm that the claim $d\hat{\boldsymbol{u}}/dt \cdot \hat{\boldsymbol{u}} = 0$ is correct if $\hat{\boldsymbol{u}}$ is $\hat{\boldsymbol{r}}$ or $\hat{\boldsymbol{\phi}}$.

7. Consider an extremely thin rod of mass M and length ℓ with a uniform mass density M/ℓ along its length. The rod rotates around a point at one of its ends. We consider the behavior in free fall, so g is effectively zero. What is the tension in the rod at the pivot point? Hint: Think of the rod as a series of differential length segments. Each segment has its own acceleration, and is connected to the two segments that are neighboring along the rod.

18.4 Solutions to the Worked Problems

1. There is a mass M climbing a beanstalk cable into geocentric orbit. It climbs at $r = \frac{1}{2}a_r t^2$. The cable rotates around the pole to pole axis at $\dot{\theta} = c$. You are asked to find $\boldsymbol{v}$ and $\boldsymbol{a}$ in cylindrical polar coordinates. The time rates-of-change of r and θ are then

$$\frac{dr}{dt} = a_r t, \tag{18.29}$$

$$\frac{d^2 r}{dt^2} = a_r, \tag{18.30}$$

$$\frac{d\phi}{dt} = c, \tag{18.31}$$

$$\frac{d^2 \phi}{dt^2} = 0. \tag{18.32}$$

We lift from this Chapter the results

$$\boldsymbol{v} = \frac{dr}{dt}\hat{\boldsymbol{r}} + r\frac{d\phi}{dt}\hat{\boldsymbol{\theta}} \tag{18.33}$$

$$\boldsymbol{a} = \left(\frac{d^2 r}{dt^2} - r\left(\frac{d\phi}{dt}\right)^2\right)\hat{\boldsymbol{r}} + \left(r\frac{d^2\phi}{dt^2} + 2\frac{dr}{dt}\frac{d\phi}{dt}\right)\hat{\boldsymbol{\theta}}, \tag{18.34}$$

and substitute the values from this problem for the derivatives, leading to

$$\boldsymbol{v} = a_r t \hat{\boldsymbol{r}} + rc \hat{\boldsymbol{\theta}} \text{ and} \tag{18.35}$$

$$\boldsymbol{a} = (a_r - rc^2)\hat{\boldsymbol{r}} + (r \cdot 0 + 2a_r tc)\hat{\boldsymbol{\theta}}. \tag{18.36}$$

18.4. SOLUTIONS TO THE WORKED PROBLEMS

2. We have another polar coordinate problem. As shown in the Chapter,

$$v = \frac{dr}{dt}\hat{r} + r\frac{d\phi}{dt}\hat{\theta} \text{ and} \tag{18.37}$$

$$a = \left(\frac{d^2r}{dt^2} - r\left(\frac{d\phi}{dt}\right)^2\right)\hat{r} + \left(r\frac{d^2\phi}{dt^2} + 2\frac{dr}{dt}\frac{d\phi}{dt}\right)\hat{\theta} \tag{18.38}$$

In this problem

$$\frac{d\phi}{dt} = \alpha t \text{ and} \tag{18.39}$$

$$r = r_o + v_r t \tag{18.40}$$

so

$$\frac{d^2\phi}{dt^2} = \alpha, \tag{18.41}$$

$$\frac{dr}{dt} = v_r, \text{ and} \tag{18.42}$$

$$\frac{d^2r}{dt^2} = 0. \tag{18.43}$$

Substituting the above five equations into the general forms for v and a,

$$v = v_r\hat{r} + (r_o + v_r t)\alpha t\hat{\theta}, \tag{18.44}$$

$$a = (0 - (r_o + v_r t)(\alpha t)^2)\hat{r} + ((r_o + v_r t)\alpha + 2v_r\alpha t)\hat{\theta}. \tag{18.45}$$

This page reserved for your notes.

Chapter 19

Angular Momentum

19.1 Introduction

We now consider angular momentum. The angular momentum L is a vector. Newton's Laws of Motion determine the time-rate-of-change of L. For an isolated group of objects, the total angular momentum is conserved. From these results, interesting problems can be solved.

19.2 Angular Momentum

We define the angular momentum L of a point object to be

$$L = r \times p. \tag{19.1}$$

If the object is extended, say if it is the bob on a pendulum, the definition requires refinement. We will return to this point. Here r is the position of the object and p is the momentum of the object. Remember that $p = mv$ refers to Cartesian coordinates. The value of L depends on the location of the origin of the coordinate system in which r is measured. If you move the origin, in general you will change the value of L. If the velocity of the object is such that the object will pass through the origin, then v and hence p are parallel to r, in which case $r \times p = 0$.

The definition of L *does not* imply or require that an object is performing circular motion or is 'going in a circle'. If the object is performing circular motion, the definition does not imply or require that the origin of r is located at the center of rotation.

We now consider several examples. We remain in SI units throughout.

General case: We have a 500 kg mass. Its velocity is $v = -10\hat{i}$ m/s. Its location relative to the origin is $r = 500\hat{i} + 10\hat{j} + 2\hat{k}$ m. *Because we are using Cartesian coordinates*, we may write $p = mv$. If we were instead using, for example, cylindrical polar coordinates, $p = mv$ would be incorrect. We therefore can write

$$L = r \times mp, \tag{19.2}$$
$$L = (500\hat{i} + 10\hat{j} + 2\hat{k}) \times 500(-10\hat{i}). \tag{19.3}$$

In the above, there are several housekeeping issues. First, the algebraic symbols were replaced with their numerical values, but no simplifications were made. Second, the units (here kilograms, meters, seconds) are suppressed until the calculation is finished, so that m-the-mass and m-the-unit-meters do not become conflated. Third, we only plugged in numbers at the end of this very short calculation. Taking the vector cross product, we obtain

$$L = 5 \cdot 10^4 \hat{k} - 1 \cdot 10^4 \hat{j} \text{ kg m}^2/\text{s}. \tag{19.4}$$

I've inserted the units for L, which are m · kg · m/s or kg · m^2/s.

There is, however, a minor inobvious detail. The multiplication that takes you from m · kg · m/s to kg · m^2/s was originally a vector cross product. The units therefore have hidden in them their vector nature. It just isn't written out. What is that vector nature?

A simple vector has a true three-dimensional direction in space. There is a consequence. Suppose we invert the direction of the $\hat{i}$, $\hat{j}$, and $\hat{k}$ basis vectors. $\hat{i}$ now points to the left. $\hat{j}$ now points down. $\hat{k}$ now points into the page. After inversion, a velocity vector $\boldsymbol{v} = a\hat{i} + b\hat{j} + c\hat{k}$, keeps its direction in space. The inversion therefore reverses the signs of the three constants a, b, and c, so after inversion the velocity vector becomes $-a\hat{i} - b\hat{j} - c\hat{k}$. The same sign inversion would affect a position vector $\boldsymbol{r}$; inversion of the direction of the basis vectors transforms $\boldsymbol{r} \Rightarrow -\boldsymbol{r}$ and $\boldsymbol{v} \Rightarrow -\boldsymbol{v}$.

Now we come to $\boldsymbol{L}$. We know $\boldsymbol{L} = \boldsymbol{r} \times \boldsymbol{p}$. If we invert the three basis vectors, $\boldsymbol{r}$ and $\boldsymbol{p}$ become $-\boldsymbol{r}$ and $-\boldsymbol{p}$. The two minus signs cancel, meaning that, under inversion, $\boldsymbol{L}$ goes onto $+\boldsymbol{L}$. The basis vectors are all being inverted, so $+\boldsymbol{L}$ points opposite in direction to the direction that $\boldsymbol{L}$ pointed before the basis vectors were inverted.

Said differently, $\boldsymbol{r}$ and $\boldsymbol{p}$ have absolute directions in space that do not change when the direction of the basis vectors is inverted. Inverting the basis vectors changes the representation, the list of coordinates, of each vector in the new coordinate system, just as changing from Cartesian to cylindrical polar coordinates changes the representation of a vector, from its (x, y, z) components to its (r, θ, z) components. In contrast, $\boldsymbol{L}$ has the same representation before and after the basis vectors are inverted. If $\boldsymbol{L}$ points straight up before the basis vectors are inverted, so that $\boldsymbol{L} = L\hat{k}$ before the inversion, after the inversion it is still true that $\boldsymbol{L} = L\hat{k}$, but now $\boldsymbol{L}$ points straight down. For this reason, $\boldsymbol{L}$ is technically not a vector; it is a *pseudovector* or *axial vector*. Is this a difficulty? If we have an object moving in a circle, its motion is confined to a plane; nothing is moving in the direction that the angular momentum vector points, so it is arbitrary whether we say that $\boldsymbol{L}$ points up or points down, so long as we stay consistent in our direction.

We now return to the dimensions of angular momentum, which are mass· length^2· time. Buried in those units is the statement that $\boldsymbol{L}$ is a pseudovector, not a scalar. A quantity X that had dimensions mass· length^2· time but was a scalar actually would not have the same dimensions as angular momentum. We may say (in a very special case) that work can be *force times distance*, while angular momentum can be *distance times momentum*, but the word *times* has completely different meanings in the two cases. In one case, *times* refers to the scalar product of two numbers; in the other case, *times* refers to the vector product of two vectors.

As another example, we will consider what happens to $\boldsymbol{L}$ when we change our choice of origin. Suppose I have a mass m on the end of a string. I am at the origin $(0,0,0)$. At the moment of interest the mass is at $(0,1,0)$ and has velocity $-2\hat{i}$. *With respect to my location*, the angular momentum of the mass is $\boldsymbol{L} = 1\hat{j} \times m(-2\hat{i})$, which multiplies out to $\boldsymbol{L} = -2m\hat{j} \times \hat{i} = 2m\hat{k}$. (Check my math. Did you get $+2m\hat{k}$? If you found $\boldsymbol{L} = -2m\hat{k}$, did you remember that the vector cross product is anticommutative?)

Now suppose I instead calculate the angular momentum around the point $(-1, 1, 0)$. That's a perfectly legitimate calculation, because in calculating the angular momentum I am free to put the origin of the coordinate system wherever I choose. (Though having made a choice, I must use the same choice for all angular momenta in the problem.) With $(-1, 1, 0)$ as the origin, the vector from the origin to the mass is $(0, 1, 0) - (-1, 1, 0) = (1, 0, 0)$. The angular momentum relative to this center is $\boldsymbol{L} = 1\hat{i} \times m(-2\hat{i})$. However, the cross product of two vectors that are parallel or antiparallel is $\boldsymbol{0}$, so with respect to the second origin $\boldsymbol{L} = -2m\hat{i} \times \hat{i} = \boldsymbol{0}$.

What happened here? How can that angular momentum have two different values? The answer is that the angular momentum of an object is only defined with respect to a choice of origin. If you change the origin, you change the value of each object's angular momentum.

19.3 Torque

How do we act on an object in order to change its angular momentum? The answer is that we apply a torque. Torque? A torque could be said, in many cases, to be a twisting force, but that's very imprecise. We begin with the fundamental definition of angular momentum $\boldsymbol{L} = \boldsymbol{r} \times \boldsymbol{p}$, and take a derivative with respect to time. Applying the chain rule for the vector cross product,

$$\frac{d\boldsymbol{L}}{dt} = \boldsymbol{r} \times \frac{d\boldsymbol{p}}{dt} + \frac{d\boldsymbol{r}}{dt} \times \boldsymbol{p}. \tag{19.5}$$

For many of you, this equation is actually a bit of a cheat. You should have seen the chain rule in calculus, but the derivation of that rule referred to the multiplication of two scalars or two scalar functions, e.g.,

19.4. CONSERVATION OF ANGULAR MOMENTUM

$df(x)g(x)/dx$. Here, however, the product is the vector product. Is the chain rule still valid? The answer is 'yes'. Proving my claim is a homework problem. The final term in the equation vanishes, because $\frac{d\boldsymbol{r}}{dt} = \boldsymbol{v}$, $\boldsymbol{p} = m\boldsymbol{v}$, and the cross product of a vector with itself vanishes.

We now recall Newton's Second Law

$$\boldsymbol{F} = \frac{d\boldsymbol{p}}{dt}. \tag{19.6}$$

Applying this form to equation 19.5, we have

$$\frac{d\boldsymbol{L}}{dt} = \boldsymbol{r} \times \boldsymbol{F}. \tag{19.7}$$

The right-hand-side of this equation is defined to be the torque $\boldsymbol{\tau}$, namely

$$\boldsymbol{\tau} = \boldsymbol{r} \times \boldsymbol{F}. \tag{19.8}$$

If the total torque on a mass is zero, $\frac{d\boldsymbol{L}}{dt} = \boldsymbol{0}$, or $\boldsymbol{L}$ = constant. For example, objects that are simply sitting in place and that have no applied forces on them have zero torque applied to them, in which case $\boldsymbol{L}$ has a specific constant value: $\boldsymbol{0}$.

19.4 Conservation of Angular Momentum

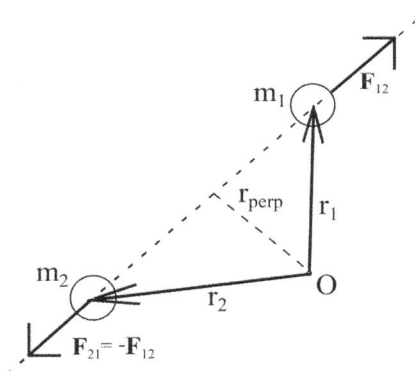

Figure 19.1: Two masses m_1 and m_2 that exert *central forces* $\boldsymbol{F}$, forces that act along their lines of centers, on each other.

We now consider a specific situation, name a pair of objects subject to no external forces, as seen in Figure 19.1. The two objects have masses m_1 and m_2, and are located at $\boldsymbol{r}_1$ and $\boldsymbol{r}_2$, respectively. The two objects also exert a force on each other. As a special case, the action-reaction pair of forces that the two objects exert on each other are *central forces*. By central forces, we mean that the two forces in the action-reaction pair of forces act parallel to the line of centers of the two objects, as shown in Figure 19.1. These forces are called central forces because there are also forces that are not central. For example, some forces only act perpendicular to the line of centers. There are also three-body forces, determined by the relative positions of three rather than two bodies. Three-body forces appear in and are important for understanding molecular motions in liquids. Here we are considering only two-body central forces.

The two forces each have one non-zero component, corresponding to the force acting along the line of centers. The two components of each force, perpendicular to the straight line connecting the centers of the two objects, are both zero.

The two objects are both moving. Their velocities are $\boldsymbol{v}_1$ and $\boldsymbol{v}_2$, respectively. Their angular momenta are therefore

$$\boldsymbol{L}_1 = \boldsymbol{r}_1 \times m_1 \boldsymbol{v}_1 \tag{19.9}$$

and

$$\boldsymbol{L}_2 = \boldsymbol{r}_2 \times m_2 \boldsymbol{v}_2. \tag{19.10}$$

Here we consider the total angular momentum $\boldsymbol{L} = \boldsymbol{L}_1 + \boldsymbol{L}_2$, in particular its time derivative. We have

$$\frac{d}{dt}(\boldsymbol{L}_1 + \boldsymbol{L}_2) = \boldsymbol{\tau}_1 + \boldsymbol{\tau}_1. \tag{19.11}$$

Substituting for the two torques, we find

$$\frac{d\boldsymbol{L}}{dt} = \boldsymbol{r}_1 \times \boldsymbol{F}_{12} + \boldsymbol{r}_2 \times \boldsymbol{F}_{21}. \tag{19.12}$$

As shown in the above Figure, $\boldsymbol{r}_1$ and $\boldsymbol{r}_2$ must be measured from the same origin. The origin may be placed in any convenient location, but the same origin must be used throughout the calculation. The force $\boldsymbol{F}_{12}$ on particle 1 due to particle 2, and the force $\boldsymbol{F}_{21}$ on particle 2 due to particle 1, are an action-reaction pair. They therefore point in opposite directions, as shown in the Figure. From the Third Law, the magnitudes of $\boldsymbol{F}_{12}$ and $\boldsymbol{F}_{21}$ must be equal.

Because the position and force vectors all lie in the plane of the page, the two torques must both be perpendicular to the page. Because the two forces point in opposite directions, one torque vector must point into the page, while the other torque must point out of the page. Furthermore, the two torque vectors must have equal magnitudes. Why? The magnitude of a torque vector may be calculated as $|\boldsymbol{r} \times \boldsymbol{F}| = r_\perp F$. The forces in the two torques are equal in magnitude. The figure shows $r_\perp$, which is the distance along the perpendicular to the line of centers to the origin. The two torques have the same $r_\perp$. $r_\perp F$ of the two torques are therefore equal. The two torque vectors have now been shown to be equal in magnitude and opposite in direction, so they add to zero.

We thus find, for objects interacting via central forces, with no external forces acting, that

$$\frac{d\boldsymbol{L}}{dt} = \boldsymbol{0} \tag{19.13}$$

or equivalently

$$\boldsymbol{L} = \text{constant}. \tag{19.14}$$

If the forces are all central and internal, the total angular momentum of the system is a constant. That's a *conservation law*. The above two equations are the *law of conservation of angular momentum*. The two equations do apply to collisions, if the forces are central.

19.5 Displacement of the Origin

The angular momentum of an object in general depends on what location $\mathcal{O}$ is chosen for the origin. If we displace the origin, say from $\mathcal{O}$ to $\mathcal{O} - \boldsymbol{a}$, then the locations of all of the objects are displaced oppositely, so that an object that had been at $\boldsymbol{r}_i$ will now be located at $\boldsymbol{r}_i + \boldsymbol{a}$. The object has not been moved; the definition of its location vector has been altered.

As a result, the object's angular momentum has been changed. Before the displacement of the origin, an object i having linear moment $\boldsymbol{p}_i$ would have had angular momentum $\boldsymbol{r}_i \times \boldsymbol{p}_i$. After the displacement, the same object would have had angular momentum $(\boldsymbol{r}_i + \boldsymbol{a}) \times \boldsymbol{p}_i$. An essentially similar argument applies to the torque on an object. If the torque before the displacement was $\boldsymbol{r}_i \times \boldsymbol{F}_i$, the torque on object i after the displacement would be $(\boldsymbol{r}_i + \boldsymbol{a}) \times \boldsymbol{F}_i$.

The total angular momentum and the total torque of a group of objects have this same property. If the total angular momentum of a group of objects is

$$\boldsymbol{L} = \sum_i \boldsymbol{r}_i \times \boldsymbol{p}_i, \tag{19.15}$$

then the effect of the displacement is to change the total angular momentum to

$$\boldsymbol{L} = \sum_i (\boldsymbol{r}_i + \boldsymbol{a}) \times \boldsymbol{p}_i. \tag{19.16}$$

If the total torque on a group of objects is

$$\boldsymbol{L} = \sum_i \boldsymbol{r}_i \times \boldsymbol{F}_i, \tag{19.17}$$

then the effect of the displacement is to change the total torque to

$$L = \sum_i (r_i + a) \times F_i. \quad (19.18)$$

However, there are two interesting special cases. If $\sum_i p_i = 0$, the total angular momentum is independent of the location of the origin. And if $\sum_i F_i = 0$, the total torque is independent of the location of the origin. These two claims are homework problems.

19.6 ω and L

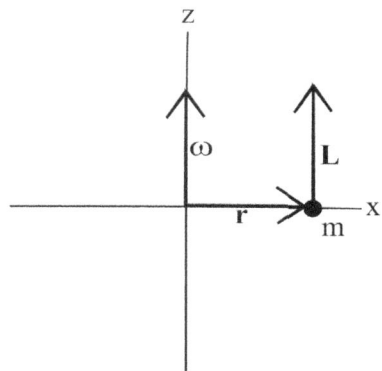

Figure 19.2: A mass m located at r. m is rotating in the x-y plane (the plane is perpendicular to the page) around the z axis. Its angular velocity is ω as indicated. Its momentum p is directly into the page. From the right-hand-rule, its angular momentum L is parallel to ω.

This section considers the relationship between the angular rotation rate vector ω and the angular momentum vector L. We begin with the simplest case, in which a point mass m is rotating around the origin. The object rotates in the x-y plane at $z = 0$. The angular rotation vector ω is taken to be parallel to the z-axis, so that $\omega = \omega \hat{k}$. The Figure shows the rotation vector ω and the location vector r of the object. As drawn, ω and r are both in the plane of the paper, so $v = \omega \times r$ is perpendicular to the plane of the paper. From the right-hand-rule, v points into the page, so $v = r\omega \hat{j}$.

The angular momentum vector is given by $r \times p$, namely

$$L = r\hat{i} \times m\omega r \hat{j}, \quad (19.19)$$

which reduces to

$$L = mr^2 \omega \hat{k}. \quad (19.20)$$

In this special case, L is parallel to the rotation vector $\omega = \omega \hat{k}$.

Minor technical issue: Does the rotation vector ω have to be shown as passing through the origin $\mathcal{O}$? ω is placed so as to define the axis around which there is rotation, meaning that the origin must be located as shown. Reminder: If we have objects rotating around an axis, we have a non-zero rotation vector ω and also a non-zero angular momentum vector L, but as shown early in this Chapter an object moving in a straight line has an angular momentum that is not equal to zero. (Exception: The angular momentum of an object moving in a straight line is zero if the straight line, when extended, passes through the origin.)

We now consider a slightly different case, in which the mass is no longer at $z = 0$, as seen in the next figure. Relative to the previous figure, the position vector r is still in the plane of the page. ω is still in the plane of the page, so once again v is perpendicular to the plane of the page. Which way does L now point? L must be perpendicular to $p = mv$, so it must be in the plane of the paper. L must also be perpendicular to r. The right hand rule gives the direction: As shown in Figure 19.3, L is not parallel to ω.

[Aside: One might ask what mathematical construction turns one vector into another, when the second vector is not parallel to the first. A general answer is provided by matrix multiplication. We introduce a 3×3 matrix I, the moment of inertia *tensor*. Its nine elements, since it is a matrix, are the I_{ij}, where i and j independently take values 1, 2, 3 corresponding to the x, y and z coordinates. In the case at hand, L and ω are related by

$$L = I \cdot \omega \equiv \sum_{j=1}^{3} I_{ij}\omega_j \qquad (19.21)$$

where here ω_j represents the three components of ω, the right-hand-most side of this equation being interpreted as a vector whose components are labelled by i.]

Is it of practical interest that ω and L are not always parallel? Yes. An example familiar to some readers will be buying a new set of tires for a car. The last step in preparing the tires to be mounted on the vehicle is to 'balance' each tire, done by attaching small weights to the rims. The purpose of balancing the tires is to ensure that L and ω are parallel, without which condition the tire will try to wobble on the axle, possibly doing all sorts of interesting and potentially expensive-to-repair damage.

However, in many interesting cases ω and L are parallel. *If the two vectors are parallel*, we can replace $L = I \cdot \omega$ with $L = I\omega$. How may we readily recognize when this special case has arisen? The next Figure supplies a useful if not complete answer.

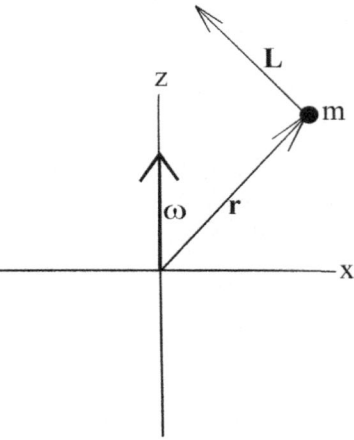

Figure 19.3: A mass m located at r, rotating around the z axis. Its angular velocity ω is indicated. Its velocity v and momentum p are directly into the page. The right-hand-rule gives the direction of its angular momentum L, as shown. This time L and ω are not parallel.

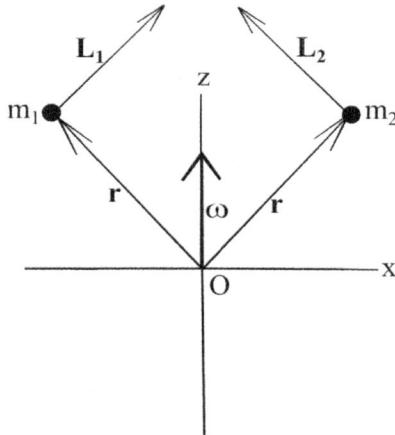

Figure 19.4: Two equal masses m_1 and m_2 placed symmetrically on opposite sides of the axis of rotation and their angular momenta L_1 and L_2. The two angular momenta are calculated with respect to the same origin O so they can be added. In $L_1 + L_2$, the x components of the two angular momentum vectors are equal in magnitude, but opposite in direction, so they sum to zero. The z components of the two angular momentum vectors add, giving a vector $L_1 + L_2$ that is parallel to ω.

Here we see a rotation vector ω, and two equal masses m_1 and m_2. The two masses are equal, so that $m_1 = m_2$. The masses, which are located at r_1 and r_2, respectively, are rotating around the axis. Their perpendicular distances from the axis are equal, so their velocities are equal in magnitude and opposite in direction, namely into the plane of the paper and out from the plane of the paper. Finally, the z-components of r_1 and r_2 are equal to each other, while the x-components of r_1 and r_2 are equal in magnitude but point in opposite directions. The angular momenta of the two masses are then as indicated in the figure. It follows that

$$L_1 = -L_x\hat{i} + L_z\hat{k} \qquad (19.22)$$
$$L_2 = L_x\hat{i} + L_z\hat{k}, \qquad (19.23)$$

The x-components of the two angular momentum vectors sum to zero. The total angular momentum is $L = L_1 + L_2 = 2L_z\hat{k}$, so here $L \| \omega$.

$L_1 + L_2$ is parallel to ω if the two masses m are mirror symmetric left and right with respect to the axis around which they are both rotating. If you have several pairs of masses, each pair must be mirror symmetric with respect to the rotation axis but do not all have to lie in the same x-z plane. For example, the wheel of a bicycle may be imagined to be composed of a very large number of pairs of masses m, each pair being mirror symmetric with respect to the rotation axis. In general any object that can be described as a *surface of revolution*, i.e., you can make it on a lathe, has the needed symmetry. A dumbbell with a rotation axis through the center, the rotation axis being parallel or perpendicular to the line of centers of the two masses, has the needed symmetry.

[Aside: Is there a more general statement than 'masses must be mirror symmetric in pairs around the desired rotation axis'. The answer is 'yes'. Any object, regardless of shape, has three axes that *pass through the center of mass* around which rotation gives $\boldsymbol{L} \| \boldsymbol{\omega}$.]

19.7 Discussion

Here we discuss several minor technical issues relating to angular momentum and torque.

First, in describing angular momentum, it is incorrect to refer to rotation as being clockwise or counterclockwise. Imagine a transparent wall with a clock on it. From the usual side, the clock hands are moving in the clockwise direction. However, if you move around to the other side of the wall, because the wall is transparent you can still see the clock hands. The hands will now be seen to be rotating in the counterclockwise direction. The rotation is the same, but the direction has changed name from clockwise to counterclockwise.

Second, the sign applied to $\boldsymbol{L}$ is arbitrary. Suppose you calculate the direction of $\boldsymbol{L}$ using your left hand rather than your right hand. The thumb still goes in the direction of $\boldsymbol{r}$. The forefinger still points in the direction of $\boldsymbol{p}$. The remaining fingers, pointing perpendicular to the plane formed by $\boldsymbol{r}$ and $\boldsymbol{p}$, point in the direction of $\boldsymbol{L}$. However, if when you used your right hand you found that $\boldsymbol{L}$ pointed up, when you repeat the calculation using your left hand you find that $\boldsymbol{L}$ now points down. Surprise! Which direction is correct?

The answer is that nothing is moving in the direction of $\boldsymbol{L}$. $\boldsymbol{L}$ is a pseudovector, not a true vector. Recall that $\boldsymbol{L} = \boldsymbol{r} \times \boldsymbol{p}$. The particle motion is in the plane perpendicular to $\boldsymbol{L}$, the plane defined by the two vectors $\boldsymbol{r}$ and $\boldsymbol{p}$. If you reverse the coordinate axes so that $\boldsymbol{r}$ and $\boldsymbol{p}$ switch signs, $\boldsymbol{L}$ has the same numerical value that it did before. Does nature care whether you use the left hand rule or the right hand rule? In the cases of interest here, the answer is in the negative. It turns out that the weak nuclear force does not, in a certain sense, 'conserve parity', i.e., there are subatomic phenomena in which nature does distinguish left and right handedness.

Third, we have discussed total angular momentum and angular momentum conservation. For objects acted on by central forces, the total angular momentum is conserved. Electrical and gravitational forces are central forces. However, in making that statement, all angular momenta must be calculated using the same origin. For example, suppose we have two wheels mounted on two parallel axes. They are rotating in the same direction. Their individual angular momenta, each calculated using its rotation axis as the origin, are $\boldsymbol{L}_1$ and $\boldsymbol{L}_2$. We may not, however, refer to their total angular momentum $\boldsymbol{L}$ as $\boldsymbol{L}_1 + \boldsymbol{L}_2$, because $\boldsymbol{L}_1$ and $\boldsymbol{L}_2$ are calculated using different points for their origins.

Fourth, we specified the right hand rule in terms of directions taken by the thumb, the index finger, and the remaining fingers. An alternative and erroneous description refers to the curvature of the four fingers as corresponding to a 'direction of rotation', with the thumb stretching perpendicular to them giving the direction of $\boldsymbol{L}$. The reason this description is erroneous is that, while rotating objects do have an angular momentum with respect to an origin, objects moving in a straight line also have an angular momentum with respect to an origin. In each case, the direction and numerical value of the angular momentum vector depends on where the origin of the coordinate system is located.

19.8 Worked Problems

1. We have three masses spaced along a rigid massless rod that initially lies on the x-axis. Mass 1 is 1 kg and starts at $x = 0$. Mass 2 is 2 kg and is $+1$ m from mass 1 along the x axis. Mass 3 is 4 kg and is $+ 1$ m along the x-axis from mass 2, meaning it is not 0 or 1 m from the origin. The masses are spun around an axis occupied by the z-axis, revolving at 2 radians/second. What is the angular momentum of the three masses and their rod?

2. A mass m is at the mobile end of a rigid, nearly massless, pivoted rod of length ℓ. The rod extends perpendicular to the axis of rotation out to the mass. The axis of rotation lies on the z-axis, from which

the angular momentum is to be calculated. The mass m incorporates an electrically driven propeller which applies a force $\mathbf{F}$ that lies in the plane of rotation, perpendicular to the rotating rod. (a) The fan is gradually sped up in speed, so that it supplies a force whose magnitude is $F_o \exp(st)$, where F_o and s are constants determined by the fan rotation. (a) What is the angular momentum of the mass relative to its rotation axis as a function of time? (b) Find the angular velocity of the mass m around the rotation axis as a function of time.

3. Prove
$$\frac{d}{dt}(\mathbf{A} \times \mathbf{B}) = \frac{d\mathbf{A}}{dt} \times \mathbf{B} + \mathbf{A} \times \frac{d\mathbf{B}}{dt} \tag{19.24}$$
by (1) writing the cross product in term of the scalar components, (2) taking the derivatives by using the rule for derivatives of products having scalar multiplication, and (3) collecting terms. Warning: You are not proving the derivative-of-product rule you learned in calculus. For example, your proof probably assumed that multiplication is commutative, and the cross product is not commutative, so you may not have proved this before.

4. A space station consists of two very heavy objects, M_1 located at $(0,0,0)$ and M_2 located at $(\ell,0,0)$. The two objects are linked by a very light boom having length ℓ. A rocket motor applies a force of magnitude a in the $-\hat{\mathbf{j}}$ direction to M_2. What is the torque applied to the space station, if the torque is calculated relative to the location of (A) M_2, (b) M_1, or (c) the space station's center of mass?

19.9 Homework

1. Prove that if the total momentum of a group of objects satisfies $\sum_i \boldsymbol{p}_i = \mathbf{0}$, the total angular momentum is independent of the location of the origin.

2. Prove that if the total force on a group of objects satisfies $\sum_i \boldsymbol{F}_i = \mathbf{0}$, the total torque is independent of the location of the origin.

3. Sometimes a seemingly general proof is not valid in a special case. For the two previous problems, is the claim true if there is only one object in the group?

4. A truck proceeds due east (the $+x$ direction) on Salisbury street. The truck has a mass of 22,000 kg, and is moving at 15 m/s. The truck's center of mass stays in the center of the lane. Three students are standing 4m south of the center of the lane (here north is the $+y$ direction). They compare notes as to the truck's angular momentum with respect to them. The first notes that the truck is moving in a straight line, so the truck's angular momentum is zero. The second announces that the angular momentum of the truck is $1.32 \cdot 10^6$ kg m^2/s. The third correctly calculates the trucks' angular momentum. Does she agree with either of the other two students? What angular momentum does she calculate for the truck?

5. You are given two vectors, $\mathbf{A} = 3\hat{i} + 5\hat{j} + 0\hat{k}$ and $\mathbf{B} = -\hat{i} + 2\hat{j} - 3\hat{k}$. Find the cross product of the vectors $\mathbf{A}$ and $\mathbf{B}$ and find the angle between the two vectors.

6. A truck proceeds down George Street in the illegal $-5\hat{i} - \hat{k}$ direction. The truck has a mass of 10,000 kg, and is moving at 20 m/s. The truck's center of mass stays in the center of the street. Three students are standing 3 m south of the center of the street (here north is the $+\hat{j}$ direction). They compare notes, as the truck passes, as to the truck's angular momentum with respect to them. The first observes that the truck is moving in a straight line, and concludes that the truck's angular momentum is zero. The second announces that the truck's angular momentum is $6.00 \cdot 10^5$ kg m^2/s. The third correctly calculates the truck's angular momentum. Does he agree with either of the other two students? What does he calculate for the angular momentum of the truck?

7. Four masses, of 1, 2, 3, and 4 kg, are located in a rectangular frame, with the corners at $(0,0,0)$, $(2,0,0)$, $(2,0,3)$, and $(0,0,3)$, respectively. The frame is rigid but massless. The frame is rotated around the z-axis with angular rotation vector $\omega = 3\hat{\mathbf{k}}$. Calculate the total angular momentum $\mathbf{L}$ of the four masses with respect to the origin. Is $\mathbf{L}$ parallel to ω?

8. Four masses, of mass 1, 2, 3, and 4 kg, momentarily lie in a line on the x-axis. Their locations are $(-2,0,0)$, $(-1,0,0)$, $(1,0,0)$, and $(2,0,0)$, respectively. They are rotating in circles around the z-axis; they all have the same rotational velocity $\omega = 3\hat{\mathbf{k}}$. Find their total angular momentum with respect to the origin.

9. Our intrepid astronauts are in a circular equatorial orbit around a distant planet. The orbital radius of the 750,000 kg spaceship is 20,000 km. The astronauts fire the main thrusters, which apply a force of 50,000 N to the spaceship. The astronauts simultaneously rotate the spaceship so that the thrusters always apply the force parallel to the surface of the planet that is directly under the spaceship, and in the plane of rotation. How long does it take for the thrusters to double the angular momentum of the spaceship?

10. Show by explicit integration that the torque exerted by the force of gravity on a *uniform* rod having mass M and length L is equal to the torque exerted by gravity on the rod by a point mass M located at the center of mass of the rod. DO NOT blindly use this result for non-uniform rods.

11. On a fine winter day (if you ignore the layer of glare ice covering everything) a pickup truck and a car having masses m and M, respectively, proceed south on Park Avenue and west on Salisbury Avenue. They travel at the same speed v; their velocity vectors are perpendicular. They enter the Park/Salisbury intersection simultaneously, and collide at its center. Following the collision, they move out of the intersection as a single piece of wreckage. (You may ignore changes in mass attendant to broken glass, etc.) (i) Is this collision elastic, inelastic, or neither of the aforementioned? (ii) List up to five distinct physical quantities that are sometimes conserved during collisions, and indicate whether each of these quantities is conserved here. (This question will be graded as right minus wrong.) (iii) Calculate the speed and direction of the wreckage after the collision. Include a sketch showing your coordinate system. Label your axes! (iv) Calculate the change in the kinetic energy of the vehicles from before to after the collision. (Of course, if something is conserved, 'calculate' can be a very short process.) (v) After the collision, the wreckage slides up the hill towards the Antiquarian Society Library. The slope of the hill measured perpendicular to its border (which is parallel to Park Avenue) is ϕ. Assuming that the ice is smooth, so that the coefficient of kinetic friction is 0.00, through what vertical distance does the wreckage slide up the hill? (vi) Suppose that instead of sliding over glare ice, the wreckage slides up the hill over grass with coefficient of kinetic friction μ_k. The sliding trajectory makes an angle θ with respect to the horizontal. Through what vertical distance does the wreckage slide up the hill? (vii) Take Park and Salisbury Avenues to have widths W and w, respectively. The vehicles are driving exactly down the center lines. Compute the total angular momentum $\mathbf{L}$ of the two vehicles before the collision, as determined by an observer located exactly on the northeast corner of the intersection. For this section of the problem, you may approximate each vehicle as being a point mass, and you may assume that the centers of mass of the two vehicles are at the height of the observer. (viii) The same collision is observed by an ant located on the ice exactly at the collision point. Calculate the angular momentum $\mathbf{L}'$ of the wreckage as it moves away from the collision point. For this part of the problem, you may assume that the wreckage is not rotating around its own center of mass, and you may approximate that the center of mass of the wreckage is at the altitude of the ant. (ix) I defined $\mathbf{L}$ and $\mathbf{L}'$ in parts vii and viii of this question. Why did you find $\mathbf{L} \neq \mathbf{L}'$? Isn't angular momentum conserved during collisions? Alternatively, if you found that $\mathbf{L} = \mathbf{L}'$, explain why these two quantities should be equal to each other.

12. A frisbee is mounted through its center to the end of a drill bit, perpendicular to the end of the bit. The drill points along the z axis, and rotates at -100 rpm. (a) What is the rotation vector of the drill? (b) The frisbee is 0.25 m in diameter. The coordinate origin is placed where the frisbee is joined to the drill. Consider a point on the perimeter of the frisbee in the $\hat{\jmath}$ direction relative to the origin. What is its coordinate vector? (c) A nut of mass $3 \cdot 10^{-2}$ kg is mounted at the perimeter point in question. What is its velocity? What is its angular momentum? (d) We now *move the origin*, so that the origin is at the base of the drill, and the frisbee is joined to the drill at a location $+0.3\hat{\mathbf{k}}$ relative to the origin. What is the location vector of the nut? What are the nut's velocity and angular momentum? (e) Use (c) and (d) to answer: Are the momentum and angular momentum vectors always parallel?

Are the rotation and angular momentum vectors always parallel? Is the angular momentum vector independent of the location of the origin?

13. Consider an extremely thin rod of mass M and length ℓ with a uniform mass density M/ℓ along its length. The rod rotates around a point at one of its ends, rotation being around an axis perpendicular to the length of the rod at angular velocity $\boldsymbol{\omega}$. We consider the behavior in free fall, so g is effectively zero. What is the tension in the rod at the pivot point? Hint: Think of the rod as a series of differential length segments. Each segment has its own acceleration, and is connected to the two segments that are neighboring along the rod.

14. A massless rod having length L is free to rotate around a pivot at one end. At the other end, it has a pointlike mass m. The rod rotates in the horizontal plane. Included as part of the mass m is a variable-speed electric fan that applies a force $\mathbf{F}$. The force is in the plane of rotation and perpendicular to the rod, so the fan tries to drive rotation in one direction or the other. The fan is gradually spun up in speed, so that it applies a force whose magnitude is $F = F_o \exp(st)$, where t is the time and where F_o and s are constants. (a) What is the angular momentum of the rod and mass, relative to the central pivot, as a function of time? (b) Find the angular velocity of the mass around the pivot point as a function of time.

15. A ride at a Florida amusement park consists of two seats, each of mass m including the rider, placed at opposite ends of a long (but very light) rod having length l. The rod is supported at its middle by a circular pipe, also very light, having radius R. The braking mechanism consists of a screw driven pair of clamps that apply a frictional (tangential) force to the outer walls of the pipe. The frictional force increases with time as $F_r = at^3$, $t = 0$ being the moment at which the brakes are applied. At $t = 0$, the ride has angular velocity ω and is located at $\theta = 0$. Obtain the angular velocity of the ride as a function of time, and calculate how long is required for the brakes to bring the ride to a stop.

16. A space station consists of two living sections of mass $m = 1 \times 10^5$ kg separated by a rigid, virtually massless beam of length 100 m. The station will rotate around the center of the beam. A pair of counter-pointing rotation thrusters are located on the beam. Each thruster is 25m from the center of the beam. While firing, the thrusters each produce a 1×10^5 N force on the beam, perpendicular to the beam. See sketch. (i) What is the moment of inertia I of the space station around its center of mass? (ii) What is the torque $\boldsymbol{\tau}$ created at the center of mass by the two thrusters, while the thrusters are firing? (iii) What is the angular acceleration $\frac{d^2\theta}{dt^2}$ of the space station while the thrusters are firing? (iv) The thrusters begin firing at $t = 3$ seconds, and continue to fire until $t = 6$ seconds. For the period while the thrusters are firing, write the angular position of the station in the form $\theta = \theta_0 + \omega_0 t + 0.5\alpha t^2$, using the definition of $t = 0$ supplied in this problem. Before the thrusters were fired, the station had angular orientation $\theta = 0$. (v) Give the angular velocity and angular orientation of the space station at $t = 6$. (vi) The thrusters cease to fire at $t = 6$ seconds. At $t = 9$ seconds, the set of gears attaching each living section to the beam are set into operation by electric motors inside each living section. Between $t = 9$ and $t = 30$, the motors and gears move the living sections in towards the center of mass. Each living section finally ends up stationary relative to the beam, and 25 m from the center of mass. What is the angular velocity of the space station after the living sections have come to rest at their new locations?

19.10 Problem Solutions

1. For each mass, $\boldsymbol{L} = \boldsymbol{r} \times \boldsymbol{p}$. In this case, for each mass

$$\boldsymbol{L} = r\hat{\boldsymbol{i}} \times m\omega r\hat{\boldsymbol{j}} = m\omega R^2 \hat{\boldsymbol{k}}. \qquad (19.25)$$

The total angular momentum is then

$$\boldsymbol{L} = 0 + 2 \cdot 2 \cdot 1^2 \hat{\boldsymbol{k}} + 4 \cdot 2 \cdot 2^2 \hat{\boldsymbol{k}} \qquad (19.26)$$

19.10. PROBLEM SOLUTIONS

or
$$\boldsymbol{L} = 36\hat{\boldsymbol{k}}. \tag{19.27}$$

Remember that $\boldsymbol{L}$ is a vector. $L = 36$ is incorrect.

2. We have a rod, length ℓ, that is pivoted at one end and is free to rotate around the z-axis. At the other end of the rod, there is a mass m that drives a fan that creates a force $F_o \exp(st)$ in the $\hat{\boldsymbol{\theta}}$ direction. The torque on the system is
$$|\boldsymbol{\tau}| = |\boldsymbol{r} \times \boldsymbol{F}| = rF\sin(\phi), \tag{19.28}$$
where ϕ is the angle between $\boldsymbol{r}$ and $\boldsymbol{F}$. Here $\phi = 90°$ so $\sin(\phi) = 1$. The torque on the system is therefore
$$\boldsymbol{\tau} = \ell F_o \exp(st)\hat{\boldsymbol{k}}. \tag{19.29}$$

The torque is related to the angular momentum by
$$\frac{d\boldsymbol{L}}{dt} = \boldsymbol{\tau}. \tag{19.30}$$

The angular momentum of the system is related to the mass's angular velocity by
$$L = m\ell^2\omega. \tag{19.31}$$

We therefore have
$$L\hat{\boldsymbol{k}} = \int dt \frac{dL}{dt}\hat{\boldsymbol{k}} = \int dt \ell F_o \exp(st)\hat{\boldsymbol{k}} \tag{19.32}$$

and therefore
$$L\hat{\boldsymbol{k}} = \hat{\boldsymbol{k}}\left(\frac{\ell F_o}{s}(\exp(st) - 1) + L_o\right), \tag{19.33}$$
where L_o is the constant of integration.

The angular velocity is then
$$\omega = \frac{L}{m\ell^2} = \frac{F_o}{sm\ell}(\exp(st) - 1) + \frac{L_o}{m\ell^2}. \tag{19.34}$$

3. First we can write the vector product in terms of the components of the vectors.
$$\boldsymbol{A} \times \boldsymbol{B} = (A_x B_y - A_y B_x)\hat{\boldsymbol{k}} + (A_y B_z - A_z B_y)\hat{\boldsymbol{i}} + (A_z B_x - A_x B_z)\hat{\boldsymbol{j}}. \tag{19.35}$$

We now take the time derivative. To make the resulting form less bulky, I introduce Newton's fluxion notation for the time derivative
$$\frac{dx}{dt} \equiv \dot{x}. \tag{19.36}$$

We then have
$$\frac{d\boldsymbol{A} \times \boldsymbol{B}}{dt} = (\dot{A}_x B_y - \dot{A}_y B_x)\hat{\boldsymbol{k}} + (\dot{A}_y B_z - \dot{A}_z B_y)\hat{\boldsymbol{i}} + (\dot{A}_z B_x - \dot{A}_x B_z)\hat{\boldsymbol{j}} \tag{19.37}$$
$$+ (A_x \dot{B}_y - A_y \dot{B}_x)\hat{\boldsymbol{k}} + (A_y \dot{B}_z - A_z \dot{B}_y)\hat{\boldsymbol{i}} + (A_z \dot{B}_x - A_x \dot{B}_z)\hat{\boldsymbol{j}}. \tag{19.38}$$

In writing the derivative, I took the time derivatives of the three basis vectors $\hat{\boldsymbol{i}}$, $\hat{\boldsymbol{j}}$, and $\hat{\boldsymbol{k}}$ to be zero, as discussed in an earlier chapter. However, the first and second lines of the above equation are $\frac{d\boldsymbol{A}}{dt} \times \boldsymbol{B}$ and $\boldsymbol{A} \times \frac{d\boldsymbol{B}}{dt}$, respectively, (write it out to check) so therefore
$$\frac{d\boldsymbol{A} \times \boldsymbol{B}}{dt} = \frac{d\boldsymbol{A}}{dt} \times \boldsymbol{B} + \boldsymbol{A} \times \frac{d\boldsymbol{B}}{dt}. \tag{19.39}$$

4. We apply a force to two masses connected by a very light boom. What is the resulting torque, as computed relative to several different centers? In general

$$\boldsymbol{\tau} = \boldsymbol{r} \times \boldsymbol{F}. \tag{19.40}$$

(a) The force is applied at the origin, so $\boldsymbol{r} = \boldsymbol{0}$. It follows that

$$\boldsymbol{\tau} = \boldsymbol{0} \times (-a\hat{\boldsymbol{j}}) = \boldsymbol{0}. \tag{19.41}$$

(b) The origin is the other mass, so $\boldsymbol{r} = \ell\hat{\boldsymbol{i}}$. It follows that

$$\boldsymbol{\tau} = \ell\hat{\boldsymbol{i}} \times (-a\hat{\boldsymbol{j}}) = a\ell\hat{\boldsymbol{k}}. \tag{19.42}$$

(c) The center of mass is at

$$\boldsymbol{R}_{\text{cm}} = \left(\frac{M_2 \ell}{M_1 + M_2}, 0, 0\right), \tag{19.43}$$

so the vector from the center of mass to the point where the thrust is applied is

$$\boldsymbol{r} = (\ell, 0, 0) - \left(\frac{M_2 \ell}{M_1 + M_2}, 0, 0\right) = \left(\frac{M_1 \ell}{M_1 + M_2}, 0, 0\right), \tag{19.44}$$

leading to

$$\boldsymbol{\tau} = -\frac{M_1 \ell a}{M_1 + M_2}\hat{\boldsymbol{k}}. \tag{19.45}$$

This space reserved for your notes.

Chapter 20

Moment of Inertia

20.1 Introduction

In the previous chapter, we discussed the motion of point masses, for which we were able to write

$$\boldsymbol{L} = \boldsymbol{r} \times \boldsymbol{p}, \tag{20.1}$$

$$\boldsymbol{\tau} = \boldsymbol{r} \times \boldsymbol{F}, \text{ and} \tag{20.2}$$

$$\boldsymbol{\tau} = \frac{d\boldsymbol{L}}{dt}. \tag{20.3}$$

If there are no external forces, and if the internal forces are all central forces, then the total angular momentum of the system is conserved.

In this chapter we discuss the rotational motion of extended objects, objects that are larger than points. We begin by considering a matter of kinematics: rotational motion involving contact without slip between two objects. We then advance to considering the rotation of a rigid body, rotation around a single axis, and rotation around a displaced center. We will often be interested in cases in which motion is confined to a plane. Angular momentum is a vector, but there are many practical cases in which $\boldsymbol{L}$ and the angular velocity vector $\boldsymbol{\omega}$ are parallel. In this case, coordinates can be chosen so that $\boldsymbol{L}$ and $\boldsymbol{\omega}$ each have only one non-zero component. These results have many engineering applications, for example in drills, crankshafts, turbines, motors, generators, and windmills.

Figure 20.1: A rotating drill bit with its angular velocity.

Suppose we have a drill bit (the part that rotates and makes holes) attached to a drill via a chuck, as seen in the Figure. The drill bit is carefully balanced, so that if it rotates around its long axis ($\boldsymbol{\omega}$ in the drawing) its angular momentum $\boldsymbol{L}$ is parallel to $\boldsymbol{\omega}$. To accelerate the drill bit from stationary up to $\boldsymbol{\omega}$, and to keep it spinning at that rate while it is cutting a hole, an external torque is supplied by the drill. We now come to an interesting bit of terminology. Torque is a vector. In addition to the torque component τ_x that lies parallel to $\boldsymbol{\omega}$, we also can have torque components τ_y and τ_z corresponding to torques that are perpendicular to $\boldsymbol{\omega}$. These components try to bend the drill.

In physics, we say that the three components τ_x, τ_y, and τ_z all represent torques. However, in mechanical engineering a different set of terms are used to describe the same τ_x, τ_y, and τ_z. In engineering, the torque is τ_x. The components τ_y and τ_z are called the *bending moments*. There is no physical difference between a torque τ_y and a bending moment τ_y. The difference is purely in the choice of words being used to describe the same mathematical structure.

20.2 Rigid Body Motion; Rolling Motion Without Slip

A specific case of interest is described as *rigid body motion*. In rigid body motion, the velocity of any point i in the body may be written as

$$\boldsymbol{v}_i = \boldsymbol{V}_{\text{CM}} + \boldsymbol{\omega} \times \boldsymbol{r}_i \tag{20.4}$$

In this equation, V_{CM} is the linear velocity of the body's center of mass, r_i is the location of point i in the body with respect to the object's center of mass, ω is the angular velocity, and $\omega \times r_i$ is the velocity of the point, taken to be simple rotational motion, with respect to the center of mass.

How does this equation for the motion of the points in a body lead to the motion being the motion of a rigid body? Consider any two points in the body, which we shall identify as points 1 and 2. Their locations are r_1 and r_2, with $r_2 = r_1 + a$. The velocities of the two points, with respect to the center of mass, are then $v_1 = \omega \times r_1$ and $v_2 = \omega \times r_2$. However, the latter can also be written $v_2 = \omega \times r_1 + \omega \times a$. The velocity difference between the two points is then $v_2 - v_1 = \omega \times a$. The velocity difference $\omega \times a$ is necessarily perpendicular to a. Because $\omega \times a$ is for every infinitesimal time step perpendicular to the current value of a, the velocity difference can change the direction of a but cannot change the distance between points 1 and 2. This result is fundamentally the same as the result that rotation cannot change the magnitude of a unit vector. For all pairs of points in the object, the distance between the two points is fixed, meaning the object is rigid and cannot change its shape.

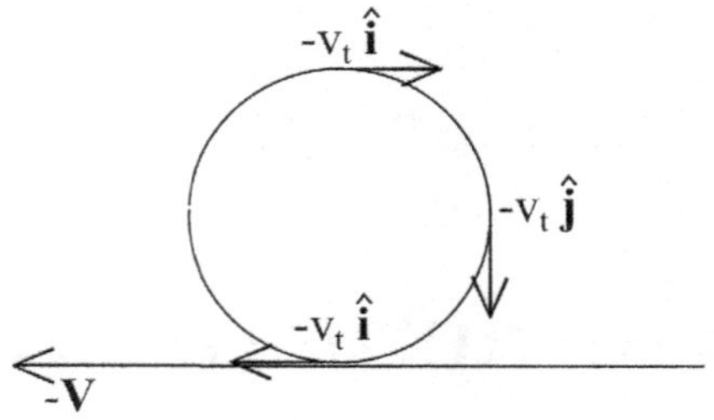

Figure 20.2: Wheel rotating without slip in a vehicle travelling at velocity V, as seen by an observer traveling with the wheel. The wheel's tangential speed is v_t. From the observer's point of view, the axle of the wheel is stationary and the road is going backward at speed $-V$.

[Aside: Simple examples of rigid and non-rigid bodies. Three points separated by fixed distances form a triangle, an object that cannot change its shape without changing those fixed distances. Four points in a square, the lengths of the four sides of the square being fixed, can change its shape, namely it can flop sideways. However, if we also fix the lengths of the two diagonals, we have transformed the square into a set of interlocking triangles; the resulting shape cannot change so long as all the distances remain fixed.]

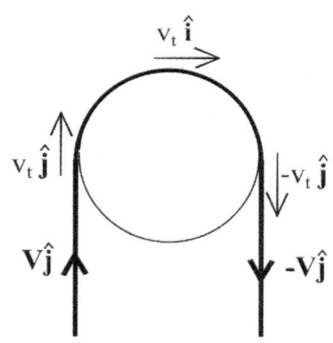

Figure 20.3: A rope (heavy line) moving at speed V passing over a pulley (light circle) moving with transverse speed v_t.

A further example of rigid body motion is provided by rolling motion without slip. The simplest example is provided by a wheel of radius R and angular speed ω rolling down a highway at velocity V. From the standpoint of someone standing on the wheel's axle, the center of the wheel is stationary, and the highway is moving backward at velocity $-V$. If the wheel is rolling without slipping, the velocity of the wheel at the point of contact, as seen by someone moving with the velocity $V\hat{\imath}$ of the car, must be $-v_t\hat{\imath}$, where v_t is the speed at which the outer surface of the wheel is rotating, with $v_t = \omega R$. The wheel is turning in a circle, with respect to the bicyclist, so the top of the wheel is moving at $+v_t\hat{\imath}$ and the forwardmost point on the wheel is moving at $-v_t\hat{\jmath}$.

As seen by an observer on the ground, the center of mass of the wheel has velocity V. The wheel is not slipping, so at the point of contact between the wheel and the ground the wheel must be stationary, i.e.,

$$V - v_t\hat{\imath} = 0. \tag{20.5}$$

The velocity of the top point of the wheel is then $V\hat{\imath} + v_t\hat{\imath} = 2V\hat{\imath}$, while the velocity of the front point of the wheel is $V\hat{\imath} - v_t\hat{\jmath} = V\hat{\imath} - V\hat{\jmath}$. Said differently, as a general form, the velocity of a point on the outer surface of the wheel is $V = V_{car} + \omega \times R$, where V_{car} is the velocity of the car as seen by the observer.

An alternative to rolling motion without slip is rolling motion with slip. A simple example is provided by an automobile attempting to go up a steep hill on wet sheet ice. The car wheel spins, so v_t may be very large, but the speed of the car with respect to the ground is zero.

[Aside: The fastest speed attained by a land vehicle as of this writing is apparently 763 miles per hour (Yes, the vehicle went supersonic; the driver survived the experience.) At that speed, the top surface of the wheel would have been travelling at 1526 miles per hour with respect to the ground, a speed accessible while flying to a very small number of experimental and combat aircraft.]

A second example of rolling without slip is provided by a rope passing over a pulley. The rope is moving at speed V. The outer surface of the pulley is moving at tangential speed v_t. If there is to be no slip, it must be the case that $V = v_t$. V and v_t are both speeds, not velocities. The tangential speed, pulley radius R, and angular rotation rate of the pulley are connected by

$$v_t = \omega R, \tag{20.6}$$

so that $\omega = v_t/R$ or $v_t = V/R$.

[Aside: For ropes passing over pulleys, the no-slip condition is not automatic, though we will systematically use it here. In 18th century sailing warships making radically sharp turns, sails were redirected at great speed, ropes slipped over pulleys that had not had time to accelerate to the needed ω, and sliding friction greatly heated the wooden pulley wheels, so that either a sailor was stationed at the pulley with a bucket of water and a ladle or the pulley block tended to catch on fire, a negative outcome for a ship made almost entirely of highly flammable materials.]

20.3 Composite Rotating Systems

As an example, we consider a shaft aligned along the z axis. The shaft rotates with $\boldsymbol{\omega} = \omega \hat{\boldsymbol{k}}$. ω is a signed number, so the shaft's rotation may be clockwise or counterclockwise. Extended out from the side of the shaft is a nearly massless staff to which masses m and M have been attached. As seen in the figure, the distance from the shaft to m is ℓ_1, while the distance from m to M is ℓ_2.

The two masses are performing circular motion, so their speeds follow from $v_t = \omega r$, namely

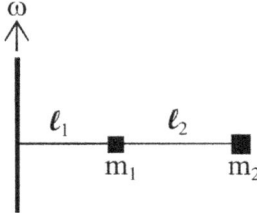

$$v_m = \omega \ell_1 \text{ and} \tag{20.7}$$
$$v_M = \omega(\ell_1 + \ell_2). \tag{20.8}$$

Figure 20.4: A rotating rod, attached to which are masses m_1 and m_2 with separations ℓ_1 and ℓ_2 as marked.

The two masses have the same angular velocity ω, but rotate around the shaft at different distances. The total kinetic energy of the two masses is therefore

$$K = \frac{1}{2}m(\omega \ell_1)^2 + \frac{1}{2}M(\omega(\ell_1 + \ell_2))^2. \tag{20.9}$$

This equation is usefully rearranged as

$$K = \frac{1}{2}(m\ell_1^2 + M(\ell_1 + \ell_2)^2)\omega^2. \tag{20.10}$$

We can also calculate the total angular momentum for the two masses, relative to an origin placed at the point where the wire intersects the shaft. The masses are then rotating purely in the horizontal x-y plane so their angular momenta $\boldsymbol{r} \times \boldsymbol{p}$ are both in the $\hat{\boldsymbol{k}}$ direction. The $\hat{\boldsymbol{k}}$ component of their angular momentum is then written

$$L_z = pr_\perp + PR_\perp, \tag{20.11}$$

with lower-case letters corresponding to m and upper-case letters corresponding to M. Here p is equal to mv or to $m\omega r$, while $r_\perp$ and $R_\perp$ are the distances from the masses to the rod, so that

$$L_z = \ell_1 \omega m \ell_1 + (\ell_1 + \ell_2)\omega M(\ell_1 + \ell_2). \tag{20.12}$$

This form for the angular momentum may be reorganized as

$$L_z = \omega(m\ell_1^2 + M(\ell_1 + \ell_2)^2). \tag{20.13}$$

If you compare equations 20.10 and 20.13, you find they have a common factor

$$I = m\ell_1^2 + M(\ell_1 + \ell_2)^2. \tag{20.14}$$

I is the moment of inertia. In the case here, I is shown as a scalar. In the general case, which we will not reach in this course, I is a 3×3 matrix, the moment of inertia tensor. We can write the total kinetic energy and the total angular momentum in terms of the moment of inertia, namely

$$K = \frac{1}{2}I\omega^2 \text{ and} \tag{20.15}$$
$$L_z = I\omega. \tag{20.16}$$

20.4 Discussion

The above two equations have an interesting relationship. Recognizing that $\omega = \dot{\theta}$, where I have used the Newtonian fluxion notation for the time derivative, namely $\dot{\theta} \equiv d\theta/dt$, and that E is the energy of the system, we may write

$$L_z = \frac{\partial K}{\partial \omega} \tag{20.17}$$
$$L_z = \frac{\partial E}{\partial \dot{\theta}} \tag{20.18}$$

Note the similarity between these two equations and the corresponding equation

$$p_z = \frac{\partial E}{\partial \dot{z}} \tag{20.19}$$

for linear motion. These two equations are not an accident or a coincidence. They are aspects of Hamilton's Theory of Classical Mechanics, as developed by the Irish mathematician William Rowan Hamilton (1805-1865) and the German mathematician Carl Gustav Jacob Jacobi (1804-1851). Hamilton's theory provides an extremely powerful mathematical tool for studying complicated systems. In particular, it provides the general formula for the momentum that replaces $\boldsymbol{p} = m\boldsymbol{v}$.

20.5 Worked Problems

1. A uniform rod lies in the x-y plane. It rotates about one end, the axis of rotation being the z axis, at ω radians per second. The rod has mass M and density M/ℓ along its length. What is the angular momentum $d\mathbf{L}$ of an infinitesimal piece of rod? By integrating, determine $\mathbf{L}$ of the entire rod.

2. Consider the rod and pivot of the previous problem. Assume $M = 2$ kg, $\ell = 3m$, $g = 10$ m/s^2. If the angular acceleration $d\omega/dt$ is 5 radians/second^2, compute the external torque on the rod.

20.6 Homework

1. A rigid rod is attached at one end perpendicular to a rotating axis. The linear density of the rod increases linearly with distance from the axis, starting at 3 kg/m at the axis, and increasing to 5 kg/m at the outermost end of the rod, which is 2 m from the axis. The rod rotates in the $x - y$ plane. A torque is applied to the rod, such that the angular velocity of the rod increases by 4π radians/second over two seconds. What is the torque τ applied to the rod, using the attachment point between the rod and the axis as the origin.?

2. The roller coaster. A roller coaster of sedate design consists of a straight, modestly sloping set of rails on top of which sits a roller coaster car. The roller coaster car has a body (mass M) and four wheels (each also mass M and moment of inertia I). The rails make an angle θ with respect to the horizontal. As a test, the roller coaster car is released at the top of the slope without carrying any passengers. Calculate its speed when it has travelled a distance s down the slope.

3. A cylindrical shell of mass M, length L, and radius R is set into rotation around an axis running the length of the cylinder. Its angular rotation rate is $\omega = A\cos(Bt^2)$. Calculate the torque on the cylinder as a function of time.

4. Consider a space station in the form of a bicycle wheel. The inner radius of the wheel is 100 m; the outer radius of the wheel is 108 m. The wheel is 5 m tall, tall being a distance perpendicular to the radius vector and the transverse velocity of the wheel, and has a density 0.1 that of water. The innards of the wheel are such that the wheel effectively has a uniform density. The wheel rotates around its central axis; the angular velocity is such that the outer surface of the wheel accelerates toward the center of the wheel at one gravity. Compute the wheel's moment of inertia, the magnitude of the wheel's angular momentum as measured around its center, and the kinetic energy of the wheel. Do not assume that the wheel is a thin sheet, all at one radius from its center.

5. We have lying on a frictionless surface a uniform meter stick having length ℓ and mass M, initially at rest. The rod is struck at one end by a hammer, its force $\boldsymbol{F}$ being perpendicular to the rod, so that $\int \boldsymbol{F}\, dt$ of the very short-lived $\boldsymbol{F}$ is a transfer of momentum $\boldsymbol{P}$ to the rod. How far will the center of mass have traveled when the meter stick has completed one revolution?

6. A uniform rod of mass M and length ℓ is pivoted at the top. The rod is initially stationary and is hanging vertically from its top end. A pellet having mass m moving horizontally with speed v strikes the rod at a distance d from the top and sticks. What speed for the pellet is required so that the rod rises through 90 degrees to be horizontal at its maximum height?

7. We consider a top spinning about its vertical axis. The top remains vertical during the entire problem. The top has moment of inertia I_o. (i) The angular momentum of the top decays with time, the time dependence being $\boldsymbol{L} = -L_o \exp(-0.3t^2)\hat{\boldsymbol{k}}$. What is the torque on the top? [Reminder: by definition $\exp(a) = e^a$.] (ii) We now change the surface under the top, so that the angular momentum of the top depends on time as $\boldsymbol{L} = -L_o \exp(-0.5t)\hat{\boldsymbol{k}}$. We describe the angular position of the top via an angle θ measured around the rotation axis. Find the most general expression for θ as a function of time.

20.7 Solutions to the Worked Problems

1. The general form is $\boldsymbol{L} = \boldsymbol{r} \times \boldsymbol{p}$. For the differential piece,

$$d\boldsymbol{L} = r\left(\frac{M}{\ell}dr\right)\omega r\,\hat{\boldsymbol{i}} \times \hat{\boldsymbol{j}}, \qquad (20.20)$$

where r is the distance from the axis of rotation. This form reduces to

$$d\boldsymbol{L} = \frac{Mr^2}{\ell}\omega\,\hat{\boldsymbol{k}}\, dr. \qquad (20.21)$$

We can integrate this form over the length of the rod, from 0 to ℓ, namely

$$\int d\boldsymbol{L} = \int_0^\ell \frac{Mr^2}{\ell}\omega\,\hat{\boldsymbol{k}}\, dr, \qquad (20.22)$$

which yields

$$\boldsymbol{L} = \frac{M\ell^2}{3}\omega\hat{\boldsymbol{k}}. \qquad (20.23)$$

2. We start with

$$\boldsymbol{\tau} = \frac{d\boldsymbol{L}}{dt} \qquad (20.24)$$

and substitute, obtaining

$$\tau = \frac{M\ell^2}{3}\frac{d\omega}{dt}, \tag{20.25}$$

$$\tau = \frac{2\cdot 3^2}{3}\cdot 5, \tag{20.26}$$

$$\boldsymbol{\tau} = 30\,\hat{\boldsymbol{k}}\ \text{kg m}^2/\text{s}^2. \tag{20.27}$$

Note that torque does not have units Joules. While energy and torque are both 'force times distance', the word 'times' has two different meanings in the definitions of the two units, namely 'times' refers to the scalar product for energy and to the vector product for torque.

This space reserved for your notes.

Chapter 21

Rigid Body Rotation

21.1 Introduction

This chapter considers the rotation of rigid bodies. A rigid body may be viewed as being composed of a vast number of objects, for example, atoms, that are connected to each other by rigid bonds. We considered a simple example of a rigid body in the last chapter. That body contained all of two masses. Here we consider bodies that when viewed sufficiently carefully would be found to be composed of huge numbers of objects.

What do we mean by a "rigid" body? When a rigid body moves, all internal distances between its parts are preserved. A rigid body has no internal motions. "Rigid body" is an approximation. If we consider the atoms in a solid, they do behave as though they were connected by tiny springs, but the springs are slightly compressible. If they were not compressible, the body would be unable to transmit sound, except that an impact on one side of the body would be felt instantaneously throughout the entire body. Real bodies do transmit sound, because their atoms are able to move very slightly with respect to each other, even before the body is cracked or sheared. For the same reason, if a body is struck on one side, the impact spreads its influence through the body at the speed of sound.

In a rigid body, all kinetic energy is present as center of mass motion and rotation of the body as a whole. No kinetic energy is stored internally in atomic motions. That's another approximation.

In a real solid, heat is in part the kinetic energy stored in the motions of the atoms with respect to each other. At least some of that energy is always present in a solid. Note that I said "in part". While part of the thermal energy in a solid is stored in the kinetic energy of the individual atoms, part of the energy is stored in the potential energy connecting the atoms together. That behavior of a solid is very different from the behavior of an ideal gas, in which all of the thermal energy is stored as the kinetic energy of the atoms. In a liquid, some of the thermal energy is stored as kinetic energy of the individual molecules, some is stored as the internal vibrations of the atoms within each molecule, and some is stored in the potential energy keeping the liquid molecules together so that the liquid does not instantly evaporate.

21.2 Example: The Physical Pendulum

In this section, we consider a simple example of rigid body motion, the uniform motion of a rod swinging from a pivot. The system is sometimes called a *physical pendulum*. The example is particularly noteworthy because it lends itself to a series of straightforward experiments. The figure shows a typical pivoted rod. The rod has length ℓ and mass M. The pivot is at the top end of the rod. The rod hangs vertically and swings back and forth. This rod happens to be uniform, but that's not a physical requirement. How do we treat all of the atoms in the rod? The answer is that we mathematically break the rod up into a large number of differential segments of some size, consider the behavior of a single segment, and then by means of integration sum the total of all of the differential segments that the rod contains. For this simple rod, an appropriate approach is to define a distance coordinate r, the distance from the pivot point to a differential segment of the rod. The differential segment of the rod has a mass dm, which is related to the density of

the rod by
$$dm = \rho\, dr. \tag{21.1}$$

In this equation, ρ is the linear density of the rod, the mass per unit length of the rod. (Density can also be mass per unit area, for example for a thin sheet of metal, or mass per unit volume, for example the water in a pitcher. Here we are talking about mass per unit length.)

How do we obtain ρ? We begin by observing that the total mass of the rod is the sum of the masses of all of the differential segments, letting us write
$$M = \int dm. \tag{21.2}$$

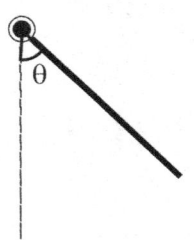

This appears to be an equation for the total mass M, but the total mass is a *known*. The unknowns are the dm, the masses of the small segments of the rod. We can rewrite the above equation in terms of the density, namely
$$M = \int_0^\ell \rho\, dr, \tag{21.3}$$

Figure 21.1: A pivoted rod, hanging from its pivot at an angle θ to the vertical.

where we have applied $dm \equiv \rho\, dr$. In this simple example, the density ρ is a constant, so the integral is
$$M = \rho r|_0^\ell, \tag{21.4}$$

giving us
$$M = \rho(\ell - 0), \tag{21.5}$$

and finally
$$\rho = M/\ell. \tag{21.6}$$

In this example, you might well have been able to find equation 21.6 by inspection, simply by looking at the problem statement. The method seen above was shown here because it will be needed in more complicated cases. It should be emphasized that the final two equations depend explicitly on the statement that the density ρ is independent of the position along the rod. If the density were $\rho(r)$ and actually depended on the position along the rod, the final integral would look quite different.

How fast are the rod differential segments moving? The rod is rigid, and straight, so if we imagine an angle θ as indicated in the figure, and measured from the vertical, all differential rod segments have the same θ. The rod segments must therefore also all have the same $\omega = d\theta/dt$, because if they did not they would be unable to keep having the same θ. The speed of a differential segment a distance r out from the origin is therefore
$$v = \omega r. \tag{21.7}$$

We are now ready to calculate the total angular momentum of the rod. The rotation of the rod is in the plane of the paper, so for each differential segment of the rod $\mathbf{r}$ is parallel to the rod, while $\mathbf{p}$ is perpendicular to the rod and also in the plane of the paper. It follows that the angular momentum $d\mathbf{L}$ of each differential segment of the rod is perpendicular to the page. We'll call the direction perpendicular to the rod the $\hat{\mathbf{k}}$ direction. The component of the angular momentum in this direction is then denoted L_z.

The total angular momentum of the rod is the sum of the angular momenta of the individual differential segments. For a single segment, $dL_z = r\, dm\, \omega r$, the $\sin(\theta)$ from the cross product disappearing because $\mathbf{r}$ and $d\mathbf{p}$ are perpendicular, so that $\theta = \pi/2$ and $\sin(\theta) = 1$. We can write an integral for L_z as
$$L_z = \int_0^\ell dL_z. \tag{21.8}$$

Replacing the differential angular momentum segment with the differential mass, we have
$$L_z = \int_0^\ell dm\, r\omega r \tag{21.9}$$

21.2. EXAMPLE: THE PHYSICAL PENDULUM

or

$$L_z = \int_0^\ell dr\, \rho \omega r^2, \tag{21.10}$$

which gives for the angular momentum

$$L_z = \frac{\omega \rho r^3}{3}\Big|_0^\ell \tag{21.11}$$

or, finally, recalling $\rho \ell = M$,

$$L_z = \omega \frac{ML^2}{3}. \tag{21.12}$$

Let us pause to note the dimensions of this quantity. On the right hand side of the equation, ω has dimensions time^{-1}. ML^2 has dimensions mass $\cdot$ length^2. In ω, what happened to the angle? Angles have dimensions unity, i.e., mass$^0 \cdot$ length$^0 \cdot$ time^0.

By comparison with the above, the moment of inertia of the rod is

$$I = \frac{ML^2}{3}. \tag{21.13}$$

The above equation refers to this particular rod and this particular motion. With a different rod, or a different motion, the moment of inertia will also be different.

We now turn to the kinetic energy of the same rod. The kinetic energy of a differential piece of rod is

$$dK = \frac{1}{2} dm (\omega r)^2. \tag{21.14}$$

The total kinetic energy is then

$$\int dK = \int_0^\ell dr\, \frac{1}{2} \rho \omega^2 r^2. \tag{21.15}$$

On performing the integral, we obtain

$$K = \frac{1}{2} \rho \omega^2 \frac{r^3}{3}\Big|_0^\ell, \tag{21.16}$$

which leads to

$$K = \frac{1}{2} \frac{M\ell^2}{3} \omega^2. \tag{21.17}$$

By comparison with the previous chapter, one infers that the moment of inertia is

$$I = \frac{M\ell^2}{3}, \tag{21.18}$$

which is the same as the moment of inertia found above for the angular momentum for this problem. We note that $K = \frac{1}{2}\omega L$ or

$$L_z = \frac{\partial K}{\partial \dot\theta}. \tag{21.19}$$

We have now examined a special case. The case was special because the mass density of the rod (here, the mass per unit length) was a constant, so that the mass density was the same everywhere along the rod. We also were examining the special case in which $\omega \parallel \mathbf{L}$. That special case occurs if the mass density has mirror symmetry around the axis of rotation. It also occurs if the motion is in a plane that passes through the origin, so that the rod always remains in a single plane as it rotates.

The above is a special case, but it was a very useful special case. For example, in the design of machinery, it is very often the case that all motions of a machine can be analyzed as two-dimensional motions, leading to the importance of four-bar linkages as design elements. That isn't absolutely always true but it is true often enough to be worth noting.

We could also write out the angular momentum and the kinetic energy of a collection of masses using vector notation. The masses are taken to be identified by an index i; their velocities are $\boldsymbol{\omega} \times \mathbf{r}_i$. We would write

$$\mathbf{L} = \sum_i m_i \mathbf{r}_i \times (\boldsymbol{\omega} \times \mathbf{r}_i) \tag{21.20}$$

$$K = \sum_i \frac{1}{2} m_i (\boldsymbol{\omega} \times \mathbf{r}_i) \cdot (\boldsymbol{\omega} \times \mathbf{r}_i) \tag{21.21}$$

For the case that $\omega \parallel \mathbf{L}$, for example, rotation around an axis, with masses placed symmetrically around that axis, we have the special case forms

$$I = \sum_i m_i r_i^2, \tag{21.22}$$

$$K_{\text{rotation}} = \frac{1}{2} I \omega^2, \text{ and} \tag{21.23}$$

$$L_{\text{rotation}} = I \omega. \tag{21.24}$$

The sum $\sum_i$ on all the masses in the system will often need to be evaluated as an integral; for physical objects volume integrals will usually be needed.

21.3 Koenig's Theorem

Let us return to the automobile wheel, whose rolling motion without slip we have discussed. The wheel has a center of mass velocity $\mathbf{V}_{\text{cm}}$. Each point on the wheel has a rotational velocity $\mathbf{v}_i = \boldsymbol{\omega} \times \mathbf{r}_i$ with respect to the center of the wheel. The velocity of each point with respect to the ground is then

$$\mathbf{V}_i = \mathbf{V}_{\text{cm}} + \boldsymbol{\omega} \times \mathbf{r}_i. \tag{21.25}$$

To get the total kinetic energy of the wheel, we add up over the kinetic energy of each of the little pieces of the wheel. If you do that yourself, you should obtain

$$K = \sum_i \frac{1}{2} m_i (\mathbf{V}_{\text{cm}} + \boldsymbol{\omega} \times \mathbf{r}_i)^2. \tag{21.26}$$

Calling the total mass of the wheel $M = \sum_i m_i$ and its moment of inertia I, on doing the sum you get

$$K = \frac{1}{2} M V_{\text{cm}}^2 + \frac{1}{2} I \omega^2. \tag{21.27}$$

Surprise! The total kinetic energy of the wheel as it rolls down the highway is the sum of its mass M moving at speed V_{cm} and its moment of inertia I rotating at angular speed ω. This result is Koenig's Theorem.

Where is there a surprise here? If you look at equation 21.26, and write out the square, you should have found cross terms like

$$\mathbf{V}_{\text{cm}} \cdot (\boldsymbol{\omega} \times \mathbf{r}_i).$$

Those terms appear to be proportional to ωV_{cm}. Where did those terms go? The answer is that when you do the sum, those cross-terms sum to zero. Proving that is a homework problem, with the hint that everything in the cross term except one factor of $\mathbf{r}_i$ is a constant.

21.4 The Rolling Cylinder

We now turn to a concrete example of solving a problem that involves rolling motion. The problem is indeed concrete, namely we have a solid concrete ramp having a length L. The top end is a distance H above the ground, the bottom end is at height h. The ramp makes an angle θ with respect to the ground. We then have a cylinder, which is also made of concrete, starting at the top of the ramp. It is released from being stationary, and rolls to the bottom of the ramp. The mass of the cylinder is M. Its moment of inertia is I. The cylinder rolls without slipping. How fast is cylinder's center of mass moving when it reaches the bottom of the ramp?

We are given positions. We want to find the velocity. The approach that typically links positions and velocities is conservation of energy, so that is a reasonable first guess on how to attack the problem. If the approach fails, we will have to try something else. In this sort of analysis "what approach should I try?" there is often not a guarantee of success, so you should be prepared to try something else if your first approach fails, but this sort of analysis gives you a starting point. We begin with energy conservation:

$$E_i = E_f. \tag{21.28}$$

The energy of the system is
$$E = K + U. \tag{21.29}$$

In the case at hand, the potential and kinetic energies may be written

$$U = Mgz, \tag{21.30}$$

$$K = \frac{1}{2}MV^2 + \frac{1}{2}I\omega^2. \tag{21.31}$$

Note that we've written these in the general form, as opposed to the form that might be correct at the top or bottom of the ramp. We have not yet applied the no-slip condition, which tells us that $V = \omega R$.

Now we find the values of the known variables at the initial and final points. At the start, $z = H$ and $V = 0$. By implication, $\omega = V/R$ is also zero. At the end of the roll, the center of mass of the cylinder is at $z = h$ and the speed of its center of mass is $V = v_f$. By implication the final angular velocity is $\omega_f = v_f/R$.

Substitute the initial and final values of the known parameters into the Law of Conservation of Energy. Here I do the substitutions, but make absolutely no simplifications, making it as easy as possible to tell if the substitutions have been done correctly. The wise procedure in doing a calculation is to take your steps as small as possible, because this makes it as easy as possible to check your work, exactly as seen here. (If you are not checking your work, you are making an extremely serious error, which in the end will lead you to an unfortunate place.)

$$\frac{1}{2}M0^2 + \frac{1}{2}I0^2 + MgH = \frac{1}{2}Mv_f^2 + \frac{1}{2}I\left(\frac{v_f}{R}\right)^2 + Mgh. \tag{21.32}$$

We can now simplify that equation. The two terms with factors of 0 vanish. Before we go further, we should ask whether we have as many equations as we have unknowns. For that matter, we should probably start by asking what the unknowns are. Here the one unknown is the final velocity v_f. Corresponding to the one unknown we have one equation. We expect to be able to solve.

A reasonable approach is to sort out the terms that do contain the unknown and the terms that do not contain the unknown. Doing that, and noting common factors Mg and $\frac{1}{2}v_f^2$, we obtain

$$\frac{1}{2}\left(M + \frac{I}{R^2}\right)v_f^2 = Mg(H - h). \tag{21.33}$$

On moving all of the constants from the left to the right side of the equation, and taking the square root, one obtains

$$v_f = \left(\frac{2MR^2g(H-h)}{(MR^2 + I)}\right)^{1/2} \tag{21.34}$$

as the final answer.

21.5 Worked Problems

1. A mass m slides down a frictionless ramp inclined at 30 degrees to the horizontal. The mass begins stationary at a height H directly above the origin. Find the block's linear and angular momenta with respect to the origin as functions of the height h of the block. Hint: $\boldsymbol{L} \neq \boldsymbol{0}$.

2. The falling pencil. This problem is short, but quite difficult if you do not see how to do it. A pencil of length L and mass M is resting vertically, sharpened point down, on a totally frictionless surface. The air current set up by a passing butterfly disturbs its equilibrium, very slightly, and the pencil falls over. What is the speed of the pencil's center of mass just as the pencil falls flat into the table? Approximate the pencil as having diameter zero.

21.6 Homework

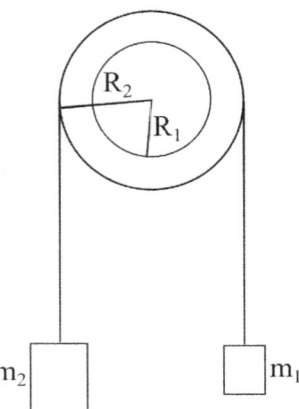

Figure 21.2: Figure for Homework problem 2.

1. Confirm Koenig's Theorem: Prove for a rigid rotating object that

$$\sum_{i=1}^{N} \mathbf{V}_{\text{cm}} \cdot \boldsymbol{\omega} \times \mathbf{r}_i = 0. \tag{21.35}$$

2. A rod hanging vertically and pivoted at its top end around a horizontal axis has length L. Denoting the pivot point by $r = 0$, the filament is massless out to $r = \ell$. Between $r = \ell$ and $r = L$, the density of the rod increases quadratically as distance from the origin, i.e., $\rho = \rho_o r^2$. Find the rod's (a) kinetic energy, (b) angular momentum, and (c) gravitational potential energy as functions of its angular velocity ω and position angle θ, with $\theta = 0$ being the vertical.

3. A cube, a sphere, and a cylinder are sent down a slope starting at the same point. The cube slides without friction. The sphere and cylinder roll without slipping. Which reaches the bottom first? How does this answer depend on the masses of the three objects? The moments of inertia of a sphere and a cylinder each having mass M and radius R are $\frac{2}{5}MR^2$ and $\frac{1}{2}MR^2$, respectively.

4. In yet another scene from the ongoing adventure movie, the heroine is hanging from the end of the world's longest flagstaff, which is attached near the top of the world's tallest building. The flagstaff is a steel girder having length L and mass M; the end attached to the building is hinged so that it is free to rotate. The other end of the flagstaff is held in place by a long wire. The heroine is hanging on to the outer end of the flagpole. The flagstaff is initially at angle $\theta_o = 90°$, with $\theta_o = 0°$ being vertical and straight down. The good news in the film is that the heroine, mass m, has just disassembled the villain's world-ending doomsday device, saving the world. The less good news is that the villain has cut the wire and the world's longest flagstaff is about to swing downward.

Find (i) The kinetic energy of the flagstaff and heroine as they swing downward, in terms of their angular velocity $d\theta/dt$. (ii) The potential energy of the flagstaff and heroine as they swing downward. (iii) The torque on the flagstaff and heroine, torque being measured relative to the location of the hinge. (iv) The angular acceleration of the flagstaff and heroine as they swing downward. Does the angular acceleration depend on the time since the wire was cut? (v) The heroine hangs on to the pole, so that she applies a force $-\mathbf{P}$ on the pole, and it applies a force $\mathbf{P}$ on her. Find the force $\mathbf{P}$ on the heroine as a function of the angle θ from the vertical. (vi) If the actress in Problem 2 hangs on as firmly as possible, she can support twice her own weight (for example, in a static vertical hold, she can support herself [including body armor], and a person of equal weight). At what angle θ does she lose her grip on the girder?

5. Consider the simplest possible yo-yo, a circular disc having a mass M, radius R, and moment of inertia $0.5MR^2$ around which a string is wrapped, the string being tied at its other end to the ceiling. The yo-yo is released from rest. How fast is the yo-yo falling (z-component of velocity) when it has fallen through a distance H?

6. A cubical block and a sphere are placed at the top of an inclined plane. The plane has length L and slope θ. The two objects each have mass M. The moment of inertia of the sphere (radius R) is $\frac{2}{5}MR^2$. The two objects are released simultaneously from the top of the ramp. (i) What is the speed of the block at the bottom of the ramp? (ii) What is the speed of the sphere at the bottom of the ramp? (iii) What is the angular momentum of the sphere at the bottom of the ramp, as measured by the ghost of Isaac Newton hovering at the position occupied by the sphere's center? (iv) Does the sphere reach the bottom of the plane before or after the block? Credit will be based on your logic, not on whether or not you happened to get the answer right, though – since the course did not cover the obvious stage play of Aristophanes – wrong logic is not worth as many points as right logic.

7. A sphere of radius R and mass M, and a uniform disc of radius r and mass m start at rest at the top of a steep incline. They then roll down the incline. The incline is a plane of length L that makes an angle θ with respect to the horizontal. There is a short shoulder curve where the ramp reaches the horizontal, sufficiently long that the objects have a smooth roll all the way along. You may take the moment of inertia of the non-uniform sphere to be $\frac{2}{5}MR^2$. Assume rolling contact without friction. (i) What are the speeds of the sphere and the disc, respectively, at the bottom of the ramp? (ii) What are the angular momenta $\mathbf{L}$ of the sphere and the disc at the bottom of the ramp, as measured with respect to their release points? (iii) The sphere and the disc next encounter an upward-bending curve in the track. Which object rolls to a greater altitude, as measured with respect to their release points, the sphere or the disc? Why? (i.e., prove your answer.)

8. Consider a space station in orbit around the sun, far from any other mass. The station consists of a nearly massless, very long rod of length L, a module of mass M in the middle of the rod, and modules of mass m at each end of the rod. (i) Calculate the location of the center of mass of the rod. (ii) What is the moment of inertia around the center of mass of the rod, for rotation through an axis perpendicular to the rod. (iii) A rocket motor is attached to the right-hand module. The motor is mounted with its main axis at an angle θ with respect to the rod. The motor can fire in either direction, and applies a force $F = F_o \sin(\omega t)$ along its main axis. What is the angular momentum of the space station, as measured around its center of mass, as a function of time? (iv) The single rocket motor is replaced by a pair of rocket motors, one mounted on each module, firing antiparallel to each other at angles θ with respect to the rod. Each rocket produces 1/2 of the thrust of the rocket in part C. By comparison with part (C), what is the angular momentum of the space station, as measured around its center of mass, as a function of time? (see sketch).

9. This problem is long and difficult. In yet another scene of the film that was a stock of old exam questions, the heroine is dropped from a helicopter onto the back of a dirigible. The studio did not have a dirigible handy. The dirigible mockup is a long plastic cylinder mounted between two towers. The cylinder rotates freely around its long horizontal axis; its vertical motion can be neglected. After a short period of time, the dirigible rotates so that the heroine is at the bottom of the dirigible. What is the angular velocity of the heroine as she passes the bottom? Put the origin of the coordinate system on the central axis of the dirigible. Do not try to simplify your final expression.

The heroine has a mass m and is traveling at speed V when she impacts the dirigible. The dirigible has mass M, radius R, and moment of inertia MR^2. Until the heroine lands on it, the dirigible is stationary. The heroine lands a distance l to the left of the center of the dirigible. (**Hints!** This is not a collision problem. A near-infinite force holds the center axis of the dirigible in place, so that the dirigible rotates freely, but its center axis does not move. The linear momentum of heroine + dirigible is therefore not conserved during her impact. However, the bearings are frictionless, so the only external torque on the actress+dirigible system is exerted by gravity. No one is making you answer the questions in the order that I am asking them!)

(a) What is the kinetic energy of the heroine just before she hits the dirigible?

(b) What is the linear momentum of the heroine just before she hits the dirigible?

(c) What is the potential energy of the heroine just before she hits the dirigible? Take the center of the dirigible to have height 0.

(d) What is the angular momentum of the heroine just before she hits the dirigible?

The heroine hangs on to the dirigible when she hits. The dirigible starts to rotate around its own central axis, which is firmly mounted and does not move, while she hangs on.

(e) What is the kinetic energy of the heroine just after she hits the dirigible?

(f) What is the linear momentum of the heroine just after she hits the dirigible?

(g) What is the potential energy of the heroine just after she hits the dirigible? Take the center of the dirigible to have height 0.

(h) What is the angular momentum of the heroine just after she hits the dirigible?

(i) Draw a force diagram for the heroine just after she lands on the dirigible.

(j) The dirigible rotates freely. What is the angular rotation rate of the dirigible just after the heroine hits it?

21.7 Solution to the Worked Problem

1. A mass slides down a ramp. We want to know its momentum and its angular momentum with respect to the specified origin. In this problem, we want to know speed or momentum as a function of position, suggesting that energy conservation or the work-energy theorem will give us the desired information. Energy conservation tells us $E_i = E_f$. The total energy of the block is $E = \frac{1}{2}mv^2 + mgh$. At the start $h = H$ and $v = 0$. After the block has slid a distance, $h = h$ and $v = V$. Energy conservation gives us

$$\frac{1}{2}m0^2 + mgH = \frac{1}{2}mV^2 + mgh,$$

where I have replaced all general algebraic symbols with their values at the start and the finish, and stopped without making any simplifications, such as eliminating terms that are equal to zero. This form can be solved for V, namely

$$V^2 = 2g(H - h)$$

which shows that the block's momentum is

$$p \equiv mV = m(2g(H - h))^{1/2}.$$

The direction is downhill, parallel to the surface of the ramp.

The angular momentum is perpendicular to the plane of the paper. L is $pr_\perp$ where from a bit of trigonometry $r_\perp = H\sin(\theta)$ so that $L = mH(2g(H - h))^{1/2}\sin(\theta)$. From the right-hand-rule, **r** goes along the thumb, **p** goes along the index finger, and therefore **L** points into the page.

2. To repeat the problem statement: We have a pencil on a frictionless tabletop, balanced on its sharpened point. The pencil is approximated as a thin, uniform rod having mass M and length L. The gentlest puff of breeze tips the pencil off center so it falls to the tabletop. The angle between the pencil and the vertical is θ. How fast is the center of mass moving when the pencil has rotated through ninety degrees and hits the tabletop?

21.7. SOLUTION TO THE WORKED PROBLEM

We want to find velocity as a function of position, suggesting that energy conservation will turn out to be a good approach. We therefore need expressions for the potential and kinetic energy of the pencil.

For the potential energy we can write $U = Mgz$, where z is the distance of the pencil's center of mass above the table top. At the start of the fall, $z = L/2$. When the pencil hits the table, $z = 0$.

To compute the pencil's kinetic energy, we have to know which way the pencil is moving. If we were to draw a force diagram of the pencil, we would see on it only two forces, namely the normal force of the table on the pencil and the force of gravity on the pencil. Both of these forces act in the vertical direction, so the pencil's center of mass never accelerates in a horizontal direction. The pencil's kinetic energy is therefore

$$K = \frac{1}{2}M\left(\frac{dz}{dt}\right)^2 + \frac{1}{2}I\left(\frac{d\theta}{dt}\right)^2. \tag{21.36}$$

The rod is uniform, so its center of mass is at the dead center of the pencil, a distance $L/2$ from either end. The moment of inertia of the pencil is then

$$I = \int_{-L/2}^{L/2} \rho r^2 dr, \tag{21.37}$$

where ρ is the mass per unit length of the pencil, i.e. $\rho = M/L$. On performing the integral, one finds $I = ML^2/12$.

We are trying to solve for dz/dt. To do that, we somehow need to find or eliminate $d\theta/dt$. However z and θ are not independent. A little trigonometry shows

$$\frac{z}{L/2} = \cos(\theta), \tag{21.38}$$

whose time derivative is

$$\frac{dz}{dt} = \frac{L}{2}\sin(\theta)\frac{d\theta}{dt}, \tag{21.39}$$

and therefore

$$\frac{d\theta}{dt} = \frac{2}{L}\frac{1}{\sin(\theta)}\frac{dz}{dt}. \tag{21.40}$$

The total energy of the system is therefore

$$E = mgz + \frac{1}{2}M\left(\frac{dz}{dt}\right)^2 + \frac{1}{2}\frac{ML^2}{12}\frac{4}{L^2}\frac{1}{\sin^2(\theta)}\left(\frac{dz}{dt}\right)^2, \tag{21.41}$$

which simplifies to

$$E = mgz + \frac{1}{2}M\left(\frac{dz}{dt}\right)^2 + \frac{1}{6\sin^2(\theta)}M\left(\frac{dz}{dt}\right)^2. \tag{21.42}$$

Conservation of energy tells us $E_i = E_f$. At the start, $\theta = 0$, $dz/dt = 0$, and $z = L/2$. At the finish, $\theta = \pi/2$ radians, $dz/dt = v_f$ (the answer), and $z = 0$. Starting by showing the substitutions with minimal simplification

$$\frac{mgL}{2} + \frac{2}{3}M0^2 = mg0 + \frac{2M}{3}v_f^2. \tag{21.43}$$

whose solution for v_f is

$$v_f = \left(\frac{3gL}{4}\right)^{1/2}. \tag{21.44}$$

This page reserved for your notes.

Chapter 22

Torque Diagrams; Pendulums

In this chapter, we consider physical systems whose behavior can be understood by applying the ideas of torque and angular momentum. I began by describing the torque diagram, which is a mnemonic tool roughly as useful as the force diagram, but appropriate for problems in which the torques on an object are the topic of interest. Pendulums provide simple examples in which torque can be used to understand the physical behavior of the system. As an alternative to pendulums, problems involving pulleys are of interest. As in all of the examples we have been considering, it is not necessarily the case that single pendulums or pulleys will be the precise topic in which you're interested, but rather that the general methods we are discussing can be applied to many other problems.

22.1 The Torque Diagram

We start by considering a simple mechanical object, shown in the accompanying sketch. We are looking at the objects of interest from the side. There are two masses, m_1 and m_2, hanging from strings, and a pulley, a large wheel on a central axle. The wheel has been machined so that it actually has two sections having two different radii, R_1 and R_2. One string is wrapped around each section. The wheel has a mass M and a moment of inertia I around the axle. There are no frictional forces in this problem; the wheel rotates freely around its axle.

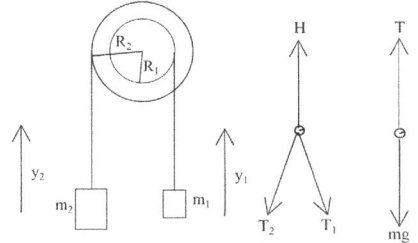

Figure 22.1: A simple machine for testing Newton's Laws of Motion.

A similar machine was used in the first precision tests of Newton's laws of motion. To understand its significance, recall that Newton's *Philosophiæ Naturalis Principia Mathematica* (Mathematical Principles of Natural Philosophy) was published in 1687. In period, people could make good bearings, permitting the construction of good pulleys. They could also weigh things reasonably accurately. Measuring time well was much more challenging; even if stop watches were available, human beings really cannot measure a time interval to better than a tenth of a second or so. A few period watches were equipped with a minute hand, reading the time in hours and minutes, the minute being a time period so tiny – minute – that it was of no consequence. The mechanical object shown here, and several variations, converts a need for precise time measurements into a need for precise weight measurements, resulting in the first successful precision tests of Newton's Laws of Motion.

We now ask what forces and torques are acting in the problem. Acting on each of the two masses is the force of gravity, leading to forces $m_1 g$ and $m_2 g$, respectively. The two strings are under tension. The tensions in the two strings are T_1 and T_2, respectively. You can immediately draw the force diagrams for the two masses. There is no reason to assume that the system is being held stationary, so it would be incorrect to claim, for example, that T_1 and $m_1 g$ are equal in magnitude.

From the force diagrams, we can immediately write the equations of motion for the two masses. The equations of motion are Newton's Second Law as written for the particular case at hand. We have

$$m_1 \frac{d^2 y_1}{dt^2} = T_1 - m_1, g \qquad (22.1)$$

$$m_2 \frac{d^2 y_2}{dt^2} = T_2 - m_2 g. \qquad (22.2)$$

The signs given to the tension and to the force of gravity were chosen based on a specific choice of coordinate system. If you had chosen the directions of y_1 and y_2 to be pointing down, then in the equations of motion the signs of the tension and the force of gravity would be reversed. As always, the sign is determined when a component of a force is inserted into an equation of motion, by comparison with the directions of the coordinates in the system.

There are several places here where you could've done things slightly differently. As I've said before, in general all choices of directions for the coordinate axes are equally valid. Some choices will make it easier to solve the problem. Some choices will make it far more difficult. However, if you are careful, no matter which choices you have made for the directions of your coordinates, the problem is still soluble. For example, you might have looked at the arrangement of the strings on the wheels. You might then have proposed that if mass one was going up, then mass two was going down, and therefore you should assign opposite directions to positive y_1 and to positive y_2. That would change some of the signs in the equations of motion, but the problem would remain soluble.

There are several reasons why this final choice of choosing directions is not a good problem solving approach. First, it is a path that makes more sense if you already think you know what the answer to the problem is. That's fine if your opinion, as to what the answer to the problem is, is correct. On the other hand, if your opinion happens to be mistaken, you can spend a great deal of time trying to force the solution to behave the way you think it should, without getting anywhere.

The main reason you will be worse off to choose a positive y_1 and a positive y_2 to point in opposite directions is that it will make it harder for you to understand your solutions. If I tell you that the acceleration of m_1 is $+0.5$ m/s^2, and you are using my coordinates, you immediately know that mass one is headed up. If you are using coordinates in which the positive accelerations for the two masses point in opposite directions, you have to think and remember whether a positive acceleration on mass one means that mass one is accelerating upward or that it is accelerating downward. Every time you insert into your solution something that you have to think about, to interpret a result, as opposed to doing things in a habitual way, you are opening yourself to error.

Note I said "habitual" and not "customary". Habit is the way you routinely do things. Custom is how people have, perhaps covertly, agreed to do things. If you have a series of routine tasks on getting up in the morning, and always carry them out in the same order, then even if you are slightly distracted everything is likely to get done in a proper manner. Here, if you always choose the upward direction to be the positive y direction, then you don't have to think as much about what sign you associate with, for example, mg.

Customs as to how you assign signs on coordinate systems are very important for communicating with other people, just as customs as to what units you are using are very important if you are communicating with other people. Customs can lead to confusion. If three of you are talking about distance for a hike, and one of you customarily uses kilometers as the distance, a second of you customarily uses miles as the distance, and the third of you customarily uses furlongs as the distance, you may end up with three very different impressions as to how long the hike is going to be. If you prepare for a 32 furlong (4 mile) hike, and the person who laid out the map and measured the distances is using miles, your preparations for your 32 mile walk are likely to be seriously inadequate. For this reason, engineering contracts often appear to be excruciatingly detailed, this being the only way to assure that everyone is in agreement as to what is to be done.

[Aside: Sometimes it is very difficult to see where an incorrect assumption has been made. Once upon a time, I spent several years of my career on and off trying to figure out why a certain alleged solution to a certain problem did not have the obviously necessary behavior, namely that the object of interest should be rotating. Only at the end did I realize that at the very start of the solution, as generated many decades ago not by me, the author had covertly assumed that the average torque on the object was zero, without bothering to mention this detail. Indeed, to further confuse the situation he had then immediately calculated a certain behavior which would only arise if the object were subject to a non-zero average torque, a torque

22.1. THE TORQUE DIAGRAM

that would cause the object to rotate. Finally he solved the problem, using his solution that said that the object did not rotate, an outcome not consistent with his calculation but consistent with his starting point. If you didn't realize that his discussion contradicted both his starting equations and his solution, and what the starting equations actually meant, you could become very confused. (Or you could, without thinking, regurgitate his final solution, most of which followed from dimensional analysis so that it had to be mostly right.) For most of those years, I knew that the author's solution was wrong. I was right about that, but until recently I couldn't see where his argument went astray.]

Returning to the problem at hand: Acting on the wheel, we have three forces, namely the two tensions pulling straight down and the force exerted by the axle pulling up, holding the wheel in place. Those three forces must sum to zero, because the wheel stays put rather than falling to the ground.

We also have acting on the wheel two torques, one due to each tension. The tension force effectively acts on the wheel at the point where the line of the string departs from the surface of the wheel. At that point, the tension force is acting perpendicular to a radius vector, namely the radius vector from the center of the wheel out to the point where the string heads off into space. The radii and the tensions all lie in a plane. Which way do the torques point? The torques calculated around the axle, at the point where the axle crosses the plane of the paper, must therefore be perpendicular to the plane of the paper. The torque vectors therefore each have only one nonzero component.

Recalling that $\boldsymbol{\tau} = \boldsymbol{r} \times \boldsymbol{F}$, the torques on the wheel due to m_1 and m_2 have magnitudes $T_1 R_1$ and $T_2 R_2$, respectively. We now introduce, as a way to describe this situation, the *torque diagram*. The torque diagram for this problem is shown in Figure 22.2. A torque diagram is a mnemonic tool; it's a sketch, not a scale drawing. Just as each mass in a Second Law problem is given its own force diagram, so also each rotating object in a torque problem is given its own torque diagram. In the Figure, the point around which the torques are calculated is indicated by a small circle. Radial lines emanating from the center represent the $\boldsymbol{r}$ vectors from the axle to the points where the forces are applied. In general it is convenient to draw all of them to be about the same apparent length. Each radial line is labelled with its associated length, here R_1 and R_2. The forces creating the torques are shown as being attached to the outer ends of the radial lines. In this case, for each mass the

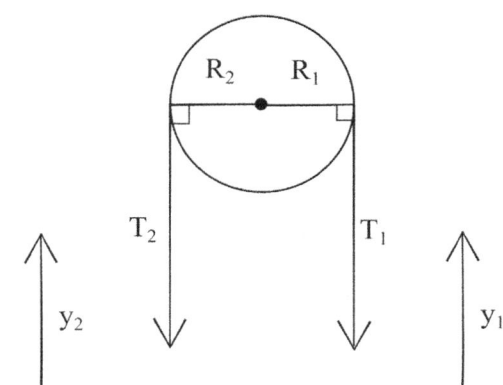

Figure 22.2: Torque diagram for the machine shown in Figure 22.1.

corresponding $\boldsymbol{R}$ and $\boldsymbol{F}$ vectors are perpendicular, so they are drawn that way in the figure. If they were not perpendicular, the Figure would need to be modified to indicate the appropriate angles. In this Figure, the 90 degree angles are indicated by the small squares. The relative positions of the $\boldsymbol{R}$ and $\boldsymbol{F}$ vectors are significant, in that they determine whether the torque tries to lead to clockwise or counterclockwise rotation. As with force diagrams and the Second Law, the sign to be associated with each term giving a torque is determined at the time the equation of motion is written.

We are now ready to write the torque equation. We begin with the general form for $\boldsymbol{L} = I\boldsymbol{\omega}$ and substitute, keeping in mind that in this problem the angular momentum and the torque each have only one non-zero component, corresponding to motions in the plane of the paper. We have

$$\frac{d\boldsymbol{L}}{dt} = \boldsymbol{\tau}. \tag{22.3}$$

For the one interesting component, we then have

$$I\frac{d^2\theta}{dt^2} = T_1 R_1 - T_2 R_2. \tag{22.4}$$

In setting up the solution to this problem, I took the vertical axis to be y rather than z, so that positive θ would have its customary direction. If the two labelled axes on the paper were x and z rather than x and y, θ would increase in the clockwise rather than the counterclockwise direction.

22.2 The Simple Pendulum

We now advance to a problem soluble with torque considerations, the simple pendulum, a mass at the end of a string. The top end of the string is attached to the ceiling. The ceiling does not move. The mass is treated as a point mass m; the string has a length r. The swing of the pendulum is described by an angle θ as shown in Figure 22.3.

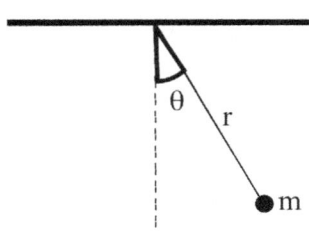

Figure 22.3: A simple pendulum hanging from the ceiling.

We now proceed to calculate how the pendulum moves as a function of time. Our starting point is equation 22.3. To advance we need to find the torque and moment of inertia of the pendulum. We advance by taking a variety of small pieces of results from earlier chapters and sections, and assemble them into a coherent whole. The objective is to write the equation of motion for the pendulum bob, this being the mass m, and then solve it to specify the position of the pendulum bob as a function of time.

A first step is to choose a center; the point where the pendulum is attached to the ceiling is an excellent but not mandatory location. Then identify the forces, as these allow us to calculate the torques in the problem. We can readily sketch a force diagram for the pendulum bob, in which the forces acting on the pendulum are seen to be a tension T and the force of gravity mg.

The force of gravity is readily broken into a component $mg_\parallel$ parallel to the string and a component $mg_\perp$ perpendicular to the string. In drawing the decomposition into components, the components are the two legs of a right triangle, while the force of gravity is the hypotenuse of the right triangle. From trigonometry we have for the magnitudes of the two components

$$mg_\parallel = mg\cos(\theta), \qquad (22.5)$$
$$mg_\perp = mg\sin(\theta). \qquad (22.6)$$

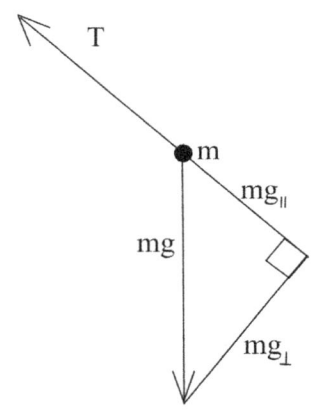

Figure 22.4: Force diagram and gravity components for the simple pendulum.

It's perfectly legal to choose these two components. The question is why we might want to do so. The answer is we are advancing toward computing the torque on the pendulum. In doing so, the component of the force of gravity that is parallel to the string exerts no torque on the pendulum bob, because $mg_\parallel$ is parallel to $\boldsymbol{r}$, and the cross product $\boldsymbol{r} \times \boldsymbol{F}$ of two parallel vectors is zero. Only the component of gravity perpendicular to the string can contribute to the torque. The magnitude of the nonzero component of the torque is $rmg_\perp$ or $rmg\sin(\theta)$. We need to assign a sign to this quantity separately. (Alternatively, we work carefully with cross products.)

Considering the swing of the pendulum, one immediately sees that $\boldsymbol{v}$, $\boldsymbol{r}$, and $\boldsymbol{F}$ are all in the plane of the paper, and therefore the torque $\boldsymbol{\tau}$ and angular momentum $\boldsymbol{L}$ are perpendicular to the plane of the paper.

The mass is performing circular motion, namely it moves at a fixed distance r from the point at which the string is attached to the ceiling. Its velocity is therefore perpendicular to the string: the string is the radius of the circle, and in circular motion the mass moves along the circumference of the circle, which is perpendicular to the circle's radius. The angular momentum therefore has magnitude

$$L = rp_\perp. \qquad (22.7)$$

Recalling circular motion, the above equation may be rewritten

$$L = rmr\frac{d\theta}{dt}. \qquad (22.8)$$

By inspection, the moment of inertia I of this pendulum bob is mr^2.

We are now almost ready to insert these results into the general equation for the time rate of change of angular momentum. The one detail we have to settle is the sign of the torque. I will first discuss how to do this by looking at the physical details of the problem. The alternative way to do this is to apply the right hand rule. We'll do that next. As indicated in Figure 22.3, the angle θ as measured from the vertical axis is positive and less than 90 degrees. Therefore, in equation 22.6, the term $\sin(\theta)$ is positive. However, as drawn, the force of gravity is attempting to induce rotation of the pendulum bob around the origin in the negative θ direction. When θ is positive, the torque is negative. (And when θ is negative, meaning that the pendulum bob is on the other side of the y-axis, the actual torque must be positive.) Both of these conditions are satisfied if we write $\tau_z = -mgr\sin(\theta)$.

The general result for $\boldsymbol{L}$ is

$$\frac{d\boldsymbol{L}}{dt} = \boldsymbol{\tau}. \tag{22.9}$$

Substituting into this equation for the one non-zero component of $\boldsymbol{L}$, one obtains

$$mr^2 \frac{d^2\theta}{dt^2} = -mgr\sin(\theta) \tag{22.10}$$

for the equation of motion of the pendulum bob. We will return to solving this equation later in the chapter.

Let us, however, first check the signs. We do this with the right-hand rule. First, let us establish the directions for the three basis vectors. The X and Y axes are shown in the original figure, so the directions of $\hat{\boldsymbol{i}}$ and $\hat{\boldsymbol{j}}$ are immediately apparent. Recalling that $\hat{\boldsymbol{i}} \times \hat{\boldsymbol{j}} = \hat{\boldsymbol{k}}$, we apply the right-hand rule to compute the direction of the final unit vector. The thumb goes in the positive x direction, the forefinger goes in the positive y direction, and therefore the three remaining fingers point in the positive z direction, which is seen to be out of the plane of the paper towards the reader.

We now use the right-hand rule to calculate the direction of positive angular momentum, and the direction of the torque. If $d\theta/dt$ is positive, the pendulum bob is moving in the counterclockwise direction. Corresponding to this, if you put the thumb of your right hand along the radius vector pointing outward from the origin, and point the forefinger in the direction of positive $d\theta/dt$, you should find that your remaining fingers are pointing up away from the page, this being the positive z direction, and therefore a positive $d\theta/dt$ does correspond to a positive value for the z component L_z of the angular momentum. What about the torque, for θ as indicated? Once again, point your thumb in the direction of the radius vector, pointing away from the origin. Point your index finger in the direction of gravity. To avoid breaking your wrist, contemplate first placing your right hand on your left shoulder. You should now find that your remaining fingers are pointing down, in the negative z direction, so that τ_z is negative. In order to cause the torque's algebraic form $rmg\sin(\theta)$ to have a negative sign, for θ being positive, you need to insert the minus sign explicitly in the expression for the torque. When you've done all this, you will discover that you have recovered equation 22.10. You have now checked the signs in that equation, and confirmed they are correct.

22.3 Pendulum with Extended Bob

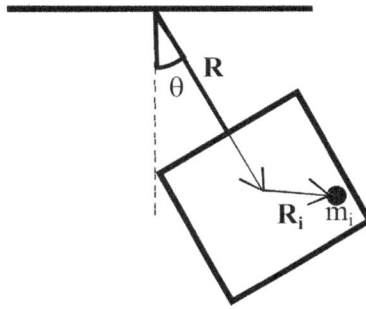

Figure 22.5: A pendulum with an extended rigid bob; size of bob enhanced for clarity of the vectors.

We now turn to a slightly more complicated problem, namely the oscillation of a pendulum with an extended bob. An extended bob is a bob that is larger than a point, so that it is meaningful to discuss the rotation of the bob with respect to its own center. I offer a sketch of an extended bob. The line from the origin at the angle θ is a stiff rod; the bob is the square at its bottom. The rod is very nearly massless relative to the bob. The bob is mounted rigidly on the rod, so it does not move with respect to the rod as the pendulum swings back and forth.

In the sketch, the vector $\boldsymbol{R}$ goes from the point where the pendulum is attached to the ceiling, to the center of mass of the bob. The vector $\boldsymbol{R}_i$ goes from the center of mass of the bob to an individual point m_i within the bob. The points, each having mass m_i, are labeled by the index i, with $\sum_i m_i = M$, M being the total mass of the bob.

A force diagram for the pendulum would show as external forces on the pendulum the tension T in the rod and the force of gravity Mg. We may readily calculate the total torque on the bob, around the point where the bob is attached to the ceiling, as a sum of the torques created by each of the individual points in the bob. We have

$$\boldsymbol{\tau} = \sum_i (\boldsymbol{R} + \boldsymbol{R}_i) \times \hat{\boldsymbol{j}}(-m_i g), \tag{22.11}$$

which may be reordered as

$$\boldsymbol{\tau} = -Mg\boldsymbol{R} \times \hat{\boldsymbol{j}} - g(\sum_i m_i \boldsymbol{R}_i) \times \hat{\boldsymbol{j}}. \tag{22.12}$$

However, by the definition of the center of mass, $\sum_i m_i \boldsymbol{R}_i = \boldsymbol{0}$, so the last term vanishes. $\boldsymbol{R} \times \hat{\boldsymbol{j}}$ has its non-zero component in the $\hat{\boldsymbol{k}}$ direction, leading to

$$\boldsymbol{\tau} = -MgR\sin(\theta)\hat{\boldsymbol{k}}. \tag{22.13}$$

We also need a form for the angular momentum $\boldsymbol{L}$. For the velocity of mass i we may write

$$\boldsymbol{v}_i = \boldsymbol{\omega} \times \boldsymbol{r}_i, \tag{22.14}$$

where the new vector $\boldsymbol{r}_i$ is the vector from the pivot point to mass i. From simple vector geometry, $\boldsymbol{r}_i = \boldsymbol{R} + \boldsymbol{R}_i$, so

$$\boldsymbol{v}_i = \boldsymbol{\omega} \times \boldsymbol{R} + \boldsymbol{\omega} \times \boldsymbol{R}_i. \tag{22.15}$$

The velocity of mass i is the sum of the center-of-mass motion of the bob plus the velocity due to rotation of the bob around its center of mass.

The total angular momentum of the bob may be found as

$$\boldsymbol{L} = \sum_i \boldsymbol{r}_i \times m_i(\boldsymbol{\omega} \times \boldsymbol{r}_i), \tag{22.16}$$

which on expansion is

$$\boldsymbol{L} = \sum_i m_i(\boldsymbol{R} + \boldsymbol{R}_i) \times (\boldsymbol{\omega} \times (\boldsymbol{R} + \boldsymbol{R}_i)). \tag{22.17}$$

Equation 22.17 can be expanded into four terms, namely

$$\boldsymbol{L} = \sum_i (m_i \boldsymbol{R} \times (\boldsymbol{\omega} \times \boldsymbol{R})) + \sum_i m_i(\boldsymbol{R}_i \times (\boldsymbol{\omega} \times \boldsymbol{R}_i)) +$$
$$\boldsymbol{R} \times (\boldsymbol{\omega} \times \sum_i (m_i \boldsymbol{R}_i)) + \sum_i (m_i \boldsymbol{R}_i)) \times (\boldsymbol{\omega} \times \boldsymbol{R}). \tag{22.18}$$

$\boldsymbol{R}$ and the $\boldsymbol{R}_i$ are both perpendicular to $\boldsymbol{\omega}$, so the two terms on the first line become $\omega M R^2$ and $\omega \sum_i m_i R_i^2$. The latter sum is the moment of inertia I relative to its center of mass of the rigid bob. On the second line, after factoring out the constants that multiply the $\sum_i$, the two terms are both proportional to $\sum_i m_i \boldsymbol{r}_i$, but by the definition of the center of mass $\sum_i m_i \boldsymbol{r}_i$ vanishes. We therefore reach

$$\boldsymbol{L} = MR^2 \boldsymbol{\omega} + I\boldsymbol{\omega}. \tag{22.19}$$

L has a center of mass term and a term describing rotation around the center of mass. Combining these with the law for the time evolution of the angular momentum, and noting that here $\boldsymbol{\omega} = \frac{d\theta}{dt}\hat{\boldsymbol{k}}$ so that $\frac{d\omega}{dt} = \frac{d^2\theta}{dt^2}$, we have

$$(MR^2 + I)\frac{d^2\theta}{dt^2}\hat{\boldsymbol{k}} = -MgR\sin(\theta)\hat{\boldsymbol{k}}, \qquad (22.20)$$

which has different constants but very much the same form as equation 22.10.

22.4 Pendulum Motion: A Solution

We have now twice derived an equation of motion for a pendulum. The equation is non-linear. The second time derivative of $\theta(t)$ is proportional to $\sin(\theta)$, not to θ. How do we solve this equation? The first part of the answer is to invoke the Taylor series expansion for $\sin(\theta)$, and use it to eliminate the nonlinearity. The Taylor series expansion indicates that for a well-behaved continuous, differentiable-many-times function $f(x)$ we can expand f(x) in terms of powers of x and the function's derivatives $d^n f(x)/dx^n$ as evaluated at $x = 0$. To be precise,

$$f(x) = \sum_{n=0}^{\infty} \frac{x^n}{n!} \left(\frac{d^n f(x)}{dx^n}\right)_{x=0}. \qquad (22.21)$$

In interpreting the derivative, it is important to remember that the derivatives of the function are taken before the function is evaluated, in this case at $x = 0$.

In particular, for $\sin(\theta)$ one has

$$\sin(\theta) = \sum_{n=1,\text{n odd}}^{\infty} \frac{(-1)^{n-1}\theta^n}{n!}. \qquad (22.22)$$

The lead terms of this expansion are

$$\sin(\theta) \approx \theta - \frac{\theta^3}{3!} + \frac{\theta^5}{5!} - \ldots \qquad (22.23)$$

At the beginning of the book, I emphasized that in calculus all angular units are in radians. In particular, the above equation is correct if θ is in radians, but completely incorrect if θ is in degrees or grads or mils.

If θ is small enough, $\theta^3/3!$ is really small, and the higher terms are even smaller, leading to the small-angle approximation

$$\sin(\theta) \approx \theta. \qquad (22.24)$$

How small is 'small enough'? To find out, you could try solving the pendulum problem while including, for example, the θ^3 term. This would be challenging analytically, possible to do numerically, and straightforward to study experimentally. However, on making the approximation, the equation of motion for the extended pendulum has the approximate form

$$-MGR\theta(t) = (MR^2 + I)\frac{d^2\theta(t)}{dt^2} \qquad (22.25)$$

Here θ is a function of time; it indicates the position of the pendulum bob. There are several ways to solve this equation. The simplest is to ask yourself which functions are, up to a constant, equal to their second derivative. Three obvious choices are the sine function, the cosine function, and the exponential. The sine and cosine are good. A homework problem lets you show that an exponential actually does not work as a solution. An alternative is to generate the solution as a Taylor series in time, as seen in Part III of the book. The proposed solutions are

$$\theta(t) = A\cos(\omega t + \phi) \qquad (22.26)$$
$$\theta(t) = A\sin(\omega t + \phi). \qquad (22.27)$$

A word of warning: The ω in the above pair of equations is the angular frequency of oscillation; it is a constant. It is not the time-dependent quantity $\omega = \frac{d\theta(t)}{dt}$ we used when we discussed rotation and angular momentum. "ω" has two different meanings. However, the notation has become standardized, customary, so you will have to live with it. A and ϕ are constants of integration appearing because we have implicitly integrated $\frac{d^2\theta}{dt^2}$ twice with respect to time. A is the *amplitude*, the size of the swing. ϕ is the *phase* of the oscillation, present so that at $t = 0$ the pendulum does not have to be at a minimum, maximum, or zero of oscillation.

Inserting the first of these equations into the equation of motion, equation 22.25, we obtain

$$-MGRA\cos(\omega t + \phi) = -(MR^2 + I)\omega^2 A \cos(\omega t + \phi), \tag{22.28}$$

which leads to

$$\omega = \left(\frac{MgR}{MR^2 + I}\right)^{1/2}, \tag{22.29}$$

and therefore to

$$\theta(t) = A\cos\left(\left(\frac{MgR}{MR^2 + I}\right)^{1/2} t + \phi\right). \tag{22.30}$$

For the simple pendulum, the pendulum bob is a point, so that $I = 0$, and

$$\omega = \left(\frac{g}{R}\right)^{1/2} \tag{22.31}$$

as found previously.

22.5 Period of a Pendulum

As an alternative to ω, it is useful to introduce the period T of a pendulum. This T is not the same as T the tension. T is usefully described as the time between pairs of zero crossings of $\theta(t)$ (points where $\theta(t) = 0$). Alternatively, T can be defined as the time between pairs of maxima or pairs of minima of $\theta(t)$. There is an interesting experimental question as to which of these three descriptions of T is to be preferred. That's a lab experiment as described later in the book. No matter which description of the period is to be preferred for measurement purposes, in one period T the phase of $\cos(\omega t + \phi)$ is increased by 2π, i.e.,

$$(\omega(t + T) + \phi) - (\omega t + \phi) = 2\pi, \tag{22.32}$$

which implies that

$$T = \frac{2\pi}{\omega}. \tag{22.33}$$

For the simple pendulum, the period is therefore

$$T = 2\pi \left(\frac{R}{g}\right)^{1/2}. \tag{22.34}$$

22.6 Discussion

Here we showed that you can sometimes solve a problem and generate a reasonably accurate solution to an uncooperative equation by identifying a variable whose values are small, and doing a Taylor series expansion in that variable. Note, however, Feynmann's warning: Just because you have a variable whose values are small, it is not necessarily a good idea to do a power series expansion in that variable and then truncate the series.

22.7 Worked Problems

1. This problem corresponds to a laboratory exercise. A uniform rod of length ℓ and mass M hangs from a pin at its top end. It makes an angle θ with respect to the vertical. Find (i) the moment of inertia of the rod, (ii) the torque on the rod, and (iii) (from (i) and (ii)) the equation of motion for the rod. (iv) Find the angular acceleration of the rod. (v) For small angles θ find the frequency and period of oscillation of the rod. (vi) A nut of mass m is now attached to the rod, a distance x from the top end; find the period of small oscillations as a function of m and x.

2. A rod of length L is pivoted at one end. At equilibrium it hangs vertically downward. At some moment, it makes an angle θ with respect to the vertical. The rod is not uniform; its density ρ depends on distance ℓ from the pivot as $\rho = \mu \ell^3$, μ being a constant. ρ is the density per unit length, so that the mass of a differential section of the rod of length $d\ell$ is $\mu \ell^3 d\ell$. Find (a) the gravitational torque on the rod, (b) the moment of inertia of the rod, (c) assuming the pivot is frictionless, the angular acceleration of the rod (a correct answer is linear in $\sin(\theta)$), and (d) for small displacements, find the frequency of oscillation.

3. Two circular disks mounted on the same axis are cemented together so that they rotate as a unit, as seen in Figure 22.1. Their combined moment of inertia is I. Their radii are R_1 and R_2, respectively. A rope is wrapped around each of them, with one end of each rope hanging down to support masses m_1 and m_2 as seen in the figure. Find the force and torque diagrams, the equations of motion, and the constraints. Find the ratio of the masses such that the disks will not accelerate in either direction. For the general case that the two masses do not balance each other, find the angular acceleration and the tension in each rope.

22.8 Homework

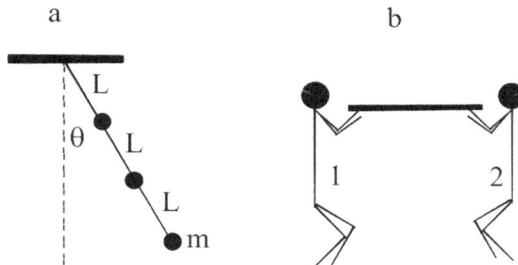

Figure 22.6: (a) A rigid rod with three attached masses m. (b) Two robots holding up a board.

1. Show that the exponential $A \exp(at)$ is not a solution of equation 22.25, if a is a real number.

2. A pendulum consists of a nearly massless rigid rod having length $3L$. As seen in Figure 22.6a, the rod is pivoted at the top end. Three masses, each having mass M, are spaced distances L apart along the rod. The rod makes an angle θ with the vertical. Find (a) the moment of inertia of the pendulum around its pivot, (b) the torque on the rod due to gravity acting on the three masses, (c) the equation of motion for the rod using θ as the time-dependent variable, and (d) the angular frequency of oscillation ω of the pendulum assuming small displacements. A reasonable guess for the time-dependent angle of the rod is $\theta(t) = \theta_o \sin(\omega t)$ in which θ_o and ω_o are constants. (e) The pendulum is released from rest at an angle θ_1 away from the vertical. What is the angular velocity of the rod around its pivot when it has fallen to an angle θ_2 from the vertical.

3. As seen in Figure 22.6b, a uniform board of length L and mass m is being supported by two robots. The robots hold the two ends of the board. Suddenly one of the robots decides to rise up against its despotic human oppressors and lets go of its end of the board. The revolutionary robot was holding

the board at end 2; the robot at end 1 steadfastly refuses to let go of the board. At the moment after robot 2 lets go, find (a) the gravitational torque on the board around end 1, (b) the angular acceleration around end 1 of the board, (c) the vertical acceleration of the board's center of mass, and (d) the vertical force exerted on the board by the robot at end 1. Robot 1 is applying a purely vertical force on its end of the board.

4. A circular pendulum consists of a string having length ℓ with a pendulum bob of mass M at the bottom end. When the pendulum is swinging, the bob goes in a circle at a uniform height, the center of the circle being where the pendulum would rest if it were stationary. Find the relationship between the period of the pendulum and the vertical distance that the bob has risen above its rest position. Find the relationship between the vertical rise and the tension in the string.

5. A physical pendulum is constructed from a *non-uniform* rod having length L. The rod is free to swing back and forth around a pivot at one end, this being the location $x = 0$. The rod is at an angle θ with respect to straight down. θ is changing at a rate $d\theta/dt = \omega$. The rod has a density that depends on length as

$$\rho(x) = A(1 + bx) \tag{22.35}$$

where A and b are constants. *Set up and evaluate the integrals* and calculate (i) the gravitational torque on the rod. (ii) the moment of inertia of the rod. (iii) The period of oscillation T of the rod.

6. A futuristic unconventional baseball bat of length L has a mass per unit length ρ that depends on position as $\rho = \lambda_0(1 + x^3/L^3)/$ Here λ_0 is a constant having units mass/length. The bat hangs downward from its $x = 0$ end, making an angle θ with respect to the downward vertical. Find by integration (a) the moment of inertia and (b) the gravitational torque on the bat. (c) Find the angular acceleration of the bat. (d) What is the bat's angular acceleration in the small-angle limit? (e) What is the bat's period of oscillation?

22.9 Solutions to the Worked Problems

1. (i) The moment of inertia of a differential segment of the rod, length dx, is

$$dI = \frac{M}{\ell} x^2 dx, \tag{22.36}$$

so the total moment of inertia of the rod is

$$I = \int dI = \int_0^\ell \frac{M}{\ell} x^2 dx = \frac{M\ell^2}{3}. \tag{22.37}$$

(ii) The interesting component of the torque on a differential segment of the rod, due to gravity, is

$$d\tau = -\frac{Mg}{\ell} x dx \sin(\theta), \tag{22.38}$$

where θ is the angle measured from the vertical. The interesting component of the total torque on the rod is

$$\tau = \int d\tau = -\int_0^\ell \frac{Mg}{\ell} x dx \sin(\theta) = -\frac{Mg\ell \sin(\theta)}{2}. \tag{22.39}$$

(iii) The equation of motion of the rod arises from

$$\tau = I \frac{d^2\theta}{dt^2}, \tag{22.40}$$

which leads to

$$-\frac{Mg\ell \sin(\theta)}{2} = \frac{M\ell^2}{3} \frac{d^2\theta}{dt^2}. \tag{22.41}$$

22.9. SOLUTIONS TO THE WORKED PROBLEMS

(iv) Rearranging the answer to (iii),
$$\frac{d^2\theta}{dt^2} = -\frac{3g}{2\ell}\sin(\theta). \tag{22.42}$$

(v) For small angles, $\sin(\theta) \approx \theta$, so
$$\frac{d^2\theta}{dt^2} = -\frac{3g}{2\ell}\theta. \tag{22.43}$$

A reasonable solution is $\theta(t) = A\cos(\omega t + \phi)$. Substituting this form in the small-angle approximation, we get
$$-A\omega^2 \cos(\omega t + \phi) = -\frac{3g}{2\ell}A\cos(\omega t + \phi), \tag{22.44}$$

and therefore
$$\omega = \left(\frac{3g}{2\ell}\right)^{1/2}. \tag{22.45}$$

The period T is $2\pi/\omega$, so
$$T = 2\pi\left(\frac{2\ell}{3g}\right)^{1/2}. \tag{22.46}$$

(vi) The nut will increase the moment of inertia by mx^2 and the torque by $mgx\sin(\theta)$, so the equation of motion becomes
$$-(\frac{M\ell}{2} + mx)g\sin(\theta) = (\frac{M\ell^2}{3} + mx^2)\frac{d^2\theta}{dt^2}. \tag{22.47}$$

The angular frequency of oscillation is then
$$\omega = \left(\frac{(\frac{M\ell}{2} + mx)g}{(\frac{M\ell^2}{3} + mx^2)}\right)^{1/2}, \tag{22.48}$$

while the period is
$$T = 2\pi\left(\frac{(\frac{M\ell^2}{3} + mx)}{(\frac{M\ell}{2} + mx^2)g}\right)^{1/2}. \tag{22.49}$$

2. (i) We begin with the differential torque (actually, the one interesting component of the torque) on a differential segment of the rod, which is
$$d\tau = -\mu\ell^3 g\ell\sin(\theta)d\ell, \tag{22.50}$$

so the total torque is
$$\tau = -\int_0^L \mu g\ell^4 \sin(\theta)d\ell = -\frac{\mu g L^5}{5}\sin(\theta). \tag{22.51}$$

(ii) The moment of inertia of a differential segment of the rod is
$$dI = dm\ell^2, \tag{22.52}$$

so the moment of inertia is
$$I = \int_0^L (d\ell\mu\ell^3)\ell^2 = \frac{\mu L^6}{6}. \tag{22.53}$$

(iii) For the one interesting component of the angular momentum vector, we may write $\tau = I\frac{d^2\theta}{dt^2}$ and hence
$$\frac{d^2\theta}{dt^2} = \frac{\tau}{I} = -\frac{\mu g L^5/5}{\mu L^6/6}\sin(\theta). \tag{22.54}$$

For small angles, this result becomes
$$\frac{d^2\theta}{dt^2} = -\frac{6g}{5L}\theta. \tag{22.55}$$

A reasonable solution to this equation has the form $\theta(t) = A\cos(\omega t + \phi)$. We have often written $\omega = \frac{d\theta}{dt}$. That definition does not apply here. Here ω is a constant, the oscillation frequency. We obtain
$$\omega = \left(\frac{6g}{5L}\right)^{1/2}. \tag{22.56}$$

3. The interesting scalar components of the Second Law for the two masses are
$$T_1 - m_1 g = m_1 \frac{d^2 z_1}{dt^2}, \tag{22.57}$$
$$T_2 - m_2 g = m_2 \frac{d^2 z_2}{dt^2}. \tag{22.58}$$

A common error is to claim at this point that $T = mg$. If that were true, which it is not, the net force on the corresponding mass would be zero, and the mass would not accelerate.

The torque equation gives
$$T_1 R_1 - T_2 R_2 = I \frac{d\omega}{dt}. \tag{22.59}$$

Can we solve this? If you look carefully, there are five unknowns, namely two tensions, two linear accelerations, and an angular acceleration. There are only three equations. We can't solve that. However, some thought shows that there are two constraints, because the speed at which the wheel turns determines how fast the two masses go up or down. If ω is positive, meaning the wheel is turning in the anticlockwise direction, then mass 1 goes down and mass 2 goes up. The rate at which the two strings are released or wound up is the tangential velocity of their wheel, leading to
$$\frac{dz_1}{dt} = -R_1 \omega, \tag{22.60}$$
$$\frac{dz_2}{dt} = R_2 \omega. \tag{22.61}$$

Corresponding to our five unknowns, we now have five independent equations, so there is a reasonable hope of solving. What if the masses all remain stationary? In that case, the accelerations are all equal to zero. The torque equation then gives
$$T_1 R_1 - T_2 R_2 = 0, \tag{22.62}$$

so $T_2 = \frac{T_1 R_1}{R_2}$. The two Second Law equations, for the special case that the accelerations are all zero, tell us
$$T_1 = m_1 g, \tag{22.63}$$
$$T_2 = m_2 g. \tag{22.64}$$

If we combine the above three equations we find
$$m_2 = \frac{m_1 R_1}{R_2}. \tag{22.65}$$

Taking the time derivatives of the two constraint equations, we have
$$\frac{d^2 z_1}{dt^2} = -R_1 \frac{d\omega}{dt}, \tag{22.66}$$
$$\frac{d^2 z_2}{dt^2} = R_2 \frac{d\omega}{dt}. \tag{22.67}$$

22.9. SOLUTIONS TO THE WORKED PROBLEMS

which lets us substitute for the accelerations of the two masses. From the Second Law equations, after multiplying by various radii,

$$R_1 T_1 - m_1 g R_1 = -m_1 R_1^2 \frac{d\omega}{dt}, \tag{22.68}$$

$$R_2 T_2 - m_2 g R_2 = +m_2 R_2^2 \frac{d\omega}{dt}, \tag{22.69}$$

$$T_1 R_1 - R_2 T_2 = I \frac{d\omega}{dt}. \tag{22.70}$$

Subtracting and adding the first two of these equations from the third, so as to eliminate the $R_i T_i$ terms, we obtain

$$m_1 g R_1 - m_1 R_1^2 \frac{d\omega}{dt} - m_2 g R_2 - m_2 R_2^2 \frac{d\omega}{dt} = I \frac{d\omega}{dt}. \tag{22.71}$$

After moving all the terms in $\frac{d\omega}{dt}$ to the same side of the equation, we find

$$\frac{d\omega}{dt} = \frac{m_1 g R_1 - m_2 g R_2}{m_1 R_1^2 + m_2 R_2^2 + I}. \tag{22.72}$$

We are also supposed to find the tensions. From the force equations

$$T_1 = m_1 g - m_1 R_1 \frac{d\omega}{dt}, \tag{22.73}$$

$$T_2 = m_2 g - m_2 R_2 \frac{d\omega}{dt}. \tag{22.74}$$

Substituting for the angular acceleration

$$T_1 = m_1 g - m_1 R_1 \frac{m_1 g R_1 - m_2 g R_2}{m_1 R_1^2 + m_2 R_2^2 + I}, \tag{22.75}$$

$$T_2 = m_2 g - m_2 R_2 \frac{m_1 g R_1 - m_2 g R_2}{m_1 R_1^2 + m_2 R_2^2 + I}. \tag{22.76}$$

This space reserved for your notes.

This page reserved for your notes.

Chapter 23

Coupled Motion Including Rotation

23.1 Introduction

In this Chapter we consider several problems involving the coupled motion of several objects, including objects that rotate but don't translate.

23.2 One Mass and a Wheel

The figure shows a pair of coupled masses. The upper mass is a spool of thread supported on a frictionless axle. The spool, moment of inertia I and radius R, is partly unwound. Hanging from the thread is a mass m_1. What is the acceleration of mass m_1?

We start by writing a torque diagram for the spool and a force diagram for the mass. We must start by choosing the origin for the torque and angular momentum of the spool. We may choose whichever point we want, but some choices are much more convenient than others. We choose the center of the axle as the origin. The torque diagram shows one torque, due to the tension T in the string. The force T is effectively applied to the spool at the point where the string departs, in a straight line, from the surface of the spool. At that point, the perpendicular distance from that point to the origin is R, so the associated torque is TR in a clockwise direction.

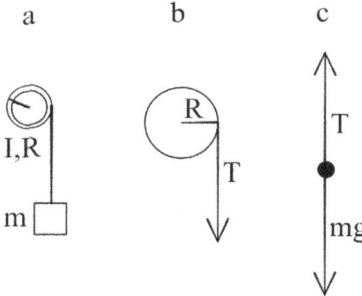

Figure 23.1: Two coupled masses, including (a) sketch, (b) torque diagram, and (c) force diagram.

The force diagram shows the mass m_1. It is acted on by two forces, namely the tension T and the force of gravity $m_1 g$, both acting in the vertical direction.

From these, we may write the equations of motion as

$$I\frac{d^2\theta}{dt^2} = -TR, \tag{23.1}$$

$$m_1\frac{d^2 y_1}{dt^2} = T - m_1 g. \tag{23.2}$$

We have here three unknowns, namely $d^2\theta/dt^2$, $d^2 y_1/dt^2$, and T. However, we only have two equations. To solve, we need a third equation. The third equation is a constraint, namely the rate at which the thread is released or winds up, this being the tangential speed of the spool, must match the speed at which the mass moves up or down. The constraint is that

$$R\frac{d\theta}{dt} = \frac{dy_1}{dt}, \tag{23.3}$$

and therefore

$$\frac{d^2\theta}{dt^2} = \frac{1}{R}\frac{d^2y_1}{dt^2}. \tag{23.4}$$

Dividing the torque equation by R, and then combining the torque equation with equation 23.4, we have

$$\frac{I}{R^2}\frac{d^2y_1}{dt^2} = -T. \tag{23.5}$$

Adding equations 23.5 and 23.2b, the tension cancels, leaving us with

$$\left(\frac{I}{R^2} + m_1\right)\frac{d^2y_1}{dt^2} = -m_1 g. \tag{23.6}$$

Rearranging terms, we finally have for the acceleration of the mass

$$\frac{d^2y_1}{dt^2} = -\frac{m_1 R^2 g}{I + m_1 R^2}. \tag{23.7}$$

We ask if this results appears to make sense. This test is part of a solution method. I and $m_1 R^2$ both have dimensions mass · length^2, so the denominator is a sum of terms having the same dimensions. The numerator and denominator each have a factor having dimensions mass · length^2, which dimensionally cancel, leaving behind a factor of g, which is indeed an acceleration, as required. Clearly $\frac{d^2y_1}{dt^2}$ is negative. The acceleration of the mass is downward, as expected. On the right hand side, in the denominator, both terms are intrinsically positive, and being added, so there is no possibility of their adding to zero and creating an infinite predicted acceleration.

Finally, we can examine the limits in which $I \ll m_1 R^2$ and $I \gg m_1 R^2$. In the former case, the very heavy mass accelerates downward at g or very close. The spool is extremely light and has next to no effect on how fast the mass falls. In the latter case, $m_1 R^2/I \to 0$; the mass and spool very nearly do not accelerate because the mass is too light to move the spool appreciably.

23.3 Two Masses and a Wheel

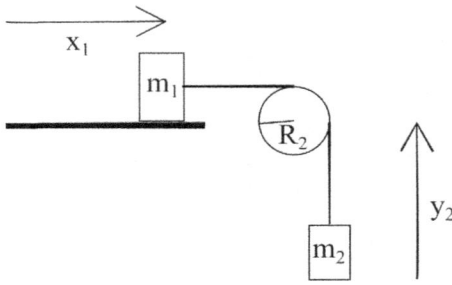

Figure 23.2: Sketch of two masses and a wheel, with m_1 resting on a table and m_2 hanging in mid-air.

Consider as an example Figure 23.2, which shows masses m_1 and m_2 connected by a piece of string. m_1 is resting on a tabletop. m_2 is hanging free in the air. The string runs over a wheel; when the masses and string move, the wheel rotates without the string slipping across the surface of the wheel. The wheel has radius R, mass M, and moment of inertia I; it is positioned with respect to mass m_1 so that the tension in the string produces a horizontal force on m_1. The horizontal distance between the center of mass of m_1 and the origin is ℓ_1; the vertical distance between the center of mass of m_2 and the origin is ℓ_2.

What is the acceleration of mass 1? We'll answer this question several times in different ways. First, suppose we set up equations of motion separately for each of the three bodies. We start by generating the needed force and torque diagrams. For the two masses, the force diagrams appear in Figure 23.3. Here N is the normal force due to the table, $m_1 g$ the force of gravity on mass m_1, and T_1 the tension in the string attached to m_1. Similarly, for mass 2 the forces are the force of gravity $m_2 g$ and the tension T_2 in the string on the other side of the wheel. The one string is continuous; it wraps around the wheel. The two tensions can be unequal because the string is applying a torque on the wheel.

23.3. TWO MASSES AND A WHEEL

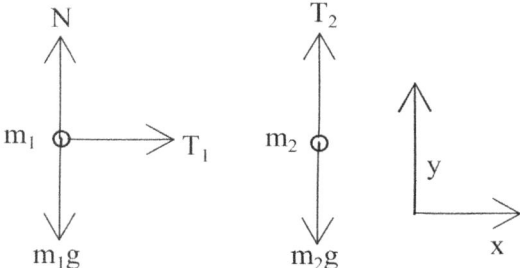

Figure 23.3: Force diagrams for the masses m_1 and m_2. They share a common set of directions for their coordinates.

Finally, for the wheel we generate a torque diagram. To do this, we must first choose an origin around which we calculate the torques. We choose to put the origin at the axle around which the wheel turns. We are free to put the origin wherever we want, but mathematically some choices are considerably more convenient than others. Note in the previous chapter our treatment of a pendulum with a rigid bob. We didn't put the origin at the center of mass of the bob; we put it at a point that was outside the bob. The bob was rotating around an outside point. Several pages of algebra were needed to compute the mechanical behavior of the bob, given that it was not rotating around its own center of mass. Here we avoid this work by choosing to put the origin at the axle.

The forces on the wheel are effectively applied to the wheel at the two points where the string is tangent to the wheel. The string tensions are T_1 and T_2. The outer perimeter of the wheel is a circle, so the perpendicular distance from the origin to the points where the forces are applied is R.

Having generated the force and torque diagrams, we now insert the forces and torques into the Second Law and the corresponding equation for time rate-of-change of angular momentum. We obtain

$$m_1 \frac{d^2 x_1}{dt^2} = T_1, \qquad (23.8)$$

$$m_1 \frac{d^2 y_1}{dt^2} = N - m_1 g, \qquad (23.9)$$

$$m_2 \frac{d^2 y_2}{dt^2} = T_2 - m_2 g, \qquad (23.10)$$

$$I \frac{d^2 \theta}{dt^2} = T_1 R - T_2 R. \qquad (23.11)$$

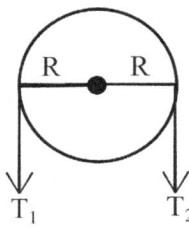

Figure 23.4: The torque diagram for the wheel, with tensions T_1 and T_2 applying torques relative to the center of the wheel.

It might be supposed that T_1 should equal T_2, or that T_2 should equal $m_2 g$. However, if either of these supposed equalities were correct, the corresponding acceleration would be zero, and the masses and wheel, once stationary, would remain stationary. In fact, they have non-zero accelerations.

There are also constraints. The tangential speed of the wheel must equal the linear speeds of the two masses, so

$$\frac{d^2 x_2}{dt^2} = R \frac{d^2 \theta}{dt^2}, \qquad (23.12)$$

$$\frac{d^2 x_1}{dt^2} = -R \frac{d^2 \theta}{dt^2}. \qquad (23.13)$$

The signs of the right hand sides of these two equations are not the same, because a positive (upward) acceleration of mass 2 corresponds to a positive (counterclockwise) angular acceleration for the wheel, while on the other hand a positive linear acceleration for mass 1 corresponds to a negative (clockwise) angular acceleration for the wheel. Also, a positive (right-ward) linear acceleration for mass 1 corresponds to a negative (downward) acceleration for mass 2, and vice versa.

There are multiple valid paths to a solution. Here we use the constraints to eliminate $d^2\theta/dt^2$ and

d^2x_2/dt^2 in favor of d^2x_1/dt^2, leading to

$$\frac{I}{R}\frac{d^2x_1}{dt^2} = T_1 R - T_2 R \qquad (23.14)$$

$$-m_2 \frac{d^2x_1}{dt^2} = T_2 - m_2 g. \qquad (23.15)$$

If we divide the first of these equations by R, we now have three equations, namely these two and equation 23.8a, which have on their right sides terms T_1, T_2 and $m_2 g$. By adding and subtracting these three equations from each other, we have

$$\left(m_1 + \frac{I}{R^2} + m_2\right)\frac{d^2x_1}{dt^2} = T_1 - T_1 + T_2 - T_2 + m_2 g \qquad (23.16)$$

On canceling the tensions, multiplying by R^2 to remove a fraction, and rearranging terms, we get

$$\frac{d^2x_1}{dt^2} = \frac{m_2 R^2 g}{m_1 R^2 + I + m_2 R^2}. \qquad (23.17)$$

We will now solve the problem of two masses and a wheel again, this time by calculating the angular momentum and the torque of the system as a whole. We will do everything from the very beginning. showing all steps.

We begin by constructing the force and torque diagrams. The force diagrams are seen in Figure 23.4. Mass m_1 is subject to a normal force $\boldsymbol{N}_1$, a force of gravity $-m_1 g \hat{\boldsymbol{k}}$, and a tension force $T_1 \hat{\boldsymbol{i}}$, as indicated in the Figure. Mass m_2 is subject a tension force $T_2 \hat{\boldsymbol{k}}$ and a force of gravity $-m_2 g \hat{\boldsymbol{k}}$. The wheel (force diagram not shown) is subject to forces $-T_1 \hat{\boldsymbol{i}}$ and $-T_2 \hat{\boldsymbol{k}}$ due to the string. It is also subject to a force $\boldsymbol{S}$ provided by the bracket holding it in place. The tensions in the string on the two sides of the wheel might not be equal, so they have been called T_1 and T_2. Indeed, if T_1 and T_2 were equal, they would put equal and opposite torques on the wheel, meaning that if the system began at rest it would not then move. It does move, namely m_2 drops, dragging along the wheel and m_1.

We now calculate all of the torques and contributions to the angular momentum of the system, using as our origin the wheel's axle, the point around which the wheel rotates. All of the objects, forces, and velocities are in the same plane, so the torques and angular momenta are all perpendicular to that plane, into or out of the plane of the paper. To calculate magnitudes of torques and angular momenta, we use the results from an earlier Chapter that their magnitudes are $F r_\perp$ and $p r_\perp$, respectively.

We'll do the calculation in parts, finding separately the torques on the system due to forces on m_1, m_2 and the wheel.

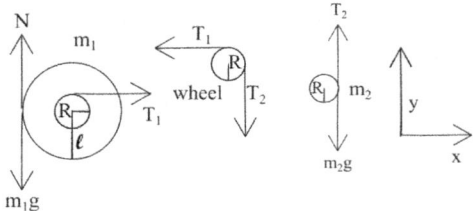

Figure 23.5: Torques on the system, calculated relative to the center of the wheel, due to forces on m_1, forces on m_2 and forces on the wheel.

Relative to our choice of origin, the center of the wheel, masses m_1 and m_2 are subject to non-zero torques. To see this point more clearly, we return to the original definitions of angular momentum and torque for a point mass, namely

$$\boldsymbol{L} = \boldsymbol{r} \times \boldsymbol{p} \qquad (23.18)$$
$$\boldsymbol{\tau} = \boldsymbol{r} \times \boldsymbol{F} \qquad (23.19)$$

These two definitions each refer to position, momentum, and force at a single instant in time, with no assumption, explicit or implicit, that the point mass is moving in a circle rather than moving in a straight line. (Similarly, in a discussion of the vector cross product based on the right hand rule, the thumb, forefinger, and remaining fingers each point in a single direction, rather than curling around some unseen axis.)

The forces on m_1 contribute to the torque as seen in Figure 23.5. The forces N and $m_1 g$ are vertical. Their perpendicular distance to the origin is therefore ℓ. There is also a torque on m_1 due to T_1. That

23.3. TWO MASSES AND A WHEEL

torque is non-zero because T_1 lies in the horizontal plane, so the perpendicular distance between the origin and the point where the string is attached to m_1 is R. The vertical distance between the tabletop and the point where the string is attached to m_1 does not enter the calculation. The forces on m_1, as shown in the torque diagram, contribute three torques as indicated.

There are three forces on the wheel, namely S, T_1, and T_2. The support force S acts on the axle at almost no distance from the origin, so $Sr_\perp$ vanishes; S creates no torque on the system. The strings that create the two tension forces are tangent to the wheel, a distance R out from the origin. The two tension forces thus create two torques, namely $T_1 R$ and $T_2 R$, pointing in opposite directions as seen in Figure 23.5.

Finally, there are two forces on m_2, namely T_2 and $m_2 g$. These two forces both act vertically, parallel to the string. The perpendicular distance between the point where they act and the origin is R. The corresponding torques on the system then have magnitudes $T_2 R$ and $m_2 g R$, but point in opposite directions.

These are all torques on the same object, the two masses and the wheel, around the same origin, so we may add them. From the torque diagram, the total torque on the system follows as

$$\boldsymbol{\tau} = \hat{\boldsymbol{k}} \left(-N\ell_1 + m_1 g \ell_1 - T_1 R + T_1 R - T_2 R + T_2 R - m_2 g R \right). \tag{23.20}$$

We separately need to calculate the angular momentum of the system. For the two masses, L can be found as $p r_\perp$. The motions of the masses are exactly parallel to the two string sections, so for each mass $r_\perp = R$. Combining the contributions of the three objects to $\boldsymbol{L}$, we find

$$\boldsymbol{L} = \hat{\boldsymbol{k}} \left(I \frac{d\theta}{dt} + R m_2 \frac{dy_2}{dt} - R m_1 \frac{dx_1}{dt} \right). \tag{23.21}$$

The force diagram for m_1 gives us a significant result, viz.

$$m_1 \frac{d^2 y_1}{dt^2} = N - m_1 g. \tag{23.22}$$

However, so long as the table does not collapse, m_1 is simply resting on a stationary table top, so $d^2 y_1/dt^2 = 0$ and $N = m_1 g$.

Most of the terms in equation 23.20 cancel each other. The result is

$$\boldsymbol{\tau} = -m_2 g R \hat{\boldsymbol{k}}. \tag{23.23}$$

Recalling $\boldsymbol{\tau} = d\boldsymbol{L}/dt$, we find

$$-m_2 g R \hat{\boldsymbol{k}} = \hat{\boldsymbol{k}} \left(I \frac{d^2\theta}{dt^2} + R m_2 \frac{dy_2^2}{dt^2} - R m_1 \frac{dx_1^2}{dt^2} \right). \tag{23.24}$$

The three time derivatives are not independent. We need to apply the constraints. The speeds of the two masses and the tangential velocity of the wheel must be equal in magnitude. If $d\theta/dt > 0$, then mass 1 is moving to the left, implying $dx_1/dt < 0$, and mass 2 must be moving up, implying $dy_1/dt > 0$. We then have as the two constraints

$$\frac{dx_1}{dt} = -R \frac{d\theta}{dt}, \tag{23.25}$$

$$\frac{dy_2}{dt} = R \frac{d\theta}{dt}. \tag{23.26}$$

Combining all there, we have

$$-m_2 g R = \frac{d^2\theta}{dt^2} \left(I + m_2 R^2 + m_1 R^2, \right) \tag{23.27}$$

leading to

$$\frac{d^2\theta}{dt^2} = -\frac{m_2 g R}{I + m_2 R^2 + m_1 R^2}, \tag{23.28}$$

and, correspondingly,

$$\frac{d^2 x_1}{dt^2} = \frac{m_2 g R^2}{I + m_2 R^2 + m_1 R^2}. \tag{23.29}$$

The accelerations are all constant.

23.4 Discussion

If one were not careful with the signs in the calculation leading to equation 23.29, one might have found for the denominator on the right hand side not $I + m_2R^2 + m_1R^2$ but instead the incorrect $I + m_2R^2 - m_1R^2$. While it takes practice to apply the following method automatically, there is a path that reveals that $I + m_2R^2 - m_1R^2$ must someplace have an error in it:

This term is a denominator. All of the algebraic characters in the denominator are positive real numbers. Therefore, if one adjusts the relative values of I, m_2, and m_1 appropriately, the negative sign on m_1R^2 lets you arrange things so that this denominator is zero, and therefore d^2x_1/dt^2 is infinite. That's not a reasonable physical outcome. An array of several masses connected by strings cannot be applying on m_1 the infinite mechanical force needed to create an infinite acceleration. Correspondingly, some of the signs in this denominator, because they are different from each other, cannot be right.

23.5 Worked Problems

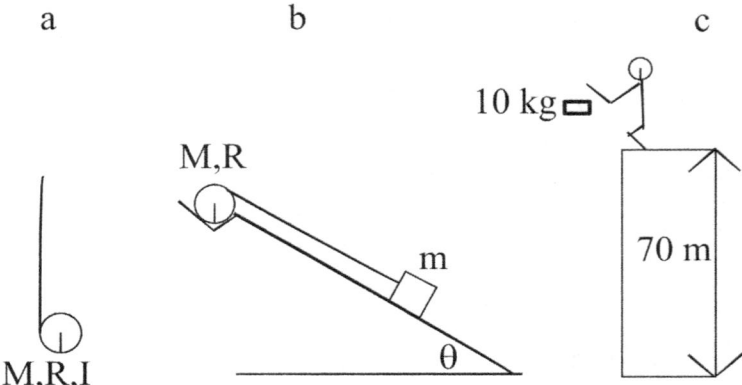

Figure 23.6: Figures for worked problems (a) problem 1, (b) problem 2, and (c) problems 3 and 4.

1. A yo-yo is a classic child's toy, as seen in Figure 23.6a. It consists of a slotted wheel at the end of a string. Part of the string is wound around inside the slot. The string makes rolling contact with the slot. If the string is held at the other end, and precisely the right tension is applied, the top end of the string remains stationary and the yo-yo simply rolls down the string. Find the tension T in the string if the top end of the string is stationary and the yo-yo is unrolling the string that has been rolled around it. Assume that the yo-yo has mass M, radius R and moment of inertia $0.5MR^2$.

2. A long cylinder (see figure) lies in a circular groove within which it rotates freely, as seen in Figure 23.6b. The uniform cylinder has mass M and radius R. A rope wrapped repeatedly around the cylinder then goes down to a block of mass m. The mass lies on an inclined plane that makes an angle θ with respect to the horizontal. Find the force and torque diagrams, the constraints, the angular acceleration of the cylinder, and the tension in the rope.

3. An outraged student at a competing university takes his 10 kg textbook to the top of his 20-story (70 m high) dormitory, walks to the edge of the roof, holds the book 0.5 m forward, and drops the book, as seen in Figure 23.6c. The book falls to earth 70 m below. "Forward" is in the x direction; "up" is the y direction. Find the angular momentum of the book relative to the student as a function of its altitude h above the ground. DO NOT try this experiment yourself without having guards to keep the impact zone clear; otherwise you might accidentally kill someone.

4. Return to the previous problem, as seen in Figure 23.6c. Calculate the torque on the book, with respect to the location of the student, due to the gravitational force on the book. Beginning with $\boldsymbol{\tau} = d\mathbf{L}/dt$,

calculate the elapsed time of the fall as a function of the height h of the book above the ground. The objective of this problem is to use the torque/angular momentum approach to solve the problem, so if you write s= -0.5 g t^2 + H you are completely missing the point of the question and will get no points.

23.6 Homework

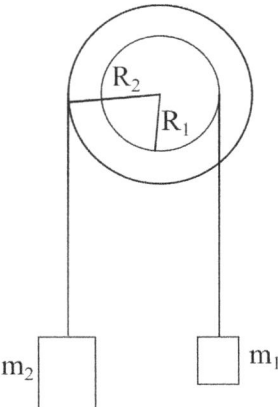

Figure 23.7: Figure for Homework problem 2.

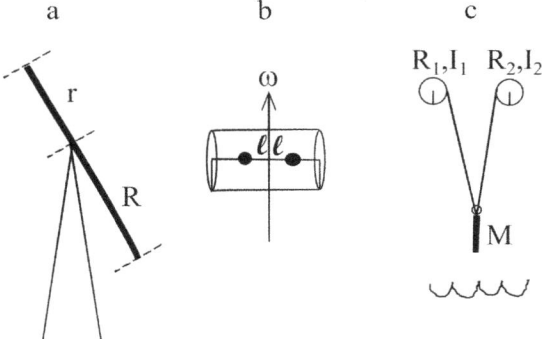

Figure 23.8: Figures for Homework problems: (a) Problem 8, (b) Problem 9, and (c) Problem 10.

1. In yet another motion picture scene, the villain Captain Goldbug and his evil henchmen are winching a 56 Mg missile from the deck of a barge onto their secret island base. When the missile is 10 m above the water, Goldbug is shown the error of his ways; solid gold is not a good structural material. The winch shaft breaks. The missile falls from rest toward the water. As it falls, the rope moves without slipping around the winch drum and over a pulley wheel, whose moments of inertia are 15,000 kg m^2 and 800 kg m^2, and whose radii are 5 and 1 meters, respectively. How fast is the missile going when it hits the water?

2. An Atwood machine consists of two masses M_1 and M_2 connected by a rope that passes over a pulley. Figure 23.7 shows this situation, Atwood's original machine being the special case $R_1 = R_2$. The pulley has moment of inertia I and radius R. Initially the two masses are each held at a height H off the floor and are stationary. $M_1 > M_2$; when the masses are released, M_1 falls floorward. How fast is M_1 going just before it hits the floor?

3. A cannon is mounted rearward-facing from an iceboat. The iceboat, mass M, is free to slide forward or backward without significant friction. The cannon fires a ball of mass m. The muzzle velocity of

a cannon is the velocity that the cannon ball has *with respect to the cannon* at the moment that the cannon ball leaves the mouth of the cannon. The italics are there for a reason. The muzzle velocity of this cannon is v_c. (i) The cannon is fired. The ball travels in the $-\hat{i}$ direction. Immediately after the cannon has fired, what are (a) the velocity, (b) the momentum, and (c) the kinetic energy of the iceboat? (ii) Compute the minimum amount of energy that must have been released when the gunpowder in the cannon was detonated. (iii) The cannonball sails a short distance through the air (effects of friction and gravity may be neglected) and strikes a block of wood of mass W hanging from a tree by a chain having length R. The cannon ball lodges in the block of wood. (iv) What is the angular momentum $\boldsymbol{L}$ of the cannonball at the instant it is fired, according to the gunner who fired the cannonball. The gunner stands directly behind the cannon when it is fired. The gunner's coordinate system has its origin along the cannon barrel just behind the cannon. (v) What is the angular momentum $\boldsymbol{L}_B$ of the block of wood at the moment after the cannon ball strikes it, according to the squirrel in the tree. The squirrel uses a coordinate system whose origin is at the point where the chains are attached to the tree. (vi) What is the angular momentum $\boldsymbol{L}_c$ of the cannon ball the moment after the cannon ball lodges in the block of wood, according to the gunner who fired the cannon ball. The gunner stands directly behind the cannon when it is fired. The gunner's coordinate system has its origin along the cannon barrel just behind the cannon. (vii) An innocent bystander safely to one side checks if angular momentum is conserved by checking if $\boldsymbol{L} = \boldsymbol{L}_B + \boldsymbol{L}_c$? Is the equality satisfied by your values of $\boldsymbol{L}$? Does your result confirm that angular momentum is conserved? Why or why not?

4. The Office of Sledding Safety of the remote kingdom of Noonynannia has issued new regulations for the protection of children taking their sleds down Noonynannia's only hill, which spends the winter covered with sheer, frictionless, ice inclined at an angle θ with respect to the horizontal. Under the new regulations, any child taking her sled down the slope is required to have a braking rope attached to the rear of her sled. The child and her sled have a combined mass m. The other end of the braking rope is wound around a large cylindrical drum that is free to rotate around its horizontal axis. The drum has mass M, radius R, and moment of inertia I around its rotation axis. (i) Give force or torque diagrams as appropriate for the sled and the drum. (ii) Give the constraints relating the motions of the sled and the drum. (iii) Find the tension in the rope. (iv) Find the accelerations, linear or rotational as appropriate, of the sled and the drum. (v) If the sled is to have a speed no larger than V when it reaches on bottom of the hill (a distance L as measured along the hill) how large must I be made? (vi) The sled begins from rest at altitude H. Find the total energy of the system (sled plus drum) after the slid has slid downhill to altitude h.

5. In yet another scene from the motion picture, a heavy steamer trunk (mass m) is being winched up from the dock onto a waiting clipper ship. The rope securing the trunk passes up over a small roller (radius r, moment of inertia i) and then back down over a large take-up drum (radius R, moment of inertia I). In the scene, the gear driving the take-up drum snaps, so the drum, roller, steamer trunk and rope are the same as before, but now the drum and roller both rotate freely. The trunk, initially stationary, falls back through a distance h onto the dock. How fast is the trunk moving just before it hits the dock?

6. In an exotic New Zealand sport, participants are strapped inside rigid plastic spheres that are allowed to roll downhill. An extreme sports advocate has found an improvement – I use that word somewhat broadly – in the sport, namely the sphere is rolled down a steep ramp and then sent around a circular loop before reaching the end of the course. The sphere starts at a height H. Its radius is r. It has moment of inertia $I = 0.4 M r^2$. The vertical loop has radius R. Find the minimum height H of the top of the ramp needed for the sphere to make its way around the loop. Hint: In order for the sphere to stay on the track when at the top of the loop, its vertical acceleration at the top of the loop must have a particular relationship to the gravitational force Mg on the sphere.

7. We have a frictionless ramp that makes an angle θ with respect to the horizontal. On the ramp is a cylinder that has mass M, moment of inertia I, and radius R. A rope is wrapped around the cylinder. The far end of the rope is attached to the top of the ramp. The cylinder slides down the ramp, unrolling the rope as it travels. The rope goes over the top of the cylinder, so that the direction of rotation of the cylinder as it goes down the ramp is the opposite of the direction that it would have had, if it

23.7. PROBLEM SOLUTIONS

rolled without slipping down the ramp. Find the center-of-mass speed of the cylinder as a function of the distance it has slid down the ramp.

8. Consider a uniform rod having mass M and length ℓ. The rod is pivoted at a distance R from its lower end, this being r from its upper end. $R > r$; it matters. The rod makes an angle θ with respect to the vertical. (a) Find the gravitational torque around the pivot of the lower section R of the rod. (b) Find the gravitational torque around the pivot of the upper section r of the rod. (c) Find the moment of inertia of the rod around the pivot (Hint: It's *not* one of the standard formulae.) (d) Write $\tau = d\mathbf{L}/dt$ explictly for this system. Assuming small oscillations, find the oscillation frequency ω. (e) How does the oscillation frequency depend on the mass or density of the rod? (f) What happens to the oscillation frequency if $r > R$? What does your answer *mean*?

9. A circular tube having mass M, length L and moment of inertia $ML^2/5$ is pivoted thorough its center, the pivot rod being perpendicular to the cylinder. The tube contains two masses m separated by a distance 2ℓ, spaced symmetrically around the pivot. The masses are held in place by a string that breaks when the cylinder is brought up to angular velocity ω. The masses then move to the outer end of the cylinder and stick to the ends. What is the final angular velocity of rotation of the system?

10. It is another scene from our motion picture. Once again, the evil Doctor Goldbug is making his escape from the heroine. The Doctor has mass M. He is hanging on to the ends of two ropes. The other ends of the two ropes are wrapped around two massive drums. The drums have masses m_1 and m_2, radii R_1 and R_2, and moments of inertia I_1 and I_2, respectively, and are both free to rotate. (i) What are the force diagrams and torque diagrams for this problem? (ii) What are the equations of motion for this problem? (iii) What are the constraints for this problem? (iv) What is the acceleration of Doctor Goldbug? (v) What are the tensions in the two ropes?

11. We have an Atwood machine apparatus, as seen in Figure 23.7. The two masses are now called M and m; the pulley wheel has mass P, radius R, and moment of inertia I. The rope connecting the masses is sufficiently light that its mass can be neglected. You have been provided with a gadget that measures the tension in the string. You use it to measure the tension in the string at the two points where the string is attached to each of the two masses. If you measure the tension correctly: What tension do you measure at the two points?

12. In the terminal scene of our motion picture, the villain is hanging from a wire directly above his swimming pool. The rope is wound at the top around a circular drum having mass M, radius R, and moment of inertia I. The villain, mass m, has a drop of length L before he hits the water. The heroine releases the brake holding the drum in place, while voicing the immortal line "Goldbug, you have one problem. You're all wet!". The drum is now free to rotate. The villain drops into the water. (a) How fast is the villain going when he hits the water? (b) How long does the villain need to fall through the distance L? (c) What was the torque on the drum due to the wire, as measured with respect to the drum's axis of rotation, while the villain was falling? (d) what is the torque on the drum due to all external forces, as measured with respect to the point where the rope first contacts the drum? Hint: While the drum is free to rotate, it is not free to fall.

13. A mass m_1 rests on a horizontal surface having kinetic friction coefficient μ_k. A rope is attached to m_1; it proceeds horizontally above the surface, is wrapped one-and-a-quarter times around a drum of mass M and radius R, and then proceeds vertically downward to a mass m_2. The moment of inertia of a drum is $\frac{1}{2}MR^2$. (i) What are the force and torque diagrams for this problem? (ii) Write the equations of motion for the two masses and for the drum. (iii) What torque must be applied to the drum to hold it and the two masses stationary? (iv) The external torque is removed. What is the acceleration of m_2? (v) Using conservation of energy, what is the velocity of mass m_2 after it has fallen from $z_2 = h$ to $z_2 = h + B$? Assume that m_1 had an initial speed V to the right, and that the ropes were taut.

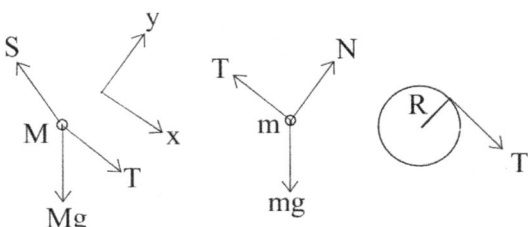

Figure 23.9: Figures for worked problem 2, including force diagrams for the roller M and the mass m, and a torque diagram for the roller.

23.7 Problem Solutions

1. We are trying to calculate the tension in the string of a yo-yo, with the constraint that the tension is such that the top of the string remains stationary as the yo-yo rotates and unrolls the string that initially was wrapped around it. We want to calculate a force, so starting with the Second Law appears to be appropriate.

   As the yo-yo descends, the release of the string can be described as rolling contact; the string is stationary at the release point. For rolling contact we can write for the magnitudes

   $$v = \omega R$$

   and therefore

   $$\frac{d^2 z}{dt^2} = \frac{dv}{dt} = -R\frac{d\omega}{dt}.$$

   We have inserted the correct sign: If ω is positive, the yo-yo is rotating in the anti-clockwise direction and is descending, meaning that a positive ω corresponds to a negative v.

   The yo-yo is subject to two outside forces, tension and gravity, so for the vertical component of its motion we can write the Second Law as

   $$M\frac{d^2 z}{dt^2} = T - Mg.$$

   The angular acceleration is non-zero. We can write a torque equation for the angular acceleration. This problem illustrates the utility of choosing cleverly the origin around which the torque is calculated. We choose the origin to be the center of mass of the yo-yo, its center, because with that choice of origin the force of gravity is applied at distance zero from the origin and does not contribute to the torque. Only the tension contributes to the torque, letting us write

   $$TR = \frac{1}{2}MR^2 \frac{d\omega}{dt}.$$

   Apply the constraint to the Second Law equation, and multiply by R, giving

   $$-MR^2 \frac{d\omega}{dt} = TR - MgR.$$

   On subtracting this result from the torque equation,

   $$\frac{3}{2}MR^2 \frac{d\omega}{dt} = MgR.$$

   Applying the constraint again, this time to replace the angular with the linear acceleration,

   $$\frac{d^2 z}{dt^2} = -\frac{2}{3}g.$$

23.7. PROBLEM SOLUTIONS

Finally, for the tension

$$T = Mg + M\frac{d^2z}{dt^2}, \tag{23.30}$$

$$T = \frac{Mg}{3}. \tag{23.31}$$

2. We can readily draw the force and torque diagrams for the mass and the roller. In addition to the tension forces T and the gravitational forces mg, there is also a support force S that holds the roller in place. The roller rotates but does not translate, so the support simply contributes whatever force is needed to hold the roller in place. An engineering issue arises: If you put enough force on a support, it ceases to hold the object it is supporting in place. Instead, it breaks. This support does not break.

The force of gravity on the mass m needs to be decomposed into its x and y components. We have done this before, so we do not repeat the decomposition here. The component of gravity parallel to the plane is $mg\sin(\theta)$. The equations of motion begin with the Second Law and the torque equation, leading to

$$-T + mg\sin(\theta) = m\frac{d^2x}{dt^2}, \tag{23.32}$$

$$-TR = \frac{1}{2}MR^2\frac{d\omega}{dt}. \tag{23.33}$$

We have two equations. Can we solve? Our list of unknowns reads T, $\frac{d^2x}{dt^2}$, and $\frac{d\omega}{dt}$. That's three unknowns, but only two equations, so we need another equation. That's the constraint equation linking ω and $\frac{dx}{dt}$. Sign check: A positive ω corresponds the anticlockwise rotation of the roller and the mass moving uphill, in the negative x direction, so the relationship between the tangential velocity of the roller and the velocity of the mass is

$$\omega R = -\frac{dx}{dt}.$$

We use the time derivative of the above equation to eliminate $\frac{d^2x}{dt^2}$ from the Second Law equation, leading after multiplying by R to

$$-TR + mgR\sin(\theta) = -mR^2\frac{d\omega}{dt}.$$

Subtracting from this result the torque equation gets

$$mgR\sin(\theta) = -mR^2\frac{d\omega}{dt} - \frac{1}{2}MR^2\frac{d\omega}{dt},$$

so that

$$\frac{d\omega}{dt} = -\frac{mgR\sin(\theta)}{mR^2 + \frac{1}{2}MR^2}$$

and hence

$$\frac{d^2x}{dt^2} = \frac{mgR^2\sin(\theta)}{mR^2 + \frac{1}{2}MR^2}.$$

The factors of R^2 cancel in the fraction. From the Second Law equation,

$$T = -\frac{m^2g\sin(\theta)}{m + \frac{1}{2}M} + mg\sin(\theta).$$

Putting the right-hand-side of the equation over a common denominator,

$$T = \frac{mMg\sin(\theta)}{2m + M}.$$

258   CHAPTER 23. COUPLED MOTION INCLUDING ROTATION

3. We want to calculate the angular momentum $\boldsymbol{L} = \boldsymbol{r} \times \boldsymbol{p}$ of the book. We have $\boldsymbol{L} = rp\sin(\theta)(-\hat{\boldsymbol{i}} \times -\hat{\boldsymbol{j}})$ or $\boldsymbol{L} = r_\perp p \hat{\boldsymbol{k}}$. The $r_\perp$ form for $\boldsymbol{L}$ is useful here because the book falls straight down, a direction perpendicular to the vector $\boldsymbol{r}$ from the student to the release point of the book. To calculate p, we need momentum and hence velocity as a function of height. Relationships between velocities and positions often arise straightforwardly from energy conservation $E_i = E_f$. For this problem, that equation becomes

$$\frac{1}{2}mv_i^2 + mgh_i = \frac{1}{2}mv_f^2 + mgh_f.$$

Here $v_i = 0$, $h_i = H$ is the height of the building, and h_f is the altitude of the book at the point where we want to know $\boldsymbol{L}$.

There is also an interesting rearrangement of the kinetic energy form, namely

$$\frac{1}{2}mv^2 = \frac{m^2v^2}{2m} = \frac{p^2}{2m},$$

leading to

$$\frac{p^2}{2m} = mg(H - h), \tag{23.34}$$

$$p = m(2g(H - h))^{1/2}. \tag{23.35}$$

The angular momentum is therefore

$$\boldsymbol{L} = r_\perp m(2g(H - h))^{1/2}\hat{\boldsymbol{k}}$$

or, substituting in numbers to one significant figure

$$\boldsymbol{L} = 0.5 \cdot 10(2 \cdot 10(70 - h))^{1/2}\hat{\boldsymbol{k}}.$$

4. We repeat the previous problem. We now want to use the torque equation $\boldsymbol{\tau} = \frac{d\boldsymbol{L}}{dt}$ to calculate time versus height. The torque on the book, calculated with respect to the student, is $\boldsymbol{\tau} = \boldsymbol{r} \times \boldsymbol{F}$. Noting $\boldsymbol{r}$ and $\boldsymbol{F}$ are perpendicular, the crossproduct direction having been calculated in the previous problem,

$$\boldsymbol{\tau} = r_\perp mg\hat{\boldsymbol{k}}.$$

The torque is independent of time, from release until the book hits the ground. A time integral of the torque equation gives us

$$\boldsymbol{L} = \int_0^{\Delta t} dt\, \boldsymbol{\tau}, \tag{23.36}$$

$$\boldsymbol{L} = \int_0^{\Delta t} dt\, r_\perp mg\hat{\boldsymbol{k}} = r_\perp mg\Delta t\hat{\boldsymbol{k}}. \tag{23.37}$$

However, we know $\boldsymbol{L}$ as a function of altitude from the prior problem, namely

$$\boldsymbol{L} = r_\perp m(2g(H - h))^{1/2}\hat{\boldsymbol{k}},$$

which gives us

$$r_\perp mg\Delta t\hat{\boldsymbol{k}} = r_\perp m(2g(H - h))^{1/2}\hat{\boldsymbol{k}},$$

and therefore

$$\Delta t = \left(\frac{2(H - h)}{g}\right)^{1/2},$$

which is the familiar answer.

Chapter 24

Statics

We now reach a topic of great engineering importance, namely the study of *statics*. In a statics problem, the objects of interest do not move with respect to each other. Linear and angular accelerations therefore both vanish.

Recall the fundamental laws of motion for individual objects.

$$\boldsymbol{F} = \frac{d^2\boldsymbol{r}}{dt^2}, \tag{24.1}$$

$$\boldsymbol{\tau} = \frac{d\boldsymbol{L}}{dt}. \tag{24.2}$$

If objects are all stationary with respect to each other, and are not accelerating as seen in an inertial reference frame,

$$\frac{d^2\boldsymbol{r}}{dt^2} = \boldsymbol{0}, \tag{24.3}$$

$$\frac{d\boldsymbol{L}}{dt} = \boldsymbol{0}, \tag{24.4}$$

leading to

$$\boldsymbol{F} = \boldsymbol{0}, \tag{24.5}$$

$$\boldsymbol{\tau} = \boldsymbol{0}. \tag{24.6}$$

That is, if we are discussing a collection of objects that do not move with respect to each other, and are not accelerating as seen in an inertial reference frame, then (i) the total force on each object must be zero, and (ii) the total torque on each object must be zero. It is worthwhile to repeat that $\boldsymbol{F} = \boldsymbol{0}$ and $\boldsymbol{\tau} = \boldsymbol{0}$ are not basic laws of nature. They are results that happen to be true in a particular, albeit interesting, special case.

A simple statics problem is provided by a lamp hanging from a tree limb. The lamp has a mass m. What is the tension in the string? We start with a sketch of the system and then a force diagram for the lamp.

From the force diagram, we see that there are no forces acting on the lamp in the x or y directions, so the components of the acceleration in those two directions must also be zero. For the vertical direction the Second Law gives us

$$m\frac{d^2z}{dt^2} = T - mg. \tag{24.7}$$

However, we are solving this as a statics problem. The lamp just sits there, so its vertical acceleration is zero. Equation 24.7 simplifies to

$$0 = T - mg, \tag{24.8}$$

or

$$T = mg. \tag{24.9}$$

One might encounter proposals that statics problems correspond to situations in which nothing is moving, so that for each object $v = 0$. If one makes measurements in a single inertial frame, and finds that all objects are stationary, then the objects correspond to a statics problem. However, there are other inertial reference frames, frames that move but do not accelerate with respect to the first reference frame, in which the objects are all moving at the same non-zero velocity, so that for these objects $v = c$, c being a constant. If the objects are still not accelerating, the forces on each object must still sum to 0. The behavior of the objects can still be described as a statics problem, because they are all moving with constant velocity..

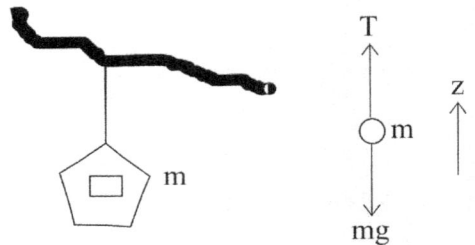

Figure 24.1: A lamp hiding from a branch and the lamp's force diagram.

Most architectural objects, buildings, e.g., do not move, so the forces on them correspond to statics problems. (During a hurricane or an earthquake, the forces on a building do depend on time, so the dynamic response of a building becomes important. Some decades ago, it was recognized that New York City is potentially subject to rare, respectably large earthquakes. There followed a significant program of architectural engineering modifications to stabilize the city's skyscrapers against this threat of non-static forces.)

24.1 Simple Statics Problem

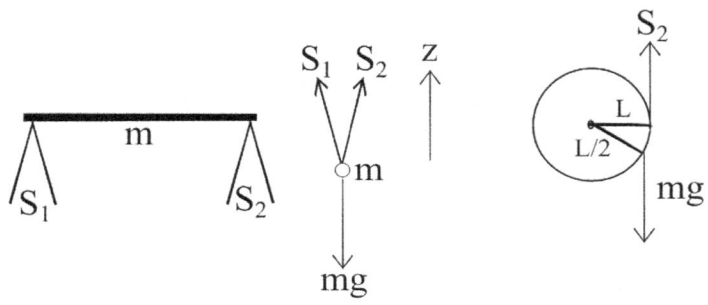

Figure 24.2: Sketch, force diagram, and torque diagram for a beam m supported by two knife edges S_1 and S_2. In the force diagram, the two support forces are drawn so that they are clearly distinct. In drawing the torque diagram, the origin for the torques was chosen to be the point where S_1 acts on the beam.

We start as seen in the sketch with a horizontal beam supported at its ends by two knife edges. The beam has length L. Its weight mg is entirely created by a single point mass located at its midpoint, a distance $L/2$ from either end of the beam. What force does the beam apply on each of the knife edges?

We can readily set up a force diagram. S_1 and S_2 are the vertical forces supplied by the two knife edges, while mg is the gravitational force on the beam. The beam has acceleration $\mathbf{0}$, so for the z component of the Second Law we can write

$$S_1 + S_2 - mg = 0. \tag{24.10}$$

We can also write an equation for the torques. As emphasized in earlier chapters, the law of conservation of angular momentum is equally correct no matter where we put the origin around which the torques are calculated. All choices are valid, but some choices are better than others. A guide for choosing the best point is the rule that a force applied at the origin creates no torque, so by placing the origin at a point where a force is applied that force is eliminated from the torque equation.

To illustrate this guide, we put the origin at the location of the left support. We start with equation 24.1b. For once we will start by writing out the torques in their full vector form. From the torque diagram, we have

$$\mathbf{0} \times S_1 \hat{\mathbf{k}} + \frac{L}{2}\hat{\mathbf{i}} \times (-Mg)\hat{\mathbf{k}} + L\hat{\mathbf{i}} \times S_2 \hat{\mathbf{k}} = \frac{d\mathbf{L}}{dt}. \tag{24.11}$$

However, the beam is stationary, so $d\mathbf{L}/dt = 0$. The vector cross product is $\hat{\mathbf{i}} \times \hat{\mathbf{k}} = -\hat{\mathbf{j}}$. The above equation simplifies to a vector equation:

$$(Mg\frac{L}{2} - S_2 L)\hat{\mathbf{j}} = \mathbf{0}. \tag{24.12}$$

Note that the x and z axes are in the plane of the paper, while $\hat{\boldsymbol{j}}$ is perpendicular to the paper. A rotation around the $\hat{\boldsymbol{j}}$ axis is therefore clockwise, not counterclockwise. You can check my claim with the right hand rule. Recalling that a vector equation is a symbol for the equations for its scalar components, equation 24.12 tell us that

$$\frac{MgL}{2} - S_2 L = 0, \tag{24.13}$$

and therefore $S_2 = Mg/2$. From equation 24.10, one must also have $S_1 = Mg/2$.

We could just as well have placed the origin at the center of the beam. In this case the weight of the beam is applied to the beam at the origin, so the weight of the beam does not contribute a torque around the new origin. Calculating the total torque around the new origin at the center of the beam, and noting that $d\boldsymbol{L}/dt = \boldsymbol{0}$ around the new origin,

$$S_1 \frac{L}{2}\hat{\boldsymbol{j}} - S_2 \frac{L}{2}\hat{\boldsymbol{j}} = \boldsymbol{0}. \tag{24.14}$$

whose solution is $S_1 = S_2$. Applying this result to equation 24.10, one finds $S_1 = Mg/2$ and $S_2 = Mg/2$. This result for S_1 and S_2 is the same as the result found by putting the origin at the end of the beam.

24.2 Ladder on a Wall with Friction

We now consider a slightly more complicated problem, namely a ladder leaning on a wall. The contact between the ladder and the wall is frictionless. The ladder, length L, makes an angle θ with respect to the horizontal. The mass of the ladder is M. The ground on which the ladder rests has a coefficient of static friction μ_s, so the ground will keep the ladder from sliding out from the wall with a force that must be less than $\mu_s N$, N being the normal force between the ground and the ladder. If the ladder is nearly vertical, it is happy to stay in place. However, as the angle θ is reduced, there is a critical angle θ_s at which the ladder's base slides sideways while the ladder falls until it hits the ground. What is the angle θ_s?

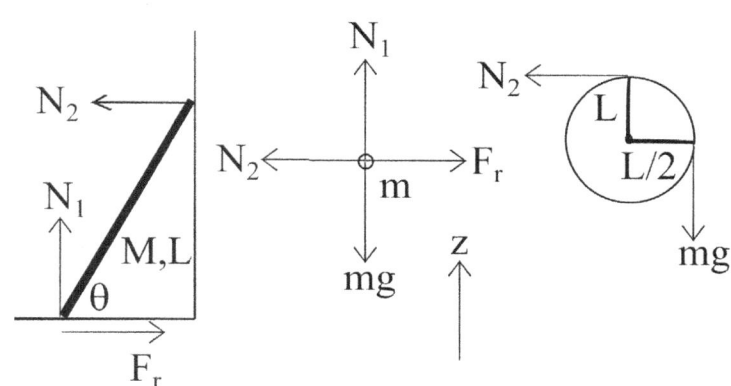

Figure 24.3: Sketch, force diagram, and torque diagram for a ladder leaning against a wall.

We begin with a sketch showing the ladder against the wall. The sketch displays all of the facts that we are given, and all the forces on the ladder and where those forces are applied. In addition to the force of static friction F_r, whose magnitude is not yet known, there are two normal forces, due to the ground and to the wall. The wall is frictionless, so it exerts no force on the ladder in a vertical direction. Figure 24.3 then shows the force diagram, including forces N_1, N_2, Mg, and F_r. At this point we have not yet gleaned from our calculation the magnitude or direction of the static friction force, though we reasonably expect it will be pointing in the $+\hat{\boldsymbol{i}}$ direction.

All forces in this image point either in the horizontal of the vertical direction. From the force diagram, we usefully write

$$N_1 - Mg = 0, \tag{24.15}$$
$$-N_2 + F_r = 0. \tag{24.16}$$

Finally, Figure 24.3 shows the torque diagram. While any location of the origin is valid, we might plausibly place the origin at the bottom of the ladder, the top of the ladder, or the ladder's center of mass. The bottom of the ladder appears to be the best choice. If the origin is at the bottom of the ladder, then two of the four forces in the problem, namely the normal and friction forces due to the ground, do not contribute to the torque. The two other plausible locations for the origin each eliminate only one force from the calculation.

The magnitude of the cross product of two vectors $\boldsymbol{A}$ and $\boldsymbol{B}$ is $|\boldsymbol{A} \times \boldsymbol{B}| = AB\sin(\phi)$, ϕ being the angle between the two vectors. Two different angles appear here. For the torque around our choice of origin, we have

$$\frac{L}{2}Mg\sin(\frac{\pi}{2} - \theta)\hat{\boldsymbol{j}} - N_2 L\sin(\theta)\hat{\boldsymbol{j}} = \boldsymbol{0}, \tag{24.17}$$

which simplifies to

$$\frac{Mg}{2}\cos(\theta) = N_2 \sin(\theta), \tag{24.18}$$

and therefore to

$$N_2 = \frac{Mg\cos(\theta)}{2\sin(\theta)}. \tag{24.19}$$

The ladder slides when the needed static friction exceeds its largest available value, namely $\mu_s mg$, and at that point from equation 24.15

$$-\frac{Mg\cos(\theta)}{2\sin(\theta)} + \mu_s Mg = 0, \tag{24.20}$$

whose solution for θ is

$$\tan(\theta) = \frac{1}{2\mu_s}, \tag{24.21}$$

or finally

$$\theta = \arctan\left(\frac{1}{2\mu_s}\right). \tag{24.22}$$

24.3 The Hinged Flagpole

Consider a flagpole. It's a horizontal rod, length L, with a flag, mass M, hanging from its end. It is supported at the wall by a hinge and by a wire, tension T, that makes an angle θ as indicated with respect to the wall. The rod is uniform and has a mass m. Find the tension T in the wire and the force that the flagpole exerts on the wall. We start by drawing a sketch indicating the known and significant unknown quantities. The force of the wall on the flagpole is $\boldsymbol{H}$; we are actually being asked for the force on the wall, which is the reaction force to $\boldsymbol{H}$.

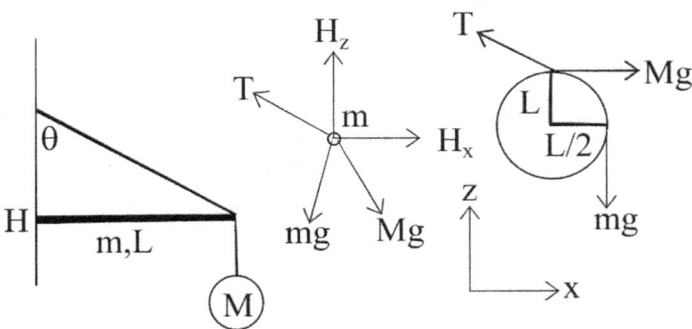

Figure 24.4: Sketch, force diagram, and torque diagram for the hinged flagpole. In the force diagram, the two gravitational forces are drawn splayed so that they may be seen as distinct forces.

We then generate a force diagram and a torque diagram for the pole. In generating the force diagram, we need to remember that $\boldsymbol{H}$ may have either a horizontal component H_x or a vertical component H_z or both, so we indicate both of them on the force diagram. We have reached a standard point where some people go slightly wrong. Instead of giving the hinge force an x and a z component, they assume that one component or the other is zero. It may of course turn out as a result of the calculation that H_x or H_z is zero, but that answer should follow from the calculation, not be put into it as an assumption. The force diagram has interesting x and z forces shown, including one force that will need to be resolved into its horizontal and vertical components. To create a torque diagram, we need to choose an origin. The general rule is to choose an origin that eliminates some of the unknown forces from the torque calculation. We choose the hinge, the point where the flagpole is attached to the wall, because with this choice of origin neither H_z nor H_x contributes a torque. The force and torque diagrams are found in Figure 24.4.

We need to decompose the tension force $\boldsymbol{T}$ into its horizontal and vertical components T_x and T_z. The force $\boldsymbol{T}$ is the hypotenuse of a triangle, while the two components are the two legs of the triangle. In making the decomposition, some minor attention is needed. The angle θ as defined is at the top of the triangle, so the horizontal component of $\boldsymbol{T}$ is $-T\sin(\theta)$ while the vertical component is $+T\cos(\theta)$.

We now write the Second Law and the corresponding equation for the torque:

$$m\frac{d^2x}{dt^2} = -T\sin(\theta) + H_x, \qquad (24.23)$$

$$m\frac{d^2z}{dt^2} = T\cos(\theta) - mg - Mg + H_z, \qquad (24.24)$$

and

$$I\frac{d^2\phi}{dt^2} = \frac{mgL}{2} + MgL - LT\cos(\theta). \qquad (24.25)$$

These equations are the equations of motion for the flagpole. In this equation, ϕ is the rotation angle of the flagpole around the hinge. You can check the signs in this equation by applying the rules for the cross product of unit vectors, noting that the vertical axis is $\hat{\boldsymbol{k}}$.

This is a statics problem, so the translational and rotational accelerations are zero. From the torque equation,

$$T = \frac{mg + 2Mg}{2\cos(\theta)}. \qquad (24.26)$$

Is this equation reasonable? The right hand side has the dimensions of a force, as required. However, there is a cosine function in the denominator. As θ goes to ninety degrees, $\cos(\theta)$ goes to zero and the calculated tension diverges. Is this plausible?

Yes! The flag is pulling down on the outer end of the flagpole. The tension in the wire, pulling the other way, keeps the flagpole from drooping. However, only the vertical component $T_z = T\cos(\theta)$ of the tension keeps the pole horizontal. As θ advances toward ninety degrees, the pull of the wire on the pole is almost purely horizontal, so a smaller and smaller part of T contributes to T_z. The required T_z does not depend on the angle, so, as $\theta \to 90°$, T must become very large.

We may now compute H_x and H_z. From the Second Law equations

$$H_x = \frac{(mg + 2Mg)\sin(\theta)}{2\cos(\theta)} \qquad (24.27)$$

and

$$H_z = \frac{mg}{2}. \qquad (24.28)$$

Finally, there is a small trick. The problem asks for the force due to the hinge on the wall, not the force due to the hinge on the flagpole. The forces from the hinge on the flagpole were H_x and H_z. The forces you were asked to find are the reaction forces to H_x and H_z, which are $-H_x$ and $-H_z$. The force $\boldsymbol{F}$ on the wall due to the flagpole is therefore

$$\boldsymbol{F} = -\left(\frac{(mg + 2Mg)\sin(\theta)}{2\cos(\theta)}\right)\hat{\boldsymbol{i}} - \frac{mg}{2}\hat{\boldsymbol{k}} \qquad (24.29)$$

24.4 Discussion

The method has a limit: Suppose I have an object resting on the ground, a footstool with three or four legs holding it up. The forces due to the three or four supports are denoted S_1, S_2, S_3, and if need be S_4. Do the methods in this chapter let us calculate the support forces?

We did this problem at the start of the Chapter, for a horizontal beam having two supports, and therefore two support forces S_1 and S_2. In that case, the Second Law gave us one equation, and the torque as calculated around some point on the beam gave us a second equation. That's two equations for two unknowns, so we are able to solve. The second equation was simply the statement that in this statics problem the torque component perpendicular to the page vanished; that's one equation no matter where the origin is chosen.

We could also consider a traditional three-legged footstool. A three-legged footstool has three supports, so there are three support forces S_1, S_2, and S_3. However, we also have three equations, namely the Second Law for motion perpendicular to the floor, and two torque equations corresponding to attempted rotation around two axes parallel to the floor and perpendicular to each other.

Suppose, however, that we have a four-legged footstool. There are now four support forces, S_1, S_2, S_3, and S_4. However, we still have only three equations, no different than we did with the three-legged footstool. We have fewer equations that we do unknowns. The statics method described in the above therefore cannot be used to determine the support forces on the legs of a four-legged footstool. Other, more advanced techniques as treated in mechanical and civil engineering are then needed.

Riddle: Why are we treating statics toward the end of the course? There are no accelerations, so the problems look very simple. Indeed, some people propose that you should teach statics first, and only advance to the Second Law later. The difficulty with this approach is that some students will interpret the special case results, e.g. $F = 0$ or $T = mg$, as the general result, and will continue to invoke the special case result when the result ceases to be true. That's why, when we discussed masses hanging from pulleys, we kept emphasizing that $T = mg$ for a hanging mass meant that the acceleration of the hanging mass had to be zero, even though the acceleration was obviously not going to be zero.

Someday, some of you will be teaching or doing research. There is an interesting distinction between using special cases to teach a problem and using special cases to understand a problem. If you are teaching about a topic, the special cases often best appear after the general case, so that it is clear that they are indeed special cases, in which general results can simplify. If you teach the special case results first, some of your students will keep using the special case rules when they have ceased to apply. On the other hand, if you are trying to find a general answer, by starting with special cases you may be able to find results in a restricted domain, and then see how to extend those results one step at a time, each step giving new insight into the nature of the general solution.

24.5 Worked Problems

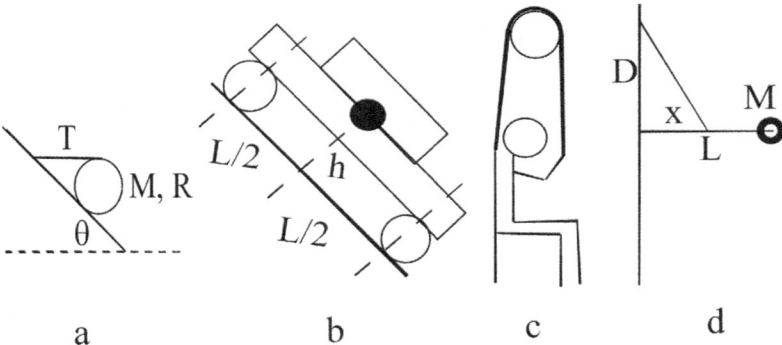

Figure 24.5: Figures for Worked Problems (a) Problem 1, (b) Problem 2, (c) Problem 3, and (d) Problem 4.

1. A large circular pipe, mass M, lies on a long ramp, angle θ with respect to the ground, as seen in Figure 24.5a. A rope is tacked to the top of the pipe and proceeds horizontally to the ramp. The coefficient of static friction is large enough that the pipe does not move in the indicated situation. Find the force applied by the rope to the pipe, the force applied by friction to the pipe, and the normal force from the ramp on the pipe.

2. As seen in Figure 24.5b, we have an automobile resting on a slope. The slope makes an angle θ with respect to the ground. The distance between the front and rear tires is L. The car has a mass M, located at the midpoint between the front and rear tires at a perpendicular distance h above the

ground. The ground exerts, perpendicular to the ground, a force N_f on the front wheels and a force N_r on the rear wheels. There are also forces on the wheels parallel to the ground. What are the forces N_f and N_r?

3. Some years ago, this was an exam problem. An impressive number of people forgot that force is a vector, not a scalar, and got no credit as a result. A painter sits in a bosun's chair, as seen in Figure 24.5c. She holds herself in position by holding the cable in one hand as shown in the sketch. The chair and painter are stationary. The painter has a mass of 100 kg; the chair has a mass of 20 kg. (a) Draw the force diagram for the painter. (b) Draw the force diagram for the bosun's chair. (c) Find the tension in the cable. (d) Find the normal force N that the painter applies to the chair.

4. The sign at a local cocoa store, as seen in Figure 24.5d, is a block of mass M attached to the end of a horizontal rod of length L. The other end of the rod is fastened to a wall with a large hinge. A rope is fastened to the wall at a distance D above the hinge; the rope is also fastened to the rod at a distance x from the wall. The mass of the rod is negligible. Find the tension in the rope and the force on the hinge.

24.6 Homework

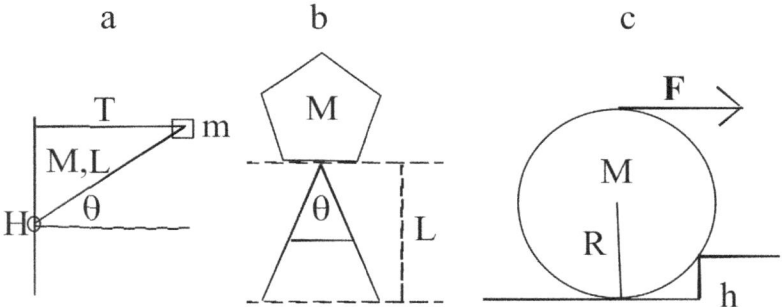

Figure 24.6: Figures for Homework Problems (a) Problem 1, (b) Problem 2, and (c) Problem 3.

1. The owners of a local store set a display pole at the front doorway of their establishment, as seen in Figure 24.6a. The pole is a rod having constant cross-section mounted at an angle θ pointing above the horizontal. The pole has a length L and mass M; fastened at its outer end is a steel ball having mass m. The pole is supported by a horizontal guy wire attached to its outer end, and by a fastener at its inner end. The guy wire has a tension T; the horizontal fastener applies a force **H** having components H_x and H_y (either or both of these may be zero) to the pole. Find:

   A) The force and torque diagrams for the pole.

   B) The tension T.

   C) The horizontal force H_x.

   D) The vertical force H_y.

2. It is wintertime in New Hampshire. We contemplate a (somewhat simplified) A-frame house under a snow load, as seen in Figure 24.6b. Fortunately, the snow has come down as a single flake of mass M that rests precisely at the peak of the house. Alas, the contractor skipped installing a foundation, so the nearly massless (relative to the snowflake) roof is resting on the frozen, icy, nearly frictionless ground. The peak of the roof is height L off the ground; the angle between the two sides of the roof is θ. At a height $L/2$ off the ground, a horizontal steel rope connects the two sides of the roof. The tension T in the rope keeps the two halves of the roof from sliding apart at their base and collapsing. What is T?

3. It is the end of the semester at a neighboring school, and the residents of a dormitory are attempting to roll a barrel of nonalcoholic root beer up the stairs to their rooms to make root beer ice cream floats, as seen in Figure 24.6c. The barrel has radius R, and is pressed against a tread of height h by a force horizontal F applied at the top of the barrel. The corner of the stair tread has a substantial coefficient of static friction with respect to root beer barrels, so the barrel does not slide across the tread; instead, it applies forces S_x and S_y on the barrel. If the students push just hard enough that the normal force of the lower tread on the barrel is zero, what are S_x and S_y?

4. You are the physics consultant for a not-so-major Victorian-epoch motion picture not actually set in England. In the next scene, the hero is to place a ladder against the wall of the heroine's house so that she may descend the ladder, following which they will elope to Scotland (1). The ladder has length L and mass M, and makes an angle θ *with respect to the vertical*. The heroine, instead of descending the ladder like a sensible person would, is to walk down the ladder with a parasol over her shoulder, doubtless to protect her skin from the rays of the full moon. (2) Thus, the ladder must rest against the wall at a fairly large angle from the vertical. The first time the scene is shot, the hero sets the ladder against the wall, and the ladder falls to the ground. Fortunately, you happen to know the coefficient of static friction μ_s between the ladder's feet and the floor of the sound stage. The director ignores your suggestion to cut the take, nail the ladder to the building, and resume. The director is a perfectionist. He orders you to calculate the range of angles from the vertical (zero degrees, ladder straight up, clearly works) at which the ladder will not fall off the wall. For your daily paycheck, calculate the allowed angles at which the ladder may be placed before the heroine steps on it. (The lead actor nods carefully when you tell him privately to keep his ankles against the bottom rung of the ladder.)

(1) No, I did not make this detail up. A peculiar feature of period English and Scottish law intervenes. See Jane Austen and Gretna Green for details.

(2) It's a horror film. She's a werewolf. He doesn't know.

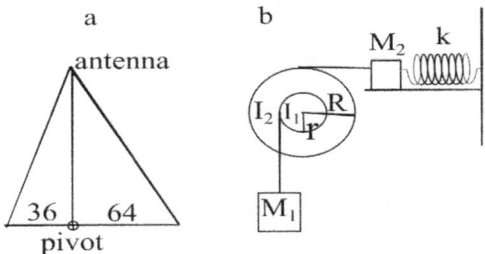

Figure 24.7: Figures for Homework Problems (a) Problem 6 and (b) Problem 7.

5. The owners of the local candy emporium have decided to erect a new sign. The sign consists of a horizontal support beam having length L and mass m, a sign having mass M hanging from the outer end of the beam, a hinge holding the beam to the wall by exerting a force on the beam, and a guy wire attached to the outer end of the beam that makes an angle θ with the beam. Compute the tension in the guy wire and the force **H** that the beam and its hinge place on the wall of the building. (Hint: The standard error is to assume in advance that you know which way **H** points, "horizontal" and "vertical" being among the more common guesses. Do not guess. Calculate.)

6. An 80 kg, 120m tall, radio antenna is stabilized by a pivot at its bottom and by two guy wires, one on each side of the pivot, attached to the antenna at its top end, as seen in Figure 24.7a. The two guy wires are attached to the ground 64m and 36m, respectively, from the pivot joint. The tension in the guy wire anchored closer to the antenna is measured to be 500 N. (a) If the radio antenna remains stationary, what is the tension in the other guy wire? (b) The antenna pivot is attached to the flatbed of a truck. What force is the antenna bottom placing on the pivot and truck?

7. A mass M_1 is hanging on a massless wire that is strung around the inner wheel of a two-wheel pulley,, as seen in Figure 24.7b. The inner wheel has radius r. The moment of inertia of the inner wheel of the pulley is I_2. The outer pulley has mass M_p and radius R. The effective moment of inertia of the outer pulley is I_1. This system is further connected to a mass M_2 through a massless wire connected to the outer wheel of the pulley. The inner and outer pulley wheels have been welded together and rotate as a unit. The mass M_2 lies on a frictionless surface; it is also attached to a spring that has spring constant k. The system is static; nothing is moving. For the system shown in this figure: A) calculate the tension in the wire connecting the pulley to M_1, B) the tension in the wire connecting the pulley to M_2, and C) the distance that the spring is stretched from its unstretched length.

8. A uniform rod of mass M is suspended horizontally by two wires hung at two different heights. Wire 1 is at an angle $150°$ from the $+x$ direction and wire 2 is at an angle θ from the $+x$ direction. The wires are firmly attached to the ceiling and to the two ends of the rod. A box of mass m_b is resting on top of the rod at a distance x from the left-most edge of the rod. In terms of M and m_b, calculate: A) the tension in wire 1, B) the tension in wire 2, and C) the angle at which wire 2 hangs.

9. The sign at a local cocoa store is a mass m_1 suspended from the middle of a rod having length L. The mass of the rod is negligible. The inner end of the rod is attached to the wall with a hinge. The outer end of the rod is supported by a rope. The rod and the rope make angles θ and ϕ with respect to the horizontal. Find the tension in the rope. Find the forces that the rope and the rod put on the wall to which they are attached.

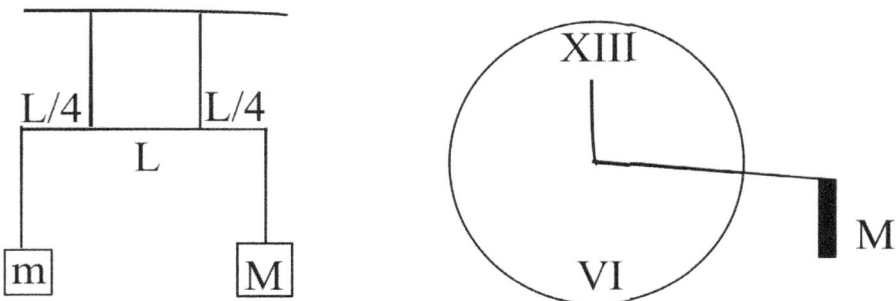

Figure 24.8: Figures for Homework Problems (a) Problem 10 and (b) Problem 11.

10. The owners of The Flying Rhino, a local coffeehouse, hang a sign above their front door. (Figure 24.8a) The display consists of two very heavy masses M and m that hang by wires from the ends of a rigid, extremely light rod. The rod has length L. The rod is supported by two wires that hang vertically. The wires are connected to the rod at distances $L/4$ from the two ends of the rod. Find the force and torque diagrams for the rod. Find the tension T in the two wires, as seen in Figure 24.8a.

11. There is a famous motion picture scene in which Charlie Chaplin, playing the role of a clock repairman, hangs from the end of the hand of a large outside clock, as seen in Figure 24.8b. Suppose that Chaplin had a mass of 70 kg, that the hand's length was 4 m, and that the clock was slightly unconventional. The hand tapered to a point; the density of the hand as a function of position was

$$\rho(x) = \rho_o + ar \tag{24.30}$$

where $\rho = 100 \text{kg/m}$, $a = -25 \text{ kg/m}^2$, and r is the distance along the arm measured from the central shaft toward Chaplin. For simplicity, the hand is currently horizontal. Find (i) The force diagrams for Chaplin and for the arm. (ii) The force exerted on the clock hand by the central shaft. (iii) The torque exerted on the clock hand by the center shaft. Caution: do not assume that you can treat the arm as a point mass at some location. For credit, you must do the appropriate integral.

12. A worker of weight M stands at the center of a ladder of length L, filling a barrel hanging from the upper end of the ladder with water, using a hose that delivers water at the rate of 10 gallons per minute. The ladder is inclined to the ground at an angle of 45 degrees. The ground is extremely damp, due to the spilled water; it supports the ladder, but has a coefficient of static friction of only μ_s. The wall is a course brick; its coefficient of static friction with the ladder is μ'_s. What is the weight of the barrel and water at the moment that the ladder begins to slide groundwards? [This is not a recommended safe construction practice.] [Hint: $\sin(45) = \cos(45) = 1/\sqrt{2}$.]

13. The 20 kg sign of a local restaurant hangs at the end of a (for practical purposes) massless beam of length 2 m. The beam points upward away from the wall; it makes an angle of 30 degrees with respect to the wall. A support wire is tied to the center of the beam; the wire also makes an angle of 30 degrees with respect to the wall. (i) Find the force and torque diagrams for the beam. (ii) Find the tension in the support wire. (iii) What is the force that the wall applies to the beam? Hint: Is this a right triangle? Really?

14. Demonstration that choosing an odd location for the origin does not make the problem insoluble. Consider the beam problem described by equation 24.10. Repeat the calculation, this time putting the origin a distance X ($X \in (0, L)$) from the left end of the beam.

15. Consider a diving board of length L and mass m. It is attached to two supports, one support at an end of the diving board and the other support a distance ℓ from the first support toward the swimming pool. Standing on the pool end of the diving board is a diver of mass M. By calculating torques around your choice of origins, calculate the force applied to the diving board by each of the support posts.

16. A 5m long diving board with total mass of 50kg is supported by two piers, one at the end and the other 1.5m in from the end. At the far end of the diving board, a 90 kg diver is at the bottom of his spring. The board and leg movements of the diver are accelerating the diver upwards at $+5$ m/s^2. The diving board at this moment is basically straight, not significantly bent. Please use my coordinates. (i) Give the free body diagram for the diving board. The diver is not part of the diving board. (ii) Find the force applied to the diving board by each pier.

24.7 Solutions to the Worked Problems

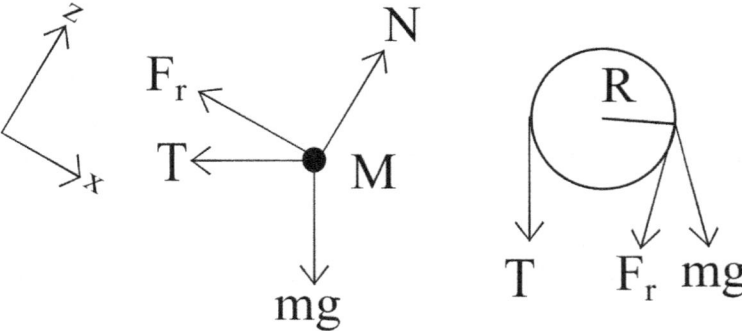

Figure 24.9: Figure for Worked Problem 1.

1. We write the force and torque diagrams for the log. A plausible set of coordinates puts the x-axis parallel to the ramp pointing uphill and the y-axis perpendicular to the ramp, pointing away from the ramp. The usual trig construction for ramps decomposes the force of gravity into its x and y components. We also have to decompose T into its x and y components.

24.7. SOLUTIONS TO THE WORKED PROBLEMS

We may write the x and y components of the Second Law as

$$F_r - Mg\sin(\theta) + T\cos(\theta) = M\frac{d^2x}{dt^2}, \tag{24.31}$$

$$N - Mg\cos(\theta) - T\sin(\theta) = M\frac{d^2y}{dt^2}, \tag{24.32}$$

and the angular momentum equation as

$$F_r R - TR = \frac{dL_z}{dt}. \tag{24.33}$$

However, this is a statics problem, so the two accelerations and the torque are always zero. The right-hand-side of these three equations all vanish. Is the problem soluble? We have three unknowns, namely F_r, T, and N, and three equations, so a solution is likely.

From the angular momentum equation, $F_r = T$. From the x-component of the forces,

$$0 = T - Mg\sin(\theta) + T\cos(\theta) \tag{24.34}$$

$$T = \frac{Mg\sin(\theta)}{1 + \cos(\theta)} \tag{24.35}$$

The normal force is

$$N = Mg\cos(\theta) + T\sin(\theta) \text{ or} \tag{24.36}$$

$$\boldsymbol{N} = \left(Mg\cos(\theta) + \frac{Mg\sin^2(\theta)}{1 + \cos(\theta)}\right)\hat{\boldsymbol{j}}. \tag{24.37}$$

Finally, the friction force is

$$\boldsymbol{F_r} = \left(\frac{Mg\sin(\theta)}{1 + \cos(\theta)}\right)\hat{\boldsymbol{i}}. \tag{24.38}$$

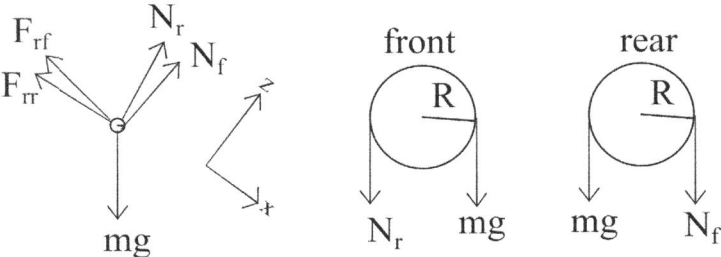

Figure 24.10: Force and torque diagrams for Worked Problem 2. We calculate the torque around two points, so there are two torque diagrams.

2. We have a car on a slope. There are two unknowns of interest, namely the two normal forces N_f and N_r that the road applies to the car at the location of the wheels. (We also do not know the frictional forces that the car applies to the road, these forces acting parallel to the road surface, but those forces are not requested as part of the solution and play no role in solving for N_f or N_r.) We now advance to solving. We begin with the force and torque diagrams. The torque is calculated twice, using as the origin each of the two points where the wheel touches the road, not the centers of the two wheels. Why are these good choices of origin? In each case, one of the two unknown forces is being applied at the origin, so it can contribute nothing to the torque equation. The two friction forces act parallel to the road surface, and therefore point directly at or away from the origin, so these two forces also contribute nothing to the torque. For each of the two choices of origin, only the weight of the car and the normal force on one of the wheels contribute to the torque.

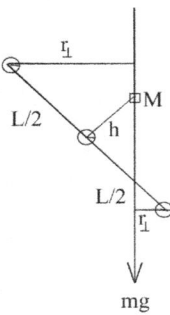

Figure 24.11: Geometric constructions for determining $r_\perp$ for gravity and each of the two origins in Worked Problem 2.

The magnitude of the torque due to each force can be written as $Fr_\perp$. N_f is perpendicular to the road surface, so its $r_\perp$ is simply the length L of the car. For the weight of the car, $r_\perp$ is indicated in Figure 24.11. Some careful trigonometry shows how $r_\perp$ may be obtained, namely as the difference of two other distances. The first is the horizontal distance (not the distance along the road surface) this distance being $(L/2)\cos(\theta)$. However, because the car is on a slope, the car's center of mass is displaced to the right, relative to the car's midpoint, the distance of the displacement being $h\sin(\theta)$. $r_\perp$ for the gravitational force relative to the rear wheel is therefore

$$r_\perp = \frac{L}{2}\cos(\theta) - h\sin(\theta). \tag{24.39}$$

For the rear wheel, the zero-torque condition becomes

$$-LN_f + Mg\left(\frac{L}{2}\cos(\theta) - h\sin(\theta)\right) = 0, \tag{24.40}$$

which gives us for the normal force on the front wheel:

$$N_f = Mg\left(\frac{1}{2}\cos(\theta) - \frac{h}{L}\sin(\theta)\right). \tag{24.41}$$

For the torque around the front wheel, $r_\perp$ for the center of mass is

$$r_\perp = \frac{L}{2}\cos(\theta) + h\sin(\theta), \tag{24.42}$$

so that the zero-torque condition becomes

$$-LN_r + Mg\left(\frac{L}{2}\cos(\theta) + h\sin(\theta)\right) = 0, \tag{24.43}$$

and the normal force on the rear wheel becomes

$$N_f = Mg\left(\frac{1}{2}\cos(\theta) + \frac{h}{L}\sin(\theta)\right). \tag{24.44}$$

Are these results reasonable? The equations are dimensionally sound. Note, however that there is a subtraction in equation 24.41. If the final term of that equation were dominant, N_f would become negative. The normal force would be pulling in the wrong direction. How is this possible?

The answer is that if θ is made sufficiently large, the center of mass of the car will be to the right of the rear wheel. The car will now roll over and do somersaults as it heads down the hill.

24.7. SOLUTIONS TO THE WORKED PROBLEMS

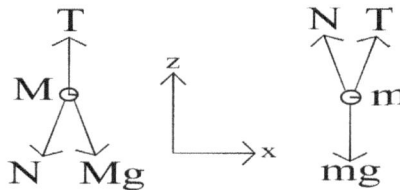

Figure 24.12: Force diagrams for the bosun's chair, the chair being M and the painter being m. The two parallel arrows on each point are drawn splayed, so that they can both be seen.

3. We first generate the force diagrams. There are two diagrams, one for the chair and one for the painter. The diagram for the chair shows the tension T in the rope, pulling up, the weight Mg of the chair pulling down, and the normal force N down of the person sitting on the chair. A standard error was to replace $-N$ with the weight of the person, which as will be seen is incorrect. The diagram for the person shows the tension T pulling up on the person, the normal force N of the chair pushing up on the person, and the force of gravity mg pulling down one the person.

The only interesting components of the Second Law for these two masses refer to the vertical direction, for which we may write the Second Law components as

$$-Mg - N + T = M\frac{d^2 z_c}{dt^2}, \tag{24.45}$$

$$N + T - mg = m\frac{d^2 z_p}{dt^2}. \tag{24.46}$$

The two accelerations are both zero, because this is a statics problem. It would be perfectly correct simply to wrote 0 as the right-hand-sides of the two equations, but doing that as part of the next step makes it easier to spot errors. We have two unknowns, T and N. If we add the two equations the N terms cancel, giving

$$2T = Mg + mg, \tag{24.47}$$

$$T = \frac{Mg + mg}{2}. \tag{24.48}$$

Either equation may now be solved for N, getting

$$N = mg - T, \tag{24.49}$$

$$N = \frac{mg - Mg}{2}, \tag{24.50}$$

and finally, on the chair,

$$\boldsymbol{N} = -\frac{mg - Mg}{2}\hat{\boldsymbol{k}}. \tag{24.51}$$

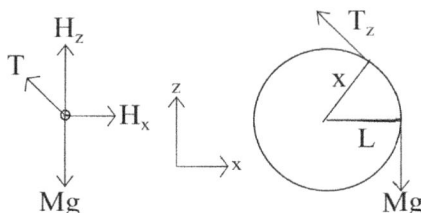

Figure 24.13: Force and torque diagrams for the flagpole problem.

4. There is a valuable meta-lesson here, namely that sometimes the order in which questions are asked is not the order in which the questions are most easily answered. We are looking for the tension in the rope and the force on the hinge.

The force diagram for the pole shows the weight Mg of the sign, the tension T of the rope, and the force components H_x and H_z due to the hinge. You do not initially know which way the force due to the hinge is pointing, so you need both force components. The tension T is usefully broken into components, these being

$$T_x = -\frac{Tx}{D^2 + x^2}, \tag{24.52}$$

$$T_z = \frac{TD}{D^2 + x^2}. \tag{24.53}$$

We can now write two force equations and one torque equation. For the x and z components of the force we have

$$H_x - \frac{Tx}{D^2 + x^2} = 0, \tag{24.54}$$

$$H_z + \frac{TD}{D^2 + x^2} - Mg = 0 \tag{24.55}$$

This time, I just set the accelerations equal to zero. After all, this is a statics problem.

We can also write an equation for the torques acting on the pole. To calculate the torques, we must choose an origin. We can put the origin anywhere we want, but a thoughtful choice of origin can simplify the calculation. A useful rule is often to put the origin where there are unknown forces being applied, because a force being applied at the origin creates no torque. For this reason, I choose to put the origin at the hinge. That choice has nothing to do with the possibility that the pole can rotate around the hinge. There is a rationale for putting the origin at the hinge, the one I just gave; it has nothing to do with whether or not is not the pole can rotate around that point. For the torque we write

$$xT_z - LMg = 0. \tag{24.56}$$

Substituting its value for T_z,

$$T\frac{Dx}{D^2 + x^2} - LMg = 0, \tag{24.57}$$

so

$$T = LMg\frac{D^2 + x^2}{Dx}. \tag{24.58}$$

For the hinge forces

$$H_x = LMg\frac{D^2 + x^2}{Dx}\frac{x}{D^2 + x^2} = \frac{LMg}{D}, \tag{24.59}$$

$$H_z = Mg - LMg\frac{D^2 + x^2}{Dx}\frac{D}{D^2 + x^2} = Mg - \frac{LMg}{x}. \tag{24.60}$$

The force *on* a hinge is the reaction force to H_x and H_z; it is

$$\boldsymbol{H}_{\text{on}} = -\frac{LMg}{D}\hat{\boldsymbol{i}} - (Mg - \frac{LMg}{x})\hat{\boldsymbol{k}}. \tag{24.61}$$

Chapter 25

Gravity

25.1 Introduction

We finally reach the force of gravity and the first physicists, people who lived three millennia ago.

Actually, I need to begin with some descriptive astronomy, what you would see if you stood outside at night before electrical lighting became common. The sights are the same now, but city lights drown out the stars and constellations, at least for many of us.

The most obvious celestial phenomenon is the sun. It rises, crosses the sky, and finally sets, once a day. At night, there are stars. If you watch them for hours and hours, the stars near the North Star Polaris are seen to travel in vast circles, the center of the circles being close to Polaris. (The location of the pole, the center of those circles, drifts very slowly with time, making a full cycle every 26,000 years.) Stars can be grouped into patterns, the constellations. There is some internal evidence that the first constellations of the zodiac were named 16,000 or so years ago. This naming, if it occurred, was perhaps not the first astronomical study. European archeologists have identified bone fragments with patterned scratches, the scratches being consistent with the interpretation that someone, perhaps 40,000 years ago, was trying to determine how many days the moon needed to cycle from full to new and back to full.

We now reach the phenomenon of First Heliacal Rising. Wait until just before sunrise. The sun is still below the horizon. The sky is bright. Perhaps ten degrees – the width of your hand held out at arms length – above the sun, stars are still visible. Some were not visible the day before. This is the first heliacal rising of those stars. If you wait a day, the same stars will become visible again, but now they will be farther up in the sky relative to the sun, and rise over the horizon earlier relative to sunrise, and new stars will be having their first heliacal rising. (Correspondingly, the day before the first heliacal rising, those stars were a degree lower in the sky, hence impossible to see against the sun's glare.) Day after day at sunrise the stars are found to have drifted in a westering direction, rising earlier in the east and setting earlier in the west.

While they cannot readily be seen directly, except during a solar eclipse, on any given day there are stars that rise with the Sun, so that we may say that the sun is in a particular part of the sky, a particular constellation. As the seasons advance the sun moves relative to the constellations, tracing out a circular path that brings it almost but not quite back to its starting point year after year.

In addition to the stars, other points of light are seen in the heavens, moving across the sky with respect to the stars. These to the ancient Greeks were the *wandering stars*, the planets. The Greeks recognized five of these, corresponding to our Mercury, Venus, Mars, Jupiter, and Saturn, though for some time they called Venus seen in the morning and Venus months later seen in the evening Hesperus and Phosphorus, respectively. The seventh planet, Uranus, actually can be seen with the naked eye under very favorable conditions, but was not recognized by the Greeks as a planet. Like the stars, every night the planets rise in the east and set in the west. Their positions drift with respect to the stars, so looking up at them at night their positions slowly drift from west to east. However, the paths of the planets are sometimes retrograde, moving relative to the fixed stars from east to west rather than from west to east. This much ancient astronomy was not original with the ancient Greeks. For example, the First Heliacal Rising of Sirius marked the start of the Egyptian year; soon thereafter the Nile began its yearly flood.

In addition to these regular and frequent astronomical phenomena, the passage of planets between the constellations of the zodiac, there are rarer events, such as eclipses when the Moon stands between the Sun and the Earth or the Earth stands between the Sun and the Moon. (Our Solar system also has those oddities the *planetary occultation* and *planetary transit*, in which our line of sight to a planet is blocked or partially blocked because another planet is in the way. The most recent of these was in the year 1818. The next two are in 2065 and 2067)

And now we reach the first physicists. The objective of physics, after all, is to reduce nature to patterns and numbers, often using the most advanced mathematics known to man. We actually know the names of some of these physicists, and when they lived, namely they lived in ancient Babylonia, served as the Scribes of *Enuma Anu Enlil*, and wrote memoranda naming who had found an event or what someone should do. The Scribes had two sorts of activity, one of which we would still view as scientific, one of which we would not.

The first activity, which is rather like modern astrology, was that astronomical phenomena were ominous; they gave omens. They warned that things (usually bad) might occur. However, Babylonian astrology was apotropaic. The positions of the stars and planets did not cause things to happen. The positions gave warnings, but rituals could be performed to avert evil events. Of course, to determine which positions warned of which events, you had to keep careful records of events and the starry signs that forewarned them. This sort of astrological activity, for obvious reasons, does not work. However, if you predicts lots of bad events, like the stock market newsletter that predicted a stock market crash like the Great Crash of 1929, except making the prediction for six dates in the last year, you have the advantage that bad events are rare. If you also make sure that the appropriate rituals are performed, the bad events almost never happen (not that they would have otherwise), and your apotropaic astrology actually appears to work, because the rituals were performed and no bad things happened. It doesn't work, but it appears to work. (Astrology will show up again in this historical note).

The second research activity, however, was modern physics, performed with the most advanced mathematics then known, notably integer factions. The objective was to predict when the planets would perform various astronomical acts, such as First Heliacal Rising, given that bad weather meant that the acts often could not be observed from the ground. The effort was challenging, and limited by the very modest quality of period instruments.

It remained for the ancient Greeks and Romans to construct mechanical models that accounted for planetary motions. They had a variety of different models, some with all planets and the Sun revolving around the earth, some with the inner planets revolving around the Sun (which in turn revolved around the earth), and some that more or less matched our modern image of the Solar system. There are several surviving Graeco-Roman books on Astronomy. The one best known, due to Claudius Ptolemaeus of Alexandria (Ptolemy) is the *Mathematike Syntaxis* (The Mathematical Arrangement), commonly titled the *Almagest*. The models were also incorporated into the first precision analog computers, as built by the ancient Greeks, one of which has survived albeit in poor condition. The *Antikythera mechanism*, recovered from an ancient shipwreck, used precision cut gears to predict the positions of the planets, phases of the moon, and eclipses, all in a box the size of a mantle clock. The Antikythera Mechanism is the supreme triumph of ancient science; nothing as complex was made anywhere in the world for the next thousand years.

How complicated was the Ptolemaic model of the Solar system, complete with planets moving in crystal spheres, spheres within spheres to describe retrograde motion, and more? Of these Philip the Wise, King of Spain ca. XIV, said "Had the Almighty had consulted with me, prior to embarking on the Creation, I might have recommended something simpler."

And now we reach a trio of medieval astronomers. The first was Copernicus, who actually did measure a modest number of planetary positions, and who revived an ancient Greek model that had the Sun in the center of the Solar system, with all the planets including the Earth revolving in circles around it. The second was a Danish nobleman, Tycho Brahe. Brahe gave the world the first 'big science' project. It consumed 2% of the Gross National product of Denmark. At the time, Denmark was one of the great powers of Europe. 2% of GNP? That's a lot. An American big science project that consumed 2% of GNP would cost $400 billion a year, dwarfing the one hundred million dollars that was spent over many years on LIGO, the Large Interferometric Gravitational Observatory.

Why spend the money? In some interpretations, the answer is "Combat Astrology". The hope was that better measurements of planetary positions would lead to the creation of more accurate horoscopes, thus predicting the behavior of national enemies and reinforcing Denmark's status as a great power. Whatever the reason, Tycho Brahe made by far the most accurate measurements performed until then of the paths of the planets across the constellations. The King of Denmark with this hope died. His successor disliked Brahe. Tycho Brahe found it expedient to move to Prague, where he became Imperial Court Astronomer to the Holy Roman Emperor.

Brahe's measurements then fell into the hands of his associate, Johannes Kepler. Some of Brahe's relatives used harsher words than 'fell into', so there were legal disputes as to the ownership of the data that delayed publication. The known challenge was the orbit of Mars, which fit Ptolemy's model poorly, especially when confronted with Brahe's data. Brahe measured planetary positions without using the not-yet-invented telescope to within a half a minute of arc. (As a scale, the moon is 30 minutes of arc across, and *Mare Crisium*, the little dark spot in the upper right corner of the moon as seen in the northern hemisphere, is about two minutes of arc across. If you have good vision, you can see that Mare Crisium has a shape.)

Kepler attempted to create more accurate models of planetary orbits, finally concluding that planetary orbits were ellipses rather than circles, and that the speed at which a planet travelled around its ellipse was not a constant, but instead followed a specific geometric rule, the *equal area* rule. Planets travel around the ellipse, cutting off equal areas in equal times. For much more on this topic, read Thomas Kuhn's *The Copernican Revolution*. At this point, planetary motion became quite complicated. The Keplerian model, which we recognize at being more or less the modern model, was no simpler than Ptolemy's earth-centered model.

25.2 Newtonian Gravity

And now we come to Isaac Newton. Newton created the differential and integral calculus, the mathematical methods that let him treat his model of the Solar system. He made a radical innovation in physical law. Gravity – things fall – has always been known. It had historically been believed that the forces that move the planets through the heavens were unrelated to the purely mundane forces that are found on our earth. Newton demonstrated that the gravitational force that the Earth exerts on an object on Earth and on the moon are quantitatively the same, and are described by his Law of Gravity and Newton's laws of motion. Doubting critics were impressed by precision measurements made by the English mathematician George Atwood. Newton also explained the nature of color and studied fluid flow. Finally, Newton attempted to explain the science of chemistry, but his efforts to reform alchemy and reliably transmute lead to gold were not successful. He appears for good reason to have kept secret his ideas about theology.

Newton's Law of Gravity, in modern terms, tells us that if we have two masses m_1 and m_2 their gravitational potential energy $U(r_{12})$ is

$$U(r_{12}) = -\frac{Gm_1m_2}{r_{12}}. \tag{25.1}$$

Here r_{12} is the scalar distance between masses 1 and 2 and G is the Gravitational Constant. $G = 6.67 \cdot 10^{-11}$ Newton·meter^2·kilogram^{-2}. Note that the same name 'gravitational constant' is sometimes applied both to G and to g. g is also sometimes called the 'gravitational acceleration', even though the book sitting on my desk has a weight mg when it is not accelerating. G is extremely difficult to measure with any degree of accuracy; G is currently known to perhaps fifteen parts per million.

The equation as written refers to point masses. It refers equally to non-overlapping spheres that are uniform or whose density depends only on the distance from the center, the distance r_{12} then being the distance between the centers of mass of m_1 and m_2.

We can use equation 25.1 to write the gravitational force between two masses. The force on mass 2 due to mass 1 is

$$\boldsymbol{F}_{12} = -\frac{Gm_1m_2}{r_{12}^2}\hat{\boldsymbol{r}}_{12}. \tag{25.2}$$

Here $\hat{\boldsymbol{r}}_{12}$ is the unit vector pointing from m_1 toward m_2.

From the Third Law, the gravitational force on m_1 due to m_2 is

$$\boldsymbol{F}_{21} = \frac{Gm_1m_2}{r_{12}^2}\hat{\boldsymbol{r}}_{12} \equiv -\frac{Gm_1m_2}{r_{12}^2}\hat{\boldsymbol{r}}_{21}. \tag{25.3}$$

Denoting the locations of m_1 and m_2 as $\boldsymbol{r}_1$ and $\boldsymbol{r}_2$, respectively, the vector $\boldsymbol{r}_{12}$ is $\boldsymbol{r}_2 - \boldsymbol{r}_1$, which may also be written as $-\boldsymbol{r}_1 + \boldsymbol{r}_2$. See Figure 25.1. From simple geometry, $\boldsymbol{r}_{12} = -\boldsymbol{r}_{21}$ and $\hat{\boldsymbol{r}}_{12} = -\hat{\boldsymbol{r}}_{21}$. However written, the Figure shows how the vector from m_1 to m_2 is related to the two vectors giving their individual locations.

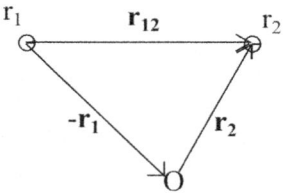

Figure 25.1: The vector $\boldsymbol{r}_{12}$ connecting the points r_1 and r_2, and its decomposition into two vectors linking to the origin, namely $-\boldsymbol{r}_1$ and $\boldsymbol{r}_2$

It is sometimes said that gravity is the weakest of the natural forces. That's based on calculating the gravitational force between the electron and the proton inside a hydrogen atom. If you calculate the force between the Earth and the Sun, or between the Earth and you, you get somewhat larger numbers for the gravitational forces. In the solar system, the gravitational forces between the sun and the planets are much stronger than the electrical and magnetic forces between the planets.

The gravitational force is a central force. We can confirm this by calculating the torque around m_1 that m_1 exerts on m_2. We find

$$\boldsymbol{\tau} = \hat{\boldsymbol{r}}_{12} \times \left(-\frac{Gm_1 m_2}{r_{12}^2} \hat{\boldsymbol{r}}_{12} \right). \tag{25.4}$$

However, $\hat{\boldsymbol{r}}_{12} \times \hat{\boldsymbol{r}}_{12} = \boldsymbol{0}$. The gravitational force due to m_1 exerts no torque on m_2. Therefore, if two masses interact with each other only via the gravitational force, their angular momentum as calculated with respect to any choice of center is conserved.

What about the gravitational force near the Earth's surface? According to equation 25.2, the gravitational force acting between two masses depends on the square of the distance between them. However, if we move around on the earth, climb a mountain, go up in an airplane, or take a submarine to the bottom of the ocean, the gravitational force is very nearly constant, so that the weight of m_1 remains very nearly $m_1 g$ throughout. Why is there no discrepancy here?

We answer this question by applying Taylor series expansions to the gravitational force and the gravitational potential energy of an object near the surface of the earth. If a function $f(x)$ is continuous and all of its higher derivatives $f^{(n)}(x) \equiv \frac{d^n f(x)}{dx^n}$ exist and are well-behaved, we may write

$$f(X + a) = \sum_{n=0}^{\infty} \frac{a^n}{n!} f^{(n)}(x)|_{x=X} \tag{25.5}$$

for the expansion of $f(x)$ around the point $x = X$, where the derivatives must be taken before being evaluated at $x = X$.

If $a \ll X$, and if the derivatives $f^{(n)}(x)|_{x=X}$ are not too large, $f(X + a)$ is dominated by its first few terms, namely

$$f(X + a) = f(X) + af^{(1)}(X) + \frac{a^2}{2} f^{(2)}(X) + \ldots \tag{25.6}$$

This equation is only a sampling of early terms from the correct Taylor series, not the series itself, but it is enough for our purposes. For the magnitude of the gravitational force, with $X = R$ being the distance from the center of the Earth to mean sea level, $a = \delta$ being the altitude from mean sea level, and M and m being the masses of the Earth and you at the surface of the Earth, respectively, we write

$$\mid \boldsymbol{F}(R + \delta) \mid = \frac{GMm}{R^2} + \delta \left(-\frac{2GMm}{R^3} \right) + \ldots \tag{25.7}$$

which may be rewritten as

$$\mid \boldsymbol{F}(R + a) \mid = \frac{GMm}{R^2} \left(1 - \frac{2\delta}{R} + \mathcal{O}((\frac{\delta}{R})^2) \right). \tag{25.8}$$

The radius of the Earth is about 6400 km. From the highest to the lowest point on the Earth's surface, $\delta \approx 20$ km. As a result, the expansion parameter $2\delta/R$ is never larger than $\approx 0.7 \cdot 10^{-2}$. Even if we take the observer into outer space, say at $\delta = 200$ km, the force of the Earth's gravity would only be reduced by $-\frac{2\delta}{R} \approx 0.07$ or 7%.

25.2. NEWTONIAN GRAVITY

If we identify $GM/R^2 = g$, we may rewrite the force of gravity as

$$|\mathbf{F}(R+\delta)| \cong mg(1 - \frac{2\delta}{R}). \qquad (25.9)$$

The Taylor series approximation may also be applied to treat the gravitational potential energy between m and M. If you do the expansion, and calculate $U(r)$ at a distance h above the ground, then for small h/R you will find

$$U(R+h) = -\frac{GMm}{R} + \frac{GMmh}{r^2}. \qquad (25.10)$$

This equation may be rewritten

$$U(h) = -\frac{GMm}{R} + gmh. \qquad (25.11)$$

On the right hand side of that equation, the first term is a constant that does not contribute to the dependence of U on altitude h. It therefore has no effect on the motions of objects. The second term is the potential energy mgh we used earlier in the course; it is an approximation appropriate if the change in h is not too large.

What if there are more than two masses present? Suppose we have four masses m_1, m_2, m_3, and m_4. What is the gravitational force on m_1 due to the other three masses? Newtonian gravity is linear; the total gravitational force $\mathbf{F}_1$ on m_1 is the vector sum of the gravitational forces $\mathbf{F}_{21}$, $\mathbf{F}_{31}$, and $\mathbf{F}_{41}$ due to the other three masses. Using the notation given above, this statement would be written

$$\mathbf{F}_1 = \mathbf{F}_{21} + \mathbf{F}_{31} + \mathbf{F}_{41}. \qquad (25.12)$$

Equation 25.12 is a vector equation. It's equivalent to three scalar equations, one for each of the three Cartesian components. Labelling the three components with additional subscripts x, y, and z, this equation is equivalent to

$$F_{1x} = F_{21x} + F_{31x} + F_{41x}, \qquad (25.13)$$
$$F_{1y} = F_{21y} + F_{31y} + F_{41y}, \qquad (25.14)$$
$$F_{1z} = F_{21z} + F_{31z} + F_{41z}. \qquad (25.15)$$

Is equation 25.12 to be understood as a mathematical result or as an additional law of nature? In our original discussion of the Second Law, we said that the total force on an object was simply the vector sum of the individual forces on the object. The separate forces on an object combine linearly to form the total force of the Second Law.

However, equation 25.12 is to be understood as a physics result, not just a mathematics result. According to this equation, the total gravitational force on an object can be decomposed into a sum of pair forces, namely the pair forces on mass 1 due separately to masses 2, 3, and 4. How can we tell that this result is a physics equation, not simply a mathematical statement? The answer is straightforward. The equation reveals physics, because there is an alternative theory of gravity in which equation 25.12 is false. The equation is true for Newtonian Gravity. However, Newton's Law of Gravity is an approximation, an excellent and accurate approximation, but nonetheless only an approximation.

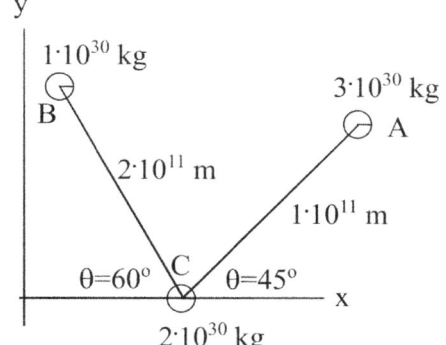

Figure 25.2: Masses and positions of the three stars in the example.

We know this because there is a more modern and accurate theory of gravity, namely gravity as described by Einstein in his Theory of General Relativity. The gravity of General Relativity is not linear. In General Relativity, equation 25.12 is not true.

In this course, we stay with Newtonian gravity, which is extremely but not perfectly accurate under almost all conditions. The planetary astronomy known to Kepler and Newton is accurately described by Newtonian gravity. More accurate measurements of the orbit of Mercury and to a much lesser extent the orbit of Venus show noticeable deviations from Newtonian gravity. Mercury's orbit deviates from Newtonian gravity by ca. 10^{-4} degrees per year, a phenomenon known as *orbital precession*. This deviation was known in the 19th century; various explanations such as the hypothesized intramercurial planet Vulcan were shown to be incorrect. The deviation is explained by General Relativity. Many readers use devices that determine their location using GPS, the Global Positioning System, based on highly accurate clocks placed into orbit around the earth. The time told by those clocks is affected not only by Special Relativity, but also by General Relativity. If you assumed that the clocks were subject to Newtonian rather than Einsteinian gravity, GPS systems would fail catastrophically, namely they would generate progressively increasing wrong reports as to their locations.

We now give an example of calculating the gravitational force in a system containing three stars. The stars all lie in the x-y plane, so the z-components of their forces on each other are all zero. As seen in the Figure, we have three stars A, B, and C having mass $3 \cdot 10^{30}$, $1 \cdot 10^{30}$, and $2 \cdot 10^{30}$ kg, respectively. The distances between the stars are $r_{CA} = 1 \cdot 10^{11}$m and $r_{CB} = 2 \cdot 10^{11}$m. The unit vectors $\hat{r}_{CA}$ and $\hat{r}_{CB}$ make angles of 45 and 120 degrees, respectively, with the positive x-axis.

On substituting the above numbers into equation 25.2, the force $\boldsymbol{F}_C$ on star C is

$$\boldsymbol{F}_C = -\left(\frac{6.67 \cdot 10^{-11} \cdot 2 \cdot 10^{30} \cdot 3 \cdot 10^{30}}{(1 \cdot 10^{11})^2}\right)(\hat{\boldsymbol{i}}\cos(45) + \hat{\boldsymbol{j}}\sin(45)) \tag{25.16}$$

$$-\left(\frac{6.67 \cdot 10^{-11} \cdot 2 \cdot 10^{30} \cdot 1 \cdot 10^{30}}{(2 \cdot 10^{11})^2}\right)(\hat{\boldsymbol{i}}\cos(120) + \hat{\boldsymbol{j}}\sin(120)). \tag{25.17}$$

The forces on star C due to stars A and B *add as vectors, not as scalars*. Thus the x-component of the total is the sum of the x-components of the terms being summed, or

$$\boldsymbol{F}_C = -\left(\frac{6.67 \cdot 10^{-11} \cdot 2 \cdot 10^{30} \cdot 3 \cdot 10^{30}}{(1 \cdot 10^{11})^2}\cos(45) - \frac{6.67 \cdot 10^{-11} \cdot 2 \cdot 10^{30} \cdot 1 \cdot 10^{30}}{(2 \cdot 10^{11})^2}\cos(120)\right)\hat{\boldsymbol{i}} \tag{25.18}$$

$$-\left(\frac{6.67 \cdot 10^{-11} \cdot 2 \cdot 10^{30} \cdot 3 \cdot 10^{30}}{(1 \cdot 10^{11})^2}\sin(45) - \frac{6.67 \cdot 10^{-11} \cdot 2 \cdot 10^{30} \cdot 1 \cdot 10^{30}}{(2 \cdot 10^{11})^2}\sin(120)\right)\hat{\boldsymbol{j}}. \tag{25.19}$$

25.3 Escape Velocity

We now turn to the notion of *escape velocity*. We have a rocket sitting on the surface of the Earth. It fires up its engines and takes off. It comes up rather quickly to a maximum velocity, and then runs out of fuel. The rocket travels more or less straight up, traveling more and more slowly as it climbs. If its starting velocity is not too large, the rocket climbs to a maximum altitude, comes to a stop, and falls back to the ground. On the other hand, if the rocket is travelling fast enough, it will still be moving upward as it heads out toward infinity. Between these two alternatives there is a velocity, the *escape velocity*, at which the rocket comes asymptotically to a stop at an infinite distance from the Earth. The definition here is formal, in that the calculation ignores air friction, which is actually an important effect. In addition, in launching a rocket from the surface of a planet, if planetary or solar escape velocity is desired, the rocket has initial velocity contributions from the tangential velocity of the planet as it spins on its axis and the orbital velocity of the planet as it rotates around the sun.

We can calculate the escape velocity using Conservation of Energy. The total energy E of the rocket satisfies $E = K + U$. The rocket has a kinetic energy and a gravitational potential energy, so for this problem we may write the total energy of the system as

$$E = \frac{1}{2}mv^2 - \frac{GMm}{r}. \tag{25.20}$$

In this equation, m is the mass of the rocket, M is the mass of the planet, and r is the distance between the rocket and the Earth's center of mass.

25.4. WEIGHTLESSNESS

At the start, $r = R$ and $v = V_e$, where R is the Earth's radius and V_e is the escape velocity. At the end, the rocket is infinitely far from the earth, so that $r = \infty$, and has come to a stop, so that $v = 0$. From conservation of energy, $E_i = E_f$, leading to

$$\frac{1}{2}mV_e^2 - \frac{GMm}{R} = \frac{1}{2}m0^2 - \frac{GMm}{\infty}, \qquad (25.21)$$

and therefore

$$V_e = \left(\frac{2GM}{R}\right)^{1/2}. \qquad (25.22)$$

25.4 Weightlessness

We note the term *weightless*. If we consider an object resting on the floor, we would say that the object is weightless if it is exerting no normal force on the floor, and, correspondingly, if the floor is exerting no normal force on it. The simplest way to achieve weightlessness is to have your object and floor extremely far from any mass, for example, well outside the orbit of Pluto. This condition is experimentally difficult to achieve over a reasonable time period, though it has been done.

Fortunately, there is an easier way to achieve the condition of weightlessness. If an object is moving in a vertical arc such that its acceleration downward is precisely one gravity, then it is weightless. Let's show this more precisely.

First, consider an object moving in a curving arc such that its acceleration vertically downward is g. This is a sketch, not a force diagram, so it is entirely appropriate to indicate velocities or accelerations. We are considering the object exactly at the top of its trajectory. The corresponding force diagram would show that the forces on the object are the force of gravity and the normal force.

The object is performing circular motion, so it has an acceleration V^2/R, where V is the object's forward velocity and where R is the radius of the circle. V and R have been chosen so that $V^2/R = g$. We may then write the Second Law for the vertical direction as

$$-m\frac{V^2}{R} = -mg + N. \qquad (25.23)$$

The negative sign on the left-hand side appears because V^2/R is the magnitude of the acceleration, but the acceleration is straight down. However, $\frac{V^2}{R} = g$, so this equation becomes

$$-mg = -mg + N, \qquad (25.24)$$

which tells us that

$$N = 0. \qquad (25.25)$$

There are several ways of obtaining this condition under practical circumstances. Some modern roller coasters duplicate this effect near the top of one or more hills. If you put a space plane into a circular orbit around the earth, the people on board experience weightlessness. Manned space probes other than the Apollo missions all remained quite close to the surface of the earth, so that the local force of gravity was not zero. However, the orbital motion of the manned probes were such that at all times the acceleration toward the center of the earth was g, leading to $N = 0$ and weightlessness. One can obtain the same effect in a drop tower, in which objects are allowed to fall vertically. The space around the falling objects needs to be pumped to vacuum in order to get an accurate weightlessness condition. Medieval gunners used shot towers in which molten lead was dropped from a respectable height, say 100 feet, into cold water to produce circular round shot. These shot towers reached some approximation of free fall. Finally, for short periods of time weightlessness can be obtained in an aircraft flying in a parabolic trajectory. NASA has used such aircraft for scientific studies.

25.5 Discussion

Let us concede that the medieval shot towers did not approximate weightlessness very well. Not counting the shot towers, in which decade was weightlessness first used to produce a successful commercial product?

It's a trick question. It's like the question "Which President of the United States was the first to make a televised address?".

Now the answers. The first commercially successful product was apparently made with the airplane method. The airplane in question was the hottest thing (at the time) in commercial aviation, the Douglas DC-3. Back around 1940, a Hollywood film-maker claimed he had had the bright idea of building a sound stage on one of these, and for one of their space opera serials filmed the intrepid fliers in fictional outer space in true zero-gee conditions. The first American president to deliver a televised address was Franklin Delano Roosevelt, who in 1939 addressed the New York World's Fair via television. The technology was quite limited by modern standards, but it worked.

25.6 Worked Problems

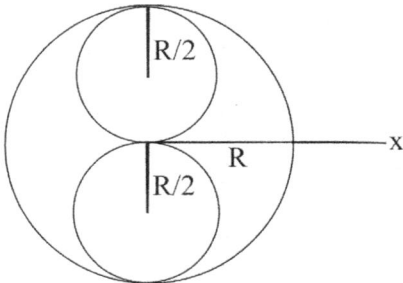

Figure 25.3: Three embedded spheres for Worked Problem 2.

1. Find the first four terms in the series expansions for the Earth's gravitational potential and the earth's gravitational force, centering the expansion at the radius R of the earth, and calling the expansion variable x. Interesting math feature: If you do the expansions correctly, the series are not convergent for $x/R > 1$, no matter what R you choose. "Why" is a sophisticated math question.

2. A solid sphere having radius R has had removed from its middle two spheres of radius $R/2$. The sphere is located at the origin. The two empty spherical volumes have their centers on the z axis at $+R/2$ and $-R/2$. Find the gravitational potential energy created by this object for a mass m located along the x-axis outside of the large sphere. Do not attempt to reduce your answer to a simpler form. Comment: There are three or four ways to work this problem. Some use more brute force. Some use more cleverness.

25.7 Homework

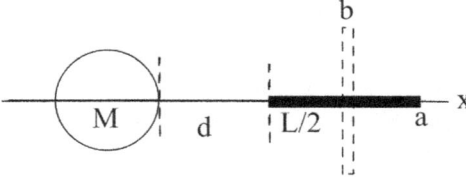

Figure 25.4: The mass M sphere and the rod, showing rod positions (a) before, and (b) after the rod is rotated through 90 degrees.

1. Two persons having masses of 50 and 100 kg, respectively stand one meter apart. What is the magnitude of the gravitational force attracting them together?

2. If you do the numerical calculations implied in equation 25.16, what is the gravitational force on star C? Reminder: If your answer is correct, it is a vector. If your answer is not a vector, something has gone wrong.

3. A uniform sphere of mass M is located at the origin. (a) Aligned along the x axis, with its center at a distance $d + L/2$ from the origin, is a uniform rod having mass m and length L. The rod does not intersect the sphere. What is the gravitational potential energy of the rod? (b) The rod is now rotated around its middle until it is parallel with the y-axis. Reduce to quadratures (that is, find a one-dimensional integral for) the gravitational potential energy of the rod. CLUE: Approximating the rod as a point mass located at its center of mass is wrong.

4. This is a variation on an exam problem from a previous year. It was easy for people who (1) remembered that force is a vector and not a scalar, and (2) did not blandly assume values for angles rather than calculating them. Recalling that the gravitational constant G is $6.67 \cdot 10^{-11}$ in SI units, calculate the gravitational force that I (M = 90kg) exert on a 75 kg student in the lecture hall. I am $(x, y, z) = (2, -1, 1)$m; the student is at $(x, y, z) = (8, -4, -3)$m. 'Zero for all practical purposes' was worth zero points as an answer.

5. Estimate the gravitational potential energy U(r) of a mass m, at a distance $r = R + x$ from a point mass M, by performing a third-order ($\mathcal{O}(x^3)$) Taylor series expansion of $-GMm/r$ around its value $U(R)$ and derivatives at R, using x as the expansion variable.

6. Estimate the gravitational force $\mathbf{F}$ on a mass m, at a distance $r = R + x$ from a point mass M, by performing a third-order ($\mathcal{O}(x^3)$) Taylor series expansion of $-GMm/r^2$ around R, using x as the expansion variable.

7. We have three point masses, all having mass m, located at $(0, L, 0)$, $(L, 0, 0)$, and $(-L, -L, 0)$. Calculate the gravitational force $\mathbf{F}$ on the third mass, the mass at $(-L, -L, 0)$, due to the other two masses. Find the gravitational field $\mathbf{g} = \mathbf{F}/m$ due to the other two masses at the location of the third mass. Find the magnitude of the gravitational field due to the other two masses at the location of the third mass.

8. Two small masses of mass m are located on the x-axis at $x = \pm a$. A mass M is located on the y-axis at $y = d$. What is the gravitational force on M due to the two small masses?

9. A sphere of mass M and radius R exerts a gravitational force on a long rod having length L and mass m. The sphere is uniform in density, so it creates the same gravitational field that would be created by a mass point. The sphere is centered at the origin. The uniform rod, total mass m, lies on the x-axis. The rod end nearer the sphere is at $x = +a$. (a) Find the gravitational force exerted on the rod by the sphere. (b) Find the gravitational force on the rod that you would calculate if you claimed that the rod acted as a point mass, mass m, located at the rod's midpoint. (c) If you did (a) and (b) correctly, are your answers to (a) and (b) the same? When we proved something like this in an earlier chapter, what assumptions did we make? Are those assumptions true here?

10. The masses of the Moon, the Earth, and the Sun are $7.35 \cdot 10^{22}$ kg, $5.97 \cdot 10^{24}$ kg, and $1.99 \cdot 10^{30}$ kg, respectively. The distances from the Earth to the Moon and from the Moon to the Sun are $3.85 \cdot 10^8$ m and $1.50 \cdot 10^{11}$ m, respectively. The constant G is $6.67 \cdot 10^{-11}$ in SI units. The Earth, Moon, and Sun lie in a right triangle with the Sun and Moon on the y-axis, and the Earth to the left of that axis. (i) Find the force on the Moon due to the Sun. (ii) Find the force on the Moon due to the Earth. (iii) Find the total gravitational force on the Moon due to the Earth and the Sun. (iv) Which is stronger, the force on the Moon due to the Sun or the force on the Moon due to the Earth? Based on your answer, is the Moon primarily in orbit around the Sun or around the Earth?

11. Three of us are (briefly) transported to deep intergalactic space, remote from all other masses. Student A is 60 kg. Student B is 80 kg. I am, alas, above 90 kg, which is the number to be used in solving this problem. In SI units the gravitational constant G is $6.67 \cdot 10^{-11}$. I am located at $\mathbf{r} = 10\hat{i} + 10\hat{j} + 0\hat{k}$. Student A is located at $\mathbf{r} = 10\hat{i} - 20\hat{j} + 0\hat{k}$. Student B is located at $\mathbf{r} = -10\hat{i} + 15\hat{j} + 0\hat{k}$. (i) Find my

gravitational potential energy U as created by the two students. (ii) Find the total gravitational force exerted on me by the two students. (iii) Student A is now returned to Earth. I am held fixed. Student B is given a speed V directly away from me that is just barely sufficient that student B keeps receding from me, more and more slowly as time goes on, almost but never quite stopping. What is the speed V?

12. Two objects of masses m_1 and m_2 are located so that the vector from m_1 to m_2 is $\boldsymbol{r} = a\hat{\boldsymbol{i}} + b\hat{\boldsymbol{j}} + c\hat{\boldsymbol{k}}$. (i) Find the gravitational force that m_1 exerts on m_2. (ii) Find mutual gravitational potential energy of m_1 and m_2. (iii) Show that the potential energy and the gravitational force are related as predicted by the work-energy theorem.

13. You have been hired as the staff physicist for the ongoing motion picture production. The Director announces that a spaceship has been transported (not consistent with the laws of physics) to another solar system, a solar system that contains three stars. Each star has the mass of the sun, $3 \cdot 10^{30}$ kg. The three stars and the spaceship lie in a flat square, with the spaceship at the coordinate origin. The sides of the square are each $3 \cdot 10^7$ km ($3 \cdot 10^{10}$ m) long. The spaceship has a mass of $1 \cdot 10^8$ kg. Hint: $G = 6.67 \cdot 10^{-11}$ m^3/kg/s. (o) Show that your formulas for the gravitational force and potential have the correct units, given that I have told you the units of G. Find (i) the gravitational potential energy of the spaceship, and (ii) the gravitational force on the spaceship.

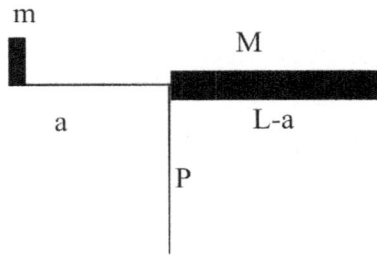

Figure 25.5: The one-person teeter-totter of problem 14. P marks the pivot rod; m and M are the masses of the person and the concrete.

14. In the world's first one-person teeter-totter, a person of mass m sits at one end of an extremely light but rigid rod of length L, at a distance a from a pivot P. The pivot is suspended a distance $2a$ above the ground. The part of the rod beyond the pivot has been uniformly coated with concrete, so over its length $L - a$ it carries a mass M of concrete. The rod is initially held in place by a brake, and is mounted at an angle θ from the horizontal, so it does not move. (i) Find the gravitational torque τ on the teeter-totter, as measured from the pivot point. (ii) The builders did not use enough concrete to balance the person in place, so when the brake is released, the teeter-totter begins to move. The person begins to fall, while the concrete rises. Calculate the angular acceleration $d^2\theta/dt^2$ of the rod at the moment the rod is released. (iii) The person continues to fall until the rod is vertical. The person's velocity will be horizontal at this point. Calculate the person's speed at this moment.

15. We have three stars in an equilateral triangle. Their masses are $1 \cdot 10^{30}$ kg, $2 \cdot 10^{30}$ kg, and $1.5 \cdot 10^{31}$ kg, respectively. The distance between each pair of stars is $2 \cdot 10^7$ km. The base of the equilateral triangle is parallel to the x-axis. Calculate the gravitational force on the $1.5 \cdot 10^{31}$ kg star, which is located at the origin. The gravitational constant G is $6.67 \cdot 10^{-11}$.

16. Recall that the gravitational constant G is $6.67 \cdot 10^{-11}$ in SI units. For yet another scene in the film, the heroine (70 kg mass, plus 50 kg body armor) is allegedly transported to outer space near a distant double star system. The director orders his staff physicist (Hollywood Directors do not usually hire staff physicists, but the staffer's uncle is Executive Producer) to calculate the gravitational force on the heroine. To keep things simple, the heroine is at the origin. One star is in the $x - y$ plane, at a distance of 1×10^8 km and an angle $+30$ degrees from the x axis, and has a mass of 1×10^{30} kg. The

other star is at **R** = $(0, -3 \times 10^7, 4 \times 10^7)$ km, and has a mass of 2×10^{30} kg. (i) Draw a clear picture, reasonably labelled, of the relative locations of the heroine and the two stars. (ii) Compute the force on the heroine. Reduce your answer to arithmetic form, but don't plug stacks of numbers into your calculator.

17. A spaceship finds itself in a distant double star system. Its engines are turned off, so it falls freely through space. The spaceship and the two stars of the system occupy three corners of an equilateral triangle whose side has length l of 2×10^8 km. The two stars have masses of 1×10^{30} kg and 2×10^{30} kg, respectively. Find the acceleration **a** of the spaceship.

18. Two masses m lie on the x axis, separated from the y axis by displacements a and $-b$, respectively. A mass M lies on the y axis, at a distance d below the x axis. a, b, and d are positive numbers. Find the gravitational force on M due to the two masses m.

    And harder bonus: Consider the mass M. It is mounted on a thin wire that allows it to slide freely back and forth along the y axis. It begins at rest a small distance δ below the x axis. Find the frequency of small oscillations of the mass M.

19. You have secured a summer job as staff physicist to the motion picture company whose film plans have appeared regularly in past exams and problem sets. The pay is good, but the questions are sometimes odd. In the forthcoming scene, the heroine and her space suit ($m = 100$ kg) supposedly find themselves in outer space, so far from any planet that there is almost no gravity. However, in the set's coordinate system the heroine is at coordinates $(1, 1, 1)$ m. In her immediate vicinity are said to be two blocks of the mythical substance "collapsed matter", block A at $(4, 1, 5)$ m having mass $1 \cdot 10^6$ kg, and block B at $(8, 13, 1)$ m having mass $13 \cdot 10^6$ kg. The blocks are rigidly mounted in place and do not move. "Collapsed matter" is so dense that the blocks are shown on the set as small spheres, diameters 1 cm. The director asks "Won't the heroine fall toward the blocks, because the blocks have gravitational fields?" "How fast does she fall?" To answer his question, (i) Calculate the vectors from the heroine to each of the two blocks. (ii) Calculate the force on the heroine due to the two blocks of "collapsed matter", assuming the blocks have the cited masses. (iii) Calculate the heroine's acceleration at the start of the scene due to these gravitational forces.

20. A student at one of our neighboring institutions, desperate for an excuse for not having submitted his final lab report in the form of the mandated series of Shakespearean sonnets, informs his Professor that the gravitational force exerted on him by his significant other had dragged him bodily from his dorm room. Astonished by the audacity of this elsewise absurd excuse, the Professor informs the student that if he can calculate the force correctly, the student will be allowed to submit the lab report late. Consultation with the campus mapping system shows that the student was located relative to the Professor at $-100\hat{\mathbf{i}} - 200\hat{\mathbf{j}} + 10\hat{\mathbf{k}}$, while the significant other was located at $200\hat{\mathbf{i}} - 100\hat{\mathbf{j}} - 20\hat{\mathbf{k}}$. The student and significant other have masses of 75 and 50 kg, respectively; you may take $G = 6.67 \times 10^{-11}$. (i) Assume SI units throughout. What was the actual force on the student due to the significant other? (ii) If the gravitational force due to the significant other was actually as large as the student's weight, what was the significant other's mass? Assume that the significant other is a mass point made of collapsed matter.

21. Hard problem: Explain the relationship between ocean tides and the closest distance at which a natural satellite (something held together by the force of its gravity, not by nuts and bolts or the like) can orbit the Earth.

25.8 Solutions to the Worked Problems

1. We are doing a series expansion for the gravitational potential energy around $r = R + x$. The potential energy itself is

$$U(r) = -\frac{GMm}{r}. \tag{25.26}$$

The series expansion is written

$$U(r) = -\left[\frac{GMm}{r} + x\left(\frac{d}{dr}\right)\frac{GMm}{r}\right. \quad (25.27)$$

$$\left. + \frac{x^2}{2!}\left(\frac{d}{dr}\right)^2 \frac{GMm}{r} + \frac{x^3}{3!}\left(\frac{d}{dr}\right)^3 \frac{GMm}{r} + \ldots\right]_{r=R}.$$

On taking the indicated derivatives

$$U(r) = -\left[\frac{GMm}{r} - x\frac{GMm}{r^2} + \frac{x^2}{2!}\frac{2GMm}{r^3} - \frac{x^3}{3!}\frac{3!GMm}{r^4} + \ldots\right]_{r=R}. \quad (25.28)$$

All terms have a common factor GMm/r followed by powers of x/r. Rearranging terms and evaluating the derivatives at $r = R$, we get

$$U = -\frac{GMm}{R}\left[1 - \frac{x}{R} + \left(\frac{x}{R}\right)^2 - \left(\frac{x}{R}\right)^3 + \ldots\right]. \quad (25.29)$$

If $x/R < 1$, that's a geometric series and convergent. On the other hand, if $x/R \geq 1$ the series is not convergent, including at $r = R + x$. Why? There is a detailed mathematical explanation. As a simple statement, the original function $\frac{GMm}{r}$ diverges as $r \to 0$. However, if $x = -R$, the series attempts to describe $U(r)$ at $r = R + x = R - R = 0$, where the function $U(r)$ is divergent. The series can't do that, which, it turns out from somewhat sophisticated mathematics, limits the range of validity of the series expansion to $-R < x < R$.

We can also do the expansion around $r = R + x$ for the radial component of the Earth's gravitational force. Our starting point is

$$F(r) = -\frac{GMm}{r^2} \quad (25.30)$$

The series expansion is written

$$F(r) = -\left[\frac{GMm}{r^2} + x\left(\frac{d}{dr}\right)\frac{GMm}{r^2}\right. \quad (25.31)$$

$$\left. + \frac{x^2}{2!}\left(\frac{d}{dr}\right)^2 \frac{GMm}{r^2} + \frac{x^3}{3!}\left(\frac{d}{dr}\right)^3 \frac{GMm}{r^2} + \ldots\right]_{r=R}. \quad (25.32)$$

On taking the indicated derivatives

$$U(r) = -\left[\frac{GMm}{r} - 2x\frac{GMm}{r^2} + \frac{3!x^2}{2!}\frac{GMm}{r^3} + \frac{4!x^3}{3!}\frac{GMm}{r^4} + \ldots\right]_{r=R}. \quad (25.33)$$

All terms have a common factor GMm/r followed by powers of x/r. Rearranging terms and evaluating the derivatives at $r = R$, we get

$$U = -\frac{GMm}{R^2}\left[1 - 2\frac{x}{R} + 3\left(\frac{x}{R}\right)^2 - 4\left(\frac{x}{R}\right)^3 + \ldots\right]. \quad (25.34)$$

This series is again divergent for $x/R > 1$.

2. We have a sphere, density ρ. The sphere is two spherical hollows in it. What gravitational force does it exert on a mass m on the x axis? The calculation can in principal be set up as a volume integral with a very complicated set of boundaries. However, sometimes it helps to be clever. The key issue is that the gravitational force is linear, so the force due to a group of masses is the sum of the forces due to each mass separately. In particular, the point where one must think well outside the box, the

25.8. SOLUTIONS TO THE WORKED PROBLEMS

force due to a mass $m = 0$ at a point is equal to the sum of the force due to two masses, $+M$ and $-M$ located at the same point. The forces due to $+M$ and $-M$ cancel and are the same in total as a mass 0 at the same point.

For the case here, we describe the problem as the sum of a spherical mass, radius R and density ρ, and two smaller spheres, radius $R/2$ and density $-\rho$. The masses of the large and small spheres are then

$$M_R = \rho \frac{4\pi R^3}{3} = \frac{4\pi \rho R^3}{3}, \tag{25.35}$$

$$M_{R/2} = -\rho \frac{4\pi (R/2)^3}{3} = -\frac{\pi \rho R^3}{6}. \tag{25.36}$$

$$\tag{25.37}$$

Since $\rho = 3M/(4\pi R^3)$,

$$M_{R/2} = -\frac{\pi R^3}{6} \frac{3M}{4\pi R^3} = -\frac{M}{8}. \tag{25.38}$$

The force of gravity due to a spherical mass, if you are outside the mass, is

$$\boldsymbol{F} = -\frac{GMm}{r^2} \hat{\boldsymbol{r}}. \tag{25.39}$$

We are interested in the force a distance d out along the x-axis. That force is

$$\boldsymbol{F} = -\frac{GMm}{d^2} \hat{\boldsymbol{i}} + \frac{G(M/8)m}{(d^2 + (R/2)^2)} \frac{d\hat{\boldsymbol{i}} - \frac{R}{2}\hat{\boldsymbol{k}}}{(d^2 + (R/2)^2)^{1/2}} + \frac{G(M/8)m}{(d^2 + (R/2)^2)} \frac{d\hat{\boldsymbol{i}} + \frac{R}{2}\hat{\boldsymbol{k}}}{(d^2 + (R/2)^2)^{1/2}}. \tag{25.40}$$

In the last two terms, the terminal fractions are the unit vectors pointing from the centers of the smaller spheres to the point of interest. That expression simplifies. The masses creating the gravitational force are placed symmetrically with respect to the x-axis, so the resulting force should be along the x-axis; forces in the y or z direction should sum to zero. The final result is

$$\boldsymbol{F} = -\hat{\boldsymbol{i}} \left(\frac{GMm}{d^2} - \frac{1}{4} \frac{GMmd}{(d^2 + (R/2)^2)^{3/2}} \right). \tag{25.41}$$

This space reserved for your notes.

This page reserved for your notes.

Chapter 26

Planetary Orbits

We now advance to the original motivation for Newton's Laws of Motion and Newton's Law of Gravity, the behavior of planets in their orbits around the Sun.

The orbit of the moon is in there someplace, too, but the theory of the orbit of the Moon with respect to the Sun and the Earth is much more complicated. For starters, neither the Moon nor the Earth is a sphere. The deviations of the Earth and the Moon from simple spherical forms are not small. The Earth is flattened at the poles. Because there are tides, the shape of the Earth is not entirely fixed. The Moon is shaped somewhat like an egg, stretched by the tides created by the Earth's gravity. While the Moon is tidally locked to the Earth, so that it always keeps more-or-less the same face toward the Earth, it in fact is subject to *libration*. It wobbles east-and-west, and north-and-south, so that with patience an observer on Earth can see around 59% of the Moon's surface. In addition, the Moon is subject to the force of gravity due to the Sun as well as the force of gravity due to the Earth. The former is as important as the latter, in the sense that the orbit of the Moon around the Sun is always concave toward the Sun. Most *orrery* models of the Earth-Moon system are for good reasons not to scale, so that the Lunar orbit appears part of the time to be concave away from the Sun. That concavity is a minor imperfection in the mechanical model, not an actual feature of the Lunar orbit.

We now consider two masses m_1 and m_2 in orbit around each other. Their locations relative to some origin are the vectors r_1 and r_2. In general, the gravitational force between them has the form

$$F = \frac{Gm_1 m_2}{r^2}\hat{r}, \qquad (26.1)$$

in which r is the distance between them, and in which the unit vector $\hat{r}$ points in the direction of the force. The distance between the two masses, r, is $r_{12} = |r_2 - r_1|$.

The Second Law equations for the two masses are

$$m_1 \frac{d^2 r_1}{dt^2} = f(r)\hat{r}_{12}, \qquad (26.2)$$

$$m_2 \frac{d^2 r_2}{dt^2} = -f(r)\hat{r}_{12}. \qquad (26.3)$$

We now introduce two new coordinates. The first is the center-of-mass coordinate R_{cm},

$$R_{\text{cm}} = \frac{m_1 r_1 + m_2 r_2}{m_1 + m_2}. \qquad (26.4)$$

The second is a separation vector r, namely

$$r = r_2 - r_1. \qquad (26.5)$$

Are these a legitimate pair of variables? R_{cm} and r are independent. You can chose any value you want for R_{cm}, and still be free to choose any value you want for r_{12}, and vice versa. We may usefully rewrite the equations of motion, equation 26.2, as

$$\frac{d^2\boldsymbol{r}_1}{dt^2} = \frac{1}{m_1}f(r)\hat{\boldsymbol{r}}, \tag{26.6}$$

$$\frac{d^2\boldsymbol{r}_2}{dt^2} = -\frac{1}{m_2}f(r)\hat{\boldsymbol{r}}. \tag{26.7}$$

where $f(r) = Gm_1m_2/r^2$.

Noting that

$$\frac{d^2\boldsymbol{r}_2}{dt^2} - \frac{d^2\boldsymbol{r}_1}{dt^2} = \frac{d^2\boldsymbol{r}}{dt^2}, \tag{26.8}$$

the difference between the two lines of equation 26.2 is

$$\frac{d^2\boldsymbol{r}}{dt^2} = -\left(\frac{1}{m_1} + \frac{1}{m_2}\right)f(r)\hat{\boldsymbol{r}}. \tag{26.9}$$

We now introduce the *reduced mass* μ, namely

$$\frac{1}{\mu} = \frac{1}{m_1} + \frac{1}{m_2}. \tag{26.10}$$

Substituting μ into equation 26.9, one finds

$$\mu\frac{d^2\boldsymbol{r}}{dt^2} = -\frac{Gm_1m_2}{r^2}\hat{\boldsymbol{r}} \tag{26.11}$$

for the equation of motion of the separation vector $\boldsymbol{r}$, where

$$\mu = \frac{m_1m_2}{m_1 + m_2}. \tag{26.12}$$

Equation 26.11 looks exactly like Newton's Second law, but it isn't. μ is not the mass of a particle. $\boldsymbol{r}$ is not the location of a particle. $\boldsymbol{r}$ is a vector linking two particles, both of which are accelerating. $\boldsymbol{r}$ is the location of particle 2 with respect to particle 1. However, m_1 is accelerating. If we take m_1 to be the origin, $\boldsymbol{r}$ is a coordinate in a non-inertial reference frame. In a non-inertial frame, the Second Law is invalid.

However, equation 26.11 has exactly the same mathematical structure as does the Second Law, so everything that is true about solutions to the Second Law is equally true about solutions to equation 26.11. That last statement is actually a very important tool in theoretical physics. If you can make two problems look like each other, so that they are described by the same mathematical equations, then everything known about the solution to each problem is equally applicable to the other problem, except of course that the symbols in the equations change their meanings.

In the case here, in equation 26.11 the force $\boldsymbol{F}$, the right-hand-side of the equation, is parallel to the line-of-centers vector $\hat{\boldsymbol{r}}$, so therefore $\hat{\boldsymbol{r}} \times \boldsymbol{F} = \boldsymbol{0}$. The mathematical object that looks like the angular momentum $\boldsymbol{L} = \boldsymbol{r} \times \boldsymbol{p}$ therefore does not depend on time. Furthermore, the only forces here are action-reaction pairs, so the total force on the two particles is zero, in which case $\boldsymbol{L}$ is equally independent of time in all reference frames. In particular, the angular momentum of a planet in orbit around the sun, as calculated using the Sun as the center around which the angular momentum is calculated, is a constant.

Some discussions introduce the concept of the centrifugal potential, which is actually a term in the total energy that arises from the kinetic energy. In the reference frame in which

$$\frac{d\boldsymbol{R}}{dt} = \boldsymbol{0}, \tag{26.13}$$

we may write a quantity that looks exactly like the total energy of the system, namely

$$E = U(r) + \frac{1}{2}\mu v^2. \tag{26.14}$$

The reduced mass is not the mass of a particle, and v is not a particle's velocity, so $\frac{1}{2}\mu v^2$ only looks like a kinetic energy. v^2 can be written in circular polar coordinates (r, θ) as

$$E = \frac{1}{2}\left((\frac{dr}{dt})^2 + (r\frac{d\theta}{dt})^2\right) + U(r). \tag{26.15}$$

In addition, the magnitude $|\boldsymbol{L}|$ of the angular momentum may equally well be written as $\mu r v_\theta$, $\mu r v_\perp$, or $\mu r^2 \frac{d\theta}{dt}$, but, however it is written, it is a constant. From $|\boldsymbol{L}| = mr^2 \frac{d\theta}{dt}$, one finds

$$\frac{L}{\mu r} = r\frac{d\theta}{dt}. \tag{26.16}$$

Substituting equation 26.16 into equation 26.15, we have

$$E = \frac{1}{2}\mu \left(\frac{dr}{dt}\right)^2 + \frac{\mu}{2}\frac{L^2}{\mu^2 R^2} + U(r). \tag{26.17}$$

The term $\frac{\mu}{2}\frac{L^2}{\mu^2 R^2}$ looks like an R^{-2} potential energy. It isn't a potential energy, because it is part of the nominal kinetic energy, but it is still called the *centrifugal potential*. Keep in mind that we are in the coordinate system defined by equation 26.11, which is not an inertial coordinate system, so the quantities that appear to be the kinetic and potential anergies are actually their analogs in these coordinates.

How are the actual masses and the reduced mass related? For the Earth-Sun system, $M_s \gg M_e$, so

$$\mu = \frac{M_s M_e}{M_s + M_e} \approx M_e. \tag{26.18}$$

If one body is much larger than the other, the reduced mass is very nearly equal to the mass of the lighter body.

What is the period of a circular planetary orbit? For a circular orbit, energy conservation says that the speed of the planet as it orbits around the Sun is a constant. The planet's acceleration is the acceleration for circular motion at constant speed, namely $v^2 r^{-1} \hat{\boldsymbol{r}}$, so the Second Law for a planet in a circular orbit under the influence of gravity may be written

$$-\frac{GMm}{r^2}\hat{\boldsymbol{r}} = -\frac{mv^2}{r}\hat{\boldsymbol{r}}. \tag{26.19}$$

From this equation we find $v^2 = GM/r$ or $v = (GM/r)^{1/2}$. The orbital period is

$$T = \frac{2\pi r}{v}, \tag{26.20}$$

which leads to

$$T = 2\pi r \left(\frac{r}{GM}\right)^{1/2} \tag{26.21}$$

or

$$T = 2\pi \left(\frac{r^3}{GM}\right)^{1/2}. \tag{26.22}$$

The predicted relationship between the period and the orbital radius is

$$T \sim r^{3/2}. \tag{26.23}$$

This result, written as $T^2 \sim r^3$, was first found by Kepler.

Finally, consider a planet in a circular or elliptical orbit around the sun. The angular momentum in the reference frame centered on the Sun is a constant. Now consider the area swept out by a planet as it orbits around the Sun. By 'swept out', we refer to the area defined during a time interval $(t, t + \delta t)$ by two lines, drawn from the Sun to the locations of the planet at times t and $t + \delta t$, and by the orbital path. Several areas swept out by a planet are shown in Figure 26.1.

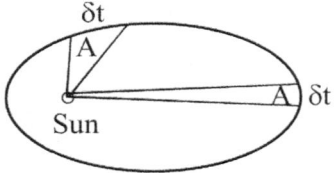

Figure 26.1: The Kepler equal-area law. In equal times δt, a planet sweeps out equal areas A.

If δt is very short, the orbital path is very nearly a straight line of length $v\,\delta t$, and the two lines from the Sun to the planetary positions have very nearly equal lengths r. The swept-out area is then a triangle having area approximately proportional to $0.5rv\,\delta t$. However, rv is proportional to the angular momentum L, the proportionality constant including a factor of m_e. The area of the triangle is therefore proportional to $L\,\delta t$. L is independent of time, so the swept-out area simply increases linearly with time. Regardless of where the planet is in its orbit, the area A that the planet sweeps out in a given time δt is the same. This result was first found by Kepler, and is his equal-area rule of planetary orbits.

26.1 Discussion

I repeat what was said earlier. Equation 26.11 has exactly the same mathematical structure as does the Second Law, so everything that is true about solutions to the Second Law is equally true about solutions to equation 26.11. That last statement is actually a very important tool in theoretical physics. If you can make two problems look like each other, so that they are described by the same mathematical equations, then everything known about the solution to each problem is equally applicable to the other problem, except of course that the symbols in the equations change their meanings.

26.2 Worked Problem

1. The Black Star Passes (literary reference). One of the cosmic catastrophes that could render the earth uninhabitable is the passage through the solar system of another star, a body that would significantly perturb the earth's orbit so that we froze or fried to death. For the sake of hypothesis, local astronomers have recently detected a star traveling at speed V whose straight-line trajectory would take it a distance $r_\perp$ from the sun. Alas, the Black Star's trajectory will be perturbed by the sun's gravity, so that it will actually pass a distance r_m from the sun. Noting that the gravitational potential is GMm/r where M is the solar mass, m is the black star mass, G is the gravitational constant, and r is the distance between the sun and the Black Star, find r_m in terms of the other variables. Note Figure 26.2.

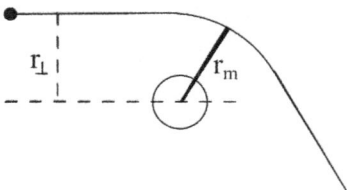

Figure 26.2: The Black Star passes. The figure shows its hyperbolic orbit as it passes by our Sun. There is currently an expectation that this will happen in 300,000 years, the star of interest having a distance of closest approach of 1/7 of a light year. The approach will probably disrupt the Kuiper belt, where comets lie sleeping in the eternal dark, but will not affect our orbit around the sun significantly.

26.3 Homework

1. A planet is held together by its gravity, not by the mechanical strength of its rock. The French astronomer Roche proved that there is a smallest orbit that a small planet can have around a large one before the small planet is torn apart by the variation in the large planet's gravity across the thickness of the small planet. Demonstrate the conditions under which his claim is correct.

26.4 Solution to the Worked Problem

1. Let us start with what we know. Angular momentum is conserved during the passage, because gravity is a central force, so
$$mVr_\perp = mv_m r_m. \tag{26.24}$$

There is no friction, so energy is conserved. The energy in this problem is
$$E = \frac{1}{2}mv^2 + \frac{GMm}{r}. \tag{26.25}$$

For the specific problem here we start out with $v = V$ and $r = \infty$ and end up at $v = v_m$ and $r = r_m$, so energy conservation becomes
$$\frac{1}{2}mV^2 + \frac{GMm}{\infty} = \frac{1}{2}mv_m^2 + \frac{GMm}{r_m}. \tag{26.26}$$

Stop! Can we possibly solve? We have two independent equations, and two unknowns, v_m and r_m, so a solution is plausibly possible.

From angular momentum conservation, we can eliminate v_m, in which we are not interested, via
$$v_m = \frac{Vr_\perp}{r_m}. \tag{26.27}$$

The m cancels out of the energy conservation equation, giving us
$$\frac{V^2}{2} - \frac{V^2 r_\perp^2}{2r_m^2} + \frac{GM}{r_m} = 0, \tag{26.28}$$

which can be simplified to
$$V^2 r_m^2 - V^2 r_\perp^2 + 2r_m GM = 0. \tag{26.29}$$

This equation is a quadratic in r_m, whose solutions are
$$r_m = \frac{-2GM \pm (4G^2M^2 + 4V^4 r_\perp^2)^{1/2}}{2V^2}. \tag{26.30}$$

The positive root is the physical answer.

This space reserved for your notes.

This page reserved for your notes.

Chapter 27

Examination 3

Third Hour Examination

This is the third hour examination. The examination lasts for 50 minutes. There are four questions, for a total of 100 points, and a bonus question. This is a closed book examination. You may not use books, notes, or reference materials during this examination. You may use a pocket calculator to perform numerical calculations.

Record your answers. Begin from first principles, e.g.,

$$F = m\frac{d^2 r}{dt^2} \tag{27.1}$$

Show all work. You are primarily graded on how you reached your answers, though in some cases there is not a lot of work to reach the correct answer. If you do not show your work, you will in general receive little or no credit for your answer.

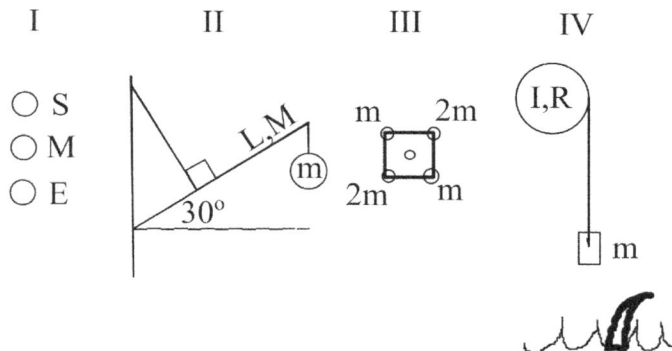

Figure 27.1: Figures for the Examination Problems.

I [Total of 25 points]

The masses of the Moon, the Earth, and the Sun are $7.35 \cdot 10^{22}$ kg, $5.97 \cdot 10^{24}$ kg, and $1.99 \cdot 10^{30}$ kg, respectively. The distances from the Earth to the Moon and from the Moon to the Sun are $3.85 \cdot 10^8$ m and $1.50 \cdot 10^{11}$ m, respectively. The constant G is $6.67 \cdot 10^{-11}$ in SI units. The Earth, Moon, and Sun lie on the y-axis, in the order Sun-Moon-Earth.

A) (9 points) Find the force on the Moon due to the Sun.

B) (9 points) Find the force on the Moon due to the Earth.

C) (7 points) Find the total gravitational force on the Moon due to the Earth and the Sun.

D) (2 points) For question I, which is stronger, the force on the Moon due to the Sun or the force on the Moon due to the Earth? Based on your answer, is the Moon primarily in orbit around the Sun or around the Earth?

II [Total of 25 points]

The owners of the coffee house The Sun And The Moon, seeing that their competitors have put up a new and larger sign, decide it is time to put up a newer and larger sign themselves. The new sign is a pole of length L and mass M, hanging from the end of which is a lantern having mass m. The poll is inclined upward at 30 degrees from the horizontal. A guy wire is attached to the pole a distance $L/3$ from the wall. The guy wire is perpendicular to the pole.

A) (10 points) Find the tension in the wire.

B) (15 points) Find the horizontal and vertical components of the force the flagpole puts on the wall.

III [Total of 25 points]

Mounted on the four corners of a very light $L \times L$ square block of wood are four heavy masses, mass m, $2m$, m, and $2m$. The Figure shows the arrangement of the four masses on the corners. A shaft that is perpendicular to the page goes through the very center of the square block; the block cannot rotate with respect to the shaft. The block is in the horizontal $x - y$ plane; the shaft lies perpendicular to the block and parallel to the z-axis.

A) (15 points) A torque having z-component $\tau_z = T_0(1 - \exp(-at))$ is applied to the shaft. Find the angular acceleration of the shaft and block.

B) (10 points) The block is stationary until $t = 0$, at which time the torque is applied. The torque follows $\tau_z = T_0(1 - \exp(-at))$ at times $t \geq 0$. Find the angular velocity ω of the block as a function of time.

IV [Total of 25 points]

In another terminal scene of the motion picture, the evil Captain Goldbug, mass m, is being winched down a cliff to his yacht, the *Gold Brick*. At the top, the rope holding him up is wound around a heavy circular drum having radius R and moment of inertia I. When Goldbug is a distance L above the water, the gearing driving the drum snaps off – solid gold is still an inopportune structural material – and the drum is free to rotate. The only torque on the drum, with respect to its line of rotation, is supplied by the rope holding up Goldbug. The drum turns more and more rapidly, unwinding the rope wrapped around it and plunging Goldbug toward his waiting, always-hungry, pet great white shark, whose fin is visible in the Figure.

A) (12 points) Find the force diagram for Goldbug and compute his acceleration as he heads toward the water. What is the angular acceleration of the drum as Goldbug falls?

B) (13 points) Using conservation of energy or the work-energy theorem, calculate how fast Goldbug is going when he hits the water.

Extra Credit (10 points)

Okay, what is *praxis*, a term that has been used in the directions to all three exams?

27.1 Examination 3 Solutions

I. The gravitational force that mass m_1 exerts on mass m_2 is

$$\boldsymbol{F}_{12} = -\frac{Gm_1m_2}{r_{12}^2}\hat{\boldsymbol{r_{12}}},$$

where r_{12} is the distance between the two masses and $\hat{\boldsymbol{r_{12}}}$ is the unit vector pointing from m_1 to m_2.

The force of the Sun on the Moon is therefore

$$\boldsymbol{F}_{SM} = -\frac{6.67 \cdot 10^{-11} \cdot 1.99 \cdot 10^{30} \cdot 7.35 \cdot 10^{22}}{(1.5 \cdot 10^{11})^2}\hat{\boldsymbol{j}},$$

while the force of the Earth on the moon is therefore

$$\boldsymbol{F}_{SM} = \frac{6.67 \cdot 10^{-11} \cdot 5.97 \cdot 10^{24} \cdot 7.35 \cdot 10^{22}}{(3.85 \cdot 10^{8})^2}\hat{\boldsymbol{j}},$$

so the total force is

$$\boldsymbol{F}_{M} = -\frac{6.67 \cdot 10^{-11} \cdot 1.99 \cdot 10^{30} \cdot 7.35 \cdot 10^{22}}{(1.5 \cdot 10^{11})^2}\hat{\boldsymbol{j}} + \frac{6.67 \cdot 10^{-11} \cdot 5.97 \cdot 10^{24} \cdot 7.35 \cdot 10^{22}}{(3.85 \cdot 10^{8})^2}\hat{\boldsymbol{j}}.$$

Note the use of vector addition. The Sun exerts a larger force on the moon than the Earth does, and, correspondingly, the Moon's orbit is always concave inward toward the Sun, a detail not obvious in some schoolbook drawings and most orreries. The Moon is thus primarily in orbit around the Sun.

II. Note that the hinge is applying to the pole a force H_x in the x-direction and a force H_z in the z-direction. We do not know if either of these is zero; the calculation will tell us that. For the torque diagram, we need to choose an origin. We could put it anywhere. We choose to put it at the hinge, because with that choice of origin two of the unknowns fall out of the torque equation. Recalling that this is a statics problem, so that the linear and angular accelerations are all zero, we can write for the force equations

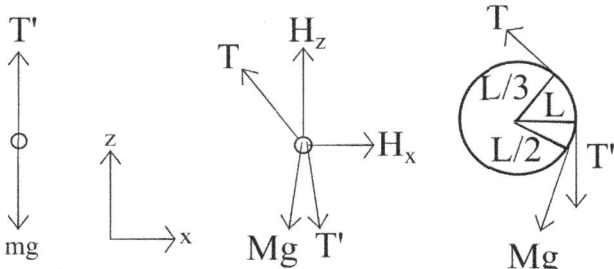

Figure 27.2: Force and torque diagrams for problem II.

$$H_z - mg - Mg + T\cos(\theta) = 0, \tag{27.2}$$
$$H_x - T\sin(\theta) = 0, \tag{27.3}$$
$$T' - mg = 0. \tag{27.4}$$

Here T' is the tension in the wire holding up the lamp. You might have skipped that last equation. For the torques around the hinge,

$$\frac{TL}{3} - \frac{MgL}{2}\cos(\theta) - T'L\cos(\theta) = 0.$$

The final equation immediately tells us that the tension in the wire is

$$T = 3\left(\frac{Mg}{2} + mg\right)\cos(\theta).$$

The forces on the wall are the reaction forces to the forces on the pole; they are $-H_x$ and $-H_z$. From the force equations, these are

$$-H_x = -3(\frac{Mg}{2} + mg)\cos(\theta)\sin(\theta), \tag{27.5}$$

$$-H_z = -mg - Mg + 3(\frac{Mg}{2} + mg)(\cos(\theta))^2. \tag{27.6}$$

III. (A) Pause: Is the square symmetric around the shaft so that $\boldsymbol{L}\|\boldsymbol{\omega}$? Yes, so we only need to think about the z components of the torque and angular momentum. We know for the z-component $\tau_z = I\frac{d^2\theta}{dt^2}$. We begin by calculating the moment of inertia of the square around the block. We have

$$I = \sum_i m_i r_i^2, \tag{27.7}$$

$$I = 2m(\frac{\sqrt{2}L}{2})^2 + 2 \cdot 2m(\frac{\sqrt{2}L}{2})^2, \tag{27.8}$$

$$I = 3mL^2. \tag{27.9}$$

Substituting for I and the torque

$$T_o(1 - \exp(-at)) = 3mL^2\frac{d^2\theta}{dt^2},$$

so

$$\frac{d^2\theta}{dt^2} = \frac{T_o(1 - \exp(-at))}{3mL^2}.$$

(B) For the angular velocity

$$\omega(t) = \int dt \frac{T_o}{3mL^2}(1 - \exp(-at)),$$

and therefore

$$\omega(t) = \frac{T_o}{3mL^2}(t + a^{-1}\exp(-at)) + \omega_0.$$

We know that $\omega(t) = 0$ at $t = 0$, which determines ω_0:

$$0 = \frac{T_o}{3maL^2} + \omega_0,$$

so

$$\omega_0 = -\frac{T_o}{3maL^2},$$

and

$$\omega(t) = \frac{T_o}{3mL^2}(t + a^{-1}\exp(-at)) - \frac{T_o}{3maL^2}.$$

IV. (A) Once again, Captain Goldbug is all wet. We begin with the force and torque diagrams. These give us a Second Law equation and a torque equation:

$$T - Mg = M\frac{d^2z}{dt^2}, \tag{27.10}$$

$$-TR = I\frac{d^2\theta}{dt^2}. \tag{27.11}$$

In the torque equation, we chose θ to increase in the counterclockwise direction. That makes the constraint

$$\frac{d^2z}{dt^2} = +R\frac{d^2\theta}{dt^2},$$

which lets us rewrite the torque equation as

$$-T = \frac{I}{R}\frac{1}{R}\frac{d^2z}{dt^2}.$$

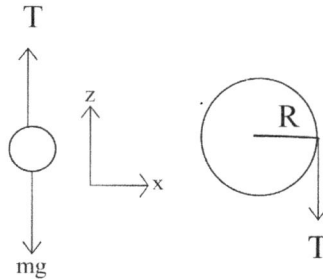

Figure 27.3: Force and torque diagrams for problem IV.

In the Second Law equation, take T to the right-hand-side of the equation and substitute

$$-Mg = (M + \frac{I}{R^2})\frac{d^2z}{dt^2},$$

so therefore

$$\frac{d^2z}{dt^2} = -\frac{MR^2g}{MR^2 + I}.$$

(B) We apply conservation of energy.

$$E = \frac{1}{2}Mv^2 + \frac{1}{2}I\omega^2 + Mgh.$$

We have the initial conditions $v = 0$, $\omega = 0$, and $h = L$. Why did I not write $v = \omega = 0$? v and ω have different dimensions and therefore cannot be equal. At the end, $v = V$, $\omega = V/R$ and $h = 0$. Energy conservation gives us

$$\frac{1}{2}M0^2 + \frac{1}{2}I0^2 + MgL = \frac{1}{2}MV^2 + \frac{1}{2}I(V/R)^2 + Mg0,$$

where I am displaying the zero terms so it is completely transparent what is happening. Simplifying,

$$2MgL = (M + \frac{I}{R^2})V^2,$$

and finally

$$V = \left(\frac{2MgL}{(M + \frac{I}{R^2})}\right)^{1/2}.$$

This space reserved for your notes.

This page reserved for your notes.

Part Two

Experimental Sequence

This page reserved for your notes.

Chapter 28

A Sequence of Experiments

In Part I of the course, we present a theoretical treatment of classical mechanics. Physics theory was the work of a number of great physicists, included Isaac Newton, James Clerk Maxwell, Josiah Willard Gibbs, Albert Einstein, Erwin Schroedinger, Werner Karl Heisenberg, Paul Adrien Maurice Dirac, and moderns to numerous to mention. There is also experiment, in which we study nature to see how it behaves. We've already mentioned Tycho Brahe, Chien-Shiung Wu, and Rainer Weiss, but there are a vast number of other people who did important experiments.

All these theorists and experimenters when successful generate new results. Newton presented his results, his Laws of Motion, by writing a book, but under modern circumstances results are reported to the scientific world in the form of short papers (typically 4-20 pages) in scientific journals. There are then a very few people whose specialty is reading all those papers, summarizing them, and presenting them to the scientific world in the form of review articles (indeed, there is a journal, *Reviews of Modern Physics*, that only publishes these review articles). As we have mentioned great theorists, and great experimentalists, it would only do to mention a great reviewer, Virginia Louise Trimble, whose reviews on the Solar neutrino problem forced recognition that there was indeed a problem.

The purpose of this experimental sequence is to give you some experience in doing real experiments. A real experiment is an experiment in which you do not know the answer in advance. Real experimental work has important parts that are not often discussed in physics courses. These experiments will introduce you to some of those parts. I therefore do not give detailed instructions as to exactly how to do your experiments. Indeed, an important part of experimental work is to realize that there may well be several different methods of measuring the same quantity, and that experiment may be needed in order to determine which experimental approach gives the best results. Real experiments have errors, some of whose sizes can be determined experimentally.

These experiments are very much not like the experiments you will encounter in some freshman physics courses, where the objective is to confirm with experiment that I was telling the truth about, e.g., the period of a pendulum, and if your experiment fails to confirm theory, you are required to go back to the lab and do the experiment again and again until it confirms the theory. (No, I did not make that up.) The experiments here are meant to show you, with very little equipment and some basic mechanics, how you would go about designing an experiment. Yes, the results are known in advance, but that's not the point of what you are doing.

You may have been told about a "scientific method" in which the sole purpose of an experiment is to test a hypothesis. In this method, before you do an experiment, you are required to make a guess (a "hypothesis") as to the correct answer. The objective of the experiment is to determine if your guess is true or false. If the experiment disagrees with your guess, your guess is wrong and must be rejected. That approach to science is not always correct. You may find it interesting to read a history of the solar neutrino problem, in which measurements of solar neutrino emissions were made, two-thirds of the neutrinos were missing, and the neutrino instrumentation, solar modeling, and nuclear physics people could each think the fault was not theirs.

Much real science does not resemble this supposed 'scientific method'. For example, when the United States set out to sequence the human genome, it was already extremely well established that *homo sapiens* had chromosomes with DNA sequences. In principle, you could have guessed the DNA sequence, and then done experiments to see if your guess was right. However, the number of possible human DNA sequences is somewhat large, say four to the 3.2 billion power, so the likelihood that any guess would be correct is exceedingly small. What was actually done was exploratory. We did not know what was there, so we set out to measure it. That's real science.

These laboratories will require a very modest amount of equipment. You are going to study a conventional pendulum, a physical pendulum, and an Atwood machine. You will also use your pendulum make a quantitative test of Newton's Three Laws for a special case.

A reasonable school laboratory will supply the equipment you need. If you are doing self-study or home schooling, you will need to create your own apparatus, just like many real scientists do. You will need a precision timer and a precision scale, but these are now incredibly inexpensive.

For the conventional pendulum, you need some strong thread and some heavy nuts from a set of nuts and bolts. The nuts will serve as the pendulum bob, the heavy mass at the bottom of the pendulum. The pendulums need to hang from something. For the conventional pendulum, a traditional chemical laboratory support stand is a good choice. The longer the support rod is, so long as it is solidly braced, the better the experiment will work. A support clamp lets you hang the pendulum a distance out from the vertical support rod, so that the pendulum does not collide with its vertical support. Some years ago, some of my cleverer students taped a plastic ruler – the sort with holes in it – to the top of a door. Some inches of the ruler stuck out beyond the end of the door. The pendulum was hung from the ruler, giving a pendulum with more-than-six-feet of swing. For the pendulums, the bottom end of the pendulum bob needs to clear the table top or floor by a small distance. The length of the string needs to be varied over a considerable range. Some ingenuity is needed here. A vertical piece of pipe, rigidly mounted, and some clamps will work. Some children's construction toys will solve the problem. Finally, at one point in these experiments a laser pointer will be convenient.

For the physical pendulum, you will need a wooden yard or meter stick, a wood drill, and some reasonably large bolts with their nuts. Many yard and meter sticks have a hanger hole, typically brass-lined, close to one end. You will also need to drill some holes in your meter stick, carefully centered relative to the width of the stick, through which you will during the experiment place some heavy nuts and bolts. A thin nail serves as a good, relatively frictionless hanger for the meter stick. Where you put the nail may require some creativity to solve.

To build a good Atwood machine you need a low-friction arrangement for the top pulley or pulleys, some thread, and some reasonably heavy weights.

Several measuring instruments are needed. To weigh pendulum bobs, a scale is required. Electronic scales good for up to 300 or 500 g with a precision of 0.01 g are now remarkably cheap, much cheaper than the older triple-beam balances. The sort of balance used to weigh jewelry, that will weigh 300 or 500 g with a precision of 0.001 g, 1 mg, is readily found on the internet. You will need a digital stopwatch. These are readily available with an accuracy of 0.01 s, but for almost the same price you can probably find a stopwatch that reports with an accuracy of 0.001 s. The more accurate watch is preferred. If someone tells you that you do not need the extra accuracy, because human reflexes are not that accurate, they are missing the point of the experiment.

[Minor aside: Particularly if you are doing this experiment in a university setting, you might very well be using a triple beam balance to measure the weight of your nuts and bolts. There is a right way to make a measurement with a triple beam balance, and a completely wrong way. A remarkable number of people will tell you to use the completely wrong method. In using a triple beam balance, you adjust the weights on the balance until they are correct. However, you have two basic choices. The balance may be swinging up and down, or it may be parked and level. In the correct approach, the balance is swinging up and down. When it is balanced, the swings have the same height up as down. If the balance is just sitting there, the static friction between the knife edge of the balance and its arm may well hold the balance in position so that it appears to be properly balanced, even though it isn't. In that case, the measured weight will be incorrect. The *parked and level* method is **wrong**.]

Finally, you will need some arrangement for recording your work. If you at school, you will be told how to do this. Otherwise, the reasonable choices are: a wire-bound large notebook, three-hole paper and a ring binder, or if all else fails a computer. The computer is dangerous. It is much easier to erase a computer file than to lose a notebook. When I taught the course, students were issued two-part carbonless paper. At the end of the lab, the student kept a copy and the TA kept a copy.

The virtue of the wire-bound notebooks is that it makes it very difficult to lose any pages...until you lose the entire notebook. Some schools, however, are very fond of this approach to recording data. If you are working on several things, with a wirebound notebook it is impossible to reorder pages to keep together results on the same topic. With a ring binder, rearranging pages is not an issue. Inserting large graphs in a wirebound notebook may also be challenging.

No matter where the page is to lurk, in a permanent notebook or a ring binder, the format of the page starts out the same. At the top, right hand corner is traditional, you place the date. Then each of the people using the page initials it. When you are done with the page, you put a line below your last written remarks, and then you initial below the line.

[Aside: In some fields of science, people are extremely picky about using either ring binders or permanently bound recording books. Other fields want the data to be recorded electronically in a computer. As you advance from this course, you will simply have to find out which method is expected in each laboratory cycle and accommodate to what you are told to do.]

To be honest: Some parts of these labs are a bit tedious. To determine statistical properties of your measurements, you need to repeat the same measurement many times. In Laboratory 2, do you really need to test all those different ways of measuring the period of a pendulum? After all, I could have told you exactly how to measure the period. Yes, the tests are needed. You are seeing real science in action, where a major part of the experiment is tuning the apparatus until it works as well as possible. With real science, you can spend weeks or months or years getting the experiment to work. Finally, you get to the point where you are taking real data. The actual data may then show up very quickly. In the experiments here, pendulums are fairly foolproof. Building a decent Atwood machine may be more challenging.

An important part of scientific experiments is determining the accuracy with which your measurements have been made. Some measurements are very crude. Some measurements are extremely accurate. You will do several straightforward experiments that demonstrate how to measure accuracy, leading to discussions of random error and statistical analysis. We then advance to experimental design. You will study different methods for measuring the period of a pendulum. You will do some physics experiments: you will study how the period of a pendulum depends on the pendulum's properties. Finally, you will try to duplicate two experiments that test Newton's Laws, namely the Atwood machine and the force balance.

This space reserved for your notes.

This page reserved for your notes.

Chapter 29

Experiment One: Measurements are Imprecise

This laboratory introduces *data analysis*. If you do measurements, writing down the numbers you obtain is only the front end of finding out what they mean. You then have to analyze your measurements, understand your results, and determine the experimental error in your results. After all, a primary difference between modern Western natural philosophy – the quantitative sciences – and competitors is that not only can you predict answers, but you can predict a minimum level of how inaccurate your answers are. Finally, you write up a report on what you found.

29.1 Experimental

In this experiment, we are going to measure the *period of a pendulum*. Here the pendulum will be some nuts tied together, hanging from a length of string. What do I mean by period? The period of the pendulum is the time it takes for the weight at the end of the pendulum, the pendulum bob, to swing back-and-forth and return to its initial position and be going in the original direction. It is often claimed that it does not matter from where you measure the period of the pendulum. You could measure the period of the pendulum starting at one end of the swing or the other, you could measure the period of the pendulum starting at the bottom of the swing, or you could measure the period of the pendulum from an arbitrary point in the middle of its swing. So long as the pendulum has swung back-and-forth, coming back to its starting point, and is going in the direction that it was going originally, the period of the pendulum should be the same. Does it matter from what point you measure the pendulum swing? The next experiment tests this claim.

Set up a pendulum that uses string and several nuts. The pendulum bob should be able to swing freely. It should clear the table top, but not by very much. You need to measure the length of the string, from its upper attachment point to the center of the bob. In this experiment, you are going to use some method, a method that you get to choose, to measure the period of the pendulum for four lengths of string.

You are now ready to start recording data in your *notebook*. At the top of a page, write the date, your initials, and a title. "Experiment 1" is a title. Draw a sketch of the apparatus, inserting numbers for measured quantities such as the length of the string and the mass of the pendulum bob. In a few words record the purpose of the experiment.

Choose the point at which the first swing of the pendulum starts. Move the pendulum bob off to the side and release it. When the pendulum passes the starting point, start your timer. After a moderate number of back-and-forth swings, say five or ten, stop the timer. You have measured the duration of the run, how long was needed for five or ten swings. Divide the duration of the run by the number of swings, and you have a number for the period of the pendulum. I used the word "run", the "duration of the run". We conventionally say that we run an experiment, a run being a single experimental trial. Keep as much precision as your timer will allow in your time measurements, even if you think that your measurements are unlikely to be that accurate. Keeping all precision is especially useful if you are fortunate enough to have a timer that reads to 1 millisecond. Now repeat the same experiment, using the same timing point, keeping the starting point of the pendulum swing as close as possible to the same. Measure the period of the pendulum ten times. That may seem like a lot, but we are going to need repeated measurements of the period of the pendulum in order to do data analysis.

There are any number of ways to record the periods that you measure. In general, in this sort of experiment, measurements are usefully recorded in a *Table*. A good table begins with "Table N" and a caption, a short paragraph specifying what was done. It then has a series of columns, one for each quantity being recorded, and a series of rows, one for each time you did the experiment. The first column is reasonably an integer 1, 2, 3, ... identifying the particular experiment. Further columns record the mass of the pendulum bob, the length of the string, the start and stop times on the timer, the time interval between start and stop, the number of swings, and finally the (calculated) time for a single swing. The first row contains symbols for each of the quantities, e.g., $M(g)$, $L(cm)$, etc, so that you can tell what the numbers in each column mean. The algebraic symbols L, M, etc. identify the quantity. The symbols in parentheses identify the units of the measurement, here g (grams) and cm (centimeters). In general, the units for the quantities are recorded only once, in the first row, unless different entries in a column are in different units. For example, a series reporting 'mass of zoo animal' might have '7 tons' and '37 grams' as entries, the units being given with the number. The caption for the table identifies the symbols being used, e.g., "L is the length of the string. M is the mass of the pendulum bob." Interpretations of the measurements, and qualitative descriptive remarks, should be in the text, not the table's caption. Rows should be labelled, sufficiently so that you can go back from your report to your data notebook and identify which set of measurements in the notebook corresponds to which line in your table.

Now repeat the whole cycle for another three lengths of the string. Record your measurements in a table, making ten determinations of the period of the pendulum for each length of string. When you change the length of the string, change the height of the top end, so that the pendulum bob still passes just just above the tabletop when the pendulum is at the bottom of swing. You have now accumulated forty measurements of the period of the pendulum for four lengths of the string.

We now advance to analyzing the measurements you've just made.

29.2 Data Analysis

You have a whole pile of numbers. What do you do with them? You could just use them as your lab report; that's what Lord Rutherford described as stamp collecting. He did not mean the description favorably. Here you will do better than that. The objective in the above section was to generate an extended series of time measurements. Now you would like to determine how accurate your measurements are, and what the likely amount of error in your measurements is. In order to go farther, something must be said about the nature of error.

There are at least four sorts of *error* in an experimental measurement There are fundamental conceptual errors, there are systematic errors, there are quantum mechanical errors, and there are random errors. 'Error' does not mean you made a mistake. It means that measurements are not perfectly accurate. 'Error' means that when you repeat an experiment several times, you will get slightly different answers each time.

A *fundamental conceptual error* is a mistake you make because you simply do not understand what is going on, and make a decision based on your lack of understanding. For example, you might note that your experiments were made at different times of day, and therefore if you cast the horoscope for each measurement you will be able to infer from astrological analysis which measurements are the most accurate. However, your analysis suffers from a fundamental conceptual error, namely that you thought astrology had some merit as a predictive tool.

Catching fundamental conceptual errors is extremely difficult. In my own professional research area, there is a fundamental conceptual error that has led to the publication of large numbers of wrong experiments based on erroneous data analysis. Another conceptual error has led to large numbers of experiments being interpreted in a way that visibly does not agree with the actual data.

You may notice that I did not tell you how to find fundamental conceptual errors. There are several choices here: You are most likely to find things if you have an extremely deep understanding of what is going on. If a book or a journal article presents you with a theoretical result, you check the calculations for yourself before accepting them. Sometimes reading a large number of different sources, tracing back through

29.2. DATA ANALYSIS

the literature, and asking yourself 'where did that assumption come from?', will help. It is sometimes useful to take an experiment and run the equipment well outside the planned range of parameters, hopefully without blowing anything up, to see what happens. Multiple approaches to measuring the same quantity are good, but may be challenging to arrange. In the end, being extremely observant and paying attention to fine detail is a useful tool for finding out what is actually going on.

Systematic errors enter because, e.g., something is wrong with the experiment. For example, if you weigh yourself on a bathroom scale, always while holding a brick in each hand, your weight will always come out high. That's a systematic error.

There can be a scatter in your measurements due to *quantum mechanics*. (This scatter will not arise here.) The quantum mechanical scatter in measurements appears because quantum mechanical processes are of their nature random. For example, suppose I have a small block of uranium, and a detector that reports that in the block, on the average, a thousand uranium atoms a minute decay and emit an alpha particle. If I look at the actual number of decays observed, minute after minute, I do not see exactly one thousand decays occurring in each minute. Instead, I see 973 decays in one minute, 1014 decays in the next, and so on. On the average, I see 1000 decays, but from minute to minute I see different numbers. The deviation from 1000 decays in one minute is completely independent of the deviation from 1000 counts in the next minute. Furthermore, if I consider a single uranium atom, I have no way to predict whether or not that atom will decay today, tomorrow, or in the far future.

Finally, *random errors* are represent the scatter in the results from measurement to measurement, in an experiment that is working completely correctly. Random error appears because measurements are not completely precise. In measuring a time interval by clicking a stopwatch or pressing a key on your computer, there will be a fluctuation from measurement to measurement, because human muscle responses fluctuate around an average time delay. If you time the pendulum swing five times, and have a very good timer, you will discover that each time you perform the experiment you will get a slightly different answer for the pendulum's period of oscillation. That's random error.

There are several paths to *reducing the random error* in a measurement. You can try to improve your measurement method. That's in the next laboratory. There are methods for calculating how various errors will affect your answer. You can use these methods, which we will not reach in this text, to identify which errors in your experiment are the most dangerous, and focus on reducing those, even if other errors become a bit larger as a result. If you have a choice of methods for making a measurement, you can think about the procedure, and try to decide from introspection which method is the best. That approach is dangerous. Many people, given the list found in the next chapter of alternative methods for measuring a pendulum's period, will choose the wrong answer as being the best. There is a method that many people think looks the best, there is a method that for most people is the best, and for most people the two methods are not the same.

So how do we measure the random error? It's considerably more difficult to find systematic errors, if you don't know how accurate your measurements might be. We therefore start by looking at the random errors. The first step is to collect all of your measurements made under nominally identical conditions. We can call an individual measurement t_1. Your collection of measurements all made with the same length of string can be viewed as a list $(t_1, t_2, \ldots t_N)$. Here N is the total number of measurements that you made under the same conditions, i.e., with the same length of string.

A reasonable first step is to *average your measurements*. You add them all up and divide by N. That average is T, calculated as

$$T = \frac{1}{N} \sum_{i=1}^{N} t_i. \tag{29.1}$$

Calculate T for each length of string that you used, and include those values of T in a second table for your lab report. For this experiment, this second table should have a column whose entries are the string lengths. The table should have a another column for the corresponding values of the average times T.

You now have a list of string lengths and the corresponding average pendulum periods. For this to be a reasonable scientific report, you need to be able to say how accurate, in some sense, your measurements of T are. Your actual measurements were recorded in the first table. The actual measurements probably do not all agree exactly. The spread in the individual measurements around the average is an estimate of the random error. How do you characterize the random error's size?

The random errors are the values of $t_i - T$, the difference between each measurement and the average of all the measurements. A simple answer would be simply to average the individual errors $t_i - T$. You may see immediately why this won't work. Whether you do or not, for one of your string lengths calculate the average of the $t_i - T$. If you did the average correctly, you should've obtained zero as the final result. Some of the individual errors were positive, some of the individual errors were negative, and when you add up all the positive and negative errors, they cancel exactly. A little algebra will confirm that, indeed, the average of the $t_i - T$ must be exactly zero. If you did not get zero for an answer, go back and check if you remembered that $t_i - T$ has a sign, plus or minus. Then check your arithmetic, because the average of the $t_i - T$ is guaranteed to be zero.

A reasonable next step is to propose that instead of averaging the $t_i - T$ you should average the magnitudes $|t_i - T|$. That approach actually does work. However, as you will see when you study statistics, in many cases it turns out to be much better to calculate the random error by averaging the quantities $(t_i - T)^2$, and then take the square root of this new average. That is, I am saying you should calculate

$$(\langle (t_i - T)^2 \rangle)^{1/2} = \left(\frac{1}{N} \sum_{i=1}^{N} (t_i - T)^2 \right)^{1/2} \tag{29.2}$$

The brackets $\langle \cdots \rangle$ denote an average of the quantity $\cdots$ between them. The quantity calculated here is called the *root-mean-square* or RMS error.

For each of your sets of measurements, calculate the root-mean-square error in measuring the time, and add it to your second table as a third column, next to the average period. If you look at your table, you should find that a single row gives a length of the string, the average period for that length, and the root-mean-square error in the period for that length. The average period will depend on the length of the string. The longer the string, the longer the period should be. On the other hand, the error in your measurements of the period perhaps will not depend very much on how long the string is, except perhaps for very short strings. If one of your determinations of the root-mean-square error in your measurements is very different from the others, a small red flag should go up in your mind. You should wonder why this is the case. That requires checking arithmetic, looking at the individual measurements, and seeing if you can figure out why that set of measurements is odd.

Why do we care about the root-mean-square error? The simple answer is that the root-mean-square error gives you a good estimate of how accurate your measurements are. The root-mean-square error is traditionally displayed on a graph of your experimental results.

29.3 Making Graphs

The next step in your work is to generate a graph of your results.

Good graphs have a correct format. To generate a graph, you will need a piece of graph paper or a computer plotting program. The graph is drawn as a square or rectangular box, with a solid line on each of the four sides. In this experiment, you reasonably want to plot the period of the pendulum against the length of the string. By convention, the horizontal axis is the variable you were changing, and the vertical axis is the response of the system to your change. That means that the length of the string is the abscissa of the plot, while the ordinate of the plot is the period of the pendulum.

You then choose where to put tick marks on the graph. *Tick marks* are short lines pointing in toward the center of the graph, starting at the outer boundaries. Tick marks let the reader estimate the numerical values corresponding to each data point. Tick marks should correspond to round numbers, e.g., 10, 20, 30, 40, not multiples of some strange number, e.g. 9.7, 19.4, 29.1, 38.8. In a scientific plot, tick marks are drawn on the inner side of the four lines that box the graph. They are drawn as a matching pairs, so if you have a set of tick marks along the lower horizontal axis, you should make an exactly matching set of tick marks, pointing the other way, along the upper horizontal axis. The same rule applies for the vertical axes. (Some graphs are much more complicated than this. Entire books have been written discussing how to cram more information, usefully, into a single graph.)

You often want to arrange your graph so that the data points fill the graph box in both the horizontal and the vertical directions. To do that, you need to choose the numerical values corresponding to the two

29.3. MAKING GRAPHS

limits of the graph so that the points to be graphed all lie inside the graph but spread from one edge of the graph to the other. That approach will work for this experiment. On the other hand, if you have varied some physical quantity, and nothing happened to the period T, then your measurements of T will appear on the graph as a series of points on a horizontal line. As a further comment, in general you want to arrange things so that the graph starts at, say, $L = 0$ and $T = 0$ as the lower left-hand corner of the graph. Graphs that start out with strange values for the minimum values for the abscissa and the ordinate are readily used to confuse interpretations of the data. Having said that, you also want to arrange things so that the root-mean-square error in the measurements is not too large relative to the size of the graph.

Now you plot your measurements on the graph. In the end, corresponding to your measurements you have a graph, with a small circular dot representing the value of T for each length of the string. You want the dot to be large enough that it is easily seen on the graph. Now go back to your table and find the root-mean-square error in your measurement for each point. The RMS errors are plotted as the *error bars* on your graph. An error bar represents graphically how accurate your measurement is. How do you plot an error bar? You go to each data point, you measure off a distance above and a distance below the data point, the two distances each being equal to your root-mean-square error, and at each of those distances you plot a short horizontal line. The two short horizontal lines are drawn to have their centers be directly above and directly below the data point. You now draw a vertical line, connecting the two short horizontal lines and passing through the data point. The actual data point needs to be large enough that you can see it after you've drawn the vertical line. (There is always a question as to whether the height of the error bars above and below the point should be the root-mean square error, twice the root-mean-square error, or some related distance. That issue is treated in statistics courses; here we will stay with the root-mean-square error.)

You will notice that I've given you a series of recommendations on how you construct a graph, and under some conditions some of my recommendations contradict each other. Constructing good graphs is an art, not a science, about which people have written entire books. One of the purposes of these laboratories is to help you think about your graphs, and how they should be drawn, because a good graph helps the reader understand what you have done. A bad graph leads to confusion.

We are not quite done. There is one additional critical step. You have to *label the graph*. In the case at hand, the first set of labels on the graph are the numerical values corresponding to the tick marks. If you have a lot of tick marks, by convention you only label a few of them. For example, for the horizontal axis, you might have a tick mark for every two centimeters of string length, but you would label only 10, 20, 30, and 40, and make the four labeled tick marks more prominent on the graph. By convention, the labels on the horizontal axis are printed with the normal orientation of letters and numbers on the page. The labels on the vertical axis are printed rotated through $90°$. Labels for each axis are only given on one side of the graph. By custom, one labels the bottom side and the left vertical side of a graph.

There is then an additional pair of *labels*, one for the horizontal axis and one for the vertical axis. These two labels identify the quantities being measured, and the units in which the quantities are measured. For the horizontal axis, you might have $L(\text{cm})$ for the length L of the string in centimeters, while for the vertical axis you might have $T(\text{s})$ for the period T of the pendulum in seconds. The first letter represents the physical quantity, in this case the length L or the period T. The letters in parenthesis stand for the units that were used in the measurement, in this hypothetical case centimeters and seconds.

Many of you will know computer software that lets you generate a whole graph including the box for the graph, the tick marks, the data points, and the error bars, without you having to do any work. In the long run, using this software is good. It saves a lot of work, and can make it very easy to change a graph if something is unsatisfactory. However, for this first experiment or two, you are encouraged to do the graphs by hand, so you have carefully focused on each of the elements in the graph. In addition, out there is a lot of not very good graphing software. For example, some sorts of software will only give you a horizontal and a vertical axis, and will fail to box the graph on all four sides, which is the physics standard. Other software sets will insist on giving you the tick marks on the outside rather than the inside of the graph, which is not acceptable. Some sets of software will need to be pounded on until they cooperate and give you a black and white graph rather than a graph with a lot of colors.

There are uses for *color* in very complicated graphs, but we are not talking about a complicated graph. You should therefore stay with black and white. If we had several sets of measurements on the graph, it would be appropriate to use *different shapes of data point* to represent each type of data. Appropriate data point shapes include ○, •, □, △ and other simple geometric forms, either open (△) or filled (▲). Using color instead of geometric shapes to distinguish different types of data point should be viewed as a desperation move for truly complicated graphs. For starters, a significant number of your fellow students do not have full human color vision, so that if you only use color they may be unable to tell your different types of data point apart. In addition, computer screens do not handle all colors equally well, yellow points on a white background being particularly unfortunate. If you print your graph, and it has colors in it, you should realize that many printers use CMYK color representation, while most video displays use RGB color representation. The conversion from one color representation to the other can not be made perfect.

29.4 The Search for Systematic Error

How do you look for systematic errors in your measurements? One search is most simply done if you are working with another student or group of students. You each make your own sets of measurements, having agreed in advance as to what lengths you will use for the string, so you have each determined the period of the pendulum for the same string lengths.

You now compare the average times T you and the other student found for each string length. If one of you consistently recorded longer or shorter average times than the other, you have good reason to suspect that there are systematic errors. A slightly more sensitive test for differences is made by plotting the individual measurements of the t_i as a histogram. The plot shows the number of experimental runs for which you and the other student found each period. To generate a histogram plot that is of any use, you want to choose the bins so that the measured times are spread over three or four or so time intervals. If the t_i are all in the same bin, or if they are spread well out, no more that one or two t_i points in a bin, the plot will be less helpful. If you measured the period for the same length of string a very large number of times, you can use more bins. Your histogram plot will then look somewhat like a smooth curve. However, you didn't measure the period that many times, so the histogram showing the distribution of the t_i will have look boxy.

Plot the other student's measurements of period on the same graph. It may be the case that your two histogram plots lie more or less on top of each other, or at least overlap. The difference between the two plots then shows the effect of random error. On the other hand, it might be the case that your histogram curve and the histogram curve for the other student group are separated from each other by a difference greater than the width of your histogram plot. That difference suggests that there is a systematic error: Your two sets of measurements were not made on quite the same system or something is slightly different between your measurement technique and theirs.

One way to test for the nature of the systematic error is for each of you to make a duplicate set of measurements using the other group's apparatus. If group A found a slightly longer time on its apparatus than group B did on its apparatus, and, when you and group B switch, you find that group B now finds the slightly longer time, a reasonable inference is that something is slightly different between the two sets of equipment. For example, the two strings might be slightly different in length. On the other hand, if, after making the switch, group B still finds a slightly longer time than group A did, a reasonable inference is something is different between the two groups in their timing procedures. Note I did not say that one group is right and the other group is wrong. It's just the two procedures are different from each other and are therefore giving slightly different answers.

I should emphasize that calling this difference "systematic error" is not to be taken in a pejorative sense. It could be the case that one group or the other has a defective stopwatch, or that both groups have defective stopwatches. It could also be the case that the systematic difference between the two sets of measurements arises because there are different ways of measuring the same quantity, and the different ways are not as equivalent as you might have thought.

Your search for systematic errors should be described in your final laboratory report.

29.5 Laboratory Report

We have now reached the point where you need to write up your results. A proper lab report begins with a title, for example "Experiment 1". Immediately below that are the names of each of the people who worked on the experiment (this may be only one person). If there are several people, you give a brief description of what each of them contributed to the work. For example, it might be the case that each of you made several time measurements, one of you always wrote down the times, and one of you did all of the write-up of the lab report. That should be noted in the report.

You then give a brief statement of what the experiment did. You should also provide, traditionally in a separate section labeled "experimental", a description of the experimental apparatus and how the measurements were performed. The description should be sufficiently detailed that someone else could duplicate the experimental work that you did. The section should include a sketch of the laboratory apparatus that you used. Note I said "sketch", not "photograph". With the advent of various electronic gadgets, it is extremely easy to generate photographs of apparatus and insert them into your report. **Almost without exception, photographs of experimental apparatus are totally and completely worthless as descriptions of the experimental equipment. There is no substitute for a line drawing.** A good sketch might well be made using straight edge and French curve or drawing software rather than drawn freehand. It might very well be convenient and is totally appropriate to do your sketch on a piece of paper, photograph or scan the paper, and then insert the image into the lab report. As a word of warning, pencil often photographs or scans poorly unless you have special pencil leads.

Having done this, you write up a description of your results, including the tables and graphs discussed above. Your report should attempt to determine how accurately you can determine the period of a pendulum.

This space reserved for your notes.

This page reserved for your notes.

Chapter 30

Experiment Two: The Search for Best Technique

In this experiment, we are going to consider different methods you might use to measure the period of the pendulum. Some of those methods are quite different from the others. Some people view it as being intuitively obvious which method is the best. In my experience, many of the people who make that claim identify a method that is wrong, often for several different reasons. What we are going to do is to use several different methods of measuring the period of the pendulum, and determine experimentally how accurate each of these methods is.

30.1 Experimental

The experimental equipment to be used in this experiment is precisely the same as the equipment that was used in the last experiment. I won't bother to repeat a description of that equipment here, but you should include it in your laboratory report. A good laboratory report is complete and self-contained.

I am now going list a number of different ways you could go about measuring the period of the pendulum. You should try each of these and include them in your laboratory report. You will need to do a lot of measurements, so you should not expect to be able to finish this laboratory in thirty minutes. The reason you are going to try each of these methods is that they are so simple that you can actually see exactly what is happening, and to some extent understand why one method might be more or less accurate than the others.

Let us now consider a series of different methods for measuring the period of a pendulum. Most of these methods are written on the assumption that two people are working together to make the measurement.

Method 1: One partner holds the pendulum at the outer end of the swing. The other partner says "Start!". The first lab partner lets go of the pendulum and the second lab partner starts the timer. After some number of swings, say ten, the first lab partner grabs the pendulum as it reaches the top of its swing, and at the same time says "Stop!". One of the lab partners then records the elapsed time. Repeat this process ten times, and record the elapsed time and the number of swings for each run of the experiment.

(I have just introduced an important physical distinction. A time, seen on a clock, is an absolute time, for example 2 in the afternoon. An elapsed time or time interval is the difference between two times reported by a clock. If I ask you what the period of a pendulum is, "two in the afternoon" is not a rational answer. If I ask you what time of day it is, "it took twenty seconds" is equally not a rational answer. These answers are mistaken because they conflate a time with a time interval. Times and time intervals are entirely different sorts of things.)

Method 2: The first partner holds and releases the pendulum. The second partner uses the stopwatch. The first partner waits until the pendulum has swung back-and-forth. When the pendulum reaches the top of its swing the first partner says "Start!". The second partner starts her stopwatch. The first partner counts off ten swings. When the pendulum again reaches the top of its swing the first partner says "Stop!". The watch is stopped. The elapsed time and number of swings are recorded.

Method 3: The partner with the stopwatch lowers her head until she is looking along the surface of the table at an object well behind the swing of the pendulum. The height of the pendulum is adjusted so that the swing of the pendulum just clears the surface of the table. You adjust the height of the pendulum by raising or lowering the bracket holding the pendulum, not by changing the length of the string, because the length of the string is a variable you will eventually be trying to study. The other partner releases the pendulum. The partner with the stopwatch times twenty passes of the pendulum bob through her line of sight to the distant object. Twenty passes? Yes. There are of course two passes of the pendulum during each period.

Method 4: Same as method three except you put an object on each side of the path through which the pendulum swings, and sight along the line defined by the two objects. The line should be the same as the line occupied by the pendulum at the bottom of its swing, when it is at rest. Those of you who have learned how to aim a firearm will find this arrangement extremely familiar as a pair of iron sights, except usually you do not place the target between the front and rear sights of your weapon.

Method 5: same as method four, except the two lab partners switch places.

Method 6: same as method one, except the same lab partner both holds the stopwatch, lets go of the pendulum, and grabs the pendulum when the ten swings are done.

Method 7: if you can find a laser pointer that is not too bright, set the laser pointer on the table (you may have to sit it on the top of a book or something to get it to exactly the right height) so the beam is interrupted when the pendulum swings by. Look at the point of light on the wall and use the shadow of the pendulum interrupting the laser pointer light beam as your time signal.

Clearly this is not an exhaustive list of different ways to do the experiment. For example you might choose to alternate which partners uses the timer, and then average over two lab partners to get a time measurement. You might choose to repeat all of these measurements (I've already indicated this for a few methods) with each lab partner taking both roles in the experiment. That will reveal systematic differences between how the two of you do timing.

30.2 Data Analysis

You now have a rather large pile of numbers. You should include all of them in your lab report. What do you do with the numbers to get interesting results out? The objective of this experiment, all of those methods described above, is experimental design. You are trying to determine which is the most accurate way to measure the period of the pendulum. There are a lot of different ways of measuring the period, and all of them are not equally good. In some cases, the difference in the accuracy of the methods may be less than the random error in your measurements, so several methods may appear to be equally good. You haven't tried all possible methods (I do not suggest doing this), but you have tried several methods.

What do I mean by "best method of measuring the period"? Until you have some decision as to what you mean by **the best way** to measure the period, you can't decide rationally which way is best. By best, I mean most accurate. A reasonable statement is that the most accurate measurement is the one with the least random error, not counting methods with known systematic errors. In the last laboratory we discussed the root-mean-square error, and how to determine the random error in your measurements.

For each of the methods you tried, calculate the average period T and the root-mean-square error for that method. Having done that, create a table in which the first column identifies the method, the second column identifies the average period, and the third column identifies the root-mean-square error for the method. You should now ask yourself: Do all of the methods give more or less the same value for the average? Or do some methods give significantly larger or small values than the others? Based on the root mean square error, are any of the methods particularly better or particularly worse than the others? In asking this question, you have to be honest with yourself. You may have had an opinion before you started the experiments as to what the right answer was, but that doesn't mean that your opinion was correct. It might have been correct, but perhaps it wasn't.

You may have a couple of methods that are clearly better than the others. In that case, if you are a determined experimentalist, you will repeat measurements using those methods, some considerable number of times each, and see if you can clarify which method is actually better. You may also decide that the two methods are about equally accurate, and that convenience or inconvenience in doing the experiment that way decides which method you should use. After all, we could have introduced a method where you sight along the line determining the bottom of the pendulum swing, while standing upside down on your hands. (Do not actually try this method!) Most people would take the position that even if this method was extremely accurate, much better than any of the others, it need not be considered further.

You're also in a position to ask whether some of the methods have systematic error, or whether there is a systematic error between the timings done by one of you and the other of you. Systematic error between methods shows up because the average times found by the different methods may not all be the same. If that's the case for your experiments, you should ask whether the differences between the times found by different methods are larger or smaller than the root-mean-square error that you determined in the measurements. If the difference is less than, say, twice the root-mean-square error, you might reasonably guess that the difference corresponds to the random error in determining T. After all, if there is a random error in the individual measurements t_i, when you average the individual measurements you make the random error smaller, but you do not make it zero. Just as there is a random error in the individual measurements t_y, so also there is a random error – a smaller one – in their average, T. On the other hand, if one method gives a very different result from the others, you might well want to try to figure out why. The fact that nine methods give one answer, and the tenth method gives a different answer, does not mean that the answer given by most of the methods is correct. **Science is based on being right, not winning a majority of the vote, though, as a practical matter in scientific debates, the side that wins over the most practitioners for the longest time very often turns out to be correct.**

As in the previous experiment, there is a test for systematic errors due to the behavior of the experimenters. Have a second group of people use your equipment to make the same measurements. You then determine whether the two sets of measurements agree or not, to within their random (root-mean-square) errors.

In your lab report, you should explain in a paragraph or so which method you have identified as the best method, and why you made that choice. Your explanation as to which method is the best has to be consistent with your measurements and data, not with your original expectation of which method is best, or anything I may have told you here. If you choose the right method, whatever that means, for wrong reasons, your analysis is incorrect.

There is a very important lesson, namely that if you are doing experiments, and the experimental data is good, *your conclusions are supposed to agree with your experimental results*, not with what you hoped to find. Of course, you can be unlucky. If you expect to find a particular answer, and thanks to the random noise in the measurements your data really fit your expectation well, you will probably conclude that you confirmed what you set out expecting to prove. Having expectations is good. Expectations can give you warnings. For example, your expectation that the zero readings do not correspond to the expected non-zero result may warn you that you forgot to turn on one of the instruments. However, expectations can also be misleading.

I am reminded of the famous paper in my field in which the extremely good researchers did a heroic pioneering experiment and announced that they had confirmed a controversial theory created by some of their friends. There are still a lot of people who believe that theory, and there is a lot of data that is interpreted — whether correctly or not — to support that theory. However, the five subsequent papers that duplicated more or less exactly the original experiment on the same system found a different result. It appears to me that these very good experimenters were fooled by extremely unfortunately placed random noise, and their very clear understanding of what they thought the answer was supposed to be.

No matter how good you are, this mistake can happen to you too.

30.3 Significant Figures

Now you have done this experiment, you are ready to step inside the data analysis and understand what significant figures actually do. Significant figures are a rote method that give a crude approximation to the random error in experimental numbers. Each of your results for the period of the pendulum can be written in the form $T \pm E$, in which T is the period of the pendulum and E is the error. You can discuss (this is a topic dealt with in statistics courses) whether you should actually show E, or $2E$, or something else after the $\pm$ to represent the range of expected errors. The important issue at this stage is that E represents how accurately you know the number T. If E is is 0.12 s, then it does not make sense to report T to an accuracy much finer than 1/100 of a second, because you only know T to within a band that is about 1/4 of a second wide. Significant figures are an extremely crude and imprecise way of representing this lack of knowledge as to the actual value of T. When we invoke significant figures, we claim that we know the value of a number to more or less exactly 1% or 0.1% or 0.01%,... of the size of the basic unit of measurement. With respect to time measurements, the significant figure approximation says you know the period to one-tenth of a second, one hundredth of a second, or one thousandth of a second, as opposed to say knowing the period to within 0.02 seconds.

Usually our error in measuring a number doesn't happen to fit exactly one of the bands assumed by significant figures, so significant figures are a very crude representation of what is going on. A detailed and exact pedagogical prescription, for calculating exactly how many significant figures you have, makes no sense, because the underlying intellectual basis for significant figures is much fuzzier than that.

Some attention to reality in estimating errors makes sense. I am reminded of a student laboratory in which students said that the temperature was known to within 1%. The temperature was being quoted in Kelvins, because we were doing physical chemistry. A 1% error is three Kelvins or five Fahrenheit degrees. There was a precision temperature control on the water bath holding constant the temperature of the system being studied, but the students were claiming that the bath worked much more poorly than a typical room thermostat.

The lesson from the above paragraph is that you should think about what you are doing and see if what you are doing makes sense. That is a difficult skill to learn, but it is extremely worthwhile to learn it.

This space reserved for your notes.

Chapter 31

Experiment Three: Properties of the Pendulum

You are now ready to do the actual experiment. You will measure the period of the pendulum as you change the length of the string or the number of nuts tied at the bottom end of the string. However, before you could do the actual experiment, you had to do the experimental design, to work out how to measure the period accurately. You have now done this, and will use your best method in the remaining work. If you really don't believe that your method actually should be the best method, for example because it's obvious that some other method is the best, repeat the above experiments using only the two methods of interest, and use each method, say, 20 times. Because you have done the experiment many more times, your determinations of T and of the root-mean-square error in determining T will be more accurate. If you choose to do the extra measurements, include in your lab report your explanation of why you did the extra measurements, what your experimental results are, and which method you will now use to measure T. Your conclusions should agree with your data. If your measurements show that method A is clearly better than method B, then you should not choose to use method B unless you can present an overwhelmingly powerful reason for doing so.

The question of interest is how the period of the pendulum depends on the length of its string, the mass of its pendulum bob, or the size of the swing. You will now search experimentally for these dependences. To test the dependence experimentally you will measure the period of the pendulum for at least four lengths of string, all with the same mass of pendulum bob, and you will measure the period of the pendulum for at least four choices of mass at the end of the string, all with the same length for the string. You will then analyze your measurements.

31.1 Data Fitting

If you look back at the dimensional analysis discussion, you will find that dimensional analysis says that T is proportional to L and to M, with L and M being raised to powers that may be integers, fractions, or real numbers, i.e.,

$$T = kM^a L^b \tag{31.1}$$

In this equation k, a, and b are unknown numbers to be determined from the experiment. A hypothetical finding that T is independent of M corresponds to a numerical result $a = 0$, and vice versa. You are about to be shown a numerical scheme for extracting numerical values for k, a, and b from your measurements. However, the method will only work for data that lies on a straight line. For a straight line following $Y = B + MX$, the method will determine the intercept B and the slope M. You can also ask how T depends on the size of the swing, the angle θ at start between the pendulum string and the vertical. θ has dimensions unity, so dimensional analysis does not tell you if there is a *theta* dependence to T. I am not telling you how to determine this, except to say that an additional objective is to find out how T depends on θ.

The dependence of T on M and L is not known to follow a straight-line linear dependence. You now have two choices. One choice is to research the use of computer fitting procedures such as the simplex method. [Warning: There are two simplex methods; they are not related to each.] With the correct computer software, you can confront your measurements with equation 31.1 and extract values for k, a, and b. The needed computer software performs non-linear least-square fits to the data, and determines values for the unknown parameters, even when the relevant equations are not linear in their parameters.

The other choice is to convert, e.g., $T = kM^a L^b$, k, a, and b being the unknowns, into a linear equation. You actually know how to do this, though you may not realize it yet. Take the logarithm of equation 31.1. You will get $\log T = \log(k) + a\log(M) + b\log(L)$, which is a linear equation in $\log T$ against $\log(M)$ and $\log(L)$. A linear fit of your measurements of T at different L and M to $\log T = \log(k) + a\log(M) + b\log(L)$ will determine k, a, and b. a and b will appear as slopes of hopefully straight lines. The proportionality constant k will appears as the intercept $\log(k)$.

The first step is to do the actual experiment, measuring how the period depends on the string length, the number of nuts, or the starting angle θ. In varying the string length, stay with a fixed number of nuts. In varying the number of nuts, keep the string length constant. As before, for each length of string make a reasonable number of measurements. As you increase the number of nuts, the load on the string, the string may stretch; you will need to adjust how the string is fastened in order to keep the string's length L constant. Your data should appear in a properly-labeled table.

Your next step is to generate the graphs of period against length, mass, and starting angle, using log scales on both axes. You should represent each measurement as a separate point, and then indicate the average as a point, and the spread around the average as error bars. In at least some cases, some of your measurements will lie outside of the error bars. That is an expected result. If all of your measurements lie inside your error bars, something is wrong with your calculations.

Your final step is to do linear-least-squares fits to the measurements and determine the exponents corresponding to the string length, the pendulum bob mass, and the starting angle. You should actually carry out the fits by hand, rather than trusting your pocket calculator, which may not have implemented precisely the algorithm that you thought it did. At the end, when you have determined the exponents of the intercepts, you should take your calculated lines and plot them on your graph to see how your calculated lines compare with your measurements.

There is a basic math problem here: We have an experiment in which we can reach in and set the value of some variable x. After each time we set x, we make one or more measurements of variable y. For example, x could be the length of the string, while y is the period of the pendulum. Then you choose different values for x, and for each x you make one or several measurements of y.

At the end of the effort, for a series of values of x, which you could call $x_1, x_2, \ldots$, you have a matching series of values of y, namely $y_1, y_2, \ldots$. The next correct step, which you could also have carried out while you were doing the measurements, is to make a scatter plot – a casual graph– of the y_i against the x_i, to see what the dependence of y on x looks like. Under many conditions, you are better off to make the plot first, and then try to analyze farther. For starters, by looking at the plot, you can detect single runs of the experiment in which you recorded the period incorrectly, for example by misplacing a decimal point, so that you have a measurement which completely disagrees with all the other measurements made under the same conditions.

In this hypothetical problem, you find from your scatter plot that y depends at least approximately linearly on x, so that

$$y = mx + b. \tag{31.2}$$

You now try to find the values of m and b that give the "best" agreement between the linear equation and your data.

Which m and b are best? A reasonable measure of the agreement between the fitting equation and the measurements is the mean square difference D between the experimental points and the fitting equation, namely

$$D = \sum_{i=1}^{N} (y_i - (mx_i + b))^2. \tag{31.3}$$

The best m and b are the values of m and b that make D as small as possible.

How do we make D as small as possible? We find where D has its minimum as a function of m and b. We can find the minimum using derivative tests, namely at the minimum

$$\frac{dD}{dm} = 0 \tag{31.4}$$

31.1. DATA FITTING

and
$$\frac{dD}{db} = 0. \tag{31.5}$$

The first derivative gives us
$$0 = \frac{dD}{dm} = 2\sum_{i=1}^{N} -x_i(y_i - mx_i - b), \tag{31.6}$$

while the other derivative yields
$$0 = \frac{dD}{db} = -2\sum_{i=1}^{N}(y_i - mx_i - b). \tag{31.7}$$

These two equations can be rewritten as
$$0 = \sum_{i=1}^{N} x_i y_i - m\sum_{i=1}^{N} x_i^2 - b\sum_{i=1}^{N} x_i \tag{31.8}$$

and
$$0 = \sum_{i=1}^{N} y_i - m\sum_{i=1}^{N} x_i - Nb. \tag{31.9}$$

Each of the sums in these equations, for example $\sum_{i=1}^{N} b = Nb$, is just a number. You now have two equations in the two unknowns m and b, so in general you can now solve for m and b.

We have now shown how to extract best values for the parameters m and b. However, the equation we care about, equation 31.1, is not linear in the parameters that we want to determine. What are we to do? There are two alternative approaches here:

First, you may be able to modify the equation of interest to make it into a linear equation, and then apply the method of linear least squares. That can be done here, namely if you take the logarithm of equation 31.1 you get
$$\log(T) = \log(k) + a\log(M) + b\log(L). \tag{31.10}$$

T is not linear in M or L, but $\log(T)$ is linear in $\log(M)$, $\log(L)$ and $\log(k)$, showing the method of linear least squares may be used to determine a, b, and k.

There is a modest caveat here. The simple method of linear least squares, as seen above, implicitly assumes that the error is entirely in T, not in M or L, and that all the error bars are about the same size. You should consider for yourself if the first assumption is true. The second assumption can get you into trouble, namely if you take the log of a variable, the effect of the log function on the error bars has to be watched carefully. If you take the log of a log (in my research there was an equation for which this step seemed to make sense), the error bars may well become very different in size for different points, and the naive linear least squares method described above will give anomalous answers.

Second, you can extract the parameters of interest via non-linear least squares, a method that we will not discuss farther here.

Outliers: Sometimes you will do a series of measurements that all more or less agree with each other, except one or two of the measurements are way off from the rest. The points that are way off are called outliers. In real science, it is understood that every so often an experiment does not work right, and therefore you should not include its outcome in further analysis. For example, if you measure the time required for five pendulum swings, you might get us answers 5.1, 5.0, 4.9, 5.0, and 6.0. Further repeats of the experiment all give numbers very close to five. What do you do with the 6.0? You might suspect, for example, that you ought accidentally let the pendulum swing six times rather than five times, but the experiment is done and gone, one with the snows of yesteryear, so you cannot prove that the pendulum was allowed to swing an extra time. The correct answer is to report the datum, put it on your graph, and explicitly say that you are not including this peculiar point in your analysis.

There is a slippery slope in outlier analysis. If you really think you know what the answer is supposed to be, you can be tempted to reject as outliers the good measurements, and keep the bad measurements that tell you what you wanted to hear. I have seen this done. You can also keep points that are really flaky, that lead you to wrong conclusions, for example you can keep them because they are your only measurements of the pendulum's period using a really long string.

The discussion in your lab report of the exponents a and b should consider the values you found for the exponents, and how your findings do or do not agree with the values you computed for a and b by means of dimensional analysis.

Finally, you have graphs showing T as a function of M or L. Add to those graphs smooth curves representing the function kM^aL^b for the values of k, a, and b that you determined. Do the curves agree with your measurements?

As an interesting addition to the experiment, determine how the period of the pendulum depends on the size of the swing, the size of the swing being how far the pendulum swings to the left or to the right.

This space reserved for your notes.

Chapter 32

Experiment Four: The Physical Pendulum

Now an experiment with much less detail. A meter stick can be turned into a physical pendulum by placing a small hole in it, close to one end, and timing the swing. You could also drill in the stick a series of small holes, add to the stick extra masses at various distances from the pivot point, and repeat your measurements of the period. For each of the physical pendulums that you assemble, the period can be calculated using techniques discussed in earlier chapters of the book. A useful graph compares your measured and calculated values of the period.

This space reserved for your notes.

This page reserved for your notes.

Chapter 33

Experiment Five: The Atwood Machine

Here we briefly consider an experiment that, with some effort, you can perform. The experimental objective is to duplicate the Atwood machine, which was the first reasonably precise test of Newton's laws of motion. Newton, of course, believed that he had a precise test, namely that he compared the time needed for an object to fall from some height with the orbit of the moon. As seen from the earth, the moon moves in a circle, because it is perpetually accelerating towards the center of the earth, but it is also moving sideways, so instead of dropping straight down it moves in a near circle. (Calculating the exact orbit of the moon is an extremely complicated problem.) The objection to Newton's test was that the Moon is not an earthly body, and might be subject to different laws.

Atwood lived in a period in which accurate time measurements were very difficult to make, especially if the times were very short. However, he also lived in a period in which Englishmen were accustomed to moving very heavy weights up and down, for example to unload goods from a merchant ship or to load cannons onto a warship. In order to do this, the English were accustomed to making very good, low friction, pulleys. They were also accustomed to weighing accurately large, heavy objects. What Atwood did was to take two heavy objects, attach a rope to the two of them, and loop the rope over pulleys, so that the two very heavy masses were suspended in midair. The masses were then allowed to fall. If one mass was heavier than the other, it would gradually sink to the ground. However, if you were clever, you could make the difference between the weights of the two masses fairly small, and the total mass of the two objects quite large, in which case the acceleration of the heavier object toward the ground was small. The time needed for the heavy object to descend to the dock was long and therefore could be timed with reasonable accuracy.

The objective in this experiment is for you to construct an Atwood machine, set it with various weights on the two ends of the rope, release the two weights, and measure how long it takes for one of the weights to descend to touch the tabletop. You should also calculate, this being a problem in coupled bodies, how long it should take for one of the two masses to descend to the tabletop. Finally, plot the observed time as a function of the calculated time. You reasonably expect as an outcome to get a more or less straight line, with the points scattered on both sides of the calculated line. This is a hard experiment to do well; you should not be surprised if your measurements only agree loosely with the predictions of Newtonian mechanics.

I am giving you very little information on how to set things up. If you do not have a good pulley or two, a steel nail is relatively frictionless. You rest the string on the nail, with the masses hanging below. You will actually need two nails spaced horizontally some distance apart, so that when you release the two masses they don't collide with each other. You will also need to measure fairly carefully the mass of the two masses you're using in your Atwood machine. As an inobvious test experiment, time the descent of the heavier mass when it has to drop several different distances before reaching the tabletop. If the acceleration is constant, and you know the exact instant when you released the mass, the drop time should go as the square root of the distance of the fall. Does it? That's a good test of your work.

If at the end you get things to work, you can reasonably say that you have duplicated one of the more important experiments in the history of physics, namely it was one of the first experiments that made a quantitative test of a mathematical theory of mechanics. Atwood got his test to work. Can you do as well?

This page reserved for your notes.

Chapter 34

Experiment Six: The Static Force Diagram

A physical model that illustrates a force diagram.

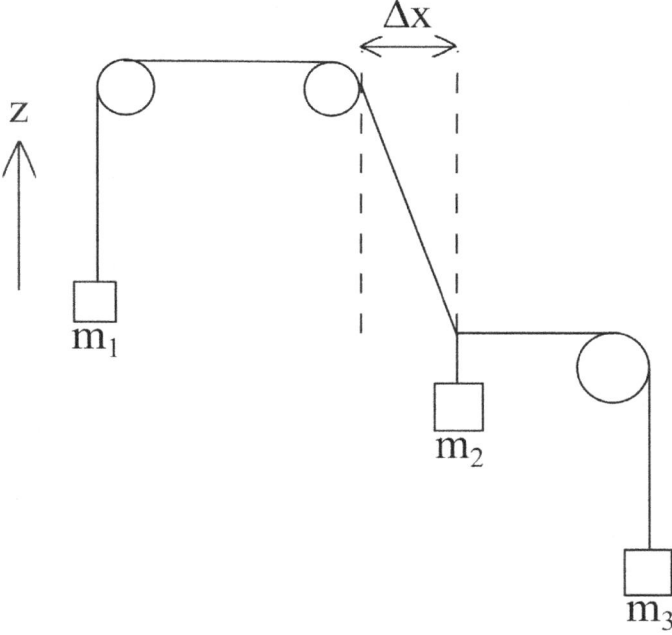

Figure 34.1: A deflected pendulum. The masses m_1, m_2, and m_3 are free-hanging, z marking the vertical direction. Circles are pulleys; straight lines are sections of twine connecting the pulleys. The objective is to measure the deflection Δx of m_2 from its unperturbed vertical hanging position, as functions of m_1, m_2, and m_3, and show that Δx and the masses are in agreement with the Second Law.

The objective of this short set of experiments is to demonstrate that the statements I've made about tension and force diagrams actually work about the way that you might expect. In a certain sense, this experiment will demonstrate that I have been telling the truth about how strings and forces work. Some of the statements that I have made may, however, have had implications that are not instantly obvious. Sometimes experiments help you to understand what a theory means.

You're going to set up a simple pendulum. You will attach to the pendulum bob an additional string allowing you to pull the pendulum bob sideways. You'll determine all the forces acting on the pendulum bob and compare those forces with sideways displacement of the pendulum.

Equipment: you will need a number of vertical rods, clamps, pulleys, masses, and string. You will set up something like the experiment in Figure 34.1, and calculate the tension in one string. The experiment has several key parts. To the ends of some strings, you need to tie hangers. These are gadgets to which you can attach weights. If the weights are nuts, a bent section of coat hanger might function as a hanger. The hangers hang more or less vertically, and provided downward forces mg determined by their weight, including the weight of any number of nuts placed over the hanger. Small pulleys or clean nails let you redirect the direction along which a string is pulling. You can use a hanger, string, and pulley to generate a horizontal force, by running the string over the pulley, having the top of the pulley level with the height at which you want to apply the horizontal force, and having enough weights on the hanger block to create the horizontal force that you want.

Your objective is to create setups similar to those shown in the sketch. Do this for several different deflecting forces. You reasonably start by creating the situation in which the pendulum simply hangs vertically. When you apply additional sideways forces to the pendulum, the pendulum will be displaced sideways, but to determine the size of that displacement you need to know exactly where the pendulum started. The picture looks very simple, but getting this set up properly is going to create some challenges. First, as the pendulum is dragged sideways, it is actually moving in a circular arc, so it's moving upward as well as sideways. As a result, if you want the force being applied to the pendulum to be perfectly horizontal, you will need to move the corresponding pulley up and down until it is at the right height relative to the central pendulum bob. A further complication arises because the pendulum is not simply attached to a top support. Instead, the string that holds up the pendulum goes over a pair of pulleys and down to another hanger. The weight on that other hanger has to be right in order to keep the system stationary. As you will discover, when you pull the pendulum sideways, you are increasing the tension in the string that supports the pendulum bob. Correspondingly, you have to increase how much force is on the hanger that holds the pendulum in place. Holding the hanger in place, and gradually adding masses to until the system is again balanced, may be relatively easy. Advancing in small steps from the simplest possible case of no sideways force on the pendulum bob may be of some advantage. As reasonable numbers, starting with 200 g as the pendulum bob, a pendulum length of half a meter or so, and sideways deflections of not more than ten or twenty centimeters may make life straightforward. How do you measure the deflection? You create a vertical reference line using clamps and rods, align the vertical reference line carefully with the line occupied by the pendulum bob when no sideways forces are being applied to it, and use the reference line to measure the horizontal distance through which the pendulum bob is deflected.

What will go in your lab report? First, you should do a simple theoretical analysis of your experiment. There are three masses in the system, so each mass gets its own force diagram. You should write the Second Law for each mass. There are two strings, so each string has a tension. The tensions in the two strings do not have to be equal. All accelerations are zero, so in this special case the sum of the forces on each mass must be zero. Calculate the deflection of the pendulum bob in terms of the three masses and calculate the tension in each string. Finally, you should compare the measured and calculated deflections. A graph of measured deflection against theoretical deflection is appropriate. Did your calculations match your measurements?

This space reserved for your notes.

Part Three

Harmonic Motion

This page reserved for your notes.

Chapter 35

Harmonic Motion

Part I of this book discussed the basic principles that enter classical mechanics, at the level of university freshman mechanics. We looked at a small number of examples showing Newton's Laws, conservation of energy, momentum, and angular momentum, tribology, rotation, collisions, statics, gravity, and planetary orbits.

Part II of this book presented a series of simple experiments demonstrating the nature of real experiments, using experiments related to topics developed in Part I of the book.

In Part III of the book, we take the basic principles developed in Part I, and apply them to a single physical problem, the existence of harmonic motion. We're going to do a somewhat deeper examination of a series of similar systems. Part III of the book is written so that it can be studied independently of Part 1, at least by students who have grasped the basics of classical mechanics. Part III requires a bit more mathematics than Parts I and II, in that we make substantial use of Euler's Identity for the complex exponential, and spend much of our time solving linear differential equations with constant coefficients. (As a practical aside, when I taught this course I found that students who had already taken a course in differential equations did not do appreciably better than students who had only had much of a year of calculus, so if you have not taken differential equations yet, do not worry.)

We turn to a specific topic that arises time and again in mechanics and other branches of physics, namely the study of harmonic motion. The word "harmonic" indeed comes from harmony, the study of tones in music. A cleanly produced musical note has a single dominant frequency. We can distinguish musical notes made by different musical instruments, from the sweet tones of the bagpipe to the brilliant glissandos of the harpsichord, because the notes produced by real musical instruments include not only the dominant frequency but also other frequencies that determine the note's timbre.

Harmonic motion, which includes motions occurring in musical instruments, refers to motions in which the system can oscillate with a single frequency. If the motion of the system is described by a variable x, in harmonic motion the variable x satisfies an equation

$$\frac{d^2x}{dt^2} = -\omega^2 x \tag{35.1}$$

Here ω – that's a Greek 'omega' – is the angular frequency of the oscillation.

To solve the harmonic motion equation, we need to integrate twice with respect to time. You could do this by assuming $x(t)$ is given by a power series in time. That assumption is certainly true; it's the same as saying you can write the solution in the form of a Taylor series [1]. We did this earlier in the course. You could also ask which of the functions you know turns back into itself, up to constants, when its second derivative is taken. If you think for a moment, you will come up with at least three, namely the sine, cosine, and exponential. The sine and cosine work here.

For example, you might guess that a solution is

$$x(t) = A\sin(\omega t + \phi). \tag{35.2}$$

Equation 35.2 arises from equation 35.1 by means of two time integrations, transforming the second derivative of $x(t)$ into $x(t)$ itself. If there are two time integrations, there must be two constants of integration *constants of integration*, two numbers whose values are not determined by equation 35.1. A and ϕ are those two constants. A is the *amplitude* of the oscillation. It reveals whether the sinusoidal motion, the oscillation, is large or small. ω is the angular frequency of the oscillation. ϕ is the *phase* of the oscillation. ϕ is needed in the solution because $\sin(\omega t)$ is zero at time $t = 0$. However, if I have a harmonic oscillator, I can choose when and how to set it swinging, so there is no requirement that at time $t = 0$ the oscillator happens to have $x(t) = 0$.

An example of harmonic motion is supplied by the swing of a simple pendulum. The pendulum swings back and forth. The horizontal coordinate of the pendulum bob is determined to good approximation by equation 35.1. It has a maximum swing, an amplitude of the oscillation. The oscillation has a well defined period, the time between pairs of zero crossings. Corresponding to the period is an angular frequency ω. Part I of this book considered a number of other systems that have similar properties. Here we bring these systems together to emphasize their similarities.

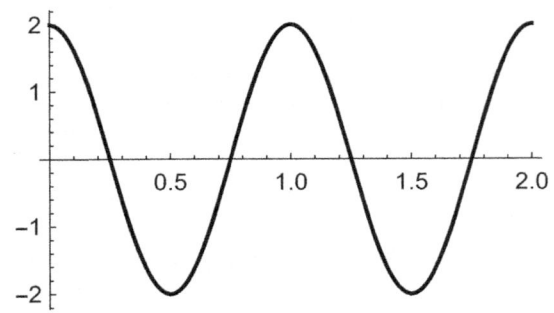

Figure 35.1: A cosine wave with $y(t) = 2\cos(2\pi t)$. The pair of zero crossings at times 0.25 and 1.25 give the period T.

A simple picture of a sine wave, Figure 35.1 with various quantities labelled, shows harmonic oscillation. We have plotted $y(t)$ against t. In the Figure, $x(t)$ has its maxima where $x(t) = +A$ and its minima where $x(t) = -A$. T is the period. The period T of the oscillation is the distance in time between pairs of non-adjacent zero crossings. Mathematically, you could equally well identify T as the time interval between pairs of maxima or pairs of minima. However, experimentally, the times at which zero crossings occur can be determined much more accurately than the times at which maxima or minima occur.

The angular frequency ω is measured in radians per second. T is the time required for the oscillation to pass through 2π radians, so

$$T = \frac{2\pi}{\omega}. \tag{35.3}$$

We also introduce the frequency ν. ν is the number of cycles in one second, so

$$\nu = \frac{1}{T}. \tag{35.4}$$

In the SI system, ν is given the units Hertz, one Hertz being one cycle per second. Hertz is abbreviated Hz. Older sources will refer to Hertz as cps – cycles per second. Angular frequency and Hertz both have dimensions 1/time, but they are not equal to each other. They differ by a factor of 2π, with $\omega = 2\pi\nu$.

35.1 Note

[1] To have a well behaved Taylor series a function must be continuous and be differentiable – have a full set of derivatives. There are functions that are not continuous. Stair steps do this. There are functions that are continuous but have an occasional point where they do not have a derivative. A conventional house roof with a peak in the middle does this. Then there are functions that are continuous everywhere but do not have even a first derivative at any point. The functions we encounter in physics almost all are well-behaved. However, you probably know a physics function that is not continuous, namely the specific heat of a material as you heat it across its melting point.

Chapter 36

Complex Numbers

As a forewarning, some of you may have already heard parts of this discussion of complex numbers.

Complex numbers arose when mathematicians came to understand how to solve quadratic equations. Once upon a time, if you studied what then passed for advanced mathematics [1], you would be taught that there were six varieties of quadratic equation, five of which had solutions, and one of which, the general equation

$$ax^2 + bx + c = 0 \qquad (36.1)$$

could not be solved for x. You may find this a bit odd, because

$$x = \frac{-b \pm (b^2 - 4ac)^{1/2}}{2a} \qquad (36.2)$$

is obviously the solution, as may be confirmed by substituting equation 36.2 into equation 36.1. However, it might readily be the case that $b^2 - 4ac < 0$, in which case the solution involved the square root of a negative number, which was viewed as not making sense. Centuries of work beginning with Girolamo Cardano were required to explain how to deal with $\sqrt{-1}$, the imaginary number.

How do complex numbers work? A typical complex number z may be written

$$z = a + bi, \qquad (36.3)$$

where a and b are familiar real numbers, and where $i = \sqrt{-1}$. a is the *real part* of z, and b is the *imaginary part* of z. (In some branches of engineering, $\sqrt{-1} = j$, because the character i is reserved for the current. In physics, the current is oft represented by the letter j.)

We define functions $\Re(z)$ and $\Im(z)$ that extract the real and imaginary parts of z, namely

$$Re(a + bi) = a, \qquad (36.4)$$
$$Im(a + bi) = b. \qquad (36.5)$$

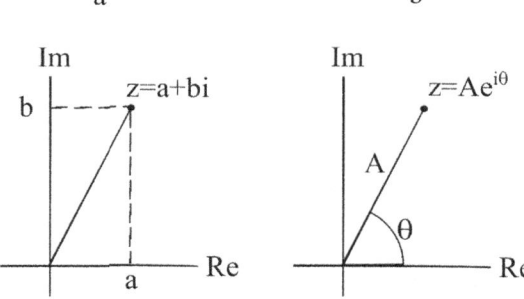

Figure 36.1: A complex number z, shown as a dot on the complex plane, and displayed (a) in terms of its real and imaginary parts a and b, or (b) in terms of its amplitude A and its phase angle Θ.

Note that the imaginary part of z is the real number b, not the imaginary number bi.

z may be viewed as a two-dimensional vector with components a and b. It may therefore be plotted as a vector (a, b) on a flat sheet of paper, the *complex plane*, as seen in Figure 36.1. The two coordinate axes are the *real axis* Re and the *imaginary axis* Im. An alternative way to write a complex number is to represent it as a *magnitude* (or *amplitude*) A and a phase angle Θ.

The relationships between a, b, A, and Θ follow from simple geometry, namely

$$A = |z|, \tag{36.6}$$

$$A = (a^2 + b^2)^{1/2}, \tag{36.7}$$

$$\Theta = \arctan\left(\frac{b}{a}\right). \tag{36.8}$$

Minor word of warning. arctan is the arctangent function. It computes an angle from the values of a and b. However, many pocket calculators have a button labelled "atn" that is in fact not the actangent function. Many calculators in fact implement Atn, the arctangent's *principal value function* Arctangent. How are these not the same? For a given ratio b/a, there are two angles having that value for the tangent. If we label the four quadrants of the plane as I, II, III, and IV, as seen in Figure 36.2, then $b/a > 0$ can refer to an angle in quadrants I or III, while $b/a < 0$ can refer to an angle in quadrants II or IV. The principal value function of the arctangent is single valued. Unlike $\arctan(b/a)$, which corresponds to two angles in opposite quadrants, 180 degrees (π radians) apart, the principal value function Arctangent(b/a) is single-valued, and corresponds to an angle in one of two quadrants. If $b/a > 0$, that angle is in the first quadrant. However, if $b/a < 0$, with some calculators the reported angle will be in quadrant II, while with other calculators the reported angle will be in quadrant IV. (There are also calculators that actually implement arctan.) There is no guarantee that the angle you want is the principal value angle. The angle you want might be the other angle that has the same value of b/a. To determine which quadrant b/a is in, the signs of a and b taken separately must be considered.

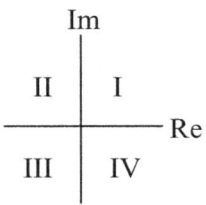

Figure 36.2: A complex plane and its four quadrants.

In equation 36.6, $|z|$ is the magnitude of z. $|z|$ is always a real, positive number. $|z|$ looks much like the familiar absolute magnitude, for which, e.g., $|-3| = 3$. In fact, the familiar absolute magnitude function is the same as $|z|$, for the special case that z is a real number.

The connection between a, b, A, and Θ can also be run in the other direction. From trigonometry,

$$a = A\cos(\Theta), \tag{36.9}$$

$$b = A\sin(\Theta), \tag{36.10}$$

$$\cos(\theta) = \frac{a}{(a^2 + b^2)^{1/2}}, \tag{36.11}$$

$$\sin(\theta) = \frac{b}{(a^2 + b^2)^{1/2}}, \tag{36.12}$$

and finally

$$z = A\cos(\Theta) + iA\sin(\Theta). \tag{36.13}$$

36.1 Arithmetic with Complex Numbers

How do we do math with complex numbers? The core answer is that $a + bi$ is a binomial, not much different from $3x + 4y$, except that $i^2 = -1$. Consider a few examples. We take $z_1 = a + bi$ and $z_2 = c + di$. Then:

$$z_1 + z_2 = (a + c) + (b + d)i, \tag{36.14}$$

$$z_1 - z_2 = (a - c) + (b - d)i, \tag{36.15}$$

and

$$z_1 z_2 = ac + bdi^2 + (ad + bc)i, \tag{36.16}$$

$$z_1 z_2 = ac - bd + (ad + bc)i. \tag{36.17}$$

Now we introduce the *complex conjugate*. If $z = a + bi$, the complex conjugate of z, denoted z^*, is $z^* = a - bi$. The complex conjugate is a path to the magnitude, namely

$$zz^* = (a + bi)(a - bi) = a^2 - b^2 i^2 + i(ab - ab) = a^2 + b^2, \tag{36.18}$$

36.2. THE EULER IDENTITY

so $zz^* = |z|^2$.

Finally we would like to compute z_2/z_1. We can write z_2/z_1 as

$$\frac{z_2}{z_1} = \frac{c+di}{a+bi}. \quad (36.19)$$

Our conventional knowledge of division only treats division by real numbers. The denominator above is a complex number. How can we transform the complex denominator to a real denominator? Complex conjugates transform complex numbers to real numbers. How to use that result here? The answer is to multiply by one – often a powerful tool – in the form $(a-bi)/(a-bi)$, leading to

$$\frac{z_2}{z_1} = \left(\frac{c+di}{a+bi}\right)\left(\frac{a-bi}{a-bi}\right), \quad (36.20)$$

which gives us

$$\frac{z_2}{z_1} = \frac{ac+bd}{a^2+b^2} + i\frac{ad-bc}{a^2+b^2}. \quad (36.21)$$

36.2 The Euler Identity

We reach a beautiful mathematical result, the Euler Identity, due to the Swiss mathematician Leonhard Euler. The identity reads

$$A\exp(i\Theta) = A\cos(\Theta) + iA\sin(\Theta) \quad (36.22)$$

In this equation, A and Θ are real numbers. I have used the exponential function notation, so that $\exp(i\Theta)$ has the same meaning as $e^{i\Theta}$. The utility of the exponential function becomes more obvious if you have the exponential of something complicated rather than the exponential of $i\Theta$. By using the exponential function, it becomes more obvious what is being exponentiated.

How is this identity to be proven? There is a Taylor series for the exponential, to wit

$$A\exp(i\Theta) = A\sum_{n=0}^{\infty}\frac{(i\Theta)^n}{n!}. \quad (36.23)$$

If you rearrange the terms on the right-hand-side of this equation, you will find that you obtain A times the Taylor series for $\cos(\Theta)$ plus iA times the Taylor series for $\sin(\Theta)$, thus proving the result.

An important feature of the result is that $\exp(i\Theta)$ really is an exponential, and has all of the properties of a conventional exponential, even though its argument $i\Theta$ is purely imaginary. In particular, if we have a complex number $z_1 = A\cos(\Theta) + iA\sin(\Theta)$, then $z_1 = A\exp(i\Theta)$. Suppose also that $z_2 = B\exp(i\phi)$. From the above, no matter what value z_2 has, z_2 can be written in this form. Then

$$z_1 z_2 \equiv Ae^{i\Theta}Be^{i\phi} = ABe^{i(\Theta+\phi)}, \quad (36.24)$$

where I have used the rule $e^a e^b = e^{(a+b)}$.

Therefore

$$z_1 z_2 = AB\cos(\Theta+\phi) + ABi\sin(\Theta+\phi). \quad (36.25)$$

You can confirm this equation by writing out equation 36.22 for z_1 and for z_2, multiplying out the sine and cosine terms, using the rules for combining sines and cosines of pairs of angles, and showing that you finally reach equation 36.25. Also, $(z_1)^n = A^n e^{in\Theta}$. In this final result, there is no requirement that n be an integer.

An important application of the Euler Identity is the proof of trigonometric identities. We begin with equation 36.22 and its complex conjugate

$$A\exp(-i\Theta) = A\cos(\Theta) - iA\sin(\Theta). \quad (36.26)$$

Combining these two equations gives us formulae for the sin and cosine functions, namely

$$\cos(\Theta) = \frac{e^{i\Theta}+e^{-i\Theta}}{2}, \quad (36.27)$$

$$\sin(\Theta) = \frac{e^{i\Theta}-e^{-i\Theta}}{2i}. \quad (36.28)$$

The other trig functions may all be written in terms of sin(Θ) and cos(Θ). To apply the Euler identity, equations 36.27 are substituted into the trig identity of interest, and simplifications are performed, until the two sides of the equation are reduced to the form $1 = 1$, proving the identity.

For example, suppose you wanted to prove that

$$(\sin(\Theta))^2 + (\cos(\Theta))^2 = 1. \tag{36.29}$$

One approach is to hunt for a plane geometry construction. An approach that tests the validity of an identity, but does not prove it, is to choose a couple of random non-trivial values, e.g 13.247955, for Θ, plug the values of Θ into the formula of interest, and see if the identity is confirmed to within the precision of the calculator. This approach is not a proof, but it will reasonably reliably detect invalid identities as being false.

Finally, there is the approach based on the Euler identity. We take the equation that we want to prove, equation 36.29. Substitute for the sine and cosine functions, using equations 36.27 and 34.28. You should find

$$\left(\frac{e^{i\Theta} + e^{-i\Theta}}{2}\right)^2 + \left(\frac{e^{i\Theta} - e^{-i\Theta}}{2i}\right)^2. \tag{36.30}$$

If you expand the squares, you find

$$\frac{e^{2i\Theta} + 2e^{i\Theta}e^{-i\Theta} + e^{-2i\Theta}}{4} - \frac{e^{2i\Theta} - 2e^{i\Theta}e^{-i\Theta} + e^{-2i\Theta}}{4}, \tag{36.31}$$

On making all the cancellations, the result is

$$\frac{4}{4} = 1, \tag{36.32}$$

confirming the identity.

36.3 Note

[1] Credit for creating imaginary numbers should perhaps be credited to the 16th century mathematician Girolamo Cardano, though it was not until the 19th century that some aspects of complex numbers were well understood. Mathematics has made great progress since the perhaps mythical ca. 15th century letter from one of the great English universities to an English lord, telling him that if his son wished to learn addition, subtraction, and multiplication, he did not need to go to Paris, because these arts were well-taught in England, but 'if he truly wishes to master the arcane art of long division, he must go to Italy, because the French don't understand it better than we do'.

36.4 Homework

1. Consider the complex numbers $z_1 = a + bi$, $z_2 = c + di$, and $Z = A + Bi$, where $Z = z_1 z_2$. By examination of a, b, c, and d, show $|z_1||z_2| = |Z|$. Do not use complex exponentials in your proof. [Hint: $|Z| = [A^2 + B^2]^{1/2}$. Brute force multiplication will work.

2. If one defines $z_1 = A_1 e^{i\theta_1}$, $z_2 = A_2 e^{i\theta_2}$, and $Z = z_1 z_2 = A_3 e^{i\theta_3}$, show $\theta_1 + \theta_2 = \theta_3$. Proceed by examining a, b, c, d, A, and B, by writing θ_1, θ_2, and θ_3 in terms of a, b, c, d, A, and B and manipulating a, b, c, and d. DO NOT use complex exponentials in your proof. [Hints: Two angles are equal if *two* of their trig functions are equal. $\cos(\theta_1 + \theta_2)$ can be written in terms of the cosine and sine of θ_1 and θ_2, which can in turn be written in terms of a, b, c, and d. (Why two trig functions? Why not one?), so show, e.g., $\cos(\theta_1 + \theta_2) = \cos(\theta_3)$ and $\sin(\theta_1 + \theta_2) = \sin(\theta_3)$.]

3. By means of a power (Taylor) series justify the formula $\cos(\theta) = e^{i\theta} + e^{-i\theta}/2$. Remark. A power series has a summation sign Σ. $e^x = 1 + x + x^2/2! + x^3/3! + ...$ is *not* a power series, only a statement of several terms. Proving that the first couple of terms of two series are equal does not prove that the two series are equal; you need to write a proof for a general term.

36.4. HOMEWORK

4. By means of a power (Taylor) series justify the formula $\sin(\theta) = (e^{i\theta} - e^{-i\theta})/2$. Remark. A power series has a summation sign Σ. $e^x = 1 + x + x^2/2! + x^3/3! + ...$ is *not* a power series, only a statement of several terms. Proving that the first couple of terms of two series are equal does not prove that the two series are equal; you need to write a proof for a general term.

5. Consider an oscillatory function $x = 5.8\cos(3t - \pi)$. a) Sketch the function. *Label* axes. b) What is the amplitude? c) What are the largest and smallest values that x attains? d) What is the angular velocity ω in radians per second? e) What is the frequency in Hertz? f) What is the period T? Sketch T on your figure from part a).

6. Consider an oscillatory function $x = 3.7\cos(5t - \pi/2)$. t is in seconds. a) Sketch the function. *Label* axes. b) What is the amplitude? c) What are the largest and smallest values that x attains? d) What is the angular velocity ω in radians per second? e) What is the frequency in Hertz? f) What is the period T? Sketch T on your figure from part a).

7. Find the amplitude and phase of $2.1 - 6.1i$, and write this complex number as a complex exponential. Calculate its third power using the complex exponential function, and rewrite the result as the sum of a real and an imaginary part. Cube $2.1 - 6.1i$ by multiplying the binomials, to verify your result. Calculate the ninth power of $2.1 - 6.1i$ by whatever means you find most convenient. Calculate the 3.4 power as well. In the last two calculations, you may write your answer in any convenient correct form.

8. Find the amplitude and phase of $1.2 + 1.6i$, and write this complex number as a complex exponential. Calculate its third power using the complex exponential function, and rewrite the result as the sum of a real and an imaginary part. Cube $1.2 + 1.6i$ by multiplying the binomials, to verify your result. Calculate the ninth power of $0.5 - 0.37i$ by whatever means you find most convenient. Calculate the 3.4 power of $0.5 - 0.37i$.

9. Find the amplitude and phase of $1 - 2i$, and write this complex number as a complex exponential. Calculate its third power using the complex exponential function, and rewrite the result as the sum of a real and an imaginary part. Cube $1 - 2i$ by multiplying the binomials, to verify your result. Calculate the ninth power of $1 - 2i$ by whatever means you find most convenient. Calculate the 3.4 power as well. In the last two calculations, you may write your answer in any convenient correct form.

10. Find the magnitudes and directions (phase angles) for $z_1 = 2 + 9i$ and for $z_2 = (2+9i)^2$.

11. For $z_1 = 2 + 3i$ and $z_2 = 2 + 3i$, find $z_1 + z_2$, $z_1 - z_2$, $z_1 z_2$, z_1/z_2, and $z_1^{z_2*}$.

12. Consider $z = a + bi$. Why is it the case that $\text{Re}(\text{Re}(z)) = a$ but $\text{Im}(\text{Im}(z)) = 0$?

13. Euler's Identity provides $\exp(i\theta) = \cos(\theta) + i\sin(\theta)$. **Use this relationship** to confirm $\sin^2(\theta) + \cos^2(\theta) = 1$.

14. Euler's Identity provides $\exp(i\theta) = \cos(\theta) + i\sin(\theta)$. **Use this relationship** to confirm $\cos(a+b) = \cos(a)\cos(b) - \sin(a)\sin(b)$.

15. Using the Euler identity, prove $\sin(A - B) = \sin(A)\cos(B) - \cos(A)\sin(B)$.

16. We have shown two standard notations for writing a complex number, namely the $a + bi$ form and the complex exponential form $A\exp(i\theta)$. In the following **show your work, which I agree is sometimes pretty simple**. I am not concerned that, if you report a value for a trig function, e.g., $\cos(0) = 1$, there was a calculator used. However, **do not** use any complex variable functions of your calculator. (i) Write $3 + 4i$ in complex exponential form. (ii) Write $2\exp(0.7i)$ in $a + bi$ form. (iii) A complex number, plotted on graph paper, is found to be 10 units from the origin, at an angle of $125°$ from the usual coordinate x-axis. Write this complex number in $a + bi$ form and in complex exponential form. (iv) Compute $(3 + 4i)^7$. You may give your answer in either format. (v) (10 points) By means of the Euler identity, either demonstrate that the equation $\cos(2\theta) = \cos^2(\theta) - \sin^2(\theta)$ is true, or demonstrate that the equation is false.

This page reserved for your notes.

Chapter 37

Harmonic Oscillation

We now turn to properties of several representative harmonic oscillators.

First, a brief aside: I present some letters of the Greek alphabet and their names. In some cases the capital and small (majascule and miniscule) forms of the letters are quite different, e.g., Γ and γ. In particular:

α - alpha
β - beta
Γ, γ - gamma
Δ, δ - delta
ϵ - epsilon
ν - nu
Ω, ω - omega
ξ - xi (pronounced 'ksy')
Σ, σ - sigma
Θ, θ - theta
Φ, ϕ - phi
ρ - rho
Ψ, ψ - psi

These are the letters we are likely to use. You will eventually need to memorize them. DO you really need to know them? If you are outside the university, look at ω, and say 'double-you', you will not have enhanced what educated people think of you.

37.1 A Mass on a Spring

A simple example of a harmonic oscillator is a mass at the end of a spring. The mass is resting on a frictionless tabletop. The other end of the spring is attached to an immobile wall. The figure shows a sketch of the mass and the corresponding force diagram. Here N and mg are the normal force from the table and the force of gravity, while $-k(x - x_0)$ is the restoring force due to the spring. The spring has a force constant k and a rest location x_0 at which it applies no force on the mass. If the spring is unstretched, the mass is at $x = x_0$. We may legitimately choose our coordinate system so that $x_0 = 0$.

Apply the Second Law

$$m\frac{d^2 \boldsymbol{r}}{dt^2} = \boldsymbol{F}. \tag{37.1}$$

Substituting for the force, and writing the Second Law as its component equations,

$$m\frac{d^2 x}{dt^2} = -k(x - x_0), \tag{37.2}$$

$$m\frac{d^2 y}{dt^2} = 0, \tag{37.3}$$

$$m\frac{d^2 z}{dt^2} = N - mg. \tag{37.4}$$

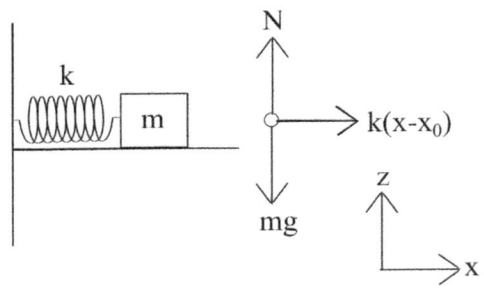

Figure 37.1: A mass m resting on a table, connected to a spring having spring constant k. The rest locations of the mass and spring are at x_0.

The accelerations in y and z vanish, so the forces in those two directions must be zero, implying $N = mg$ in this problem. We have chosen our coordinate origin so that $x_0 = 0$, leading to the equation of motion, a canonical equation, for a harmonic oscillator.

$$m\frac{d^2x}{dt^2} = -kx. \tag{37.5}$$

There are several ways to find a solution to these equations. One could try a power series. Alternatively, one might guess a solution, and then prove that it is a solution by plugging it into the differential equation, equation 37.5, and showing that the substituted equation is true. Separate from finding a solution is showing that your solution is unique. You may have found a solution to a problem, but if there are also completely different solutions life can become interesting. The series approach with some effort shows that there are only two solutions.

Here we use the second approach to finding a solution. As it turns out,

$$x(t) = A\cos(\omega t + \phi) \tag{37.6}$$

is a guess that leads to the solution. As before the amplitude A and the phase ϕ are constants, while ω is the oscillation frequency. One might note that $A\sin(\omega t + \phi)$ also appears to be a solution, but in fact the sine and cosine are the same function, differing only in their values for the constant ϕ. To move from the second time derivative of the function to the function itself, we had to do two time integrals, introducing two constants of integration. One of the trio A, ϕ, ω therefore cannot be a constant of integration. It must somehow be determined by the problem.

Inserting equation 37.6 for $x(t)$ into equation 37.5, one finds

$$m\frac{d^2}{dt^2}(A\cos(\omega t + \phi)) = -kA\cos(\omega t + \phi). \tag{37.7}$$

However,

$$m\frac{d^2}{dt^2}(A\cos(\omega t + \phi)) \equiv -m\omega\frac{d}{dt}A\sin(\omega t + \phi) \equiv -m\omega^2 A\cos(\omega t + \phi),$$

leading to

$$(k - m\omega^2)A\cos(\omega t + \phi) = 0. \tag{37.8}$$

The left hand side is a product of three terms, the product being equal to zero, so at least one of the three terms must be 0. $A = 0$ corresponds to the mass remaining at its rest position indefinitely. In this solution, there is no motion. There are occasional instants at which $\cos(\omega t + \phi)$ vanishes because its argument is zero modulo π, so at those instants A or $k - m\omega^2$ are not obliged to be zero. However, the product is zero at all times, so A or $k - m\omega^2$ must be zero at those times when the cosine is not zero. A or $k - m\omega^2$ are constants, so therefore one or the other must be zero at time when the cosine function is also zero.

By a process of elimination, one must have $k - m\omega^2 = 0$, and therefore

$$\omega = \pm\sqrt{\frac{k}{m}}. \tag{37.9}$$

As predicted several paragraphs ago, one of the trio A, ϕ, ω cannot be a constant of integration; it must be determined by the problem. We have found that ω is determined by the problem, so the constants of integration are A and ϕ. The final solutions for the position and velocity are is

$$x(t) = A\cos\left(\sqrt{\frac{k}{m}}t + \phi\right) \tag{37.10}$$

$$v(t) = A\sqrt{\frac{k}{m}}\sin\left(\sqrt{\frac{k}{m}}t + \phi\right). \tag{37.11}$$

We may plot $x(t)$ as a function of t. The phase angle determines where the maximum of $x(t)$ lies relative to $t = 0$, a ϕ that is positive but less that π indicating that $x(t)$ has its maximum that is closest to $t = 0$ for $t < 0$.

It is also possible to discuss a harmonic oscillator by writing its energy. The oscillator will have a kinetic energy due to its mass and a potential energy due to the spring, letting us write

$$E = \frac{1}{2}m\left(\frac{dx}{dt}\right)^2 + \frac{1}{2}kx^2. \tag{37.12}$$

This equation is readily shown to be consistent with the Second Law form treated above, namely if we require that E is a constant, and take a derivative with respect to time, we get

$$m\frac{dx}{dt}\frac{d^2x}{dt^2} + kx\frac{dx}{dt} = 0. \tag{37.13}$$

Dividing out the common factor of dx/dt and reordering terms, we have

$$m\frac{d^2x}{dt^2} = -kx, \tag{37.14}$$

which is the Second Law form for the mass on a spring.

Equations 37.10 gives us the position and velocity of the mass as a function of time. Inserting those forms into equation 37.12 gives

$$E = \frac{1}{2}m\left(-A\sqrt{\frac{k}{m}}\sin(\sqrt{\frac{k}{m}}t + \phi)\right)^2 + \frac{1}{2}k\left(-A\cos(\sqrt{\frac{k}{m}}t + \phi)\right)^2, \tag{37.15}$$

which simplifies to

$$E = \frac{1}{2}kA^2\left(\sin^2(\omega t + \phi) + \cos^2(\omega t + \phi)\right). \tag{37.16}$$

A standard trig identity shows that the sine and cosine terms sum to 1. The energy is finally found to be

$$E = \frac{1}{2}kA^2. \tag{37.17}$$

37.2 Canonical Forms for a Harmonic Oscillator

By 'canonical', one means 'most fundamental', 'standard' or 'simplest'. The harmonic oscillator is defined by two equations that, equivalently, describe its motions, and the one consequent equation that gives its total energy. The canonical forms are

$$m\frac{d^2x}{dt^2} = -kx \tag{37.18}$$

as the equation of motion, and

$$E = \frac{1}{2}kx^2 + \frac{1}{2}m\left(\frac{dx}{dt}\right)^2 \tag{37.19}$$

for the dependence of the energy on the position and velocity. Here k and m are constants whose meaning depends on the problem at hand, while x is a position coordinate. Any system whose dynamics can be described by either of these equations, no matter the physical meanings of k, m, and x, has the properties of a harmonic oscillator. For example, if we consider the electric and magnetic fields inside an empty container, the electrical and magnetic fields $\mathbf{E}$ and $\mathbf{B}$ can be put in this form, the two fields taking the roles of a coordinate and its time derivative.

These equations have canonical solutions, viz.,

$$x(t) = A\cos\left(\sqrt{\frac{k}{m}}t + \phi\right), \tag{37.20}$$

$$E = \frac{1}{2}kA^2. \tag{37.21}$$

The first of these is the solution to the equation of motion, equation 37.18. The second of these follows from the first, and the general form for the total energy. The total energy is determined by the amplitude of oscillation A and the spring constant k, but does not depend on the mass of the oscillating object.

37.3 The Pendulum

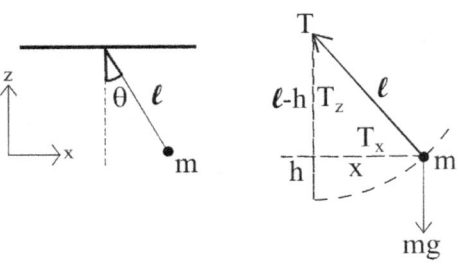

Figure 37.2: A pendulum, bob of mass m, on a string of length ℓ, and the force diagram. Dashed lines are construction lines for resolving the tension T in the string into its components $T_x = -T\sin(\theta)$ and $T_z = T\cos(\theta)$. The bob is a distance h above its rest position $\theta = 0$. The dashed curve marks the swing of the pendulum.

We now consider a second example of a harmonic oscillator, the pendulum. We considered the pendulum in Part I of the text. As seen in Part II of the text, the pendulum is particularly convenient as a choice for laboratory experiments, because almost no specialized equipment is needed to make measurements on it. It's also convenient as a place to introduce a solution technique having very broad general applications, namely *linearization*.

We begin with an analysis of the simple pendulum. Figure 37.2 shows an example. A mass m hangs from a string of length ℓ. It swings back and forth, making an angle θ with respect to the vertical. The Figure also shows the force diagram, including the forces mg and T on the mass, and the geometric construction that decomposes T into its components T_x and T_z. Note that the sine term is associated with the x component of the force, and the cosine term is associated with the z component of the force. I inserted a minus sign in the horizontal component of the force in order to agee with the coordinate system. From the force diagram, we may readily write the equations of motion for the three components of the acceleration, namely

$$-T\sin(\theta) = m\frac{d^2x}{dt^2}, \tag{37.22}$$

$$0 = m\frac{d^2y}{dt^2}, \tag{37.23}$$

$$T\cos(\theta) - mg = m\frac{d^2z}{dt^2}. \tag{37.24}$$

There is no motion perpendicular to the plane of the paper, so we will not consider farther the y acceleration. The trig functions are given as functions of θ, while the Cartesian coordinates are x and z. In a plane, there are only two independent coordinates, so we have a surplus of coordinates. We deal with this by rewriting the trig functions in terms of the Cartesian coordinates, thereby eliminating θ, namely

$$\cos(\theta) = \frac{\sqrt{\ell^2 - x^2}}{\ell} \equiv \left(\frac{\ell^2 - x^2}{\ell^2}\right)^{1/2}, \tag{37.25}$$

$$\sin(\theta) = \frac{x}{\ell}. \tag{37.26}$$

These equalities are the algebraic forms of the standard definitions of the trig functions, so that the cosine function is the adjacent over the hypotenuse, while the sine function is the opposite over the hypotenuse. The first of the above equations obtains the adjacent side of the triangle in terms of x and ℓ via the theorem of Pythagoras. We finally apply a series expansion for $\cos(\theta)$, obtaining for small x/ℓ

$$\cos(\theta) \approx 1 - \frac{1}{2}\frac{x^2}{\ell^2}, \tag{37.27}$$

terms of order $\mathcal{O}((x/\ell)^4)$ and higher being dropped. We see here an example of solution via expansion in a small parameter.

37.4. COMPLEX VARIABLE METHOD

We now say that we are talking about small oscillations, small meaning that $x/\ell \ll 1$. In each series, we keep the leading term in x/ℓ, these terms being 1 and x/ℓ, respectively, letting us rewrite the equations of motion in their approximate forms

$$T - mg = m\frac{d^2z}{dt^2}, \tag{37.28}$$

$$-T\frac{x}{\ell} = m\frac{d^2x}{dt^2}. \tag{37.29}$$

It is not automatically the case that the lowest order term is 1 or x/ℓ; for example, it might be $(x/\ell)^2$.

From the first line, the acceleration in the z direction is a constant. The pendulum does not fall through the floor or take off into space, so the acceleration in the z direction must be zero. It follows that $T \approx mg$ and

$$m\frac{d^2x}{dt^2} = -\frac{mg}{\ell}x. \tag{37.30}$$

The factors of m will cancel.

Behold! We linearized the equations of motion, so that the force is a restoring force that is linear in x, the variable of interest. You can't always do this, but in this problem you can. Consider equation 37.30. It's in a particular form, namely it is one of the two canonical forms for the harmonic oscillator. The other canonical form and the energy form, as discussed at the end of the last chapter, must also be true. Comparing with the canonical form, equation 37.18, we can immediately write

$$x(t) = A\cos\left(\sqrt{\frac{g}{\ell}}t + \phi\right) \tag{37.31}$$

for the lateral motion of the pendulum bob,

$$E = \frac{1}{2}\frac{mg}{\ell}A^2 \tag{37.32}$$

for the energy of the pendulum, and, for the period $T = 2\pi/\omega$, we write

$$T = 2\pi\sqrt{\frac{\ell}{g}}. \tag{37.33}$$

37.4 Complex Variable Method

We now demonstrate an alternative solution method. We have a variable $x(t)$ that is the solution to the problem. What we do is to replace $x(t)$ with a complex variable $z(t)$ of which $x(t)$ is the real part. Important notation issue: By convention z is used both for the vertical direction and also for a complex number. We try to keep clear which meaning is in use in any given equation.

We replace $A\cos(\omega t + \phi)$ with $A\exp(\imath(\omega t + \phi))$, because just as

$$x(t) = Re(z(t)) \tag{37.34}$$

so also

$$A\cos(\omega t + \phi)) = Re(A\exp(\imath(\omega t + \phi))). \tag{37.35}$$

[Aside: This approach is an opening to the "Laplace transform method" of solving differential equations, but Laplace transforms do considerably more than what you will see here.]

We return to equation 37.30 and substitute z for x, giving

$$m\frac{d^2z}{dt^2} = -\frac{mg}{\ell}z. \tag{37.36}$$

Replacing z with the complex exponential

$$m\frac{d^2 A\exp(\imath(\omega t + \phi))}{dt^2} = -\frac{mg}{\ell}A\exp(\imath(\omega t + \phi)). \tag{37.37}$$

There are two time derivatives. Each of them brings down from the exponential a factor of $\imath\omega$, leading to

$$-m\omega^2 A\exp(\imath(\omega t + \phi)) = -\frac{mg}{\ell}A\exp(\imath(\omega t + \phi)). \tag{37.38}$$

There are a lot of factors common to both sides of the equation. When they are all cancelled out, one obtains

$$\omega^2 = \frac{g}{\ell} \tag{37.39}$$

and therefore $\omega = \sqrt{g/\ell}$. A and ϕ have again disappeared during the solution process; the equations of motion for the pendulum do not bind either of them. (The string, however, restricts A, namely $A < \ell$ must be true.) The final solution is

$$x = Re\left(A\exp(\imath\left(\sqrt{\frac{g}{\ell}}t + \phi\right))\right) \equiv A\cos\left(\sqrt{\frac{g}{\ell}}t + \phi\right). \tag{37.40}$$

37.5 Energy of a Pendulum

We now consider the energy of a pendulum. In general we may write for the total energy $E = K + U$. The kinetic energy K is $\frac{1}{2}mv^2$. To obtain the potential energy we use the construction lines seen in Figure 37.2. The length of the pendulum's string is ℓ, no matter what the angle of the pendulum with respect to the vertical is. At some point, the pendulum bob is a distance x to one side or the other of the vertical. It's simplest to put the zero of the gravitational potential energy at the midpoint, where the pendulum bob is at the lowest point of its swing. When the pendulum bob is a distance x out to one side, it is also a distance h above its minimum height, so its gravitational potential energy at that point is $U = mgh$.

We obtain h from the theorem of Pythagoras. The right triangle seen in the Figure 37.2 has hypotenuse ℓ and leg x, so the other leg of the triangle is $(\ell^2 - x^2)^{1/2}$. That side is also $\ell - h$, so we may write

$$\ell - h = \ell\left(1 - \left(\frac{x}{\ell}\right)^2\right)^{1/2} \tag{37.41}$$

where a factor of ℓ has been extracted from the radical. The radical is now expanded by applying the binomial theorem, which, as first recognized by Newton, is valid for an arbitrary fractional power as the exponent. One obtains

$$\ell - h \approx \ell(1 - \frac{1}{2}\left(\frac{x}{\ell}\right)^2 + \ldots), \tag{37.42}$$

which simplifies to

$$h \approx \ell\frac{x^2}{2\ell^2}. \tag{37.43}$$

Combining these results, we have

$$E = \frac{m}{2}\left(\frac{dx}{dt}\right)^2 + \frac{1}{2}\frac{mg}{\ell}x^2. \tag{37.44}$$

This form is the same, except for some of the constants, as the canonical form

$$E = \frac{m}{2}\left(\frac{dx}{dt}\right)^2 + \frac{1}{2}kx^2 \tag{37.45}$$

for the energy of a harmonic oscillator. The solutions are therefore the same, once the constant k is replaced with the constant mg/ℓ, leading to a predicted motion

$$x(t) = A\cos\left(\sqrt{\frac{g}{\ell}}t + \phi\right) \tag{37.46}$$

and a predicted total energy

$$E = \frac{mg}{\ell}A^2. \tag{37.47}$$

37.6 Pendulums: Torque Approach

Having introduced linearization, I am reminded of an observation by the late Richard Feynmann, who remarked that the fact that you have a small parameter in your problem does not mean that it is necessarily a good idea to expand in that small parameter. As a second observation, note that in some systems linearization works very well, so that your errors increase linearly with the time over which you observe the system. In other systems, linearization fails catastrophically, so that the error in the linear solution increases much more rapidly than linearly with the time over which you observe the system. The prior two sentences are an opening to a discussion of the role of chaos in mechanics.

37.6 Pendulums: Torque Approach

This section is based on the general relationship

$$\boldsymbol{\tau} = \frac{d\mathbf{L}}{dt} \tag{37.48}$$

between torque and angular momentum, and its special form for in-plane motion

$$\tau_\perp = I \frac{d^2\theta}{dt^2}. \tag{37.49}$$

We consider again a mass m on a string having length ℓ, as seen in the sketch of the pendulum and its associated force diagram. The string is attached above to a support beam. The vector from the attachment point to the mass is $\mathbf{r}$. The forces on the mass are the tension $-T\hat{\mathbf{r}}$ and the force of gravity $-mg\hat{\mathbf{k}}$.

The torque on the mass, as calculated around the attachment point of the string to the ceiling, is $\boldsymbol{\tau} = \mathbf{r} \times \mathbf{F}$. Here $\mathbf{F}$ is the total force on the mass. We are, of course, free to chose whatever point we want as the origin for calculating the torque and angular momentum, but using the attachment point gives clear results, while at least one other point does not lead to a solution. The tension in the string does not contribute to the torque, namely it is $\boldsymbol{\tau} = \mathbf{r} \times (-T\hat{\mathbf{r}})$ and the vector cross product of two parallel vectors is $\mathbf{0}$. The torque is therefore

$$\boldsymbol{\tau} = \ell\hat{\mathbf{r}} \times (-mg\hat{\mathbf{k}}) \equiv -\ell mg \sin(\theta)\hat{\boldsymbol{\tau}} \tag{37.50}$$

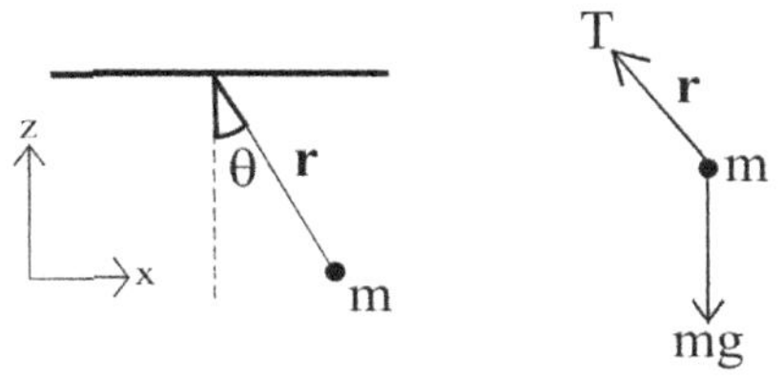

Figure 37.3: A pendulum, bob of mass m, on a string of length r hanging from the ceiling. Note the force diagram. The vector $\mathbf{r}$ points from the ceiling to the mass, opposite to the direction indicated in the force diagram as the direction of the tension.

The vector product gives the direction $\hat{\boldsymbol{\tau}}$ of the torque vector. From the sketch, we can use the right-hand-rule for the cross product to calculate the direction of the product, namely the thumb will point in the $\mathbf{r}$ direction, the forefinger will point in the $-\hat{\mathbf{k}}$ direction, and the remaining fingers will point in the $\hat{\boldsymbol{\tau}}$ direction, this being the direction pointing into the page.

The angular momentum is determined by the pendulum's moment of inertia $I = \sum_i m_i \ell_i^2$. The string is taken to be massless. The moment of inertia is therefore simply $m\ell^2$. Which way does the angular momentum vector point? Applying the right-hand-rule to the indicated mass, and taking the direction of motion to be in the direction of increasing θ, one finds that $\mathbf{L} = \mathbf{r} \times \mathbf{p}$ points out of the page. The torque therefore acts as a restoring force, seeking to drive the mass back to the $\theta = 0$ position.

Combining these results, we find for the $\hat{\boldsymbol{\tau}}$ component of equation 37.48, this being the vector component perpendicular to the page, that

$$-\ell mg\sin(\theta) = m\ell^2 \frac{d^2\theta}{dt^2}. \tag{37.51}$$

This result is a nonlinear equation for the motion of the pendulum. Nonlinear equations are often difficult to solve. However, if θ is not too large, we can linearize the equation by invoking the Taylor series expansion

$$\sin(\theta) \approx \theta - \frac{\theta^3}{3!} + \frac{\theta^5}{5!} - \ldots \tag{37.52}$$

of $\sin(\theta)$ around $\theta = 0$. Keeping only the lead term in the expansion, so that $\theta^3/3!$ is taken to be negligible, the equation of motion for the mass on a string becomes

$$-mg\ell\theta = m\ell^2 \frac{d^2\theta}{dt^2}. \tag{37.53}$$

This last equation is one of the canonical equations for the harmonic oscillator, namely except for the choice of constants this equation has the same form as

$$-kx = m\frac{d^2x}{dt^2}, \tag{37.54}$$

which corresponds to a sinusoidal oscillation having angular frequency $\omega = \sqrt{k/m}$.

In the case here, the angular frequency is predicted to be

$$\omega = \left(\frac{mg\ell}{m\ell^2}\right)^{1/2} \equiv \sqrt{\frac{g}{\ell}}. \tag{37.55}$$

The predicted motion is then

$$\theta(t) = A\cos\left(\sqrt{\frac{g}{\ell}}t + \phi\right). \tag{37.56}$$

Here, once again, A and ϕ are the amplitude and phase of the oscillation.

We may also solve equation 37.53 by using the complex variables approach, taking as the proposed solution

$$\theta(t) = \theta_0 \exp(i(\omega t + \phi)), \tag{37.57}$$

the actual solution being the real part of $\theta(t)$. Here ϕ is the phase and θ_0 is the amplitude of oscillation. Substituting in equation 37.53 and simplifying, one finds

$$-\frac{g}{\ell}\theta_0 \exp(i(\omega t + \phi)) = (i\omega)^2 \theta_0 \exp(i(\omega t + \phi)). \tag{37.58}$$

Cancelling factors found on both sides of the equation, one finds

$$\frac{g}{\ell} = \omega^2. \tag{37.59}$$

Taking the real part of equation 37.57, one again finds

$$\theta(t) = \theta_0 \cos\left(\sqrt{\frac{g}{\ell}}t + \phi\right) \tag{37.60}$$

For the x and y coordinates of the mass, we have, to first order in θ,

$$x \equiv \ell \sin(\theta) \approx \ell\theta = \ell A \cos(\sqrt{\frac{g}{\ell}}t + \phi) \tag{37.61}$$

and

$$y = \ell - \ell\cos(\theta) \approx \ell - \ell(1 - \frac{\theta^2}{2} + \ldots \approx 0. \tag{37.62}$$

[Interesting Aside: What happens if θ is not small? Try the experiment.]

37.7 Physical Pendulum: Energy Approach

We now return to the physical pendulum, considered with the energy approach. What is its period of oscillation?

Figure 37.4 shows a physical pendulum. You might imagine it as a meter stick with a hole drilled through one end, so that the meter stick can hang from a nail driven into a wall and swing back and forth. The pendulum has a length ℓ and makes an angle θ with respect to the vertical. The pendulum has a cross section A and a density ρ per unit length, so that its mass M is $\rho A \ell$. We choose a distance x to measure the distance along the pendulum starting at the pivot point, so that $x \in (0, \ell)$.

The total energy is the sum of the kinetic energy and the potential energy. We compute each of these by finding the kinetic and potential energies of a differential segment dx of the rod, and then summing over the differential segments. The section dx has a mass

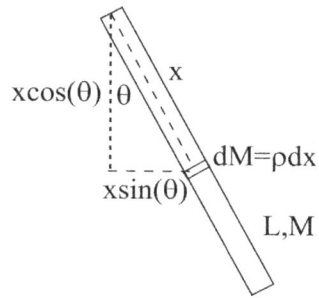

Figure 37.4: A physical pendulum, length L and mass M, at an angle θ from the vertical, showing a differential segment dM a distance x from the pivot point and its vertical distance $x\cos(\theta)$ from the pivot point.

$$dM = \rho A dx. \tag{37.63}$$

To obtain the total mass, we sum (by integration) over all differential segments. The rod is rotating around its axis, so the differential segment has speed

$$v = x \frac{d\theta}{dt} \tag{37.64}$$

and therefore a differential kinetic energy

$$dK = \frac{1}{2} \rho A \, dx \left(x \frac{d\theta}{dt} \right)^2. \tag{37.65}$$

The total kinetic energy of the whole rod is

$$\int_0^\ell dK = \frac{1}{2} \int_0^\ell \rho A \, dx \left(x \frac{d\theta}{dt} \right)^2. \tag{37.66}$$

On performing the integration, we have

$$K = \frac{1}{2} \rho A \left(\frac{d\theta}{dt} \right)^2 \frac{x^3}{3} \Big|_0^\ell. \tag{37.67}$$

One factor of ℓ disappears when we replace $\rho A \ell \to M$, leading to

$$K = \frac{1}{2} \frac{M \ell^2}{3} \left(\frac{d\theta}{dt} \right)^2. \tag{37.68}$$

The mass dM is taken to have potential energy 0 when the pendulum is vertical at $\theta = 0$. When it is at angle θ, it is also a distance x away from the pivot point. The mass's vertical distance below the pivot point, as indicated in the figure, is $x \cos(\theta)$. It has therefore moved upward from its $\theta = 0$ position by a distance $x - x \cos(\theta)$. The differential potential energy of the differential mass is then

$$dU = \rho dx \, Agx(1 - \cos(\theta)). \tag{37.69}$$

Summing the potential energy of all the differential length segments, we have

$$\int_0^x dU = \int_0^\ell dx \, \rho Agx(1 - \cos(\theta)) = \frac{Mg\ell}{2}(1 - \cos(\theta)). \tag{37.70}$$

However, for small θ we can write $\cos(\theta) = 1 - \theta^2/2! + \theta^4/4! - \ldots$. To lowest non-zero order in θ, $1 - \cos(\theta) = \theta^2/2$, leading to

$$U = \frac{Mg\ell\theta^2}{4}. \tag{37.71}$$

The total energy is then

$$E = \frac{1}{2}\frac{M\ell^2}{3}\left(\frac{d\theta}{dt}\right)^2 + \frac{1}{2}\frac{Mg\ell}{2}\theta^2. \tag{37.72}$$

By comparison with the canonical equation for the total energy of a harmonic oscillator, we have $k \to Mg\ell/2$ and $m \to M\ell^2/3$, so the oscillation frequency becomes

$$\omega = \left(\frac{Mg\ell/2}{M\ell^2/3}\right)^{1/2} \tag{37.73}$$

or

$$\omega = \left(\frac{3g}{2\ell}\right)^{1/2}. \tag{37.74}$$

Finally, this physical pendulum's period of oscillation is

$$T = 2\pi \left(\frac{2\ell}{3g}\right)^{1/2}. \tag{37.75}$$

37.8 Calculating Constants of Integration

Having repeatedly demonstrated that the motion of a harmonic oscillator is described by

$$x(t) = A\cos(\omega t + \phi), \tag{37.76}$$

we now turn to the constants of integration A and ϕ. The frequency ω is determined by the physics of the system, but A and ϕ are contingent variables, determined by the particular details of the situation. A common phrasing is to say that A and ϕ are determined by the *initial conditions* provided in the problem, the modifier *initial* referring to the common problem that one knows the starting point of the motion, and then must use the equations of motion to determine where the oscillator has gone at later times. In the phrase *initial conditions*, the adjective *initial* does not actually require that the conditions be reported for the earliest time in the problem. For example, if we are trying to calculate when there was an eclipse in Babylon in 1600 B.C., the initial conditions are the positions, velocities and masses of the planets and sun over the past hundreds or more years, not to mention all of the datable eclipses and planetary occultations.

A general approach to determining A and ϕ is numerical. The presence of the cosine function means that recourse must be had to some non-linear least squares method, e.g., the simplex method (caution: There are two simplex methods. They are completely different.) or other numerical fitting approach. As there are two unknowns, one must have at least two pieces of information in order to be able to solve, least-squares often being appropriate when there are more than two pieces of information available to help determine A and ϕ. As a minor word of warning, some pocket computing devices have what are described as 'root finders' for determining unknowns. Many of these do not work, at least not the way an innocent user might have expected. As a simple test of a root finder, present it with a quadratic such as $x^2 - 3x + 2 = 0$ and see whether it presents you with more than one root. Quadratics, after all, have two.

We here consider as an example a very special case in which a largely algebraic solution can be made. At some time T, we have for the position and velocity of the oscillator $x = X$ and $v = V$. From these three pieces of information, we can write

$$X = A\cos(\omega T + \phi) \tag{37.77}$$
$$V = -\omega A\sin(\omega T + \phi). \tag{37.78}$$

Squaring and rearranging lets us eliminate the trig functions, namely we can write

$$X^2 + \left(\frac{V}{\omega}\right)^2 = A^2\cos^2(\omega T + \phi) + A^2\sin^2(\omega T + \phi). \tag{37.79}$$

The identity on the squares of the sines and cosines eliminates them from the equation, so that

$$A = \left(X^2 + \left(\frac{V}{\omega}\right)^2\right)^{1/2}. \tag{37.80}$$

Taking the ratio of the two parts of equation 37.77,

$$\frac{\sin(\omega T + \phi)}{\cos(\omega T + \phi)} = \frac{-V/\omega}{X} \tag{37.81}$$

The ratio of the sin to the cosine is the tangent, here $\tan(\omega T + \phi)$, so taking the arctangent of both sides leads to

$$\omega T + \phi = \arctan\left(\frac{-V}{\omega X}\right) \tag{37.82}$$

and therefore to

$$\phi = \arctan\left(\frac{-V}{\omega X}\right) - \omega T. \tag{37.83}$$

As in all cases in which inverse trig functions are used numerically, care must be take to identify in which quadrant the arctangent function's answer lies.

37.9 Homework

1. Test if $\exp(-\Gamma t)$, Γ being a real number, is a solution to equation 35.1. Reminder: If A^2 is negative, then A cannot be a real number.

2. Confirm that equation 35.2 is a solution to equation 35.1 by substituting equation 35.2 into equation 35.1.

3. A mass at the end of a spring oscillates with a frequency of 2π Hertz and an amplitude of 11 cm. At $t = 0$ the mass is at its equilibrium position ($x = 0$), and is headed upwards. a) Find the possible equations describing the position of the mass as a function of time, in the form $x = a\cos(\omega t + \alpha)$, giving numerical values for a, ω, and α. b) What are the numerical values of x, $\frac{dx}{dt}$, and $\frac{d^2x}{dt^2}$ at $t = 5/3$ second? c) How would the answer change if you did not know, at $t = 0$, whether the mass was headed up or down?

4. A mass on the end of a spring oscillates with an amplitude of 3 cm at a frequency of π Hertz. a) At $t = 0$ the mass is at $x = +3$ cm. Describing the mass's position as $x = A\cos(\omega t + \alpha)$, what are A, ω, and α? b) At $t = 4$ s the mass is at $x = 3$ cm. What are A, ω, and α? [Hint: not all same as in a.] c) At $t = 2$ s the mass is at $x = 1.5$ cm. What are A, ω, and α? Hint: you should have 2 answers for α. Why? d) If in part c we had $\frac{dx}{dt} < 0$ at $t = 2$ s, what is α?

5. A mass at the end of a spring oscillates with a period of $1/4\pi$ seconds and an amplitude of 15 cm. At $t = 0$ the mass is at its equilibrium position ($x = 0$), and is headed downward. a) Find the possible equations describing the position of the mass as a function of time, in the form $x = a\cos(\omega t + \alpha)$, giving numerical values for a, ω, and α. b) For one solution from (a), what are x, $\frac{dx}{dt}$, and $\frac{d^2x}{dt^2}$ at $t = 4/3$ second. c) How would the answer change if you did not know, at $t = 0$, whether the mass was headed up or down?

6. A point moves in a circle at a constant speed of 30 cm/s. The period of one complete journey around the circle in 5 seconds. At $t = 0$ a line from the center of the circle to the point makes an angle of $+30°$ to the x axis. a) Why does the above information <u>not</u> reveal whether the point is travelling in a clockwise or counterclockwise direction? b) Obtain the equation for the x coordinate of the particle as a function of time, in the form $x = A\,\text{Re}(\exp[i(\omega t - \alpha)])$, giving the numerical values for A and ω. c) Find the values of x, $\frac{dx}{dt}$, and $\frac{d^2x}{dt^2}$ at $t = 3$ seconds. [Hint: Think about part (a). How many answers should you get in parts (b) and (c)?]

7. A mass on the end of a spring oscillates with an amplitude of 5 cm at a frequency of $+2.0$ Hertz. a) At $t = 0$ the mass is at $x = +1$ cm. Describing the mass's position as $x = A\cos(\omega t + \alpha)$, what are A, ω, and α? b) At $t = 3$ s the mass is at $x = 4$ cm. What are A, ω, and α? [Hint: not all same as in a.] c) At $t = 1$ s the mass is at $x = 1.5$ cm. What are A, ω, and α? [Hint: you should have 2 answers for α. Why?] d) If in part c we had $\frac{dx}{dt} < 0$ at $t = 2$ s, what is α?

8. a) Show $x = Ae^{i\omega t}$ is a solution of $\frac{d^2 x}{dt^2} + \gamma x = 0$. b) If ω is to be a real number, what must be the sign of γ? c) Suppose $x = Ae^{i(\omega t + \alpha)}$, $\alpha \neq 0$. Is x a solution of the same differential equation? d) Are A and α determined by the differential equation? Why?

9. A mass m hangs from one or more springs. In this problem all springs have the same force constant k. a) Find the mass's frequency of oscillation if it hangs from one spring. What would be the mass's period of oscillation if: b) It were hung from three springs side-to-side, all three springs being attached both to the mass and to the ceiling? c) It were hung from three springs in a line? d) For case (c), what would be the energy of the system, if the maximum displacement of the mass from equilibrium was L?

10. A mass of 4 g is hung from a spring and set in oscillatory motion. At $t = 0$ the displacement is 35 cm and the acceleration is -2.40 cm/s^2. Find the free body diagram, the equation of motion, and the motion as a function of time. What is the numerical value of the spring constant?

11. A mass of 3 kg is hung from a spring and set in oscillatory motion. At $t = 0$ the displacement is 0.4 m from the equilibrium position and the acceleration is -1.20 m/s^2. Find the free body diagram, the equation of motion, and the motion as a function of time. What is the spring constant? Hint: At the equilibrium position, what two forces are acting on the mass? Do they cancel? Why?

12. Consider a floating $L \times L \times L$ cube having mass m. The density of the water is ρ. A floating cube is subject to a downward force (its weight) and an upward force. The upward force is the buoyant force. The buoyant force is determined by the volume V_s of the part of the cube that is underwater. The buoyant force is numerically equal to the weight of a volume V_s of water. What is the energy of the system if the maximum displacement upward is Z? [Hints: $W = \int dx\, F_x(x)$. You'll need to find the equilibrium position, and take this as the zero point.]

13. Consider a simple pendulum, consisting of a mass and string. (i) Set up the free body diagram. (ii) Find the equation of motion for the horizontal displacement x of the pendulum bob. (iii) Show, for small x, that the pendulum is a harmonic oscillator.

14. A rod of length L is hung from a pivot. (See Figure 37.4 for an example of pivoting.) The density of the rod (mass per unit length) increases linearly with distance from the pivot point, so that the density at the extremity of the rod is twice the density at pivot point. What is the frequency of small oscillations of the rod? What would happen to the frequency of oscillation if the rod density at the pivot point were twice the density at the outer end of the rod, the rod density varying linearly with position along the rod? [Hints: This *is not* a pendulum with a mass at one end. To do the problem, integrals are unavoidable.]

15. A uniform rod having mass M is pivoted 1/4 of the way from its top to its bottom. (See Figure 37.4 for an example of pivoting.) It is free to rotate in a vertical plane. What is its oscillation frequency? Solve using the energy approach. Hint: If the rod rotates, the bottom goes up, but the top goes down.

16. A conical pendulum consists of a mass hanging at the end of a string. The mass goes in circles centered on its suspension point; the string's motion describes the surface of a cone. Find the force diagram and the differential equations of motion for the mass. Solve them. Hint: At any one instant, you can put the x-axis through the mass and the z-axis through the center of rotation of the mass.

17. A five-kilogram mass rests on a frictionless table. It is connected to two springs that lie along the x-axis. The far end of each spring is connected to a wall. Each spring has a spring constant $k = 500$ Newtons/meter. The mass is set into oscillation along the x axis. (a) Find the force (free body) diagram for the mass. (b) Write the differential equation of motion (Newton's Second law) for the

mass. (c) Solve the differential equation of motion. Find the mass's frequency of oscillation in Hertz. (d) Find the equation describing the motion of the mass as a function of time (Hint: Not the same equation as found from part (b).) Put this equation into the general form $A\sin(\omega t + \alpha)$. Hint: Take the $x = 0$ position to be the equilibrium position of the mass.

18. (a) A mass on a spring is set into vibration. Its motion follows $y(t) = A\sin(\omega t + \alpha)$. The frequency of oscillation is 3π Hertz. The oscillations take the mass between $y = 0.3$ m and $y = -0.3$ m. At $t = 8$ one finds that the displacement from equilibrium is zero. Solve for A, ω, and α.

19. We have a long, very light rod with total length L. A mass m_1 is at the top end of the rod, a distance L_1 above a pivot point. (See Figure 37.4 for an example of pivoting.) A second mass m_2 is attached to the other end of the rod, a distance $L_2 = L - L_1$ below the pivot point. $m_1 < m_2$ and $L_1 < L_2$. The angle between the orientation of the rod and the vertical is θ. (a) Find the kinetic energy of the two masses when the rod is swinging back and forth. (b) Find the potential energy of the two masses when the rod is swinging back and forth. (c) Find the total energy of the two masses. What is their frequency of oscillation? Open ended: (a) Suppose $m_1 L_1 = m_2 L_2$. What happens to the oscillation frequency? What does this odd result mean? (b) Suppose $m_1 L_1 > m_2 L_2$. What happens to the oscillation frequency? What does this odd result mean?

20. Consider an oscillatory function $x = 3\cos(2t + \pi/2)$. t is in seconds. a) Sketch the function. *Label* axes. b) What is the amplitude? c) What are the largest and smallest values that x attains? d) What is the angular velocity ω in radians per second? e) What is the frequency in Hertz? f) What is the period T? Sketch T on your figure from part a). g) What is the phase ϕ?

21. A mass at the end of a spring oscillates with a period of 2π seconds and an amplitude of 10 cm. At $t = 0$ the mass is at its equilibrium position ($x = 0$), and is headed downward. a) Find the possible equations describing the position of the mass as a function of time, in the form $x = A\cos(\omega t + \alpha)$, giving numerical values for A, ω, and α. b) For one solution from (a), what are x, $\frac{dx}{dt}$, and $\frac{d^2 x}{dt^2}$ at $t = 5/3$ second. c) How would the answer change if you did not know, at $t = 0$, whether the mass was headed up or down?

22. A mass on the end of a spring oscillates with an amplitude of 20 cm at a frequency of $+4.0$ Hertz. a) At $t = 0$ the mass is at $x = +2$ cm. Describing the mass's position as $x = A\cos(\omega t + \alpha)$, what are A, ω, and α? b) At $t = 5$ s the mass is at $x = 10$ cm. What are A, ω, and α? [Hints: Not all same as in a. You should have 2 answers for α. Why?] c) If in part b we had $\frac{dx}{dt} < 0$ at $t = 5$ s, what would α be?

23. A mass m hangs from a spring of spring constant k. The spring has unstretched length ℓ_o. The presence of the mass causes the spring to be stretched by Δ when the mass remains stationary. What is Δ? The spring now oscillates through displacement $z(t)$ around the equilibrium stretch Δ. (a) Write the force diagram and the second law for m. Show the forces due to Δ and $z(t)$ separately. (b) What is the value of Δ? (Hint: What is $z(t)$ if the mass *remains* stationary.) Find the frequency of oscillation of m. What would be the mass's period of oscillation if it were hung from two identical springs in a line? c) For case (b), what would be the energy of the system, if the maximum displacement of the mass from equilibrium was L?

24. A mass of 12 kg is hung from a spring and set in vertical oscillatory motion. At $t = 0$ the displacement is 1.0 m from the equilibrium position and the acceleration is -0.15 m/s^2. Find the free body diagram, the equation of motion, and the motion as a function of time. What is the spring constant? Hint: At the equilibrium position, what two forces are acting on the mass? Do they cancel? Why?

25. Consider two unequal masses M and $3M$ fixed at opposite ends of a rigid, nearly massless rod having length L. The rod is free to rotate, in the vertical plane, around a point P located a distance $L/4$ from the mass M. The rod's equilibrium position has the mass M vertically *above* the point P. Using the energy approach, find the frequency of small oscillations of the rod: a) The energy E of the system has two parts, a kinetic energy K and a potential energy P. Write K and P in terms of the angle θ between the long axis of the rod and the vertical, and find E. [Hint: $v = r\frac{d\theta}{dt}$ for a mass rotating at

the end of a rod; $U = mgh$ for gravitational potential energy.] b) Rewrite the energy E, putting it into the approximate form appropriate for small values of θ. c) Show for small oscillations of θ that the system behaves as a harmonic oscillator. What is the angular frequency of the harmonic oscillations?

26. Same physical system as the previous question. Consider two unequal masses M and $3M$ fixed at opposite ends of a rigid rod of length L. The rod is free to rotate around a point P located a distance $L/4$ from the mass M. The rod's equilibrium position has the mass M vertically *above* the point P. Using the *torque* approach, find the frequency of small oscillations of the rod. Hint: Your answers to to this problem and the previous problem should agree with each other.

27. A non-uniform rod hanging vertically is pivoted around its top end. (See Figure 37.4 for an example of pivoting.) The variable that labels the distance from the pivot is x. The density of the rod varies with x as $\rho = bx^2$. (a) Find the mass of the rod. (b) Using whatever method you prefer, find the frequency of small oscillations of the rod. Hint: The density depends on position, so the mass of a small section of length dx is $dm = bx^2 dx$. Find the kinetic and potential energies dK and dU of differential segments, and total them by integrating.

28. We have an oscillation with angular frequency $\omega = 10$ rad/s. At $t = 5$, $x = 2.917$ and $v_x = 6.95$. Find the amplitude of oscillation.

29. A pendulum consists of a very light rod having total length L and pivoted at $L/5$ from one end. (See Figure 37.4 for an example of pivoting.) Masses are attached to each end of the rod. The mass that is close to the pivot has mass M; the mass that is further from the pivot point has mass $5M$. The angle between the rod and the vertical is θ. (a) Find the total potential energy of the rod and masses as a function of θ. Take the potential energy to be zero when $\theta = 0$. (b) The rod is swinging back and forth, and at some moment has angular velocity $\Omega = d\theta/dt$. What is the total kinetic energy of the system? (c) Find the total energy of the system. Find the total energy of the system if θ is small. Find the frequency of oscillation of the two masses.

This space reserved for your notes.

Chapter 38

Harmonic Motion, Damped or Driven

In this chapter, we consider two special cases of harmonic motion with external forces. First, we consider harmonic oscillators in which there is also a frictional damping force that causes the amplitude of the oscillation to decay toward zero. Second, we consider harmonic oscillators subject to an external driving force that is a sinusoid in time. These two special cases are preparations for the next chapter, in which we consider a harmonic oscillator subject both to a damping and to a driving force. By treating the special cases first, and then advancing to the general case, we illustrate a fundamental approach to solving a difficult problem, namely one examines special case after special case until the difficult problem ceases to be refractory.

38.1 Harmonic Oscillation with Damping

If you set up a pendulum and set it swinging, you will observe if you watch long enough that the pendulum gradually swings less and less, and finally comes so far as can be seen to a stop. As it comes to a stop, did it swing more and more slowly? Did the periods get longer and longer as the pendulum gradually ended its motions? Answering those questions could be done with careful experimental measurements. Here we take a different approach, namely we do analytic calculations.

The friction we are discussing is a simple linear drag force

$$\mathbf{F}_f = -f \frac{d\mathbf{r}}{dt}. \tag{38.1}$$

In this equation, $\mathbf{F}_f$ is the drag force, f is the drag coefficient, and $\mathbf{r}$ is the position of the moving object. We call this linear friction, because the drag is proportional to the first power of the velocity. There are systems like this, but in many systems the drag force depends much more strongly on the speed, for example as v^3. Observe that this frictional force is not the same as the static, kinetic, rolling, and tractive friction forces discussed in an earlier chapter. Unlike those forces, the magnitude of this force is proportional to the object's speed.

We consider a mass resting on a table and attached to a spring, the mass seen in Figure 38.1. The tabletop is horizontal and non-accelerating, so the normal force and the force of gravity simply cancel. The Second Law is readily written

$$m \frac{d^2 x}{dt^2} = -f \frac{dx}{dt} - kx. \tag{38.2}$$

This equation is readily rearranged to read

$$m \frac{d^2 x}{dt^2} + f \frac{dx}{dt} + kx = 0. \tag{38.3}$$

To solve we replace x with a complex variable z of which x is the real part. There are other solution methods, but this one is valid. For the solution, we try

$$z = A \exp(\imath(\Omega t + \phi)). \tag{38.4}$$

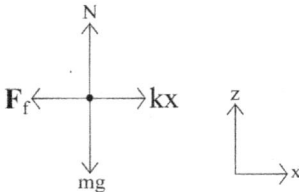

Figure 38.1: Force diagram for a mass m resting on a table, connected to a spring having spring constant k. The other end of the spring is anchored in place and does not move. The rest location of the mass is $x = 0$. A drag force $\mathbf{F}_f$ is also applied. The diagram does *not* indicate the actual directions of the spring or drag force. These both oscillate in time, not in phase with each other. Depending on where the mass is in its oscillations, the spring and drive force may be pointing in the indicated direction or the direction opposite to that indicated, and may be antiparallel or parallel.

At this stage, the best reason for using that guess is that it turns out to work. A course in differential equations will treat better rationales for trying this function. Our reason for writing the friction as Ω rather than ω will soon be apparent. Making the replacement, we find

$$m(\imath\Omega)^2 A\exp(\imath(\Omega t + \phi)) + f\imath\Omega A\exp(\imath(\Omega t + \phi)) + kA\exp(\imath(\Omega t + \phi)) = 0. \tag{38.5}$$

After extracting a common factor $A\exp(\imath(\Omega t + \phi))$, this equation becomes

$$(-m\Omega^2 + \imath\Omega f + k)A\exp(\imath(\Omega t + \phi)) = 0. \tag{38.6}$$

In all non-trivial cases, the factor $A\exp(\imath(\Omega t + \phi))$ is not zero, so it may be divided out, leaving

$$-m\Omega^2 + \imath f\Omega + k = 0. \tag{38.7}$$

This equation is a quadratic in the unknown quantity Ω. Applying the standard solution to the quadratic form,

$$\Omega = \frac{-\imath f \pm ((\imath f)^2 - 4(-m)k)^{1/2}}{-2m}, \tag{38.8}$$

which simplifies to

$$\Omega = \imath\frac{f}{2m} \pm \left(\frac{k}{m} - (\frac{f}{2m})^2\right)^{1/2}. \tag{38.9}$$

Applying this solution for Ω to equation 38.4, we have

$$z(t) = A\exp\left(\imath(\imath\frac{f}{2m} \pm \left(\frac{k}{m} - (\frac{f}{2m})^2\right)^{1/2})t + \phi\right), \tag{38.10}$$

whose real part is

$$x(t) = Ae^{-\gamma t/2}\cos\left(\left(\frac{k}{m} - (\frac{f}{2m})^2\right)^{1/2}t + \phi\right). \tag{38.11}$$

I introduced the notation $\gamma = f/m$ for the damping constant. The oscillation frequency of the cosine is $\omega = (\frac{k}{m} - (\frac{f}{2m})^2)^{1/2}$. That quantity is always less than the undamped frequency $\sqrt{k/m}$, so damping has the effect of reducing the frequency of oscillation. In saying equation 38.11 has a cosine term, the implicit assumption is made that $\frac{k}{m} - (\frac{f}{2m})^2 > 0$. If $\frac{k}{m} - (\frac{f}{2m})^2$ is less than zero, $(\frac{k}{m} - (\frac{f}{2m})^2)^{1/2})$ would be the square root of a negative number, so it would be imaginary. There would be decay but not repetitive oscillation.

Denoting $(\frac{k}{m} - (\frac{f}{2m})^2)^{1/2}) = \varpi$, our solution for $x(t)$ reads

$$x(t) = A\exp(-\gamma t/2)\cos(\varpi t + \phi). \tag{38.12}$$

38.1. HARMONIC OSCILLATION WITH DAMPING

The amplitude of the oscillation is $A\exp(-\gamma t/2)$, meaning that the amplitude of oscillation decays exponentially in time. The oscillation frequency is ϖ, meaning that the period T is $2\pi/\varpi$ or

$$T = \frac{2\pi}{\left(\frac{k}{m} - (\frac{\gamma}{2})^2\right)^{1/2}}. \tag{38.13}$$

The period of oscillation does not depend on the time t. As the mass oscillates, the amplitude of its oscillations become smaller and smaller, but the period of the oscillation, the spacing between pairs of zero crossings, does not change, because $\cos(\varpi t + \phi)$ has its zeros in the same places, no matter the initial amplitude.

That's seen in the following plot of a decaying oscillation, in which we have chosen the phase ϕ to be zero. Aside: To generate this figure, we used the parameters $\gamma/2 = 0.08$ and $\varpi = 1$.

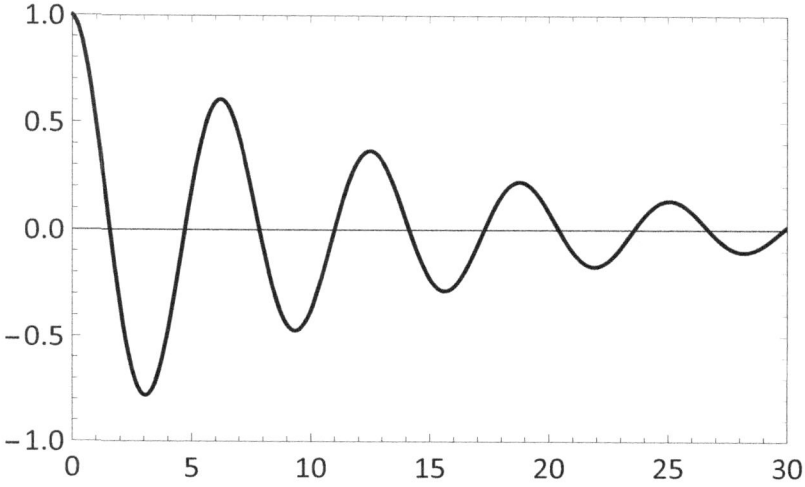

Figure 38.2: A damped relaxation described by $y(t) = \exp(-0.08t)\cos(t)$.

What is the energy of this system? We have a mass on a spring, so in general we may write

$$E = \frac{1}{2}m\left(\frac{dx}{dt}\right)^2 + \frac{1}{2}kx^2 \tag{38.14}$$

For the damped mass, the motions are described by the equations

$$x(t) = Ae^{-\gamma t/2}\cos(\varpi t + \phi), \tag{38.15}$$

$$\frac{dx}{dt} = -A\varpi e^{-\gamma t/2}\sin(\varpi t + \phi) - \frac{A\gamma}{2}e^{-\gamma t/2}\cos(\varpi t + \phi).$$

Substituting these into the general form for the energy, we obtain the cumbersome result

$$E = \frac{1}{2}kA^2 e^{-\gamma t}\cos^2(\varpi t + \phi) + \frac{1}{2}m(A^2\varpi^2 e^{-\gamma t}\sin^2(\varpi t + \phi) + \left(\frac{A\gamma}{2}\right)^2 e^{-\gamma t}\cos^2(\varpi t + \phi)$$
$$+ A^2\gamma\varpi e^{-\gamma t}\sin(\varpi t + \phi)\cos(\varpi t + \phi)). \tag{38.16}$$

Some simplifications are possible. In particular, there is a common factor $A^2 e^{-\gamma t}$. You now see why I chose to say that the amplitude decreases as $\exp(-\gamma t/2)$, namely that choice means that the energy decreases as $\exp(-\gamma t)$. If I had instead chosen to write $\gamma = f/2m$, the amplitude would indeed have decreased as $\exp(-\gamma t)$ but then the energy would decrease as $\exp(-2\gamma t)$. You can move that factor of 2 around, but you can't make it go away.

This form is a bit cumbersome. Fortunately, in many cases there is a resolution, namely *weak damping*. Under weak damping, the mass oscillates many times before the amplitude of the oscillation has decreased substantially. That's a somewhat qualitative statement. The more nearly quantitative statement is

$$\frac{k}{m} \gg \left(\frac{\gamma}{2}\right)^2. \tag{38.17}$$

However, $\omega = \left(\frac{k}{m} - \left(\frac{\gamma}{2}\right)^2\right)^{1/2}$, so $\omega \approx \sqrt{k/m}$, in which case the velocity becomes

$$\frac{dx}{dt} \approx -A\omega \exp(-\gamma t/2) \sin(\omega t + \phi). \tag{38.18}$$

If you look carefully, you see that this approximation can't actually be correct. There will assuredly be moments in time at which the sin function $\sin(\omega t + \phi)$ is exactly zero, at which moment the other term in equation 38.15b, the term in $\cos(\omega t + \phi)$, must be dominant. However, we are only going to use this approximation to calculate the energy of the system. At the moments when the cosine term is dominant, $x(t)$ is at a maximum or a minimum, so almost all the energy of the system is potential energy; the fact that the kinetic energy term is inaccurate does not matter, because that term is very nearly zero.

In this approximation, the energy under weak damping is

$$E \approx \frac{1}{2}kA^2 e^{-\gamma t} \cos^2(\omega t + \phi) + \frac{1}{2}mA^2\omega^2 e^{-\gamma t} \sin^2(\omega t + \phi). \tag{38.19}$$

There are common factors here. The covert result $m\omega^2 = k$ and a trig identity lead to

$$E \approx \frac{1}{2}kA^2 e^{-\gamma t}. \tag{38.20}$$

With weak damping, the energy of the system decreases exponentially in time

38.2 Energy Storage–The Quality Parameter

This section considers the storage of energy in a harmonic oscillator, and its disappearance due to friction. In a certain sense, all we are going to do is to introduce some new notation. It's 'only notation', but it is *very useful* notation. You've actually seen the utility of good notation before. Early in the course I mentioned the priority dispute between Newton and Leibniz as to which of them had invented calculus, Each of them advanced a notation for integrals and derivatives. Whatever the truth as to who had invented calculus, there is no doubt that we use Leibniz's notation, because Leibniz spent time thinking about his notation, making it accessible to mortals, while Newton's notation was adequate for supreme geniuses such as himself.

The notation we introduce is the quality parameter Q. Q is defined as

$$Q = \frac{\omega_0}{\gamma}. \tag{38.21}$$

Q is the ratio of the system's undamped frequency ω_0 to the relaxation rate γ for the energy. The utility of Q arises from three equations from the previous chapter. First, for weakly damped systems

$$E \approx \frac{1}{2}kA^2. \tag{38.22}$$

Second, in the presence of damping the energy of the system decreases with time approximately as

$$E \sim \exp(-\gamma t), \tag{38.23}$$

while the amplitude of oscillation of the system decreases as

$$A \sim \exp(-\gamma t/2). \tag{38.24}$$

Let us focus on γt. We rewrite it by multiplying it by one, one being in the form a/a. We have

$$\gamma t \equiv \frac{\gamma}{\omega_0}(\omega_0 t) \equiv \frac{2\pi\gamma}{\omega_0}\frac{\omega_0 t}{2\pi}. \tag{38.25}$$

Several of these quantities have simple physical descriptions. ω_0 gives the oscillation rate in radians per second, so $\omega_0/2\pi$ gives the oscillation rate in periods (cycles) per second. Its inverse $2\pi/\omega_0$ describes the relaxation in terms of seconds per period, which we denote t_0, with

$$t_0 = \frac{2\pi}{\omega_0}. \tag{38.26}$$

Correspondingly, t/t_0 is the number of oscillatory periods during a time t. We then have the notational result

$$\gamma t \equiv \left(\frac{2\pi}{Q}\right)\left(\frac{\omega_0 t}{2\pi}\right) \equiv \left(\frac{2\pi}{Q}\right)\left(\frac{t}{t_0}\right). \tag{38.27}$$

We may then write the decay of the energy as

$$E \sim \exp(-\gamma t) \sim \exp(-\frac{2\pi}{Q}\frac{t}{t_0}). \tag{38.28}$$

The larger Q is, the slower the energy relaxes out of the system due to the action of friction.

The relaxation of the energy may also by written

$$E \sim \exp\left(-\frac{t/t_0}{Q/2\pi}\right). \tag{38.29}$$

The energy relaxes by a factor of $1/e$ in $Q/2\pi$ periods.

38.3 Harmonic Oscillation with an External Driving Force

As an alternative to the harmonic oscillator with friction, we consider the response of a harmonic oscillator to an external driving force. The driving force we discuss is a cosine wave, so that

$$F_x(t) = F_o \cos(\omega t). \tag{38.30}$$

There are clearly a very large number of different driving forces that might be considered. Why do we choose this one? The first part of the answer, as was shown by Jean-Baptiste Joseph Fourier, is that any time-dependent physical force can be written as a sum of forces that depend on time as cosine waves, the amplitudes and phases of the cosines being chosen so as match the form of the particular force being described. (This statement is an extension of the familiar assertion that any sound can be described as a sum of pure tones, each tone having the needed amplitude and phase.)

The second part of the answer is that the equation of motion of an undamped harmonic oscillator

$$m\frac{d^2x}{dt^2} + kx = F_x(t) \tag{38.31}$$

is a linear equation. Other than the driving force, which is independent of x, each term in the equation of motion contains only one factor of $x(t)$, either $x(t)$ itself or one of its derivatives. As a consequence, the response of an oscillator to an arbitrary time-dependent force can be calculated from the responses of the oscillator to cosine wave driving forces, if the responses are known for all frequencies of the driving force.

We guess that the solution to the equation of motion is

$$x(t) = A\cos(\omega t + \delta). \tag{38.32}$$

Here δ is a phase factor. Inserting the guess into the equation of frictionless motion, we have

$$m(-\omega^2 A\cos(\omega t + \delta)) + kA\cos(\omega t + \delta) = F_o \cos(\omega t). \tag{38.33}$$

In order for the two sides of the equation to have zeros at the same time, δ must be 0 or π. We take $\delta = 0$ and return to this wuestion later. Deleting common factors, we have

$$-A\omega^2 + \frac{k}{m}A = \frac{F_o}{m}, \tag{38.34}$$

which leads to

$$A(\omega) = \frac{F_o/m}{-\omega^2 + \omega_o^2}. \tag{38.35}$$

We could do the same calculation using the complex variable approach. For the position and the applied force we write

$$x(t) = A\exp(i(\omega t + \delta)), \tag{38.36}$$
$$F_x(t) = F_o \exp(i(\omega t)). \tag{38.37}$$

Inserting these forms into the equation of motion, we have

$$-A\omega^2 m \exp(i(\omega t + \delta)) + kA\exp(i(\omega t + \delta)) = F_o \exp(i(\omega t)), \tag{38.38}$$

which can be factored as

$$A(-\omega^2 + \frac{k}{m})\exp(i(\omega t + \delta)) = F_o \exp(i(\omega t)). \tag{38.39}$$

Dividing by $m\exp(i(\omega t))$, and applying Euler's identity to the $\exp(i\delta)$,

$$A(-\omega^2 + \frac{k}{m})(\cos(\delta) + i\sin(\delta)) = F_o \tag{38.40}$$

The right-hand-side of the equation is a real number; its imaginary part is zero. The only way that the left-hand-side of the equation can have a zero imaginary part is if the phase δ is 0 or π. In that case

$$A(\omega) = \frac{F_o}{m}\frac{\cos(\delta)}{-\omega^2 + k/m}. \tag{38.41}$$

If δ is 0 or π, then $\cos(\delta)$ is 1 or -1. If we require $A > 0$, then $\cos(\delta) = 1$ requires that

$$-\omega^2 + k/m = -\omega^2 + \omega_0^2 > 0, \tag{38.42}$$

which means that if $\omega^2 < \omega_o^2$ then $\delta = 0$.

On the other hand, if

$$-\omega^2 + k/m = -\omega^2 + \omega_0^2 < 0, \tag{38.43}$$

then in order for A to be positive we must have $\cos(\delta) = -1$, which means that if $\omega^2 > \omega_o^2$ then $\delta = \pi$.

It is worthwhile to consider how $A(\omega)$ and $\cos(\delta)$ depend on ω. There are several interesting limits here. If $\omega \to 0$, then

$$A \to \frac{F_o/m}{k/m} = \frac{F_o}{k}. \tag{38.44}$$

That is, when we approach the static limit, we see the static response, how far a spring stretches under an external force.

On the other hand, if $\omega \to \infty$, then

$$A \to \frac{F_o/m}{\omega^2} \sim \omega^{-2}. \tag{38.45}$$

At large frequencies, $A(\omega)$ goes to zero roughly as ω^{-2}. Finally, as $\omega \to \omega_o$, $A(\omega) \to \infty$. A becomes extremely large, the phenomenon known as *resonance*. A does not actually become infinite. Instead, as $\omega \to \omega_0$ and A becomes large, it is no longer a valid approximation to ignore friction.

38.4 Discussion

In our treatment of the damped, not-driven harmonic oscillator, we introduced a useful general method for solving problems, an approximation that is good when it matters, and wrong when it has no effect. Sometimes you can find an approximation that is accurate when it matters, even though it is seriously incorrect at points where it has little effect, and that can be good enough.

38.5 Homework

1. An object of mass 50 kg is hung from a spring whose spring constant is 400 N/m. The object is subject to a hydrodynamic friction force $-bv$, where v is its velocity. a) Set up the differential equation of motion for free oscillations of the system. b) Solve the equations of motion. c) If the damped frequency is 0.85 times the undamped frequency, what is the value of the constant b? d) What is Q for this system? By what factor is the oscillation amplitude reduced after ten cycles? By what factor is the system's total energy reduced after five cycles?

2. A fine spring is used to suspend a large sphere in water. The spring constant k of the spring is 80 dyne/cm. For a slowly moving sphere, the drag coefficient is $b = 6\,\pi\eta a$, where a is the radius and η is the viscosity of the water. [$\eta = 0.01$ Poise in cgs units at 20^0 C]. The sphere, of $a = 0.5$ cm, has a mass of 5 g. What are the undamped frequency in Hertz, the damped frequency in Hertz, and the Q of the system? Set up the equation of motion, and *solve from first principles*.

3. An object of mass 3.0 kg is hung from a spring whose spring constant is 400 N/m. The body is subject to a resistive force $-bv$ with $b = 60$ in SI units. Do all calculations algebraically and plug in numbers at the end. a) What are the dimensions of b? b) Set up the equation of motion for the mass. c) Solve explicitly the equation of motion. Find the oscillation frequency and the position as a function of time.

4. An object of mass 100 kg. is hung from a spring whose spring constant is 200 N/m. The object is subject to a hydrodynamic friction force $-bv$, where v is its velocity. a) Set up the differential equation of motion for free oscillations of the system. b) Solve the equations of motion. c) If the damped frequency is 0.9 times the undamped frequency, what is the value of the constant b?

This space reserved for your notes.

This page reserved for your notes.

Chapter 39

The Damped, Driven Harmonic Oscillator

In the previous chapter, we discussed a damped harmonic oscillator, in which the natural ringing of the oscillation gradually fades out due to friction or some other dissipative mechanism. The amplitude of the oscillation was shown to fade exponentially; the quality parameter Q was introduced as a useful description of the exponential decay rate. We also discussed a harmonic oscillator that was subject to an external sinusoidal driving force, but is not subject to friction or other dissipative resistance. As the driving frequency approached the natural oscillation frequency of the undriven oscillator, the amplitude of oscillations increased without apparent limit. Infinite response amplitudes do not appear to be physical behaviors, so what is going on? The answer is that in real systems there is friction. The amount of frictional dissipation depends on the system, but in all systems there is some sort of dissipation, so the system response never becomes infinitely large.

We now reach the general case from the previous Chapter, the driven, damped harmonic oscillator. Having said that the amplitude never becomes infinite, we reasonably ask how a driven, damped harmonic oscillator does behave. The response of the system depends on the driving frequency. We can characterize the system's response in terms of the amplitude of oscillation, the velocity with which the driven particle moves, and the amount of power absorbed by the system and then lost to friction. Each of these characterizations is described by a figure, the size of the response as a function of frequency. These *response curves* show the amplitude, the velocity, and the power absorption as a function of frequency. The curves have a more or less sharp maximum, characterized by a frequency at the maximum and a width, the width being a range of frequencies near the maximum frequency over which the amplitude, velocity, or power absorption is substantially non-zero.

39.1 Basic Calculation

As an example, we consider a mass m on a spring that has spring constant k. The mass is subject to an additional external force $F_d = F_0 \cos(\omega t)$, and a frictional force $-f \frac{dx}{dt}$. The sinusoidal force as written has acted at all past times, and will continue to act at all times into the future. If we instead had a harmonic oscillator that was initially at rest, and at some time turned on the driving force, the response of the harmonic oscillator would be a bit different than what we describe here, namely there would be *transient responses* that would damp out exponentially. Details of the transient responses are beyond the scope of this volume, but you should remember that they are there. If you start driving a damped harmonic oscillator, the system takes a while to settle down into the responses discussed here.

In this diagram, the external (driving) force points alternately in one direction and in the opposite direction; the friction force has the same property. The arrows in the force diagram thus indicate the axis along which the force points but not which way each force points at any particular time. From the force diagram, we can write the equation of motion as

$$m\frac{d^2x}{dt^2} = -kx - f\frac{dx}{dt} + F_o\cos(\omega t). \tag{39.1}$$

The equation is now rearranged to put all terms depending on x on the same side of the equation. The external force is the same no matter where the particle is, so it is independent of x and remains on the other side of the equation, giving

$$m\frac{d^2x}{dt^2} + f\frac{dx}{dt} + kx = F_o\cos(\omega t). \tag{39.2}$$

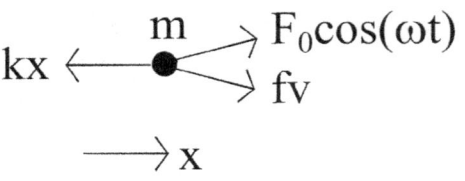

We now replace the real variable x with a complex variable z, taking

$$x = \Re(z), \tag{39.3}$$

with

$$z = A\exp(\imath(\omega t - \delta)). \tag{39.4}$$

Figure 39.1: Force diagram for a damped, driven harmonic oscillator in which a mass m is subject to a spring force kx, a friction force fv, and a driving force $F_0\cos(\omega t)$. All forces and motions lie along the x-axis (indicated), with some force vectors being splayed slightly so that they are cleanly visible.

That is, the response to the external force is at the same frequency as the driving force but the response and driving force may differ in phase by an angle $-\delta$. Why the minus sign? Rather soon, we are going to multiply $\exp(-\imath\delta)$ to the other side of the equation, where it becomes $\exp(+\imath\delta)$. Having the plus sign after this point becomes convenient. We rewrite equation 39.2 as

$$-m\omega^2 A\exp(\imath(\omega t - \delta)) + f\imath\omega A\exp(\imath(\omega t - \delta)) + kA\exp(\imath(\omega t - \delta)) = F_0\exp(\imath\omega t). \tag{39.5}$$

The time derivatives in the starting equation were applied to the complex exponential to bring down factors of $\imath\omega$. As written, the two sides of the equation have a common factor $\exp(\imath\omega t)$.

Implicit in this equation is the claim that the response and drive frequencies must be equal. Suppose that the two frequencies were not equal so that the response frequency was some ω' and not ω. Divide the above equation by $\exp(\imath\omega' t)$. On doing so, the right-hand-side of the equation would depend on time as $\exp(\imath(\omega - \omega')t)$, while the left-hand-side of the equation would be independent of time. How is this possible? The only possible answer is that $\omega = \omega'$, because in that case the right-hand-side of the equation would depend on time as $\exp(\imath(\omega - \omega)t) = \exp(0) = 1$, so that its time dependence would match the time dependence of the left-hand-side of the equation. If $\omega' \neq \omega$, there would be a contradiction.

Now that we have shown that $\exp(\imath\omega t)$ is indeed a common factor to both sides of the equation, divide it out, leading to

$$(-m\omega^2 + f\imath\omega + k)A = F_0\exp(\imath\delta). \tag{39.6}$$

In general each side of this equation corresponds to a complex number. We now apply the rule that if

$$a + b\imath = c + d\imath, \tag{39.7}$$

in which a, b, c, and d are all real numbers, then it must be the case that

$$a = c, \text{ and} \tag{39.8}$$
$$b = d. \tag{39.9}$$

Recalling the Euler identity $\exp(\imath\delta) = \cos\delta + \imath\sin(\delta)$, the above three equations give us

$$(-\omega^2 + \frac{k}{m})A = \frac{F_0}{m}\cos(\delta) \tag{39.10}$$

and

$$\frac{f}{m}\omega A = \frac{F_0}{m}\sin(\delta). \tag{39.11}$$

39.1. BASIC CALCULATION

The ratio k/m is the frequency at which the undamped, non-driven mass oscillates. I now replace $k/m \to \omega_0^2$, ω_0 being the angular frequency of the corresponding undamped, undriven harmonic oscillator. If at the top of the calculation we had taken $\cos(\omega t + \delta)$ rather than $\cos(\omega t - \delta)$ to be the time dependence of the response, then when we reached equation 39.11 the right-hand-side of the equation would have an extra minus sign to remember. The choice we made, to use $\cos(\omega t - \delta)$ as the response, means that for most of the calculation a plus sign appears. You see here a minor trick for improving the accuracy of manual algebraic calculations, namely on occasion a careful choice of variables reduces how many details need to be tracked at the same time. (With computer algebra, this constraint is less interesting.)

How large is δ? If we take the ratio of these two equations, we have

$$\frac{\frac{F_0}{m}\sin(\delta)}{\frac{F_0}{m}\cos(\delta)} = \frac{f\omega/m}{-\omega^2 + \omega_0^2}. \qquad (39.12)$$

The left-hand-side of the equation is $\tan(\delta)$. Taking the arctangent of the equation, one has

$$\delta = \arctan\left(\frac{f\omega/m}{-\omega^2 + \omega_0^2}\right). \qquad (39.13)$$

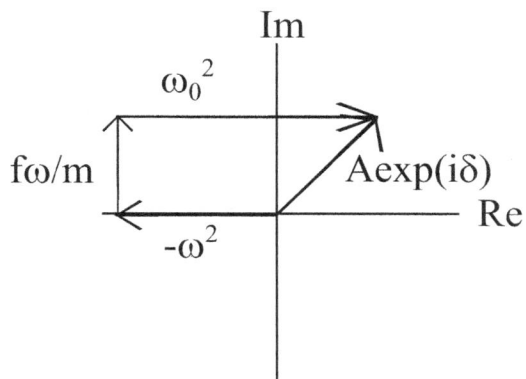

Figure 39.2: Figure for determining the quadrants in which δ can lie, namely it may lie in quadrant I or quadrant II.

We now have δ except for one minor detail. In which quadrants can δ lie? The answer is shown by a simple picture, Figure 39.2, which shows how equation 39.6 is to be represented in the complex plane. The three terms on the left side of the equation are $-\omega^2$, $f\iota\omega/m$, and $+k/m \equiv \omega_0^2$, each appearing as a vector parallel to the real or imaginary axis. The diagonal arrow is $\frac{F_0}{A}\exp(\iota\delta)$. The term $f\omega/m$ is always positive, so δ is always an angle in the first or second quadrant. The response $\cos(\omega t - \delta)$ therefore lags the driving force, but by an angle of π or less.

Is this result consistent with our discussion of the driven, undamped harmonic oscillator, made in the precious chapter? We certainly expect that it is. If we reduce the damping toward zero, we have $f \to 0$, in which case $\delta = \arctan(\pm 0)$. I say ± 0 because the sign of the argument of the arctangent could be positive or negative, depending on the sign of $-\omega^2 + \omega_0^2$. In fact:

a) If $\omega^2 < \omega_0^2$, then, if $f \to 0$, $\exp(\iota\delta)$ approaches the origin from the first quadrant, meaning that $\delta = \arctan(0(+)) = 0$.

b) If $\omega^2 > \omega_0^2$, then if $f \to 0$, $\exp(\iota\delta)$ approaches the origin from the second quadrant, meaning that $\delta = \arctan(0(-)) = \pi$.

These results agree with the previous calculation.

We also need to calculate $A(\omega)$. To do this we again invoke the trig identity $\cos^2(\theta) + \sin^2(\theta) = 1$, as applied to the squares of equations 39.10 and 39.11. Those squares are

$$((-\omega^2 + \omega_0^2))^2 A^2 = \left(\frac{F_0}{m}\right)^2 \cos^2(\delta) \qquad (39.14)$$

and

$$\left(\frac{f}{m}\omega\right)^2 A^2 = \left(\frac{F_0}{m}\right)^2 \sin^2(\delta). \qquad (39.15)$$

Observe that on the left-hand-side, these two equations have a common factor A^2, while on the right-hand-side, these two equations have a common factor $\left(\frac{F_0}{m}\right)^2$ and the squares of the sine and cosine. We again introduce the notation

$$\gamma = \frac{f}{m} \qquad (39.16)$$

for the damping constant.

Adding the two equations and extracting common factors,

$$((((-\omega^2 + \omega_0^2))^2 + (\gamma\omega)^2)A^2 = \left(\frac{F_0}{m}\right)^2. \tag{39.17}$$

Rearranging terms and taking the square root of both sides, we obtain a final form:

$$A(\omega) = \frac{F_0/m}{((-\omega^2 + \omega_0^2)^2 + (\gamma\omega)^2)^{1/2}}. \tag{39.18}$$

Figure 39.3 shows a sample amplitude curve. For this curve, $\omega_0 = 10$, $\gamma = 0.333$, and therefore $Q = 30$.

39.2 Form of the Amplitude Curve

It is worthwhile to consider properties of the function $A(\omega)$. As $\omega \to 0$, $A(\omega)$ becomes

$$A(\omega) = \frac{F_0/m}{\omega_0^2}, \tag{39.19}$$

but, recalling $\omega_0^2 = k/m$, this result reduces to

$$A(0) = \frac{F_0}{k}. \tag{39.20}$$

At near-zero frequency, the spring is simply stretched by $F_0 \cos(\omega t)/k$.

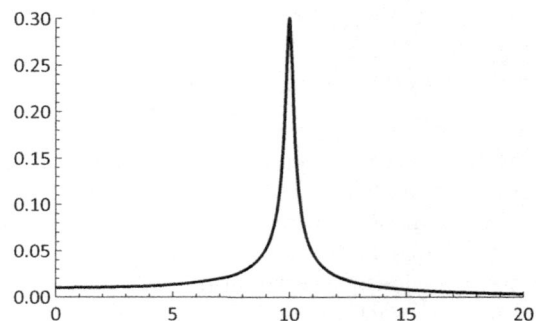

Figure 39.3: The amplitude of the response of a harmonic oscillator, equation 39.18 with $\omega_0 = 10$, $\gamma = 0.333$, and therefore $Q = 30$; the plotted amplitude is in arbitrary units.

What happens in the limit of infinite frequency? If you go to the right-hand-side of equation 39.18 and expand out the squares, you find the lead term in ω is ω^4, other terms containing ω have ω to the second or zeroth power, and $\omega \gg \omega_0$ or γ, so at large ω the denominator goes to $(\omega^4)^{1/2} \approx \omega^2$. That dependence of the denominator on ω is the only dependence of the right-hand-side on ω, so, at large ω, $A(\omega) \sim \omega^{-2}$ and $A(\omega) \to 0$. At very large frequency, the harmonic oscillator has next to no time to respond before the direction of the force reverses, so it does not move appreciably.

The maximum is found where

$$\frac{dA(\omega)}{d\omega} = 0. \tag{39.21}$$

However, the derivative test also finds minima and saddle points, and is invalid at end points, so some care is required in applying it. Taking the derivative of equation 39.18 with respect to ω, we have

$$\frac{dA(\omega)}{d\omega} = -\frac{1}{2}\frac{F_0}{m}\frac{2(-\omega^2 + \omega_0^2)(-2\omega) + 2\omega\gamma^2}{((-\omega^2 + \omega_0^2)^2 + \gamma^2\omega^2)^{3/2}}. \tag{39.22}$$

Where does this function have its zeros? Start by checking the two endpoints, ∞ and 0. The denominator takes the function to zero when $\omega \to \infty$, but in that limit $A(\omega)$ also goes to zero, so $\omega \to \infty$ represents a minimum for $A(\omega)$. At $\omega = 0$, the denominator is $\sim \omega_0^6$ while the numerator has a common factor of ω, which is zero at $\omega = 0$. At $\omega = 0$, $dA(\omega)/d\omega = 0$, so the function is flat and, from the above, equal to F_0/k. Because $\omega = 0$ is an end point of $A(\omega)$, it could be the case that $A(\omega)$ has its maximum at $\omega = 0$, no matter what the derivative test indicates, but we shall show in a bit that the maximum of $A(\omega)$ is elsewhere.

At all finite ω, the denominator is not infinite. Any remaining zeros of $dA(\omega)/d\omega$ must arise from the numerator of equation 39.22. When is the numerator equal to zero? Setting the numerator equal to zero, one has

$$4\omega(\omega^2 - \omega_0^2) + 2\omega\gamma^2 = 0. \tag{39.23}$$

39.2. FORM OF THE AMPLITUDE CURVE

We've already identified the zero of this function, the zero at $\omega = 0$. The other zeros satisfy

$$2\omega^2 - 2\omega_0^2 + \gamma^2 = 0, \tag{39.24}$$

whose solutions are

$$\omega_m = \pm(\omega_0^2 - \frac{1}{2}\gamma^2)^{1/2}, \tag{39.25}$$

so the frequency at the maximum is

$$\omega_m = \pm\omega_0(1 - \frac{1}{2}\left(\frac{\gamma}{\omega_0}\right)^2)^{1/2}. \tag{39.26}$$

One of the values of ω_m refers to the non-physical domain $\omega < 0$, which we do not consider further. Recalling that the quality parameter is $Q = \omega_0/\gamma$, we may also write this expression as

$$\omega_m = \omega_0(1 - \frac{1}{2Q^2})^{1/2} \tag{39.27}$$

The effect of damping is to reduce the frequency ω_m at which the maximum of $a(\omega)$ occurs from its undamped value ω_0.

What is the amplitude at maximum? In answering this question, we also confirm that $A(0)$, which is equal to F_0/k, is not the maximum. We advance by setting $\omega = \omega_m$, with ω_m in the form of equation 39.25, into equation 39.18. We obtain

$$A(\omega_m) = \frac{F_0/m}{((-(\omega^2 - \gamma^2/2) + \omega_0^2)^2 + \gamma^2(\omega^2 - \gamma^2/2))^{1/2}}. \tag{39.28}$$

The denominator can be simplified. Expanding and collecting terms, this equation becomes

$$A(\omega_m) = \frac{F_0/m}{((\gamma^4/4) + \gamma^2\omega_0^2 - \gamma^4/2))^{1/2}} \tag{39.29}$$

which may be rearranged as

$$A(\omega_m) = \frac{F_0}{m} \frac{1}{(\gamma^2\omega_0^2 - \gamma^4/4)^{1/2}}. \tag{39.30}$$

Extracting on the right-hand-side a factor of $\gamma\omega_0$ from the denominator's square root,

$$A(\omega_m) = \frac{F_0}{\gamma\omega_0 m} \frac{1}{(1 - \frac{\gamma^2}{4\omega_0^2})^{1/2}}. \tag{39.31}$$

To extract the factor of $\gamma\omega_0$ from the denominator's $\gamma^4/4$ term, where initially no ω_0^2 was present, we multiplied this term by one in the form of ω_0^2/ω_0^2. We now take advantage again of multiplying by one, namely we multiply $F_0/\gamma\omega_0 m$ by k/k. The lead fraction on the right hand side of the last equation transforms to

$$\frac{F_0}{m\gamma\omega_0} \equiv \frac{F_0 k}{km\gamma\omega_0} \equiv \frac{F_0\omega_0^2}{k\gamma\omega_0} \equiv \frac{F_0}{k}\frac{\omega_0}{\gamma} \equiv \frac{F_0 Q}{k}. \tag{39.32}$$

The larger Q is, the larger $A(\omega_m)$ is. Also, the denominator term may be rewritten as

$$(1 - \frac{\gamma^2}{4\omega_0^2})^{1/2} \rightarrow (1 - \frac{1}{4Q^2})^{1/2}, \tag{39.33}$$

leading to

$$A(\omega_m) = \frac{F_0 Q}{k}\left(\frac{1}{(1 - \frac{1}{4Q^2})^{1/2}}\right). \tag{39.34}$$

In the final form F_0/k is the amplitude at $\omega = 0$, so the amplitude at the maximum response frequency ω_m is larger than the amplitude at $\omega = 0$ by a factor $Q/(1 - \frac{1}{4Q^2})^{1/2}$. For a system with weak damping, and therefore large Q, $\frac{1}{4Q^2} \to 0$, so the denominator term is very nearly unity and the peak height at maximum is $QA(0)$, Q times larger than the height at zero.

We have calculated how tall the peak is. We might also ask how wide the peak is. There are a variety of ways to characterize the width of a peak. A conventional approach is to calculate the width of the peak at half of its maximum height, the so-called full width at half maximum, oft abbreviated FWHM. Recall that the amplitude as a function of frequency is

$$A(\omega) = \frac{F_0/m}{((-\omega^2 + \omega_0^2)^2 + (\gamma\omega)^2)^{1/2}}. \tag{39.35}$$

All of the frequency dependence in this equation is in the denominator. In order for the amplitude to be half of its maximum value, the denominator must twice as large as it is at ω_m, the frequency at which the amplitude is a maximum. Taking ω to be the frequency at the amplitude half-maximum, we therefore have

$$((-\omega^2 + \omega_0^2)^2 + (\gamma\omega)^2)^{1/2} = 2((-\omega_m^2 + \omega_0^2)^2 + (\gamma\omega_m)^2)^{1/2}. \tag{39.36}$$

As seen above $\omega_m^2 = \omega_0^2 - \gamma^2/2$. Substituting in equation 39.36 this form for ω_m, we reach

$$((-\omega^2 + \omega_0^2)^2 + (\gamma\omega)^2)^{1/2} = 2(\gamma^2\omega_0^2 - \frac{\gamma^4}{4})^{1/2}. \tag{39.37}$$

The right-hand-side of the above equation is the same as the denominator on the right hand side of equation 39.31. To find ω, we need to simplify. A sound first step is to eliminate the square roots. We square both sides of the equation, leading to

$$(-\omega^2 + \omega_0^2)^2 + (\gamma\omega)^2 = 4\gamma^2\omega_0^2 - \gamma^4. \tag{39.38}$$

Expanding powers and resorting terms, the above form is seen to be a quadratic in the variable ω^2, to whit

$$(\omega^2)^2 + \omega^2(-2\omega_0^2 + \gamma^2) + (\omega_0^4 - 4\gamma^2\omega_0^2 + \gamma^4) = 0, \tag{39.39}$$

whose solution is

$$\omega^2 = \frac{1}{2}\left(2\omega_0^2 - \gamma^2 \pm ((-2\omega_0^2 + \gamma^2)^2 - 4(\omega_0^4 - 4\gamma^2\omega_0^2 + \gamma^4))^{1/2}\right). \tag{39.40}$$

That's not a very attractive solution. We now try a mathematical solution technique: On ugliness, try expand and cancel. Sometimes this technique works. Sometimes this technique makes things worse. On expanding under the radical, terms in $4\omega_0^4$ cancel, leading to

$$\omega^2 = \omega_0^2 - \frac{\gamma^2}{2} \pm \frac{1}{2}\left(12\omega_0^2\gamma^2 - 3\gamma^4\right)^{1/2}. \tag{39.41}$$

On the right-hand-side, the first two terms combine to equal ω_m^2, ω_m being the location of the peak. The amplitude curve is therefore symmetric around the peak position, the two half-maxima occurring at equal distances in frequency from the maximum. From the radical, we can extract a factor of ω_0^2, leading to

$$\omega^2 = \omega_m^2 \pm \frac{1}{2}\omega_0^2\left(12\frac{\gamma^2}{\omega_0^2} - 3\frac{\gamma^4}{\omega_0^4}\right)^{1/2}. \tag{39.42}$$

Recalling that the quality parameter is $\omega_0/\gamma = Q$, this equation becomes

$$\omega^2 = \omega_m^2 \pm \frac{\omega_0^2}{2}\left(\frac{12}{Q^2} - \frac{3}{Q^4}\right)^{1/2}. \tag{39.43}$$

Introducing the quality parameter, which tells us how rapidly the ringing of the undriven oscillator decays, significantly reduces the size of the equation.

What does this equation tell us about the width? If Q is large, meaning that damping is weak, $12/Q^2 \gg 4/Q^4$, so the width term is approximately proportional to $1/Q$. At large Q, the width of the amplitude curve

39.3. POWER ABSORPTION

is narrow. As Q becomes smaller, the width curve becomes wider. For large Q, we can make these remarks more quantitative: For large Q, $Q^{-2} \gg Q^{-4}$, so that $\frac{12}{Q^2} - \frac{3}{Q^4} \approx \frac{12}{Q^2}$ and therefore

$$\omega^2 = \omega_m^2 \pm \frac{\sqrt{3}\omega_0^2}{Q}. \tag{39.44}$$

To obtain the actual frequency at each of the two half-maxima, we need to take the square root, which we can do for weak damping (and, hence, large Q) by applying for small ϵ the approximation $\sqrt{1+\epsilon} \approx 1 + \epsilon/2$.

$$\omega = \omega_m(1 \pm \frac{\sqrt{3}}{2Q} \frac{\omega_0^2}{\omega_m^2}). \tag{39.45}$$

The width of the response curve is inversely proportional to Q. The weaker the damping, the narrower the range of drive frequencies to which there is a large response. At large Q, to first order $\omega_m \approx \omega_0$ and the frequency at half-maximum becomes

$$\omega = \omega_0(1 \pm \frac{\sqrt{3}}{2Q}). \tag{39.46}$$

We are reach an interesting outcome of resonant response, the *radar problem*. The first effort to build a radar, a device to measure direction and distance using reflected radio waves, would take us back before World War I. Period technology and design plans had some success, but an effective radar requires that you generate a short, intense pulse of radio waves, send them out to a target, detect the reflected waves to measure their travel time, and have arrangements to determine a direction.

The reflected pulses are very weak. They fall off as 1/range^2 on the way to the target, and an additional factor of 1/range^2 on the way back from the target, for a total loss of 1/range^4. To detect these pulses, one might choose to construct a detector with a very large Q, because the response of the circuit to the driving force of the reflected wave is proportional to Q. Now we encounter a minor difficulty. The outgoing pulse is at the same frequency, almost, as the reflected pulse. It acts a source more-or-less necessarily close to the detector, and it is much more powerful than the reflected pulse. When the outbound pulse passes the detector on the way to the target, it sets the detector ringing. Because the detector has a very large Q, the detector will continue to ring for a long time, hiding the response of the detector to the reflected wave. All the detector sees is the outbound wave. A simple-minded radar system therefore appears not to work. Beating this obstacle, so as to build a working radar system, required clever inventiveness.

39.3 Power Absorption

We now come to the final major topic in our treatment of damped driven oscillators. As the oscillator moves, it is subject to a frictional force $-fv$ that pulls energy out of the system. However, over long periods of time the motion $x(t)$ of the oscillator is described by $A(\omega)\cos(\omega t - \delta)$. The oscillator moves back and forth, but over long periods of time its amplitude of oscillation, and therefore its total energy, does not decrease with time. The energy being lost to friction must therefore be being replaced by the energy supplied to the system by the driving force.

In contrast, if there were no friction, a driven oscillator would still move back and forth, but averaged over a complete cycle the driving force must do no work on the system. After all, when a driven oscillator returns to its start position after a full cycle of motion, it has the same speed as it did before, so its total energy has not changed. There is no friction, so there are no losses of energy due to friction, so the driving force must in total over a full cycle of oscillation have done no work on the oscillator.

We now consider the work done on a driven oscillator by the driving force. We will calculate the power

$$P = \mathbf{F} \cdot \mathbf{v}, \tag{39.47}$$

where $\mathbf{F}$ is the driving force and $\mathbf{v}$ is the oscillator's velocity, and then average P over a cycle of oscillation. In the discussion above, the oscillator is one-dimensional, and only the component of the driving force parallel to the oscillator's motion is of any consequence, so the interesting component of the driving force is

$$F = F_0 \cos(\omega t). \tag{39.48}$$

The movement of the oscillator, as calculated above, is given by

$$x = A\cos(\omega t - \delta), \tag{39.49}$$

where A and δ are both functions of frequency, and the mass, friction factor, and spring constant of the oscillator. The oscillator's velocity component along the x-axis is therefore

$$v \equiv \frac{dx}{dt} = -A\omega \sin(\omega t - \delta). \tag{39.50}$$

The power dissipation is therefore

$$P = -F_0 A\omega \cos(\omega t)\sin(\omega t - \delta). \tag{39.51}$$

For this calculation it is simplest to apply the trig identity

$$\sin(a+b) = \sin(a)\cos(b) + \cos(a)\sin(b), \tag{39.52}$$

leading to

$$P = -A\omega F_0 \cos(\omega t)(\cos(\omega t)\sin(-\delta) + \sin(\omega t)\cos(-\delta)). \tag{39.53}$$

This is the power.

However, in terms of the original expression $P = \mathbf{F} \cdot \mathbf{v}$, we have the complication that $\mathbf{F}$ and $\mathbf{v}$ both depend on time in an oscillatory manner, so that sometimes $\mathbf{F}$ and $\mathbf{v}$ both point in the same direction, and sometimes $\mathbf{F}$ and $\mathbf{v}$ point in opposite directions. Sometimes $\mathbf{F}$ is supplying energy to the oscillator, and sometimes $\mathbf{F}$ is abstracting energy from the oscillator. How do we deal with this time dependence of P? The direct answer is that we integrate P over a complete cycle of the system, so as to determine the average work done on the system by the external force.

If you are thoughtful, you ask why we should average P over exactly one cycle instead of averaging it over, for example, π cycles. The answer is that we are calculating the average power $\Delta E/T$, where ΔE is the total work done by the driving force (or, with sign reversed, by friction) and T is the time over which we are averaging. If we integrate over exactly one cycle, we find that the work done is some amount of energy E_1 during a time T_1. If we integrate only over part of a cycle, we find that the work done is δE during some time δt. Now suppose we average over a large number N cycles, plus some fraction of a cycle. We find that the average power is

$$P = \frac{NE_1 + \delta E}{NT_1 + \delta t}. \tag{39.54}$$

We may rewrite this be multiplying by one in the form $\frac{1/N}{1/N}$, obtaining

$$P = \frac{E_1 + (\delta E/N)}{T_1 + (\delta t)/N}. \tag{39.55}$$

As we integrate over longer and longer times, the terms in $\delta E/N$ and $\delta t/N$ become negligible relative to E_1 and T_1, so at long times $P \approx E_1/T_1$ to high accuracy.

Of course, if for some reason you want to know the average power over π oscillations, beginning at some specified time, there is no fundamental difficulty with doing the average, but we do not do that average here.

We therefore calculate the average over precisely one cycle. We need the calculus results

$$\frac{1}{T}\int_0^T dt\ \cos^2(\omega t) = \frac{1}{2} \tag{39.56}$$

and

$$\frac{1}{T}\int_0^T dt\ \cos(\omega t)\sin(\omega t) = 0, \tag{39.57}$$

where $\omega T = 2\pi$.

39.3. POWER ABSORPTION

Applying these results to equation 39.53, we find

$$\langle P \rangle = \frac{1}{2} A(\omega) F_0 \omega \sin(\delta). \tag{39.58}$$

I've introduced a new, but extremely standard bit of notation here. The angle brackets $\langle \cdots \rangle$ stand for the average of $\cdots$, where $\cdots$ is whatever is between the two angle brackets.

How large is $\sin(\delta)$? From equations 39.11 and 39.18

$$\sin(\delta) = \frac{\gamma \omega}{((-\omega^2 + \omega_0^2)^2 + (\gamma \omega)^2)^{1/2}}, \tag{39.59}$$

where from prior discussion δ always lies in quadrant I or quadrant II.

The average power is then

$$\langle P \rangle = \frac{1}{2} \left[\frac{F_0/m}{((-\omega^2 + \omega_0^2)^2 + (\gamma \omega)^2)^{1/2}} \right] F_0 \omega \left[\frac{\gamma \omega}{((-\omega^2 + \omega_0^2)^2 + (\gamma \omega)^2)^{1/2}} \right], \tag{39.60}$$

so the average power becomes

$$\langle P \rangle = \frac{F_0^2 f}{2m^2} \frac{\omega^2}{(-\omega^2 + \omega_0^2)^2 + (\gamma \omega)^2}. \tag{39.61}$$

What is the behavior of this form? In the limit $\omega \to 0$, $\langle P \rangle$ goes to $0/\omega_0^4 = 0$. In the limit $\omega \to \infty$, $\langle P \rangle$ is proportional to ω^2/ω^4, so, at large ω, $\langle P(\omega) \rangle$ goes to zero as ω^{-2}.

Where are the maxima? $\langle P \rangle$ is composed of quantities that are all positive or zero, so $\langle P \rangle$ is either zero or a positive number. $\langle P \rangle$ goes to zero at its end points $\omega = 0$ and $\omega \to \infty$, so its endpoints must be minima. We can find the maximum with the derivative test. It's convenient to multiply $\langle P \rangle$ by one in the form $(1/\omega^2)/(1/\omega^2)$, giving

$$\langle P(\omega) \rangle = \frac{F_0^2 f}{2m^2} \left[\frac{1}{(-\omega + \frac{\omega_0^2}{\omega})^2 + \gamma^2} \right]. \tag{39.62}$$

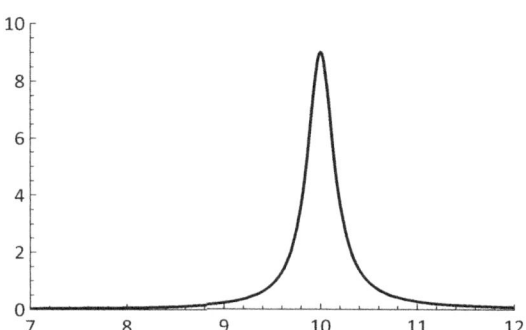

Figure 39.4: The absorbed power as a function of frequency of a damped, driven harmonic oscillator, as calculated from equation 39.61 with $\omega_0 = 10$, $\gamma = 0.333$, and therefore $Q = 30$. The plotted amplitude is in arbitrary units.

In certain ways this is a rather ugly expression, in which the simple polynomials in the numerator and denominator have been replaced by a form containing a fraction in the denominator, but for our specific question it is a very useful form. Let us first do a qualitative analysis as to where $\langle P \rangle$ will be the largest. The term containing ω is the square of a real number, so it is always zero or positive, while no other term depends on ω, so the largest that the term in square brackets can be is $1/\gamma^2$, that occurring where

$$(-\omega + \frac{\omega_0^2}{\omega})^2 = 0, \tag{39.63}$$

which occurs where $\omega = \omega_0$. The frequency at maximum power absorption must therefore be $\omega = \omega_0$.

Remark: These handwaving approaches to finding maxima or minima, while sometimes sweet to the ear, can be deceptive. They are not substitutes to finding the maxima and minima by applying the derivative test.

How much power is absorbed at ω_m? At this frequency, here denoted ω_m,

$$\langle P(\omega_m) \rangle = \frac{F_0^2 f}{2m^2} \frac{1}{(f/m)^2} \tag{39.64}$$

A series of reductions puts this equation in a particularly interesting form, namely

$$\langle P(\omega_m) \rangle = \frac{F_0^2}{2f} = \frac{F_0^2}{2m\omega_0} \frac{m\omega_0}{f} \tag{39.65}$$

But $f/(m\omega_0) = 1/Q$, so

$$\langle P(\omega_m) \rangle = \frac{F_0^2}{2m\omega_0} Q. \tag{39.66}$$

The maximum of the power absorption curve is proportional to the quality parameter Q. At the maximum of the power absorption curve, the smaller the friction coefficient f is, the larger the power absorption is.

Just as we found the frequencies where the amplitude $A(\omega)$ is half of its maximum value, so also we can find the frequencies ω_2 where $\langle P(\omega) \rangle$ is half of its maximum value. At those frequencies,

$$\frac{F_0^2 f}{2m^2} \frac{1}{(-\omega_2 + \frac{\omega_0^2}{\omega_2})^2 + \gamma^2} = \frac{1}{2} \frac{F_0^2 f}{2m^2} \frac{1}{\gamma^2}. \tag{39.67}$$

After cancellations between the two sides of that equation, and inverting, one finds

$$(-\omega_2 + \frac{\omega_0^2}{\omega_2})^2 + \gamma^2 = 2\gamma^2. \tag{39.68}$$

On rearranging and taking the square root of this equation, we have

$$-\omega_2 + \frac{\omega_0^2}{\omega_2} = \pm\gamma, \tag{39.69}$$

which on multiplying by ω_2 becomes the quadratic

$$-\omega_2^2 \pm \gamma\omega_2 + \omega_0^2 = 0, \tag{39.70}$$

whose solutions are

$$\omega_2 = \frac{\pm\gamma \pm (\gamma^2 + 4\omega_0^2)^{1/2}}{-2}, \tag{39.71}$$

which reduces to

$$\omega_2 = \pm\left(\omega_0(1 + \frac{1}{4Q^2})\right) \pm \frac{\gamma}{2}. \tag{39.72}$$

What can we say about our final solution? First, the original equation for ω_2, that we have here solved, was a quartic in ω_2, meaning that we should find four roots, four solutions for ω_2. Noting the two $\pm$ points, we indeed did so. Second, only the positive roots are of physical interest here. So long as damping is not extremely strong, meaning so long as $\omega_0 > \gamma$, the two roots of physical interest are the two roots with positive frequency, these being the roots $\omega_0(1 + \frac{1}{4Q^2})^{1/2} \pm \gamma/2$. The half-maximum points are displaced on opposite sides of the midpoint by $\pm\gamma/2$, so that the full width of the curve at half of its maximum amplitude is γ. The midpoint is $\omega_c = \omega_0(1 + \frac{1}{4Q^2})^{1/2}$. The midpoint frequency ω_c, half-way in frequency between the two frequencies at the two half-maximum points, is not the same as the frequency $\omega_m = \omega_0$ at the maximum of the power absorption curve. Instead, ω_c lies at a somewhat larger frequency than ω_0. The power absorption curve is not symmetric about its maximum; the curve lies more to higher frequencies.

If we look at this as a description of a radio receiver, we can arrange to turn a weak signal into a strong response by using a damped, driven harmonic oscillator in the form of a radio receiver circuit having a very large Q. However, there is a price. If Q is large, than the range of frequencies $\omega_c \pm \gamma$ to which the receiver is most sensitive (the bandwidth, the rate at which you can send information) is narrow, meaning the receiver will only have a slow response to changes in the signal. Of course, in an engineering design you can consider whether the bandwidth should be defined as $\omega_c \pm \gamma$ or $\omega_c \pm 3\gamma$ or whatever, but the limitation is still there. A large-Q system, large because damping is weak, responds slowly to changes in the driving force.

39.4 Homework

1. Return to equations 39.51 and 39.52. Instead of invoking the trig identity of equation 39.52, use the Euler identity to replace the sines and cosines with their complex exponential forms and take the time average. Do you get to the same final answer?

2. "...we do not do that average here..." But we do it here. Returning to equation 39.52 and following, calculate the power absorbed between $\omega t = \theta_1$ and $\omega t = \theta_2$. Assume that $\theta_2 > \theta_1$ and that $\theta_2 - \theta_1 < 2\pi$.

3. We said following equation 39.64 that "They are not substitutes to finding the maxima and minima by applying the derivative test." Apply the derivative test directly to the right-hand-side of equation 39.61 and find the maxima and minima of $\langle P(\omega) \rangle$.

4. A mass suspended on a spring moves as a damped driven harmonic oscillator with an equation of motion
$$m\frac{d^2x}{dt^2} + f\frac{dx}{dt} + kx = F_o \cos(\omega_D t). \tag{39.73}$$
The motion of the mass is given by
$$x = A\cos(\omega_D t - \delta), \tag{39.74}$$
with
$$A = \frac{F_o/m}{[(-\omega^2 + \omega_o^2)^2 + \gamma^2\omega^2]^{1/2}} \tag{39.75}$$
and
$$\delta = \arctan\left[\frac{\gamma\omega}{-\omega^2 + \omega_o^2}\right] \tag{39.76}$$
for $\gamma = f/m$. Find the major features of $A(\omega)$, namely (a) its values at zero and infinite frequency, (b) the frequency ω_m at which it is a maximum, and (c) the amplitude of oscillation at the frequency at which the oscillations are a maximum. Explain (you will need at least a few equations, I think) why the frequency at which the power absorption is a maximum is larger than the frequency at which the amplitude of driven motions is a maximum.

5. A mass M is suspended from a spring whose unstretched length is ℓ_0 and whose spring constant is K. The top end of the spring is stationary. The mass's height is $x(t)$. The mass sits in a fluid bath, which oscillates up and down at frequency Ω and amplitude A. The position of the bath is thus $y = A \cos(\Omega t)$. The fluid in the bath moves with the bath, at velocity $v_f = \frac{dy}{dt}$. Denote the displacement of the mass from its equilibrium position by x. The motion of the bath then creates a frictional force $f\left(\frac{dy}{dt} - \frac{dx}{dt}\right)$ on the mass. i) Give the free body diagram for the mass. [Hint: here y and x are parallel distances along a single axis.] ii) Find the equation of motion for the mass. iii) Solve the equation of motion and obtain the steady state solution for the mass's movements. iv) Write the amplitude A and phrase angle δ as functions of ω_0, ω, and such other variables as you find necessary. Find the frequency ω_m at which $A(\omega)$ has a maximum. (You'll have to use the derivative test to find the maximum.)

6. An alternative approach to calculating the mean power needed to maintain a constant amplitude of a forced vibration begins with the friction force $-fv$. a) Why is the power dissipated by friction $-fv^2$? b) If the amplitude of vibrations is a constant, the energy E stored in the system must on the average also be a constant. Why does $\frac{dE}{dt} = 0$ imply that the power due to the applied force $\mathbf{F}$ must on the average equal fv^2? c) Using $x = A\cos(\omega t - \delta)$ show that the average power dissipation $\langle P \rangle$ due to friction is $f\omega^2 A^2/2$. d) From the results in part c, and the known form for A as a function of ω, obtain $\langle P \rangle$ as a function of ω.

7. Consider a damped harmonic oscillator with $m = 1.0$ kg, $f = 12$ N sec/m, and $k = 300$ N/m. Suppose the driving force is $F = F_0\cos(\omega t)$ with $F_0 = 10$ N and $\omega = 40$ s^{-1}. a) Set up and solve the equations for steady state motion. b) What are the values for A and δ if steady state response is written $x = A\cos(\omega t - \delta)$? c) How much energy is dissipated by friction during one cycle? d) What is the mean power input required to maintain A constant? *Do all calculations algebraically and plug in numbers at the end.*

8. Consider a seismometer whose equation of motion is

$$m\frac{d^2y}{dt^2} + b\frac{dy}{dt} + ky = -m\frac{d^2\eta(t)}{dt^2}. \qquad (39.77)$$

i) Here $\eta(t)$ is the ground position; the ground oscillates up and down as $\eta_o \cos(\omega_E t)$. From the equation of motion, solve for $y = A\cos(\omega t - \delta)$, obtaining values for A, δ, and ω. ii) Recalling that the average power into the system must be dissipated by the frictional force $-b\frac{dy}{dt}$, compute the average $\langle P \rangle$ supplied to maintain constant amplitude of vibrations. iii) Calculate the frequencies at which the vibration amplitude and the power dissipation have maxima. [Hint: The derivative test is needed. The answers are not the same as the ones seen in previous problems.]

9. An object of mass 0.5 kg serves as the bob of a pendulum having length $\ell = 2$m. The body is subject to a resistive force $-fv$ with $f = 50$ in SI units. The object is also subject to a horizontal driving force $F_x(t) = F_0 \cos(\omega t)$ with $F_0 = 20N$ and $\omega = 50s^{-1}$ Do all calculations algebraically and plug in numbers at the end. a) What are the dimensions of f? b) Find the force diagram and set up the differential equations of motion for the mass. c) Solve explicitly the equation of motion; find the amplitude of the forced oscillation. d) Find the phase of the oscillation relative to the phase of the driving force. e) At what frequency is the driven amplitude a maximum? f) What is the amplitude of the oscillation at the frequency at which the amplitude is a maximum?

10. The power $P(\omega)$ adsorbed by a damped, driven harmonic oscillator depends on frequency as

$$\langle P(\omega)\rangle = \frac{F_o^2 f}{2m^2} \frac{\omega^2}{(-\omega^2 + \omega_o^2)^2 + \gamma^2\omega^2}. \qquad (39.78)$$

Confirm with a derivative test that the maximum power absorption is at $\omega = \omega_o$. What is the power absorbed at this frequency?

11. The power $P(\omega)$ adsorbed by a damped, driven harmonic oscillator depends on frequency as

$$\langle P(\omega)\rangle = \frac{F_o^2 f}{2m^2} \frac{\omega^2}{(-\omega^2 + \omega_o^2)^2 + \gamma^2\omega^2}. \qquad (39.79)$$

Find the width of the power absorption curve, namely the locations of the two points at which $P(\omega)$ falls to half of its maximum value, and the frequency separation of these two points. Are these points centered on the frequency at which the maximum power is adsorbed?

12. For a driven, damped harmonic oscillator the amplitude of oscillation depends on the oscillation frequency as

$$A(\omega) = \frac{F_o/m}{((-\omega^2 + \omega_o^2)^2 + \gamma^2\omega^2)^{1/2}}. \qquad (39.80)$$

where $x = A\cos(\omega t - \delta)$ gives the displacement. a) Find the time-dependent velocity of the harmonic oscillator. b) If the drive frequency is varied while other parameters, e.g., m and f, are kept the same, at what drive frequency is the velocity a maximum. What is the maximum velocity at that frequency?

13. Prove that the frequency at which a driven harmonic oscillator oscillates is equal to the frequency of the applied driving force.

14. An object of mass m serves as the bob of a pendulum having length ℓ. The body is subject to a resistive force $-fv$. The object is also subject to a horizontal driving force $F_x(t) = F_0 \cos(\omega t)$. The equation of motion of the mass is

$$m\frac{d^2x}{dt^2} + f\frac{dx}{dt} + \frac{mgx}{\ell} = F_0 \cos(\omega t). \qquad (39.81)$$

(a) Solve explicitly the equation of motion; find the amplitude of the forced oscillation. (b) Find the phase of the oscillation relative to the phase of the driving force. Identify the quadrants in which this phase may lie, and prove that your answer is correct. (c) At what frequency is the driven amplitude a maximum? (d) What is the amplitude of the oscillation at the frequency at which the amplitude is a maximum? (e) If the drive frequency is held fixed, for what length of string is the driven amplitude a maximum?

15. A very weakly damped system vibrates at 500 Hertz. With the damper applied, the system is much more strongly damped than it was before. The $1/e$ time for the damping of its stored energy is measured to be 0.1 seconds. An external driver is now applied to put the system in forced oscillation. On tuning the frequency of the driver until the maximum power is being absorbed, it is found that the power absorbed is 0.05 Watts. Draw the power absorption curve of the system. Label numerically all of the curve features that you can. For example, if you think that the power absorption maximum occurs at 314 Hertz (Hint: It doesn't), you should label the maximum in the power absorption curves as '314 Hertz').

16. An object of mass 3.0 kg is hung from a spring whose spring constant is 400 N/m. The body is subject to a resistive force $-bv$ with $b = 60$ in SI units. Do all calculations algebraically and plug in numbers at the end. a) What are the dimensions of b? b) Set up the equation of motion for the mass. c) Solve explicitly the equation of motion; find the oscillation frequency and the position as a function of time. d) The object is subject to a sinusoidal driving force $F(t) = F_0 \sin \omega t_1$ with $F_0 = 30N$ and $\omega = 40s^1$. In the steady state, what is the amplitude of the forced oscillation? What is the phase of the oscillation relative to the phase of the driving force?

This space reserved for your notes.

This page reserved for your notes.

Chapter 40

Coupled Harmonic Oscillators

So far in our discussion, we've considered a single harmonic oscillator attached to an immobile, typically effectively infinitely massive, object. In this Chapter we turn to considering the behavior of several coupled harmonic oscillators: a pendulum bob hanging from the bob of another pendulum, a line of masses connected by springs to each other as well as to walls, and the like. We begin with a mathematical interlude that considers the solution of several equations that turn out, as will be seen, to be the equations that describe the motion of coupled harmonic oscillators. We then set up and solve a particular harmonic oscillator problem, namely the double pendulum, one pendulum hanging from the other.

40.1 Mathematical Interlude

First, we consider the pair of equations

$$M\frac{d^2x}{dt^2} = -ax + by$$
$$N\frac{d^2y}{dt^2} = bx - cy. \tag{40.1}$$

You can view this as the Second Law written for two masses M and N. M and N are one-dimensional harmonic oscillators. Each has a single coordinate of interest, which I've here called x and y, respectively, with no implication that the coordinates point in particular directions. x and y are just the symbols for the two coordinates. a, b, and c are positive constants. In terms of a Second Law description, the four terms $-ax$, by, bx, and $-cy$ all represent forces on M or N. The forces on M and N are all linear in the displacements; each force resembles a spring force in that the force depends on some mass's displacement from its rest position, the displacement being raised to the first power. If you have reached a course in differential equations, these are coupled linear differential equations with constant coefficients.

Note the two minus signs on the right-hand-side of these equations. The signs ensure that each mass is subject to a restoring force. If M moves away from $x = 0$ or N moves away from $y = 0$, they are subject to forces $-ax$ and $-cy$, respectively. No matter whether x or y is positive or negative, the resulting force on M or N drives each mass back towards its origin. Each mass, if it displaces from its origin, puts a force on the other mass. The corresponding pair of forces have the same proportionality constant b. The statement that the two proportionality constants are both b corresponds to the statement $\partial^2 U(x,y)/\partial x \partial y = \partial^2 U(x,y)/\partial y \partial x$, $U(x,y)$ being the potential energy of the system. For reasonably well-behaved functions U, including those encountered in most of physics, the order of taking the derivatives with respect to x and with respect to y does not matter.

We now consider how to solve these two equations. A straightforward approach is to guess that the solutions will be much like the solutions we have seen before, namely

$$x(t) = A\exp(\imath(\omega t + \alpha)),$$
$$y(t) = B\exp(\imath(\omega t + \alpha)). \tag{40.2}$$

That guess has one overt and two covert assumptions. The overt assumption is that the solutions are sinusoids, which will be tested when equations 40.2 are substituted into equations 40.1. The covert assumptions, namely that the two masses will move with the same frequency and the same phase, will be confirmed by calculation.

You are seeing a 'try it and see if it works' approach, an approach whose uncertainty is somewhat tempered by the practical issue that I already know that the assumed answer works. The time derivatives are both straightforward, namely

$$\frac{\partial^2}{\partial t^2} \exp(\imath(\omega t + \alpha)) = -\omega^2 \exp(\imath(\omega t + \alpha)). \tag{40.3}$$

If we substitute the guesses for $x(t)$ and $y(t)$ into equations 40.1, we obtain

$$-MA\omega^2 \exp(\imath(\omega t + \alpha)) = -aA \exp(\imath(\omega t + \alpha)) + bB \exp(\imath(\omega t + \alpha)), \tag{40.4}$$

$$-NB\omega^2 \exp(\imath(\omega t + \alpha)) = bA \exp(\imath(\omega t + \alpha)) - cB \exp(\imath(\omega t + \alpha)). \tag{40.5}$$

These equations both have in every term the common factor $\exp(\imath(\omega t + \alpha))$. That factor can be uniformly divided out, leaving us with

$$-M\omega^2 A = -aA + bB \tag{40.6}$$

$$-N\omega^2 B = bA - cB. \tag{40.7}$$

We now have two equations and three unknowns (one frequency and two amplitudes), reminiscent of the earlier discussion of a single harmonic oscillator, in which we had one equation and two unknowns (one frequency and one amplitude). In the earlier case, the frequency was fixed by the problem, but the amplitude was not. The pendulum had a fixed period, but the experimenter was free to make the range of the pendulum's swing larger or smaller. Here with two equations and three unknowns there is still some one number not fixed by the equations of motion; we will in the end see what that number is.

The above equations lead to a demonstration that equation 40.2 is correct in its assumption that the two masses oscillate at the same frequency and with the same phase. After all, suppose that we simply assumed without checking that the two masses oscillate at the same frequency and with the same phase. Suppose that we march through the analysis, and show that if the two masses oscillate at the same frequency and with the same phase then we have a solution to our starting equations. How do we know that there aren't other equations, equations in which the two masses oscillate at different frequencies or with different phases, that are also solutions? The answer is that we do not. Constructing an actual solution is very important, but it does not show that the solution we have found is the only solution.

A famous example of non-unique solutions is found in the rotation of Mercury as it orbits around the Sun. Mercury is very difficult to observe with classical telescopes, because it never is many degrees away from the sun, so twilight haze tends to obscure its visibility. In addition, it is most readily seen only at certain points in its orbit. However, it was observed, carefully, leading to the conclusion that Mercury was tidally locked. Just as the Moon always presents more or less the same face to the Earth, so also Mercury was believed always to present more or less the same face to the Sun. It was at some point noted that the observations were alternatively consistent with a resonance in which three rotational periods of Mercury take about as much time as two orbits of the planet around the Sun. This alternative turned out to be correct. The simple tidal lock solution was a solution to the observations, but it was not the only solution.

To demonstrate that the two masses must oscillate with the same frequency and phase, we assume that the claim is false, and show that our new assumption leads to a contradiction. You are seeing here a different technique for proving that a result is true, namely you consider the alternative and show that the alternative can't be made to work. This technique, showing that the contrary of a supposed result leads to a contradiction, is the basis of a considerable part of modern mathematics. (The technique has at its heart an assumption, namely that I can correctly say "I have an alternative, I choose an example that matches the alternative and has no special conditions, and any other example that matches the alternative is also false if my example is false." This assumption, the Axiom of Choice, is more complicated than it sounds.)

Suppose that the two frequencies and the two phases were not the same, so that

$$y(t) = B \exp(\imath(\omega' t + \beta)) \tag{40.8}$$

40.1. MATHEMATICAL INTERLUDE

with $\omega \neq \omega'$ and $\alpha \neq \beta$. Equation 40.8 is a hypothesis being tested. As will be seen, it cannot be correct. Inserting our new hypothetical form for $y(t)$ into the first of equations 40.1, we obtain

$$-MA\omega^2 \exp(\imath(\omega t + \alpha)) = -aA\exp(\imath(\omega t + \alpha)) + bB\exp(\imath(\omega' t + \beta)). \quad (40.9)$$

If we then divide out the complex exponential multiplying A, we obtain

$$-MA\omega^2 = -aA + bB\exp(\imath((\omega' - \omega)t))\exp(\imath(\beta - \alpha)). \quad (40.10)$$

We are almost done with the proof. One and only one term in the above equation depends on time, which is impossible. To demonstrate this claim, take the time derivative of equation 40.10, which is

$$(\omega' - \omega)bB\exp(\imath((\omega' - \omega)t))\exp(\imath(\beta - \alpha)) = 0 \quad (40.11)$$

This equation can only be true if $\omega = \omega'$, contrary to our original assumption. Therefore, ω must be equal to ω'.

Now consider equation 40.10 with $\omega = \omega'$, and take the imaginary part of the equation. The imaginary part is

$$0 = bB\sin(\beta - \alpha), \quad (40.12)$$

which can only be true if $\beta - \alpha = 0$ (modulo π), contrary to our original assumption. Therefore, α must be equal to β.

We have now shown that our assumption that $x(t)$ and $y(t)$ oscillate with the same frequency and the same phase must be true. The equalities of frequency and phase are forced by the problem.

Now to solve. Return to equations 40.1 and cancel the complex exponentials. The results are

$$-MA\omega^2 = -aA + bB, \quad (40.13)$$
$$-NB\omega^2 = bA - cB. \quad (40.14)$$

We may solve the second of these two equations for B

$$B = \frac{bA}{-N\omega^2 + c}, \quad (40.15)$$

and use this result to eliminate B from the first of these equations, to obtain

$$(-M\omega^2 + a)A = \frac{b^2 A}{-N\omega^2 + c}. \quad (40.16)$$

Cross-multiplication gives us

$$(-M\omega^2 + a)(-N\omega^2 + c)A = b^2 A, \quad (40.17)$$

This equation has three solutions. The trivial solution is $A = 0$, which is equally true for any ω, namely there is no motion. The more interesting solutions are the solutions for ω^2 with $A \neq 0$, so that

$$(-M\omega^2 + a)(-N\omega^2 + c) - b^2 = 0 \quad (40.18)$$

The above equation is a quadratic in ω^2. There are therefore two solutions, ω_1^2 and ω_2^2, each of which is a function of N, M, a, b, and c. The solutions have an interesting property. Suppose you substitute either ω_1 or ω_2 into equations 40.13. Taking ω_1 to generate an example, and substituting, one obtains

$$-MA\omega_1^2 + aA = bB, \quad (40.19)$$
$$-NB\omega_1^2 + cB = bA. \quad (40.20)$$

You appear to have here two equations connecting A and B. Can you use the two equations to solve both for A and for B? No! If you have solved correctly for ω_1, these two equations are *degenerate*. They give the same value for A/B. [Remark: This equality is a test for the correctness of your solutions. If you plug a solution into both of the above equations, and you get two different values for A/B, your solution is wrong.] If these two equations are both graphed for B as a function of A, they do not describe two lines with an intersection at the point that marks a solution for A and B.

Can two equations in two unknowns actual not have solutions for the two unknowns? As examples of two equations in two unknowns that do not have solutions, consider

$$x + y = 3,$$
$$2x + 2y = 6, \tag{40.21}$$

or

$$x + y = 3,$$
$$x + y = 6. \tag{40.22}$$

Here the first pair of equations create two lines that lie on top of each other, while the second pair of equations represent two parallel lines that never meet.

Instead of finding a solution, you will discover that the two equations and ω_1 predict the same value for B/A. If you instead substitute ω_2 for ω, you get a second value for B/A. There are finally two pairs of solutions, one pair of solutions giving $x(t)$ and $y(t)$ corresponding to ω_1, and the other pair of solutions giving $x(t)$ and $y(t)$ corresponding to ω_2. These pairs of solutions are

$$x_1(t) = S_1 \exp(\imath(\omega_1 t + \alpha_1))$$
$$y_1(t) = \frac{-M\omega_1^2 + a}{b} S_1 \exp(\imath(\omega_1 t + \alpha_1)) \tag{40.23}$$

and

$$x_2(t) = S_2 \exp(\imath(\omega_2 t + \alpha_2))$$
$$y_2(t) = \frac{-M\omega_2^2 + a}{b} S_2 \exp(\imath(\omega_2 t + \alpha_2)). \tag{40.24}$$

In these solutions, the subscripts 1 and 2 correspond to the frequencies ω_1 and ω_2. Here S_1 and S_2 are amplitude factors; their sizes are not determined by the original equations. The original equations had four time derivatives, two attacking $x(t)$ and two attacking $y(t)$, so to reach the solutions we must have integrated with respect to time, implicitly, four times. Corresponding to four integrations there must be four constants of integration; these are seen to be S_1, S_2, α_1, and α_2. As shown above, for a single pair of solutions the frequencies and phases must be the same. Indeed, in the first pair of solutions the frequency is ω_1 and the phase is α_1 both for $x(t)$ and for $y(t)$. In the second pair of solutions, the frequencies and the phases are again the same for $x(t)$ and for $y(t)$, the frequency being ω_2 for both and the phase being α_2 for both.

In these solutions, I happen to have used S as the amplitude for $x(t)$ and then used equations 40.25 to compute the corresponding amplitude for $y(t)$. I could equally well have used S as the amplitude for $y(t)$ and then used equations 40.25 to calculate the amplitudes for $x(t)$. The solutions would have been physically the same, only the notation having been changed.

The two pairs of solutions each have a frequency ω_i, with $i \in (1, 2)$. The ω_i are variously termed the *characteristic frequencies*, *normal mode frequencies*, or *eigenfrequencies*; these being three different ways to say the same thing. The corresponding solutions for $x(t)$ and $y(t)$ are variously termed the *modes*, *characteristic modes*, *normal modes*, or *eigenvectors* of the problem, these being four different ways to say the same thing. The *general solutions* for $x(t)$ and for $y(t)$ are the sums of the characteristic solutions. The general solutions for $x(t)$ and for $y(t)$ are

$$x_1(t) = S_1 \exp(\imath(\omega_1 t + \alpha_1)) + S_2 \exp(\imath(\omega_2 t + \alpha_2)), \tag{40.25}$$
$$y_1(t) = \frac{-M\omega_1^2 + a}{b} S_1 \exp(\imath(\omega_1 t + \alpha_1)) + \frac{-M\omega_2^2 + a}{b} S_2 \exp(\imath(\omega_2 t + \alpha_2)). \tag{40.26}$$

We close with an aside, shown by pedagogical experiment to be useful for an alarming number of your fellow students. We have a class with 25 students and one professor. Let P be the number of professors and S be the number of students. Choose one of my answers, then keep reading. Should I write

$$P = 25S$$

or
$$S = 25P?$$

Which did you write? The latter answer is correct. If you are not sure, replace P and S with 1 and 25, respectively. If you do that with the second of these equations, you get the reasonable 25=25. If you do the replacement with the first of these equations, you get $1 = 625$.

40.2 An Example

We will now work a numerical example demonstrating all of the points made in the above discussion. Our starting point will be

$$\frac{d^2x}{dt^2} = -2x + y, \tag{40.27}$$

$$\frac{d^2y}{dt^2} = x - 2y. \tag{40.28}$$

These two equations have all the features we discussed above. The second time derivatives of x and y are linearly proportional to x and y. Viewing the coefficients on the right-hand side of these equations as a 2×2 matrix, the matrix components on the main diagonal are -2 and -2; they are both negative. The matrix is symmetric across the main diagonal, so if we write a component of the matrix as a_{ij}, then $a_{ij} = a_{ji}$. In the example here, we have $a_{12} = a_{21} = 1$.

For x and y, we now insert the general forms

$$x(t) = A \exp(\imath(\omega t + \alpha)), \tag{40.29}$$
$$y(t) = B \exp(\imath(\omega t + \alpha)). \tag{40.30}$$

These two solutions, for $x(t)$ and for $y(t)$, use the same frequency ω and the same phase α, for reasons shown above.

If we substitute these forms into equations 40.27, we obtain

$$-A\omega^2 \exp((\imath(\omega t + \alpha)) = -2A \exp(\imath(\omega t + \alpha)) + B \exp(\imath(\omega t + \alpha)), \tag{40.31}$$
$$-B\omega^2 \exp((\imath(\omega t + \alpha)) = A \exp(\imath(\omega t + \alpha)) - 2B \exp(\imath(\omega t + \alpha)). \tag{40.32}$$

The complex exponentials cancel, leading to

$$-A\omega^2 = -2A + B, \tag{40.33}$$
$$-B\omega^2 = A - 2B. \tag{40.34}$$

The first of these equations tells us that $B = -A\omega^2 + 2A$, while the second of these equations tells us $A = -B\omega^2 + 2B$. Substituting for B in the second of these equations gives us

$$A(-\omega^2 + 2)(-\omega^2 + 2) = A, \tag{40.35}$$

which on canceling the factors of A and simplifying leads to

$$(\omega^2)^2 - 4\omega^2 + 3 = 0. \tag{40.36}$$

This form is a quadratic in ω^2. Yes, we can treat ω^2 as a variable. The equation is readily factored, namely it gives us

$$(\omega^2 - 3)(\omega^2 - 1) = 0. \tag{40.37}$$

The quadratic has two roots. ω^2 is 1 or 3.

The normal mode frequencies are then

$$\omega_1 = 1 \tag{40.38}$$
$$\omega_2 = \sqrt{3} \tag{40.39}$$

The subscripts 1 and 2 identify the frequency that belongs to each of the two modes, which I have arbitrarily numbered 1 and 2.

The normal modes are specified by the ratios A/B. To obtain this ratio for mode 1, we substitute ω_1 into either of the equations 40.33. Either substitution gives the same value for A/B. Similarly, to obtain A/B for mode 2, we substitute ω_2 into either of the equations 40.33.

For mode 1, $\omega_1^2 = 1$. Substituting into equations 40.33, we find

$$-A = -2A + B \tag{40.40}$$
$$-B = A - 2B \tag{40.41}$$

These two equations have the same solution, namely $A = B$.

On the other hand, for mode 2, equations 40.33 become

$$-3A = -2A + B \tag{40.42}$$
$$-3B = A - 2B. \tag{40.43}$$

These two equations give the same solution, namely $B = -A$.

For either frequency, the two equations relating A and B are the same, both being either $A = B$ or $A = -B$. We have two equations relating A and B, but we cannot use them to solve for A and B as numbers, because at either frequency the two equations are *degenerate*, like the pair $x + y = 3$ and $2x + 2y = 6$.

The particular relationships found above for A and B, namely $A = B$ and $A = -B$, are accidents, outcomes of the particular coefficients that appear on the right-hand-side of equations 40.27. If we had a different set of coefficients in those two equations, we would get a different pair of normal mode frequencies and a different pair of normal modes.

For this problem, the two normal modes are

$$x_1(t) = S_1 \exp(\imath(t + \alpha_1)), \tag{40.44}$$
$$y_1(t) = S_1 \exp(\imath(t + \alpha_1)), \tag{40.45}$$

and

$$x_2(t) = S_2 \exp(\imath(\sqrt{3}t + \alpha_2)), \tag{40.46}$$
$$y_2(t) = -S_2 \exp(\imath(\sqrt{3}t + \alpha_2)). \tag{40.47}$$

Here S_1 and S_2 are the amplitudes for $x(t)$ in the first and second normal modes. In the first normal mode, $A = B$ was replaced by S_1. In the second normal mode, $A = -B$ was replaced by S_2. The phases are subscripted because there is no requirement that the phase in the first mode must be equal to the phase in the second mode. There is a requirement that within either mode the two phase angles must be the same, so in the first mode the two phases are both α_1 and in the second mode the two phases are both α_2.

The above four equations are the two normal mode solutions. There are also the *general solutions* describing everything that $x(t)$ and $y(t)$ can do, given that they follow equations 40.27. The general solutions are the sum of the normal modes, so

$$x(t) = S_1 \exp(\imath(t + \alpha_1)) + S_2 \exp(\imath(\sqrt{3}t + \alpha_2)), \tag{40.48}$$
$$y(t) = S_1 \exp(\imath(t + \alpha_1)) - S_2 \exp(\imath(\sqrt{3}t + \alpha_2)). \tag{40.49}$$

The final solutions for harmonic oscillator problems are given by found by taking the real parts of the above equations. For this problem, the two normal modes are

$$x_1(t) = S_1 \cos(t + \alpha_1), \tag{40.50}$$
$$y_1(t) = S_1 \cos(t + \alpha_1), \tag{40.51}$$

and

$$x_2(t) = S_2 \cos(\sqrt{3}t + \alpha_2), \tag{40.52}$$
$$y_2(t) = -S_2 \cos(\sqrt{3}t + \alpha_2), \tag{40.53}$$

while the physical general solution is the real part of equations 40.48, namely

$$x(t) = S_1 \cos(t + \alpha_1) + S_2 \cos(\sqrt{3}t + \alpha_2), \tag{40.54}$$
$$y(t) = S_1 \cos(t + \alpha_1) - S_2 \cos(\sqrt{3}t + \alpha_2). \tag{40.55}$$

40.3 Discussion

We have referred several times to eigenvectors and eigenvalues. *eigen* is a somewhat peculiar part of a word. It comes from the German. We encounter German loanwords into English with some frequency because until the approach of World War II physics was in fair part dominated by German physicists.

40.4 Homework

1. Consider two coupled variables x_1 and x_2, united by

$$m\frac{d^2 x_1}{dt^2} = -\frac{4mg}{\ell}x_1 + \frac{mg}{\ell}x_2, \tag{40.56}$$
$$2m\frac{d^2 x_2}{dt^2} = \frac{mg}{\ell}x_1 - \frac{mg}{\ell}x_2. \tag{40.57}$$

Here m, g, and ℓ are constants. Find (a) the characteristic frequencies (normal mode frequencies), (b) characteristic mode amplitudes, (c) the solutions for x_1 and x_2, and the general solution.

2. Two masses connected by springs have as their equations of motion in the x direction

$$m\frac{\partial^2 x_1}{\partial t^2} = -4kx_1 + 2kx_2. \tag{40.58}$$
$$2m\frac{\partial^2 x_2}{\partial t^2} = 2kx_1 + -8kx_2. \tag{40.59}$$

Find the normal mode frequencies, the normal modes of oscillation of the system, and the general solution.

3. Two masses connected by springs have as their equations of motion in the x direction

$$m\frac{\partial^2 x_1}{\partial t^2} = -3kx_1 + 2kx_2, \tag{40.60}$$
$$2m\frac{\partial^2 x_2}{\partial t^2} = 2kx_1 + -6kx_2. \tag{40.61}$$

Find the normal mode frequencies, the normal modes of oscillation of the system, and the general solution.

4. The Second Law equations for a pair of coupled harmonic oscillators are

$$5m\frac{d^2 x_1}{dt^2} = -2kx_1 + 3k(x_2 - x_1), \tag{40.62}$$
$$4m\frac{d^2 x_1}{dt^2} = -kx_2 - 3k(x_2 - x_1) \tag{40.63}$$

Find: (a) The normal mode frequencies, (b) the normal mode solutions, and (c) the general solution(s) for these oscillators. (d) What are the synonyms for "normal mode" as in "normal mode frequencies"?

This page reserved for your notes.

Chapter 41

The Double Pendulum

In this chapter, we treat a concrete example of a pair of coupled harmonic oscillators. Our model system will be the *double pendulum*, in which the upper of two pendulums is attached to the ceiling, and the lower of two pendulums hangs from the bob of the upper pendulum. The sketch shows what a double pendulum looks like. The two pendulum bobs each have mass m. The two strings connecting the bobs to their upper support points each have length ℓ; they make angles θ_1 and θ_2 with respect to the vertical. The strings are approximated as being massless. The tensions in the upper and lower strings are T_1 and T_2, respectively. From this sketch, we may immediately construct the force diagrams for the two masses.

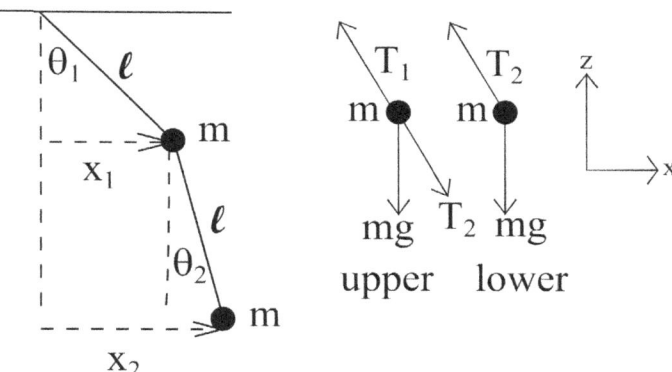

Figure 41.1: Sketch and force diagrams for a double pendulum, with two strings ℓ connecting an upper mass m to the ceiling and to a lower mass m.

Acting on the upper mass are three forces, namely the force of gravity $-mg\hat{\mathbf{k}}$ and the tension forces T_1 and T_2. Acting on the lower mass are two forces, namely the tension T_2 in the lower string and the force of gravity $-mg\hat{\mathbf{k}}$ on the lower mass. The angles θ_1 and θ_2 as indicated in the figure show the deviations of strings 1 and 2 from the vertical. To advance, we need to decompose the tension forces into their components, using the same methods previously used for the single pendulum.

For T_1, the vertical component of the force on the upper mass is $T_1 \cos(\theta_1)$. Analysis based on similar triangles shows that the horizontal component is $-T_1 x_1/\ell$. For T_2, the vertical component of the force on the lower mass is $T_2 \cos(\theta_2)$ and the horizontal component of the force on the lower mass is $-T_2(x_2 - x_1)/\ell$. The strings are massless, so T_2 acts on the upper mass with components exactly opposite in sign to the force components with which it acts on the lower mass. We now apply the small-angle approximation $\cos(\theta) \approx 1 - \theta^2/2 \approx 1$. This approximation reduces the vertical components of the two forces to T_1 and T_2, respectively.

It might appear tempting to introduce a coordinate x_2', the horizontal distance between the two masses, in which case $x_2 - x_1 \to x_2'$. This replacement of x_2 with x_2' is not literally wrong, but it has a problem. The origin of x_2' is the current location of the upper mass. The upper mass may well be accelerating in the horizontal direction, so the origin of x'_2 is accelerating, so x'_2 is the location of the lower mass in a non-inertial reference frame. In a non-inertial reference frame, Newton's Second Law of motion is invalid, so it would be incorrect to write

$$m\frac{d^2 x'_2}{dt^2} = F_x,$$

F_x being the horizontal component of the force on the lower mass. You could introduce the coordinate x'_2, but you can't use it to solve the problem, because the acceleration of x'_2 is not determined by the forces on the lower mass.

We now write the Second Law for the vertical accelerations of the two masses, namely

$$m\frac{d^2 x_1}{dt^2} = T_1 - T_2 - mg, \tag{41.1}$$

$$m\frac{d^2 x_2}{dt^2} = T_2 - mg. \tag{41.2}$$

From simple geometry, the vertical distances of the two masses from the ceiling are $-\ell\cos(\theta_1)$ and $-\ell\cos(\theta_1) - \ell\cos(\theta_2)$. In the small-angle approximation, $\cos(\theta_1)$ and $\cos(\theta_2)$ are both equal to one, so the vertical positions of the two masses are approximately $-\ell$ and -2ℓ. These distances are time-independent constants, so their second time derivatives satisfy

$$\frac{d^2 x_1}{dt^2} = 0, \tag{41.3}$$

$$\frac{d^2 x_2}{dt^2} = 0. \tag{41.4}$$

Combining the four above equations, in the small-angle limit we obtain

$$T_2 = mg, \tag{41.5}$$
$$T_1 = T_2 + mg = 2mg. \tag{41.6}$$

We emphasize that the equalities between the tensions and mg are only correct because the masses are not, in the small-angle approximation, accelerating in the vertical direction.

What about the horizontal direction? We have decomposed the tensions into their horizontal and vertical components, so we have for the equations of motion of a double pendulum

$$m\frac{d^2 x_1}{dt^2} = -\frac{T_1 x_1}{\ell} + \frac{T_2(x_2 - x_1)}{\ell}, \tag{41.7}$$

$$m\frac{d^2 x_2}{dt^2} = -\frac{T_2(x_2 - x_1)}{\ell}. \tag{41.8}$$

From equations 41.5, we know the values of the tensions, so these equations reduce to

$$m\frac{d^2 x_1}{dt^2} = -\frac{3mg}{\ell}x_1 + \frac{mg}{\ell}x_2, \tag{41.9}$$

$$m\frac{d^2 x_2}{dt^2} = \frac{mg}{\ell}x_1 - \frac{mg}{\ell}x_2. \tag{41.10}$$

On the right hand side of this equation, the coefficients multiplying x_1 and x_2 show the expected behavior. The coefficients on the main diagonal are both negative, corresponding to net restoring forces on the two pendulum bobs. The off-diagonal components that are symmetric with respect to the main diagonal are equal to each other.

We now substitute the assumed solutions

$$x_1(t) = A\exp(i(\omega t + \alpha)), \tag{41.11}$$
$$x_2(t) = B\exp(i(\omega t + \alpha)), \tag{41.12}$$

into the equations of motion, finding

$$-m\omega^2 A \exp(\imath(\omega t + \alpha)) = -\frac{3mg}{\ell} A \exp(\imath(\omega t + \alpha)) + \frac{mg}{\ell} B \exp(\imath(\omega t + \alpha)), \quad (41.13)$$

$$-m\omega^2 B \exp(\imath(\omega t + \alpha)) = \frac{mg}{\ell} A \exp(\imath(\omega t + \alpha)) - \frac{mg}{\ell} B \exp(\imath(\omega t + \alpha)). \quad (41.14)$$

The complex exponentials cancel, leading to

$$(-m\omega^2 + \frac{3mg}{\ell})A = \frac{mg}{\ell} B, \quad (41.15)$$

$$(-m\omega^2 + \frac{mg}{\ell})B = \frac{mg}{\ell} A. \quad (41.16)$$

From the first of these equations, after cancelling factors of m,

$$B = \frac{-\omega^2 + 3g/\ell}{g/\ell} \quad (41.17)$$

Substituting for B in the second of these equations,

$$(-\omega^2 + \frac{g}{\ell})(-\omega^2 + 3\frac{g}{\ell}) = \left(\frac{g}{\ell}\right)^2. \quad (41.18)$$

On expansion, this form becomes

$$(\omega^2)^2 - 4\omega^2 \frac{g}{\ell} + 2\left(\frac{g}{\ell}\right)^2 = 0. \quad (41.19)$$

We have here a quadratic in ω^2. Its solutions are

$$\omega^2 = \frac{1}{2}(4\frac{g}{\ell} \pm (16\left(\frac{g}{\ell}\right)^2 - 8\left(\frac{g}{\ell}\right)^2)^{1/2}, \quad (41.20)$$

which simplifies to

$$\omega^2 = \frac{g}{\ell}(2 \pm \sqrt{2}). \quad (41.21)$$

The two mode frequencies are therefore

$$\omega = \sqrt{\frac{g}{\ell}}(2 \pm \sqrt{2})^{1/2}, \quad (41.22)$$

so

$$\omega = a\sqrt{\frac{g}{\ell}}, \quad (41.23)$$

for $a \approx 0.77$ and for $a \approx 1.55$.

From our values for ω^2, we can calculate the form of the two modes. A reasonable starting point is the second of equations 41.15. For the lower frequency, $\omega^2 = \frac{g}{\ell}(2 - \sqrt{2})$ and

$$(-\frac{g}{\ell}(2 - \sqrt{2}) + \frac{g}{\ell})B = \frac{g}{\ell} A, \quad (41.24)$$

or

$$A = (\sqrt{2}) - 1)B \approx 0.41 B. \quad (41.25)$$

For the higher frequency,

$$(-\frac{g}{\ell}(2 + \sqrt{2}) + \frac{g}{\ell})B = \frac{g}{\ell} A, \quad (41.26)$$

and therefore

$$A = (-\sqrt{2}) - 1)B \approx -2.41 B. \quad (41.27)$$

We obtain one set of normal modes for each value of ω. The normal modes are therefore

$$x_1(t) = 0.41 B_1 \exp(\imath(0.77\sqrt{\tfrac{g}{\ell}}t + \alpha_1)), \tag{41.28}$$

$$x_2(t) = B_1 \exp(\imath(0.77\sqrt{\tfrac{g}{\ell}}t + \alpha_1)), \tag{41.29}$$

and

$$x_1(t) = -2.41 B_2 \exp(\imath(1.85\sqrt{\tfrac{g}{\ell}}t + \alpha_2)), \tag{41.30}$$

$$x_2(t) = B_2 \exp(\imath(1.85\sqrt{\tfrac{g}{\ell}}t + \alpha_2)). \tag{41.31}$$

The periods of these two modes are readily stated in terms of the period of a single pendulum containing a string of length ℓ in a gravitational field g. The slow mode has a period $0.77^{-1}T \approx 1.3T$, while the fast mode has a period $1.85^{-1}T \approx 0.54T$.

What are the appearances of these two modes? Of course, with two pieces of thread and some washers you can set up a double pendulum and watch them yourself. However, in the low frequency mode, the upper mass swings left to right, and the lower mass swings in the same direction, but covers a considerably larger left-to-right distance. In the high frequency mode, the upper mass swings left to right, and the lower mass swings in the opposite direction. In the high frequency mode, the swings of the upper mass are much wider than the swings of the lower mass. When I taught this course in lecture, I set up a single pendulum to determine g/ℓ, then set up a double pendulum, moved masses to match the outer extent of swing for each of the two modes, let go, and had the class time the swings. On at least one occasion, the class numbers were so close to the calculated values that my students asked me how I had faked things. If you try this yourself, you will find that the lower frequency mode is much easier to set up properly.

41.1 Discussion

There is at this point a minor math trick, not terribly useful until it is really helpful: You can represent the modes by writing their numerical coefficients. In the case here, the two modes would be $(0.41, 1)$ and $(-2.41, 1)$. To see why these numbers represent the modes, look at the numerical coefficients for the amplitudes in the mode equations. $(0.41, 1)$ and $(-2.41, 1)$ are lists of numbers, i.e., they are vectors. We may denote them as $\mathbf{v}_1 = (0.41, 1)$ and $\mathbf{v}_2 = (-2.41, 1)$ If you take their dot product, you get $\mathbf{v}_1 \cdot \mathbf{v}_2 = (0.41, 1) \cdot (-2.41, 1) \approx 0$. (The dot product is not exactly zero; there is round-off error.) That's a general result; we say that the modes are *orthogonal*, orthogonal meaning 'perpendicular'.

Furthermore, the right-hand-side of equation 41.13 can be viewed as a constant $C = mg/\ell$ and a matrix

$$\mathbf{M} = \begin{pmatrix} -3 & 1 \\ 1 & -1 \end{pmatrix}$$

times (scalar product) a vector (x_1, x_2). $\mathbf{M}$ and the $\mathbf{v}$ vectors are clearly not independent. Indeed, if you write the unit vectors of the $\mathbf{v}$ vectors as $\hat{\mathbf{v}}_i$ and the corresponding frequencies as ω_i, then

$$C\mathbf{M} = \sum_i \omega_i^2 \hat{\mathbf{v}}_i \otimes \hat{\mathbf{v}}_i, \tag{41.32}$$

where $\otimes$ is the vector outer product mentioned early in the book.

41.2 Homework

1. Consider the double pendulum as seen in Figure 41.2a. A mass m hangs at the end of a string having length ℓ. Dangling from it is a second mass of mass $2m$ at the end of a string having length 2ℓ. The

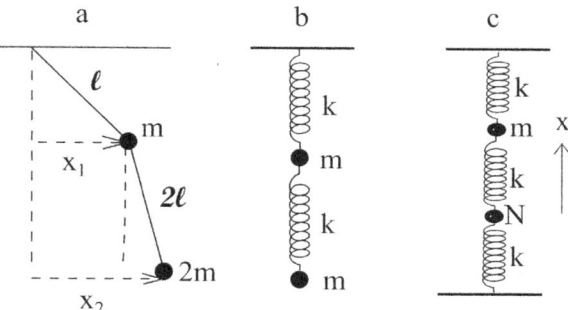

Figure 41.2: Sketches for homework problems (a) 1, (b) 2, and (c) 3.

equations of motion for the two masses are

$$m\frac{d^2 x_1}{dt^2} = -\frac{4mg x_1}{\ell} + \frac{mg x_2}{\ell} \tag{41.33}$$

$$2m\frac{d^2 x_2}{dt^2} = \frac{mg x_1}{\ell} - \frac{mg x_2}{\ell} \tag{41.34}$$

Find (a) the normal mode frequencies, (b) the normal modes, and (c) the general solution to the equations of motion.

2. Consider a pair of masses m hanging from springs having spring constant k as seen in Figure 41.2b. The masses oscillate in the vertical direction; call their displacements z_1 and z_2. a) What are the force diagrams for the system? b) What are the equations of motion of the system? c) What are the characteristic frequencies (the normal mode frequencies) of the system? d) What are the characteristic solutions (normal modes) of the system? (A correct solution will open $z_1 = \ldots, z_2 = \ldots$. e) What is the general solution for the motions of the system? (A correct solution will open $z_1 = \ldots, z_2 = \ldots$

3. Consider a set of springs and masses lying on a frictionless table between two motionless walls, as seen in Figure 41.2c. Note the orientation of the x axis in this problem. Call their displacements from equilibrium x_1 and x_2. The masses can only move parallel to the x-axis. The springs all have the same spring constant k. The two masses have mass m and N, respectively. m and N are *not* equal; instead, $N = 2m$. No external forces are applied to the system (no gravity!). a) What are the force diagrams for the system? b) What are the equations of motion of the system? c) What are the characteristic frequencies (the normal mode frequencies) of the system? d) What are the characteristic solutions (normal modes) of the system? (A correct solution will open $x_1 = \ldots, x_2 = \ldots$ e) What is the general solution for the motions of the system? (A correct solution will open $x_1 = \ldots, x_2 = \ldots$)

4. A double coupled pendulum consists of two masses m hanging from strings having equal lengths ℓ, and coupled by a spring of spring constant k, as seen in Figure 41.3a. The pendulum oscillations are sufficiently small that you can approximate the spring as remaining horizontal. Find the force diagrams and the differential equations of motion for the two masses. Solve the differential equations of motion to find the characteristic frequencies, the normal modes of vibration, and the general solution for the motion of the two masses. Assume small oscillations so that the tension T in each string is approximately mg.

5. Consider a double pendulum in which the two strings have length ℓ and the two masses m and N are **NOT** equal, as seen in Figure 41.3b. N is the lower mass. Find the force diagrams and the differential equations of motion for the two masses. For the case $N = 3m$, solve the differential equations of motion to find the characteristic frequencies. Find the normal modes of vibration. Assume small oscillations so that the tension T in each string is approximately mg.

6. Two masses m_1 and m_2 resting on a flat frictionless table – vertical is perpendicular to the page – are connected to three springs as shown in Figure 41.3c. The springs have equal spring constants k,

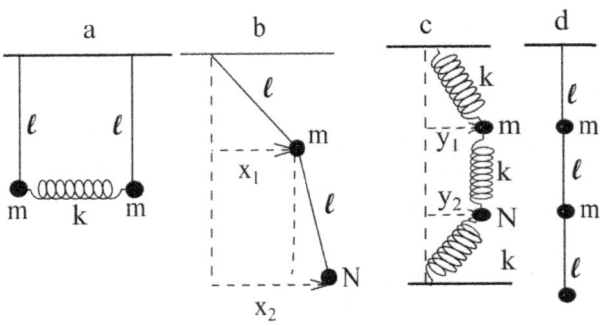

Figure 41.3: Sketches for homework problems (a) 4, (b) 5, (c) 6, and (d) 7 and 8.

and at rest are all stretched equally by a modest distance Δ, so they each supply a restoring force having magnitude $k\Delta$ along their length. The outer two springs are also attached at their outer ends to motionless walls. For small movements along the y-axis, find the force diagrams and the differential equations of motion for the two masses. Hint: a small y displacement does *not* change the length of the spring, only the axis along which it sits. For the case $m_1 = m_2 = m$, find the normal mode frequencies, the normal mode amplitudes, the normal modes, and the general solution.

7. A triple pendulum consists of three masses m_1, m_2, and m_3 linked by strings having equal lengths ℓ, as seen in Figure 41.3d. Find the force diagrams and the differential equations of motion for the three masses. You are not asked to solve for the frequencies or modes.

8. Free extra credit. Using Mathematica, Maple, or any other symbolic integration program, solve the triple pendulum problem of the previous problem and Figure 41.3d for the mode frequencies and normal modes. Include in your solution a printout of the code you used to cause the software to do the work. Remark: In the phrase "Free Credit", the use of the word "free" is debatable. Some work required.

This space reserved for your notes.

Chapter 42

The Oscillating String – Standing Waves

The mathematical techniques used to treat two coupled harmonic oscillators remain valid if three or more harmonic oscillators are coupled. At some point, computational limitations make it far more practical to calculate normal mode frequencies and normal modes if algebriac coefficients in the force constant matrix can be replaced with their numerical values. While computers can now do algebra effectively – more effectively than human beings can – the memory requirement for a large computer algebra calculation can become huge, and the purely symbolic answers become difficult to understand. Inordinate calculation times and round-off errors do not arise for numerical calculations until the number of coupled harmonic oscillators becomes very large indeed.

In this chapter, we consider a specific sort of harmonic oscillator – the string under tension – that contains a respectable fraction of Avogadro's number ($6 \cdot 10^{23}$) harmonic oscillators. With such a large system, the methods of the previous chapters are ineffective. We therefore introduce different methods for treating the system.

42.1 Normal Modes of a String

We begin by considering a string. It has a length ℓ and a mass m, and therefore a density (mass per unit length) $\mu = m/\ell$. The string is under a tension T. You can imagine the tension being generated by having one end of the string go over a pulley, with a mass M hanging on the end of the string, as seen in Figure 42.1. The other end of the string is attached to a wall and does not move. The string itself is suspended in midair, so that segments in the middle of the string are free to move, horizontally or vertically, in directions perpendicular to the string.

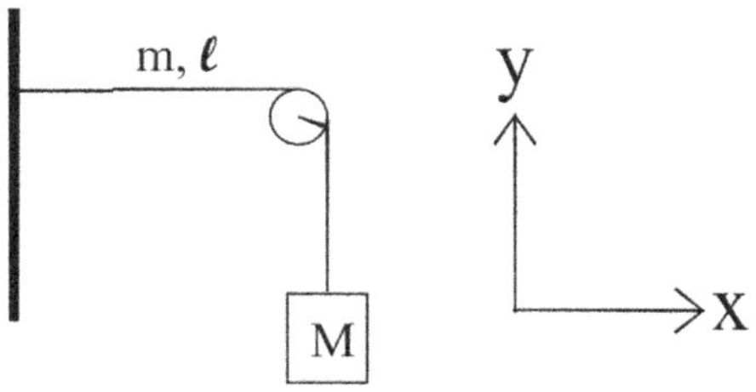

The Figure also shows the force diagram for the mass M. M is acted on by the force of gravity and the tension in the string, these both pointing in the vertical direction. For M, the Second Law gives us

$$M\frac{d^2z}{dt^2} = T - Mg. \quad (42.1)$$

Figure 42.1: A string, mass m and length ℓ (between the wall and the pulley, indicated as a circle) under a tension T due to the mass M. Where the string reaches the wall or the pulley the string cannot move.

In this problem, M is stationary, so its acceleration is zero. For this particular case, we therefore find $T = Mg$. This tension is the same everywhere along the string.

We now consider a very short piece of string. This very short segment is not a point mass; it has some very short length Δx. Treating objects by treating their differential pieces of some size could lead us to several other topics in physics, including hydrodynamics (how liquids move) and continuum mechanics (how solids move). Hydrodynamics and continuum dynamics describe forces and motions over distances sufficiently large that the existence of individual atoms – point masses – can be ignored. Here we stay with the oscillating string.

If we consider our differential segment of string, two significant forces act on it, namely the two tension forces applied to the two ends of the string. (We are ignoring gravity, which is much weaker than the tension force and can be neglected here.) If the string is not vibrating, it simply lies along the line connecting the pulley and the wall attachment point. The two tension forces shown in the force diagram, Figure 42.2 are therefore equal in magnitude and antiparallel, so they sum to zero. The differential string segment therefore does not accelerate if the string is straight.

The string does exert forces on the hanging mass M and on the wall. The force of the string on the wall is $+T\hat{\mathbf{i}}$; the reaction force to that force is the force of the wall on the string, which is $-T\hat{\mathbf{i}}$. The string exerts two forces on the pulley, namely a force $-T\hat{\mathbf{i}}$ in the horizontal direction and a force $-T\hat{\mathbf{k}}$ in the vertical direction. The pulley does not move because it is also subject to a force from its support, that force being $+T\hat{\mathbf{i}} + T\hat{\mathbf{k}}$. Finally, T is equal to the weight Mg of the hanging mass.

What if the string has been plucked, so that it is vibrating? In that case, the string is deflected to the side, where it will, in general, have some curvature. Figure 42.3 shows a sketch of this situation.

Once again, the only forces on the differential segment of the string are the tensions at each end. We now see the importance of treating a differential length of string rather than a point mass. At each end of the differential length segment, the tension must point parallel to the string. If we were treating a point segment of the string, those two forces would necessarily point antiparallel to each other, so that there would be no net force on the string segment. However, because we are treating a segment of the string of some length, albeit a differential length Δx, over the distance Δx the string has had some opportunity to curve, so that the tension forces on the two ends can point in non-parallel directions, leading to a net force on the string segment.

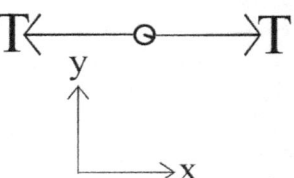

Figure 42.2: Force diagram for a differential segment of a string under tension.

[Aside: We are treating the string as having zero thickness. If it had a non-zero thickness, then the string ends would have some extent perpendicular to the direction in which the string was pointing. The tension force would be perpendicular to the end of the string, but might have different values at different points in the end of the string, so that a net torque (calculated around the end of the differential segment) would be exerted on the differential string segment. Because we treat the string as having no thickness, the net torque on the differential segment, calculated across the surface at either end, must be zero.]

In order to treat string vibrations quantitatively, we need an appropriate set of coordinates. We take x to be the coordinate parallel to the rest position of the string, and y to be the sideways displacement of the string due to the vibration, so that the sideways string motion can be written as

$$y = y(x,t). \tag{42.2}$$

However, we are going to make a radical change in the interpretation of x. In a simple function $\tilde{y}(x,t)$, we would interpret $\tilde{y}(x,t)$ as the y-coordinate of the string, the sideways displacement of the string, at location x and time t. Here, however, we interpret x as a field variable, namely it labels a particular group of atoms by their rest coordinate x. Even if the string segment moves in the x direction as the string oscillates, the segment is still labelled by its rest position x. Having said that, we then approximate the slope $\frac{dy}{dx}$ to be small. We also approximate the displacements of the string to be exactly perpendicular to the string's rest position, so that $y(x,t)$ and $\tilde{y}(x,t)$ are the same. We make the distinction between $y(x,t)$ and $\tilde{y}(x,t)$ so that you realize that there is an approximation being made here.

We now consider a sequence of three length segments $y(x-\Delta x,t)$, $y(x,t)$, and $y(x+\Delta x,t)$. Each of these segments has a differential length Δx, so the distances between their centers are also Δx. The equations of

42.1. NORMAL MODES OF A STRING

motion for the segments will begin with the Second Law forms

$$m\frac{d^2x}{dt^2} = F_x, \tag{42.3}$$

$$m\frac{d^2y}{dt^2} = F_y. \tag{42.4}$$

The mass of a differential string segment is $m = \mu \Delta x$.

Figure 42.3 lets us break the two forces T into their components. The geometric drawing shows T due to the forces from the leftward string segment as the hypotenuse of a triangle whose horizontal component is Δx and whose vertical component is $\Delta y = y(x,t) - y(x - \Delta x, t)$. We are assuming that sidewise motions are small, meaning that $\Delta y \ll \Delta x$ and $\left(\frac{\Delta y}{\Delta x}\right)^2 \approx 0$. The length of the triangle's hypotenuse is therefore

$$\Delta \ell = (\Delta x^2 + \Delta y^2)^{1/2} = \Delta x \left(1 + \left(\frac{\Delta x}{\Delta y}\right)^2\right)^{1/2} \approx \Delta x. \tag{42.5}$$

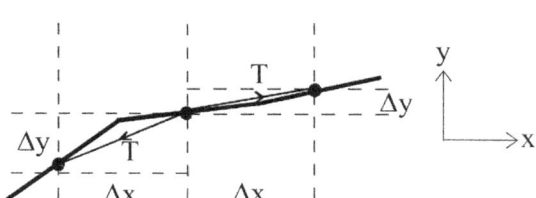

Figure 42.3: Sketch of a differential slice of a string under tension, including forces due to adjacent differential slices. Heavy lines are differential segments, with black dots marking their centers of mass. Tension forces T act from center of mass to center of mass. Δx and Δy are the displacements along the two coordinate axes from one center of mass to the next.

The vertical and horizontal components of the tension are therefore

$$T_y = -T\frac{\Delta y}{\Delta \ell} \approx -T\frac{y(x,t) - y(x - \Delta x, t)}{\Delta x}, \tag{42.6}$$

$$T_x = -T\frac{\Delta x}{\Delta \ell} \approx -T\frac{\Delta x}{\Delta x} \approx -T. \tag{42.7}$$

The same rationale, with T pointing to the right rather than the left, gives the forces on the string segment at $y(x,t)$ due to the tension exerted by the segment at $y(x + \Delta x, t)$. For these two segments, the components of the forces are

$$T_y = T\frac{\Delta y}{\Delta \ell} \approx T\frac{y(x + \Delta x, t) - y(x, t)}{\Delta x}, \tag{42.8}$$

$$T_x = T\frac{\Delta x}{\Delta \ell} \approx T\frac{\Delta x}{\Delta x} \approx T. \tag{42.9}$$

We can now combine the forces from the two neighboring segments, and substitute into the Second Law. We have one minor complication, namely $y(x,t)$ is a function of two variables, so we have to specify what we are keeping constant when we take various derivatives. Simple derivatives d/dz are therefore replaced with partial derivatives, the quantities being held constant being specified in the notation. We find

$$\mu \Delta x \left(\frac{\partial^2 x}{\partial t^2}\right) = -T + T = 0, \tag{42.10}$$

$$\mu \Delta x \left(\frac{\partial^2 y}{\partial t^2}\right)_x = T\frac{y(x + \Delta x, t) + y(x - \Delta x, t) - 2y(x,t)}{\Delta x}. \tag{42.11}$$

Our approximations lead to the outcome that F_x, the x component of the total force on the string segment, vanishes. The first of these equations thus confirms that a single differential segment of a piece of string does not accelerate in the x-direction. In the second equation, taking the first Δx to the other side of the equation, the fraction turns into the second derivative. We have as Δx becomes small that

$$\mu \left(\frac{\partial^2 y}{\partial t^2}\right)_x = T \left(\frac{\partial^2 y}{\partial x^2}\right)_t \tag{42.12}$$

For reasons that will soon become clear, equation 42.12 is known as a wave equation. Reminder: $\frac{d^2x}{dt^2} \neq \left(\frac{dx}{dt}\right)^2$.

How are we to solve this equation? We have the interesting statement that the second derivatives of $y(x,t)$ with respect to x and t are equal up to a multiplicative constant. One way to achieve this object is to consider functions whose second derivatives are equal to the original function multiplied by a constant. We know a series of these: exponential, sine, and cosine should come to mind. There are also two ways to arrange the temporal and spatial dependences, as we will see here and in the next chapter.

We try as a solution

$$y(x,t) = A\sin(kx + \phi)\cos(\omega t + \theta). \tag{42.13}$$

In this equation, A is the amplitude, the maximum value of $y(x,t)$. ϕ and θ are phases, present because we might not want $y(x,t) = 0$ at $x = 0$ or at $\omega t = \pi/2$.

What happens if we substitute this form for $y(x,t)$ into equation 42.12? The partial derivatives give the important results

$$\left(\frac{\partial x}{\partial t}\right)_x = 0 \tag{42.14}$$

$$\left(\frac{\partial t}{\partial x}\right)_t = 0. \tag{42.15}$$

As a result, the partial derivative with respect to t treats the sine function as a constant, and the partial derivative with respect to x treats the cosine function as a constant.

Making the substitution,

$$-\omega^2 \mu A \sin(kx + \phi)\cos(\omega t + \theta) = -k^2 T A \sin(kx + \phi)\cos(\omega t + \theta). \tag{42.16}$$

This equation constrains the relative values of ω and k. Dividing out the terms that are the same on both sides of this equation, and taking a square root,

$$\frac{\omega}{k} = \left(\frac{T}{\mu}\right)^{1/2} \equiv v. \tag{42.17}$$

At this point v is just a symbol. The suggestive notation turns out to be meaningful.

We now return to the original problem. The string is tied in position at both of its ends, the ends being at $x = 0$ and $x = \ell$. It does not move at either end, so we must have

$$y(0,t) = 0 \tag{42.18}$$

$$y(\ell,t) = 0. \tag{42.19}$$

In order for these equations to be true at all times, it must be the case that

$$\sin(k \cdot 0 + \theta) = 0, \tag{42.20}$$

$$\sin(k \cdot \ell + \theta) = 0. \tag{42.21}$$

$$\tag{42.22}$$

For the first of these equations to be true, we must have

$$\theta = 0 \pm n\pi. \tag{42.23}$$

We can simply ignore the $\pm n\pi$ part and say $\theta = 0$; the $n\pi$ term doesn't change anything significant.

For the second of those equations to be zero, we must have

$$k\ell = n\pi, \tag{42.24}$$

because, with $\theta = 0$, only with this equality will it be the case that $\sin(kx + \theta)$ has a zero at $x = \ell$. Said physically, in order for $y(x,t)$ to be zero at its two end points, we must have between the two end points an integer or half-integer number of waves. A single half wave between the two end points is a valid outcome. It follows that k can then only have certain values, namely

$$k = \frac{n\pi}{\ell}. \tag{42.25}$$

42.1. NORMAL MODES OF A STRING

for n an integer that is greater than zero.

If k can only have certain values, then from equation 42.17 ω can only have certain values, to be precise

$$\omega = \frac{n\pi}{\ell}\sqrt{\frac{T}{\mu}}. \tag{42.26}$$

Combining these results, the normal mode form for the wave equation, for a string pinned in place at its two ends, is

$$y_n(x,t) = A_n \cos\left(\frac{n\pi}{\ell}\sqrt{\frac{T}{\mu}}t + \phi_n\right)\sin\left(\frac{n\pi x}{\ell}\right). \tag{42.27}$$

n is the mode number. It labels the mode. n is an integer, one or greater. Equation 42.27 shows a single mode of an oscillating string. The subscripts n on y, A, and ϕ are there because each mode y_n has its own amplitude A_n and phase ϕ_n. A given string may have several modes excited at the same time.

This equation is the solution to the wave equation, equation 42.12. The wave equation has four derivatives in it. Two of them lead to the two constants of integration seen in the above equation, these being the amplitude A_n and the phase ϕ_n. The other two derivatives did not lead to constants of integration; they were taken up by the two requirements that the $y(x,t)$ is zero at the two ends of the string. Those two requirements are examples of *boundary conditions*, requirements that the solutions must have some specific value at some particular location. Boundary conditions can lead to *quantization*, the outcome seen here in which the allowed solutions are characterized by an integer n rather than by a variable that can have any of a continuous range of values. Precisely the same mathematics, wave equations and boundary conditions, leads to quantization in quantum mechanics, quantization giving us the atomic property that a given atom can only absorb or emit light at a list of allowed frequencies.

We say that a mode has been excited if it has a non-zero amplitude. There is no requirement that only one mode can be excited at a time. In general, any or all modes can be excited at the same time, so the behavior of the string is described by the general solution to the wave equation:

$$y(x,t) = \sum_{n=1}^{\infty} A_n \cos\left(\frac{n\pi}{\ell}\sqrt{\frac{T}{\mu}}t + \phi_n\right)\sin\left(\frac{n\pi x}{\ell}\right). \tag{42.28}$$

The general solution is *complete*, in the sense that any $y(x,t)$ of the string can be written in the form of equation 42.28. That's Fourier's Theorem. The A_n and ϕ_n are the *expansion coefficients* and phases that tell us how $y(x,t)$ can be written in terms of ("expanded in terms of") a sum with amplitudes and phases for the individual normal modes.

In writing this result, we have taken advantage of a consequence of the fact that the wave equation is *linear*, i. e., the terms of the wave equation are all proportional to $y(x,t)$ to the first power, never to $y(x,t)$ to some higher power. Consider two normal modes of the wave equation, $y_n(x,t)$ and $y_m(x,t)$. They each satisfy the wave equation, so

$$\mu\left(\frac{\partial^2 y_n(x,t)}{\partial t^2}\right)_x = T\left(\frac{\partial^2 y_n(x,t)}{\partial x^2}\right)_t, \tag{42.29}$$

$$\mu\left(\frac{\partial^2 y_m(x,t)}{\partial t^2}\right)_x = T\left(\frac{\partial^2 y_m(x,t)}{\partial x^2}\right)_t. \tag{42.30}$$

If we add these two equations we get

$$\mu\left(\frac{\partial^2 y_n(x,t)}{\partial t^2}\right)_x + \mu\left(\frac{\partial^2 y_m(x,t)}{\partial t^2}\right)_x = T\left(\frac{\partial^2 y_n(x,t)}{\partial x^2}\right)_t + T\left(\frac{\partial^2 y_m(x,t)}{\partial x^2}\right)_t. \tag{42.31}$$

However, for well-behaved functions the sum of the derivatives is the derivative of the sum, so that

$$\mu\left(\frac{\partial^2 (y_n(x,t) + y_m(x,t))}{\partial t^2}\right)_x = T\left(\frac{\partial^2 (y_n(x,t) + y_m(x,t))}{\partial x^2}\right)_t. \tag{42.32}$$

Because the wave equation is a linear equation, if $y_n(x,t)$ and $y_m(x,t)$ are solutions of the wave equation then $y_m(x,t) + y_m(x,t)$ is also a solution of the wave equation. This outcome is a powerful general result for linear equations, but it is only true because the wave equation is linear. If the were equation were not linear, then in general the fact that $y_n(x,t)$ and $y_m(x,t)$ are solutions would not mean that $y_n(x,t) + y_m(x,t)$ is a solution.

Equation 42.28 shows how we can write an arbitrary $y(x,t)$ in terms of normal modes. For oscillations on a string, the normal mode solutions are not unique, in the sense that there are other ways to expand $y(x,t)$, in terms of functions that are not cosine waves. An important alternative to cosine waves, and the Fourier transforms that underlie them (we shan't say more about Fourier transforms here) is provided by the study of wavelets. The simplest example of a wavelet-like description is provided by the study of musical notes. A note happens over a narrow range of frequency and time. Because a note does not go on forever in time, it cannot be composed of a single frequency; it must contain a range of frequencies in order that it starts and stops. Nonetheless, as known to any musician, notes have dominant frequencies. Wavelets also play an important role in image compression. Like cosine waves, there are wavelet families that form complete sets, so that any signal may be represented as a sum over those wavelets and their expansion coefficients. Some wavelet families are *overcomplete*, so that the same $y(x,t)$ can be represented as a sum of wavelets in more than one way.

We have now reached the physical basis for how string instruments, including not only electric guitars but also harpsichords, actually work. The instrument contains strings made of steel, synthetic polymers, or animal intestines. The strings have a certain mass and length, and are subject to an external tension. When struck or plucked or stroked they oscillate at frequencies determined by equation 42.26. The string oscillation then drives a sounding board or microphone, leading to audible notes heard by the musician and audience.

The same result describes half of how a laser works, namely why a laser only emits light at a given frequency. A gas laser is a long tube filled with a material, say Argon ions, that emits light over a range of frequencies. At the ends of the tube are a pair of mirrors (one of which is not totally reflective so that light can get out). The mirrors create boundary conditions, so the light waves inside the cavity can only have a certain short list of frequencies. In addition, when the light strikes an Argon ion, the light the ion emits comes out with the same frequency and phase as the incident light. That light leads out through one of the mirrors. We thus get laser emission at one or a few nearly-equal frequencies.

A few notational points: ω gives us the angular frequency in radians per second at which the string oscillates. It can also be of interest to note the frequency in cycles per second or periods per second (the named unit is the Hertz: One Hertz is one cycle per second). If the angular frequency ω follows equation 42.26, then the frequency ν (that's a Greek letter 'nu') follows

$$\nu = \frac{\omega}{2\pi} = \frac{n}{2\ell}\sqrt{\frac{T}{\mu}}. \tag{42.33}$$

The period T, the duration of a single cycle, is the inverse of the frequency, so that

$$T = \frac{2\ell}{n}\sqrt{\frac{\mu}{T}}. \tag{42.34}$$

Important note: In the above equation T has one meaning on one side of the equation, and a different meaning on the other side of the equation, but these are standard symbols. On the left-hand-side of the equation, T is the period of oscillation; on the right-hand-side of the equation, T is the tension in the string.

42.2 Energy in a Vibrating String

It is of some interest to ask how much energy is stored in an oscillating string. In this section we consider the energy stored in an oscillating string whose ends are fixed. The total energy in a string has two parts, the kinetic energy and the potential energy. It is fruitful to begin by asking what energy dE is stored in a differential length dx of string, for which we can write $dE = dK + dU$ or

$$dE = \frac{1}{2}\mu dx \left(\frac{dy}{dt}\right)^2 + \Delta U, \tag{42.35}$$

42.2. ENERGY IN A VIBRATING STRING

μdx being the mass of a differential length of string and ΔU being the potential energy of that differential length. The total energy of the string is then

$$E = \int_0^L dE, \qquad (42.36)$$

or

$$E = \int_0^L dx \left[\frac{\mu}{2} \left(\frac{dy}{dt} \right)^2 + dU \right]. \qquad (42.37)$$

In order to advance farther, we need a useful expression for the potential energy change due to the vibrations of the string. Our general approach is to note that if the string is vibrating, rather than being at rest, it is now a smooth curve rather than a straight line. Local segments of the string have therefore been stretched from their rest length dx to their final length ds. How much potential energy is stored in the stretch? We answer by applying the work-energy theorem, which gives us

$$W = -\Delta U = \int \vec{F}(s) \cdot d\vec{s}. \qquad (42.38)$$

In the case at hand, the force $\vec{F}(s)$ is the tension force. It pulls along the string, antiparallel to the stretch, and is very nearly independent of how much the string is stretched by the oscillatory motion. Because the tension force is very nearly constant, we can write

$$\Delta U = -F \Delta s \qquad (42.39)$$

to good approximation, where Δs is the stretch. F and Δs are antiparallel, so $\vec{F} = -\vec{T}$, and $\Delta s = ds - dx$, letting us write for the work done in stretching the differential segment of string

$$\Delta U = T(ds - dx). \qquad (42.40)$$

One sees that $ds = (dx^2 + dy^2)^{1/2}$, and therefore for $dy \ll dx$ that

$$ds = dx \left(1 + \left(\frac{dy}{dx} \right)^2 \right)^{1/2}. \qquad (42.41)$$

If dy/dx is small, we can usefully do a series expansion of the square root, namely for $\epsilon \ll 1$ one has $\sqrt{1 + \epsilon} \approx 1 + \epsilon/2$, giving in this case

$$ds = dx(1 + \frac{1}{2} \left(\frac{dy}{dx} \right)^2 + \ldots). \qquad (42.42)$$

Applying this result to equation 42.40 gives us for the potential energy

$$\Delta U = T(dx(1 + \frac{1}{2} \left(\frac{dy}{dx} \right)^2 + \ldots) - dx) \approx dx \frac{T}{2} \left(\frac{dy}{dx} \right)^2. \qquad (42.43)$$

This is the change in the potential energy for an infinitesimal string section dx. To obtain the full change in the potential energy, we must integrate over the length of the string, leading to

$$U = \int_0^L dx \frac{T}{2} \left(\frac{dy}{dx} \right)^2. \qquad (42.44)$$

The total energy of the system is therefore

$$E = \int_0^L dx \left[\frac{\mu}{2} \left(\frac{dy}{dt} \right)^2 + \frac{T}{2} \left(\frac{dy}{dx} \right)^2 \right]. \qquad (42.45)$$

If only a single normal mode of the system is excited, there is a trick for simplifying the calculation of the energy. We will write the normal mode as

$$y(x,t) = A \sin\left(\frac{m\pi x}{L}\right) \sin\left(\frac{m\pi}{L}\sqrt{\frac{T}{\mu}}t + \phi\right). \tag{42.46}$$

In writing this form, it was convenient as will be seen to change the phase of the time dependence by $\pi/2$ so that the time dependence is a sine rather than a cosine. The solution looks exactly the same. We can calculate the energy at any time that we choose. Conservation of energy, and an absence of friction, guarantee that the energy of the system does not change with time. We choose to calculate the energy at a moment at which $\sin\left(\frac{m\pi}{L}\sqrt{\frac{T}{\mu}}t + \phi\right)$ and hence $y(x,t)$ are both equal to zero. If $y(x,t)$ is zero and the string is vibrating, the velocity of the string must be non-zero. However, if $y(x,t) = 0$ everywhere along x is equal to zero, then $\frac{dy}{dx}$ is zero along its length, so the potential energy term is zero.

In equation 42.45, the total energy is found as a sum of two terms, a kinetic energy term proportional to $\frac{dy}{dt}$ and a potential energy term proportional to $\frac{dy}{dx}$. At the moments at which $y(x,t) = 0$ along the entire string, the potential energy term therefore vanishes; the total energy of the system is present as kinetic energy. To obtain the kinetic energy, we need $\frac{dy}{dt}$, which from equation 42.46 is

$$\frac{dy(x,t)}{dt} = A\frac{m\pi}{L}\sqrt{\frac{T}{\mu}} \cos\left(\frac{m\pi}{L}\sqrt{\frac{T}{\mu}}t + \phi\right) \sin\left(\frac{m\pi x}{L}\right). \tag{42.47}$$

However, if we are at a moment at which $\sin\left(\frac{m\pi}{L}\sqrt{\frac{T}{\mu}}t + \phi\right) = 0$, then we are equally at a moment at which $\cos\left(\frac{m\pi}{L}\sqrt{\frac{T}{\mu}}t + \phi\right) = \pm 1$. The energy may then be written

$$E = \int_0^L dx \frac{\mu}{2} \frac{m^2 \pi^2 A^2}{L^2} \frac{T}{\mu} \sin^2\left(\frac{m\pi x}{L}\right). \tag{42.48}$$

The time-dependent cosine term of equation 42.47 has been replaced its numerical value, unity, at this instant.

There is a useful basic calculus result. $\int dx \sin^2(ax)$ over any integer number of half-waves is $L/2$. Applying this result to do the integral, we finally obtain

$$E = \frac{\mu L}{4} \frac{m^2 \pi^2}{L^2} \frac{T}{\mu}, \tag{42.49}$$

which simplifies to

$$E = \frac{1}{2} \frac{\pi^2 m^2 T}{2L} A^2. \tag{42.50}$$

Having reached this point, we reasonably ask if our solution for $y(x,t)$ and the integral for the energy actually satisfy energy conservation. They certainly should, because these results all followed from Newton's Second Law. Testing whether these results conserve energy provides a check on the validity of the calculation. Suppose we write a mode of the string as

$$y(x,t) = A\cos(\omega t)\sin(kx). \tag{42.51}$$

The angular frequency and wave number are related by $\omega^2 = k^2 T/\mu$. The two interesting derivatives of $y(x,t)$ are

$$\left.\frac{\partial y}{\partial x}\right)_t = Ak\cos(\omega t)\cos(kx), \tag{42.52}$$

$$\left.\frac{\partial y}{\partial t}\right)_x = -A\omega\sin(\omega t)\sin(kx). \tag{42.53}$$

Substituting these two derivatives into equation 42.48, the energy of the system is

$$E = \int_0^\ell dx \left(\frac{Tk^2 A^2}{2} \cos^2(\omega t) \cos^2(kx) + \frac{\mu A^2 \omega^2}{2} \sin^2(\omega t) \sin^2(kx) \right). \tag{42.54}$$

However,

$$\int_0^L dx \cos^2(kx) = \int_0^L dx \sin^2(kx) = \frac{L}{2},$$

and $\mu\omega^2 = k^2 T$, leading to

$$E = \frac{1}{2} \frac{TLk^2}{2} A^2 \cos^2(\omega t) + \frac{1}{2} \frac{TLk^2}{2} A^2 \sin^2(\omega t)), \tag{42.55}$$

which immediately simplifies by means of a trig identity to

$$E = \frac{1}{2} \frac{TLk^2}{2} A^2, \tag{42.56}$$

at all times, confirming that the energy of the string is conserved, as required if all of the above calculations are correct.

42.3 Homework

1. Consider a differential string segment in an oscillating string. Without making the small-displacement approximation, calculate the net force on the segment in the x direction. Is it zero?

2. A uniform string of length 5.0 m and mass 0.05 kg is placed under a tension of 50 N. i) Set up the free body diagram for a section of the string. ii) Find the equations of motion for a section of string. iii) Solve the equations of motion, and find the modes of vibration. iv) What is the frequency of the fundamental mode? v) If the string is plucked transversely, and is then touched at a point 2.0 m from one end, which modes survive? Any mode that does not have a zero where the string is touched will not survive.

3. A violin string of mass 0.1 g and length 0.5 m is stretched to a tension T. A note of frequency 550 Hz is obtained as the second harmonic. What is the tension in the string?

4. Consider a string having mass M and length L fastened at both ends under tension T to immobile walls. (a) Find the force diagram for a short piece of string. (b) Using the small-displacement approximation, write the equation of motion (Second Law, for this problem) for a short piece of string (Hint: on the left, there will be a $\partial^2 y/\partial t^2$.) (c) Solve the equation of motion, and find the allowed frequencies of the waves. (d) I place my finger in contact with the string, 1/3 of the way from one end to the other. Modes in which the string is stationary at this point are not disturbed; other modes are suppressed. What are the allowed frequencies of the waves?

5. The string in Problem 3 is now subject to a damping force $-f\frac{dy}{dt}$. Note that this force is of different size at different points along the string. Discuss (equations are always better) the motion of the string after it is initially set into motion.

6. Solve symbolically. Insert numbers at the end. A uniform string of length 2.0 m and mass 0.002 kg is placed under a tension of 100 N. i) Set up the free body diagram for a section of the string. ii) Using the small-displacement approximation, write the equations of motion for a section of string. iii) Solve the equations of motion, and find the modes of vibration. iv) What is the frequency of the fundamental mode (the mode with half a wave between the two end points)? v) If the string is plucked transversely, and is then touched at a point 0.4 m from one end, which modes survive? What if the string is instead touched 0.3 m from on end? (A general solution of the latter is difficult. Find an answer.)

7. Solve symbolically. Insert numbers at the end. A uniform string of length 1.0 m and mass 0.004 kg is placed under a tension of 50 N. i) Set up the free body diagram for a section of the string. ii) Using the small-displacement approximation, write the equations of motion for a section of string. iii) Solve the equations of motion, and find the modes of vibration. iv) What is the frequency of the fundamental mode? v) If the string is plucked transversely, and is then touched at a point 0.2 m from one end, which modes survive? What if the string is instead touched 0.15 m from on end?

8. A harpsichord string of mass 0.03 g and length 0.4 m is stretched to a tension T. A note of frequency 400 Hz is obtained as the second harmonic. What is the tension in the string?

9. A set of strings having various weights per unit length, various lengths, and various tensions are all tuned until they sound the same note, concert A, when plucked, as their fundamental. Concert A is currently 440 Hz. [Note that over the years notes have drifted up in frequency, so that what a Baroque-period composer called a C we would now call approximately a B.] One string at a time is changed. (a) The tension on string 1 is increased by 50%. (b) The length of string 2 is halved. (c) Spring 3 is wrapped in nylon to double its mass density. (d) Spring 4 is increased in length by 1/4, and its tension is reduced by 1/4, without changing its mass density. What frequencies do these strings play, after these changes, as their fundamentals?

10. A stretched string is plucked, giving it a vibration that is a sum of two normal modes, namely

$$Y(x,t) = 0.3\sin(2\pi x/L)\cos((T/\mu)^{0.5}\frac{2\pi t}{L}) + 0.5\sin(4\pi x/L)\cos((T/\mu)^{0.5}\frac{4\pi t}{L}). \tag{42.57}$$

Find the energy stored in the string.

11. Hard problem: A laser may be constructed by placing a plasma tube in a cavity that has mirrors at each end. The mirrors act on light waves the way rigid walls act on strings, namely the light waves (which are cosines in space and in time) must have zeros at the mirrors at all times. The plasma tube acts as a source of energy, exciting those light waves that are normal modes of the cavity. The math describing the normal modes looks exactly like the math describing the normal modes of a string, except that v is now the speed of light, not $[T/\mu]^{1/2}$. a) What are the normal mode frequencies of the resonant cavity, expressed as functions of the speed of light c and the distance L between the cavity mirrors? b) Suppose the plasma tube emits light centered at $\nu_0 = 4.5 \cdot 10^{14}$ Hz with a spectral width $\Delta\nu$. $\Delta\nu$ is such that all modes whose frequency is within $\pm 1 \cdot 10^9$ Hz of ν_0 will be excited.

1) How many modes will be excited if $L = 2.0$ m? [Hint: an easy error would be to assume that one mode lies precisely at ν_0.] 2) What is the largest L such that only <u>one</u> normal mode will be excited? [$c = 3 \cdot 10^8$ m/s]. [Hint: the mode frequencies are tunable by displacing the mirrors. At what frequencies, relative to ν_0 and $\Delta\nu$, would you place the modes so that the laser is *just barely* able to excite two modes?].

Chapter 43

The Oscillating String – Traveling Waves

43.1 Introduction

In the previous Chapter, we considered the equations of motion for a uniform string. For small oscillations, we showed that the equations of motion were given by the wave equation

$$\mu \left(\frac{\partial^2 y(x,t)}{\partial t^2} \right)_x = T \left(\frac{\partial^2 y(x,t)}{\partial x^2} \right)_t. \tag{43.1}$$

Here $y(x,t)$ is the amplitude of vibration, the lateral displacement of the string at time t, x is a field variable labelling a differential segment of the string by its location when $y(x,t) = 0$, μ is the mass density of the string, assumed to be the same everywhere along the string, and T is the tension in the string.

We considered strings whose ends are fastened in place, so that $y(0,t) = 0$ and $y(L,t) = 0$ at all times. For these strings, the normal mode solutions to equation 43.1 are standing waves, waves having the general form

$$y_n(x,t) = A_n \sin(\frac{n\pi}{L} x) \cos(\frac{n\pi}{L} \sqrt{\frac{T}{\mu}} t + \phi_n). \tag{43.2}$$

In this equation n is a positive integer, A_n and ϕ_n are constants of integration, and $y_n(x,t)$ satisfies the boundary conditions

$$y_n(0,t) = 0, \tag{43.3}$$
$$y_n(L,t) = 0 \tag{43.4}$$

at the two ends of the string. We took it to be true, but did not actually prove (it's Fourier's Theorem) that any waveform of the string can be written as a sum of normal mode standing waves, with properly chosen amplitudes and phases for each normal mode wave.

43.2 Travelling Waves

If the string is infinitely long, so that it does not have ends at which $y(x,t)$ is forced to equal zero, the wave equation has another set of solutions, namely

$$y(x,t) = A \cos(kx - \omega t + \phi) \equiv Re(A \exp(\imath(kx - \omega t + \phi))). \tag{43.5}$$

These solutions are called *travelling waves*.

First, let's determine when equation 43.5 is a solution to equation 43.1. We advance by substitution, with

$$\mu \frac{\partial^2}{\partial t^2} (A\cos(kx - \omega t + \phi)))_x = T \frac{\partial^2}{\partial x^2} (A\cos(kx - \omega t + \phi)))_t \tag{43.6}$$

as the starting point. Because x is held constant in the time derivative, and time is held constant in the x derivative, the derivatives with respect to t and x only act on t and x, respectively, leading to

$$-\mu\omega^2 A \cos(kx - \omega t + \phi) = -Tk^2 A \cos(kx - \omega t + \phi). \tag{43.7}$$

After cancelling factors that are the same on the two sides of his equation, we find

$$\mu\omega^2 = Tk^2, \tag{43.8}$$

showing that

$$\frac{\omega}{k} = \sqrt{\frac{T}{\mu}} \equiv v. \tag{43.9}$$

The travelling waves are only solutions of the wave equation if their wave vector k and angular frequency ω satisfy equation 43.9. If k and ω are not in the ratio shown in equation 43.9, the travelling wave is not a solution to the wave equation.

We have previously seen this form for ω/k in the previous chapter as an outcome for standing waves. We now call v the *velocity* of the travelling waves. Calling v the velocity does not make v a velocity; you should confirm for yourself that v has the right units, length/time, to be a velocity. v is a scalar, not a vector, because the wave is taken to be travelling along the x-axis.

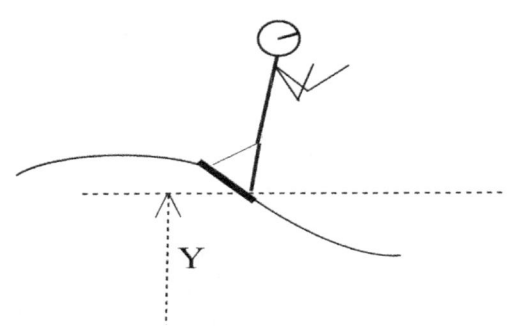

Figure 43.1: A surfer on a wave, the surfer's height being Y.

Why do we call the $y(x,t)$ of equation 43.5 a *travelling* wave? To answer this, consider this sketch of a real wave, including the surfer near the top of the wave. The surfer is so great that she can keep a constant height Y on the wave as it does whatever it is doing, so that

$$A \cos(kx - \omega t + \phi) = Y \tag{43.10}$$

at all times. The argument of the cosine function must therefore be a constant, here called Θ, so that

$$kx - \omega t + \phi = \Theta. \tag{43.11}$$

How can this be true? As time goes on, t changes, so the x of the surfer must change to match, namely

$$x = \frac{\omega}{k}t + \frac{\Theta - \phi}{k}, \tag{43.12}$$

Here x is no longer a field variable labeling a specific differential segment of the string. Instead, here x is the location of the surfer, a location that moves progressively left or right. This equation has the same form as the equation $x = vt + x_0$ for motion in the x direction at constant velocity. The location x for any constant value of $kx - \omega t + \phi$ is therefore time-dependent, changing as $\frac{dx}{dt} = v$ so that

$$v \equiv \frac{dx}{dt} = \frac{\omega}{k}. \tag{43.13}$$

v is thus a travelling velocity. It is the x-wards velocity of a point, on the wave, that keeps a fixed height y in the wave.

k and ω may each be either positive or negative, so v may be either positive or negative. We follow the convention $\omega > 0$, so that the x-component v of the wave's true velocity is positive or negative depending on whether k is positive or negative.

Our travelling wave solution does not satisfy any boundary conditions. It does describe a wave on an infinitely long string. A wave as described by equation 43.5 cannot have the feature that y is a constant at some fixed x, because as time advances $kx - \omega t + \phi$ at fixed x will change. However, we can get a fixed-y boundary condition if we allow that the wave is *reflected*. A combination of two waves

$$y(x,t) = A\cos(kx - \omega t + \phi) + B\cos(-kx - \omega t + \phi') \tag{43.14}$$

is also a solution to the wave equation.

43.3 Terminology

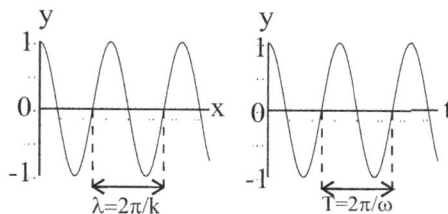

Figure 43.2: A travelling wave plotted as a function of position (left) or of time (right), with the wavelength $\lambda = 2\pi/k$ and the period $T = 2\pi/\omega$ marked.

We now have two sets of solutions for the wave equation, namely

$$y(x,t) = A\cos(kx - \omega t + \phi), \tag{43.15}$$
$$y(x,t) = A\sin(kx + \theta)\cos(\omega t + \phi). \tag{43.16}$$

The first of these equations describes travelling waves. The second of these equations describes standing waves. In the second of these equations, the value of θ and the list of allowed values of k and ω are determined by the boundary conditions.

The various parts of these equations have names:

k is the *wave number* of the wave. It gives the number of radians in a unit length. Radians have dimensions unity, so the dimensions of k are 1/length. In SI units, the dimensions would be 1/meters. If the wave were travelling in space in some direction not along the x axis, in general we replace kx with $\mathbf{k} \cdot \mathbf{r}$. In this equation $\mathbf{r}$ is a position in space and $\mathbf{k}$ is now a vector, the *wave vector*. For a travelling wave, the *wave vector* points in the direction that the wave is travelling.

ω is the *angular frequency*. It gives the number of radians in a unit time. Radians have dimensions unity, so the dimensions of ω are 1/time. In SI units, ω has units 1/seconds.

ν is the *frequency*, which gives the number of complete cycles of a wave in unit time. Its units are Hertz, which is a named unit having units 1/seconds. ω and ν have the same units, but they are not equal to each other. They instead differ by a factor of 2π. If N cycles have passed, $2\pi N$ radians have passed. One cycle is 2π radians, so $\omega = 2\pi\nu$.

T is the *period*, the time for one complete cycle, so T gives seconds per period. The SI units of T are seconds. The period and the frequency are inverses of each other, so $T = 1/\nu$. Recalling ω is larger than ν by a factor of 2π, $T = 2\pi/\omega$.

Corresponding to T, but as a description of the spatial dependence of the wave, we have the *wavelength* λ. λ is the length of one wave, the number of meters in 2π radians of change in the phase of the wave. In SI units, k has units radians per meter, while λ has units meters, so the number of meters in 2π radians change of phase is

$$\lambda = \frac{2\pi}{k} \tag{43.17}$$

which leads to the result

$$k = \frac{2\pi}{\lambda}. \tag{43.18}$$

We also introduced the velocity v of the wave. From $v = \omega/k$ as seen above, we can write $v = \frac{2\pi\omega}{2\pi k}$ which leads to the important form

$$v = \nu\lambda. \tag{43.19}$$

We now come to one of those minor terminological issues that is of practical importance. We have taken the *wave number* and wave vector to give us the number of radians in one meter. However, especially in chemistry, there are analytical spectroscopic techniques known as *infrared spectroscopy* and *Raman spectroscopy*. In these spectroscopy techniques, the relevant wavelengths are conventionally measured in wave numbers, except that the chemistry wavenumber is $1/\lambda$, which is not the physics wavenumber $2\pi/\lambda$.

[Aside: We capitalize *Raman* because the technique is named for the Indian physicist Chandrashekhara Venkata Raman, who in 1930 received the Nobel prize for discovering the Raman effect on which Raman spectroscopy is named. Infrared and Raman spectra of chemical molecules are characterized by spectral lines at different frequencies. The spectral frequencies reveal, in first approximation, the frequencies at which the molecule in question can vibrate.]

It may be helpful to display these quantities as a series of figures. First, note that a travelling wave, plotted as a function of x at some fixed moment in time, looks exactly like the same wave plotted as a function of t at some fixed location in space. Both plots are sinusoids. In Figure 43.2, I plot a wave $y(x,t)$ as a function of position x at a fixed time. I've indicated the wavelength λ, the length of a single wave, shown as the distance between pairs of zero crossings. I've also plotted $y(x,t)$ for a travelling wave, at a single point x in space, as a function of time. The period T during which the wave passes through a full cycle is indicated in the figure. T is shown in its correct experimental form, as the measurement of the time between pairs of zero crossings of the wave. I mentioned earlier the surfboarder staying at constant height on the wave. If we had plotted the surfboarder's $y(x,t)$ as a function of time, we would see a horizontal line.

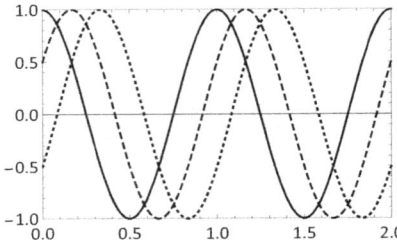

Figure 43.3: A travelling wave, moving right, as indicated sequentially by the solid, dashed and dotted lines.

It is perhaps worthwhile to discuss the appearance of standing and travelling waves. At a single moment in time, $y(x,t)$ of a travelling wave looks exactly like $y(x,t)$ of a standing wave. If the wave is a travelling wave, $y(x,t)$ has a cartoonish similarity to a line of ocean waves. Figure 43.3 shows a travelling wave at three times. The zeros of the travelling wave seen here are moving from left to right. A single traveling wave is actually infinitely long; the figure only shows a short segment of it.

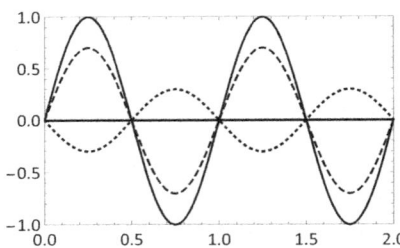

Figure 43.4: A standing wave as a function of position at three times, indicated sequentially by the solid, dashed and dotted lines.

In contrast, Figure 43.4 shows a standing wave at several times. The two end points of the wave are the points where the wave is pinned to $y = 0$. As the wave oscillates in time, it becomes larger or smaller. The solid, dashed, and dotted lines show wave at three successive times. However, this wave is not traveling. It gets larger or smaller, but it is fixed in space, so the points where it has zeros do not move as time goes on. Some readers will recognize the oscillation pattern seen here as an oscillation pattern for a jump rope.

43.4 Longitudinal and Transverse Velocities

How fast are the waves moving? That depends on exactly what we measure. There are two different velocities here, not to mention that the waves have a slope $\frac{\partial y}{\partial x})_t$ at each point.

43.4. LONGITUDINAL AND TRANSVERSE VELOCITIES

One velocity we can readily define is the *transverse velocity* v_t, the rate at which the waves are moving transversely, perpendicular to their direction of motion. That velocity is

$$v_t \equiv \left.\frac{\partial y(x,t)}{\partial t}\right)_x = -\omega A \sin(kx - \omega t + \phi). \tag{43.20}$$

To measure v_t, we sit at a fixed point x (whencefrom the $)_x$ in the definition) and measure how y changes as time goes on.

We next introduce the *longitudinal velocity* v_ℓ. v_l gives us the rate of change of x, if we sit at some fixed height y and watch the wave move. If we are standing on an ocean beach watching the waves roll in, v_ℓ is the speed at which the waves are rolling toward us before they break on the sand. We could define v_ℓ by

$$v_\ell = \left.\frac{\partial x}{\partial t}\right)_y. \tag{43.21}$$

How can we calculate this odd derivative? For a cosine wave, if the wave is observed at a constant height y, then the phase $kx - \omega t + \phi$ of the wave must remain constant, meaning

$$\left.\frac{\partial(kx - \omega t + \phi)}{\partial t}\right)_y = 0, \tag{43.22}$$

and therefore

$$k\left.\frac{\partial x}{\partial t}\right)_y - \omega = 0, \tag{43.23}$$

or

$$\left.\frac{\partial x}{\partial t}\right)_y \equiv v_\ell = \frac{\omega}{k}. \tag{43.24}$$

The transverse and longitudinal velocities are not independent. They are related by a result you may or may not have seen elsewhere, the *cyclic permutation identity*

$$\left.\frac{\partial x}{\partial t}\right)_y \left.\frac{\partial y}{\partial x}\right)_t \left.\frac{\partial t}{\partial y}\right)_x = -1 \tag{43.25}$$

A derivation of the identity is suggested by Figure 43.5.

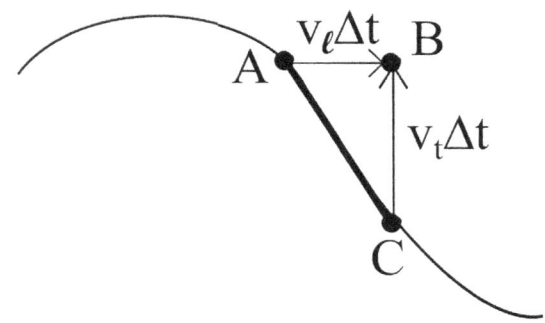

Figure 43.5: Derivation of the cyclic permutation identity. During the time Δt, the wave moves horizontally from A to B and vertically from C to B, at the speeds v_ℓ and v_t, respectively.

During a time Δt the wave moves forward from its original position. Consider a point on the wave staying at fixed y. During the time Δt the point moves from the point labelled A to the point labelled B. The point has moved a distance $\Delta x = v_\ell \Delta t$ while staying at fixed y. Now consider instead an observer observing the wave at a fixed x, the observer's x being chosen to line up with point B. The observer initially sees the wave at point C. During the time Δt, from the perspective of the observer the wave moves upward a distance $\Delta y = v_t \Delta t$ upward, while staying at fixed x.

The two lines AB and CB are the legs of a right triangle, whose hypotenuse is the line AC. The line AC connects two points that were measured at the same time t. What is the slope of AC? Its horizontal is Δx. Its vertical change is $-\Delta y$, the minus sign appearing because in moving from B to C one moves to a smaller value of y. The slope is therefore

$$\frac{\Delta y}{\Delta x} = \frac{-v_t \Delta t}{v_\ell \Delta t}. \tag{43.26}$$

If Δ is significantly larger than zero, the cosine wave curves between points A and C. However, in the limit of small Δt, $\frac{\Delta y}{\Delta x}$ becomes the slope $\left.\frac{\partial y}{\partial x}\right)_t$, leading to

$$v_\ell \left.\frac{\partial y}{\partial x}\right)_t = -v_t. \qquad (43.27)$$

Let's consider an example showing the two velocities and the slope. Our travelling wave is a propagating Gaussian, a bell-shaped curve that migrates across the page. Our example satisfies

$$y(x,t) = A\exp(-b(x+bt)^2). \qquad (43.28)$$

A and b are constants. In this example, they are both positive numbers.

The longitudinal velocity is determined by the requirement that $y(x,t)$ remains constant as time passes. In order for that to occur, $x + bt$ must remain constant, so that the time derivative of $x + bt$ must be zero. That requirement gives us

$$\left.\frac{\partial x}{\partial t}\right)_y = -b, \qquad (43.29)$$

or

$$v_\ell = -b. \qquad (43.30)$$

The transverse velocity is

$$\left.\frac{\partial y}{\partial t}\right)_x = \left.\frac{\partial A\exp(-b(x+bt)^2)}{\partial t}\right)_x. \qquad (43.31)$$

Taking the indicated partial derivative, under which $\left.\frac{\partial x}{\partial t}\right)_x = 0$ because x is being held constant, we have

$$v_t = -2b^2(x+bt)A\exp(-b(x+bt)^2). \qquad (43.32)$$

Finally, we need the slope at fixed t, which is

$$\left.\frac{\partial y}{\partial x}\right)_t = \left.\frac{\partial A\exp(-b(x+bt)^2)}{\partial x}\right)_t. \qquad (43.33)$$

Taking the indicated partial derivative, under which $\left.\frac{\partial t}{\partial x}\right)_t = 0$, because t is being held constant, we have

$$\left.\frac{\partial y}{\partial x}\right)_t = -2b(x+bt)A\exp(-b(x+bt)^2). \qquad (43.34)$$

If we insert these results into equation 43.27, we get

$$-b \cdot 2b(x+bt)A\exp(-b(x+bt)^2) = -2b^2(x+bt)A\exp(-b(x+bt)^2), \qquad (43.35)$$

which is indeed true. Also, if we calculated any two of the three quantities $dy/dx)_t$, v_t, and v_ℓ, we could obtain the third with simple division or multiplication. Using two to compute the third will often be useful for finding the longitudinal velocity, since it will not always be obvious what is to be done to x in order to hold $y(x,t)$ constant as t changes.

43.5 Homework

1. Show that equation 43.27 actually does rearrange into equation 43.25.

2. An isosceles triangular wave pulse of height 1.0 m is moving in the negative x direction on a string on which the wave speed is 20 m/s. At $t = 0$ the ends of the pulse are at $x = 0$ and $x = -2$. Draw, at $t = 1$, the pulse. Label coordinates. Draw a graph of the transverse velocity, as a function of time, at $x = +1.0$ m. [Hint: $t < 0$ is valid for t.]

3. Consider a travelling pulse whose shape as a function of position and time is
$$Y = \frac{1}{1+(3x+7t)^2}. \tag{43.36}$$
Find the transverse velocity, the longitudinal velocity, and the slope of Y, and show that these three quantities satisfy the cyclic permutation identity.

4. From information given you in each part, write the corresponding travelling wave in the form $Y(x,t) = A\cos(kx - \omega t)$. Take the $+x$ direction to be to the right. (a) The wave has an amplitude of 3 volts, with $\lambda = 4$ m and $\nu = 2$ Hertz. The wave is moving to the right. (b) The wave has a peak-to-trough height of 4 m, with $\lambda = 2\pi$ m and a speed of 5 m/s. The wave is going to the left. (c) The wave has an amplitude of 0.5 Pascals with $\omega = 3$ s^{-1}. The wave is moving at 7 m/s to the right.

5. A tightly stretched extremely long cable is under tension T. The cable has mass density M and length L; it is driven into small transverse (sideways) oscillations by being struck with small hammers. A travelling wave is set up in a long part of the cable. Take x to be the direction along the cable and Y to be the direction perpendicular to the cable. (a) Find the force diagram for an infinitesimally long segment of cable. (b) Find the differential equation of motion for the cable segment. (c) Show that a travelling wave $Y(x,t) = A\cos(kx - \omega t + \phi)$ is a solution to the differential equations of motion. (d) Find the speed at which a point of constant phase moves along the cable.

6. Consider a travelling wave of the form $Y(x,t) = A\cos(kx - \omega t + \theta)$. Here θ is the constant of integration that specifies the phase the wave would have, at $x = 0$ and $t = 0$, if the wave continued to that position and time. The phase of the wave at a given position and time is $\Phi = kx - \omega t + \theta$. (a) Suppose the wave has amplitude $A = 3$ and angular frequency $\omega = 40$ radians/second, with $\theta = \pi/6$. The wave moves to the right at 200 m/s. Write the wave in the form $Y(x,t) = A\cos(kx - \omega t + \theta)$. (b) What is the difference in the phase Φ between two points 5 m apart in space, the phases being determined at the same time? (c) Repeat part (b) but this time the constant of integration is $\theta = \pi/3$. (d) What is the difference in the phase Φ between two points at the same location along the x axis but separated in time by 7 seconds? (e) What is the difference in the phase between two points, the second point being 3 m to the right and 0.1 second later than the first point? (f) What happens to the phase of the wave at a given point in space and time if the amplitude A of the wave is doubled?

7. A pulse travels along a string with wave form $y = A\cos((x+at)^2)$. a) Sketch y against x at $t = 0$. b) Calculate the slope $\frac{\partial y}{\partial x})_t$ of the curve. c) Calculate the transverse velocity $v_t = \frac{\partial y}{\partial t})_x$. d) Using the cyclic permutation identity, calculate the longitudinal velocity v_l of the wave. e) Can you determine whether the pulse is travelling to the left or to the right? Why?

8. A travelling wave pulse has the form
$$Y(x,t) = \frac{Y_0}{\ell^2 + (ax+bt)^2}. \tag{43.37}$$
(This is a *Lorentzian* wave form.) Here Y_0, ℓ, a, and b are all constants.

(a) Without using the cyclic permutation identity, find (i) the transverse velocity, (ii) the longitudinal velocity, and (iii) the slope of the wave pulse.

(b) show that your answers to part (a) satisfy the cyclic permutation identity.

9. The equation of a transverse wave travelling along a wire is given by $y = 0.3\sin(\pi(0.5x - 50t))$, where x and y are in centimeters and t is in seconds. (a) Find the amplitude, wavelength, wave number, frequency, period, and velocity of the wave. (b) Find the maximum transverse speed of any segment of the string. c) Is the equation of motion for a segment of string having mass density μ and tension T
$$\mu\frac{\partial^2 y(x,t)}{\partial t^2} = -T\frac{\partial^2 y(x,t)}{\partial x^2}? \tag{43.38}$$
To answer this question derive the correct equation of motion for the string and identify errors (there may not be any) in the above equation. An answer "yes" or "no" is worth NO points.

10. From information given you in each part, write the corresponding travelling wave in the form $Y(x,t) = A\cos(kx - \omega t)$. Take the $+x$ direction to be to the right. (a) The wave has an amplitude of 2 volts, with $\lambda = 3$ m and $\nu = 7$ Hertz. The wave is moving to the right. (b) The wave has a peak-to-trough height of 15 m, with $\lambda = 1\pi$ m and a speed of 3 m/s. The wave is going to the left. (a) The wave has an amplitude of 5.0 Pascals with $\omega = 7$ s^{-1}. The wave is moving at 5 m/s to the right.

11. Consider a travelling wave of the form $Y(x,t) = A\cos(kx - \omega t + \theta)$. Here θ is the constant of integration that specifies the phase the wave would have, at $x = 0$ and $t = 0$, if the wave continued to that position and time. The phase of the wave at a given position and time is $\Phi = kx - \omega t + \theta$. (a) Suppose the wave has amplitude $A = 5$ and angular frequency $\omega = 60$ radians/second, with $\theta = \pi/3$. The wave moves to the right at 400 m/s. Write the wave in the form $Y(x,t) = A\cos(kx - \omega t + \theta)$. (b) What is the difference in the phase Φ between two points 5 m apart in space, the phases being determined at the same time? (c) Repeat part (b) but this time the constant of integration is $\theta = \pi/8$. (d) What is the difference in the phase Φ between two points at the same location along the x axis but separated in time by 7 seconds? (e) What is the difference in the phase between two points, the second point being 3 m to the right and 0.1 second later than the first point? (f) What happens to the phase of the wave at a given point in space and time if the amplitude A of the wave is doubled?

12. A pulse travels along a string with wave form $y = A\sin((x - at)^3)$. a) Sketch y against x near $t = 0$. b) Calculate the slope $\frac{\partial y}{\partial x})_t$ of the curve. c) Calculate the transverse velocity $v_t = \frac{\partial y}{\partial t})_x$. d) Using the cyclic permutation identity, calculate the longitudinal velocity v_l of the wave. e) Does the given information determine whether the pulse is travelling to the left or to the right? Why?

13. A trapezoidal (quadrilateral, narrower at top) wave pulse of height 3.0 m is moving in the positive x direction on a string on which the wave speed is 5 m/s. At $t = 0$ the front end of the pulse rises in a straight line from $y = 0$ at $x = 0$ to $y = 3$ at $x = -2$, continues level until $x = -8$, and then declines in a straight line to $y = 0$ at $x = -10$. Draw, at $t = 1$, the pulse. Label coordinates. Find the slope and the transverse velocity as functions of position.

14. The equation of a transverse wave travelling along a wire is given by $y = 0.3\sin(\pi(0.5x - 50t))$ where x and y are in centimeters and t is in seconds. (i) Find the amplitude, wavelength, wave number, frequency, period, and velocity of the wave. (b) Find the maximum transverse speed of any segment of the string.

15. The equation of a transverse wave travelling along a wire is given by

$$y = 0.3\exp(-(0.4x + 40t)^4). \tag{43.39}$$

(i) Find the transverse velocity.
(ii) Find the slope of the function.
(iii) Without using the cyclic permutation identity, find the longitudinal velocity.
(iv) Use your results from (i)-(iii) to confirm the cyclic permutation identity.

Chapter 44

About the Author

George Phillies is Professor of Physics Emeritus from the Worcester Polytechnic Institute. He now lives in retirement in Worcestér, Massachusetts, with his cat Pounce. At WPI he did research and taught physics courses for three decades. He also taught for the WPI Associated Faculty on Interactive Media and Game Development, developing two nationally unique courses in board game design, based on five textbooks that he wrote for those courses. His degrees (S.B., M.S., D.Sc., all in Physics, and S.B., Life Science) are from the Massachusetts Institute of Technology. He also did research at the University of California, Los Angeles.

Physics One is his second physics textbook, the first being *Elementary Lectures in Statistical Mechanics*. His third physics book is a research monograph, *Phenomenology of Polymer Solution Dynamics*. Two additional volumes on polymer dynamics are in preparation.

His hobbies include politics and science fiction. He ran for Federal office twice; one of his debates with his two opponents was carried on national television. He has written four books on campaign finance and political strategy.

A published science fiction author, Phillies has a half-dozen published novels and short story collections. He has also published 170 scientific papers, two separate short stories, and one article in a law journal.

Books by George Phillies include:

Physics Books

1. George D. J. Phillies, *Elementary Lectures in Statistical Mechanics*, Springer-Verlag: New York (2000) [textbook].

2. George D. J. Phillies, *Phenomenology of Polymer Solution Dynamics*, Cambridge University Press, Cambridge, U.K. (2011) [research monograph].

3. George D. J. Phillies, *Complete Numerical Tables for Phillies' Phenomenology of Polymer Solution Dynamics*, Third Millennium Publishing, Tempe, AZ (2011) [research monograph].

4. George D. J. Phillies, *Physics One*, Amazon (2020) [textbook].

Novels and Short Stories

1. George D. J. Phillies, *This Shining Sea*, Third Millennium Publishing, Tempe, AZ (2000) [novel].

2. George D. J. Phillies, *Nine Gees*, Third Millennium Publishing, Tempe, AZ (2000) [short story collection].

3. George D. J. Phillies, *The Minutegirls*, Third Millennium Publishing, Tempe, AZ (2006) [novel].

4. George Phillies, *Mistress of the Waves*, Amazon.com Kindle, Createspace.com (2012) [novel].

5. George Phillies, *The One World*, Smashwords, Amazon.com Kindle (2012) [novel].

6. George Phillies and Jefferson Swycaffer, Editors, *A Sea of Stars Like Diamonds*, Kindle, Smashwords (2016).

7. George D. J. Phillies, *Minutegirls, Second Edition*, Kindle, Smashwords (2017) [novel].

8. George D. J. Phillies, *Against Three Lands*, Kindle, Smashwords (2018) [novel].

9. George D. J. Phillies, *Eclipse - The Girl Who Saved the World*, Kindle, Smashwords (2019) [novel].

10. George Phillies, *Airy Castles All Ablaze*, Kindle, Smashwords (2019). [novel]

Politics

1. George D. J. Phillies, *Stand Up for Liberty!*, Third Millennium Publishing, Tempe, AZ (2000) [political manual].

2. George D. J. Phillies, *Funding Liberty*, Third Millennium Publishing, Tempe, AZ (2003) [historical monograph].

3. George Phillies *Libertarian Renaissance*, Kindle, Smashwords, Third Millennium (2014).

4. George Phillies, *Surely We Can Do Better*, Kindle, Smashwords, Third Millennium (2016).

Game Design

1. George D. J. Phillies and Tom Vasel, *Contemporary Perspectives on Game Design*, Third Millennium Publishing, Tempe, AZ (2006) [textbook].

2. George D. J. Phillies and Tom Vasel, *Design Elements of Contemporary Strategy Games*, Third Millennium Publishing, Tempe, AZ (2006) [text].

3. George Phillies and Tom Vasel, *Designing Modern Strategy Games* (Studies in Game Design - 1), Smashwords, Kindle (2012) [textbook; second edition of "Design Elements of Contemporary Strategy Games"].

4. George Phillies and Tom Vasel, *Modern Perspectives on Game Design* (Studies in Game Design - 2), Smashwords, Kindle (2012) [textbook; second edition of "Contemporary Perspectives in Game Design"].

5. George Phillies, *Stalingrad for Beginners* (Studies in Game Design - 3), Kindle, Smashwords (2013).

6. George Phillies, *Stalingrad Replayed* (Studies in Game Design - 4), Kindle (2013).

7. George Phillies, *Designing Wargames - Introduction* (Studies in Game Design - 5), Kindle, Smashwords, Third Millennium (2014).

Index

K
    the kinetic energy, 143
ℓ_0
    unstretched spring length, 121
$\exp(ax)$, 24
ω
    angular frequency, 329
g, 59
k, force constant, 121
m
    meaning of, 4
ansatz, 73

absolute magnitude, 332
acceleration
    force of, 78
    in circular polar coodinates, 199
acceleration vector, 46
action-reaction pair, 59, 82, 120, 132
    properties of, 58
advanced topic
    abstract algebra, 16
    canonical partition function, 38
    chaos, 120, 343
    complex fluids, 102
    computer algebra, 386
    conservation law, 135
    continuum mechanics, 388
    de Sitter precession, 50
    engineering, 263
    general relativity, 50, 75
        displacement of a vector in, 192
    hydrodynamics, 388
    kinetic theory, 103
    Laplace transform, 341
    Lense–Thirring effect, 50
    mechanical engineering, 102
    nonlinear equations, 343
    polymer physics, 39
    quantum mechanics, 39, 391
    radar problem, 365
    radioactive decay, 39
    shear banding, 103
    simple liquids, 102
    special relativity, 75
    statistical mechanics, 38, 39
    statistics, 308
    three-body forces, 58
    viscoelasticity, 95
    wavelets, 392
alchemy, 96
amplitude, 240, 330
anecdote, 30
angular acceleration, 190
angular frequency, 392, 399
angular momentum, 205, 250
    and rotation vector, 209
    choice of origin, 206
    conservation of, 207
    definition, 250
    dimensions of, 206
    direction of
        meaningless, 211
    displacement of the origin, 208
    independent of origin, 209
angular velocity, 47
Antikythera mechanism, 274
arctangent function, 332
aside
    angles, 5
astrology, 274
atomic motion, 103
Atwood machine, 253
average, 7, 37
    index, 37
    mass-weighted, 131
    space-weighted, 39
    time-weighted, 39

unweighted, 37
weighted, 37
average acceleration, 131
average velocity, 31, 37, 131

bagpipe, 329
ballistic motion, 48, 53
basis vector
    time derivative of
        non-zero, 198
basis vectors, 15
    complete, 15
beam balance, 99
bending moment, 217
binomial theorem, 342
boundary condiions, 391

calculus, 23
    standard functions, 23
calculus-based, v, vii
Cartesian coordinates, 3, 45, 55, 73, 197
    orientation of, 78
center of mass, 131
center of mass acceleration, 132
center of mass velocity, 132
central force, 207, 211, 276
centrifugal potential, 288
characteristic frequencies, 376
characteristic modes, 376
circular motion, 45, 46, 189, 190, 200
    in an arbitrary plane, 191
    vector description, 191
circular pendulum, 86
circular polar coordinates, 46, 190, 199
    basis vectors, 191
    motion in, 197
college-level, v, vii
collision, 133
    three-dimensional, 179
collision ideon, 175
collisions, 131, 175
    examples, 176
complex arithmentic, 332
complex conjugate, 332
complex exponential, 329
complex number
    amplitude, 331

phase angle Θ, 331
complex numbers, 331
    division by, 333
    equality of, 360
complex plane, 331
    quadrants of, 332
component, 13
composite system, 61
computer algebra, 387
concepts, 9
conservation law, 141, 159, 208
    angular momentum, 207
conservation of energy, 227
conservative forces, 157
constant acceleration, 23
constant angular acceleration, 190
constantc of integration, 24
constants of integration, 240
    calculation of, 346
    meaning, 27
    solving for, 29
constraint, 62, 82, 99, 100, 128, 247
constraints, 186
contact force, 59
contingent variable, 346
coordinate
    center-of-mass, 287
coordinate inversion, 206
coordinate systems, 3
coordinates
    circular polar, 46
coupled harmonic oscillators
    example, 377
coupled masses, 80, 99, 163
coupled motion, 247
    three bodies, 248
    two bodies, 247
    with rotation, 247
coupled pendulums
    and outer product, 384
cross product, 187
    of basis vectors, 188
Curie Principal, 20
cycles per second, 330
cyclic permutation identity, 401

damped oscillator

INDEX

amplitude, 353
energy, 353
energy relaxation, 355
period, 353
damping constant, 352
definite integral, 24
derivative, 24
dimensions of, 25
finding maxima, 25
of an integral, 144
derivatives of vectors, 45, 53
dimensional analysis, 5, 9
dimensions, 3
direction, 13
direction cosine, 19
displacement, 30
double pendulum
normal modes, 384
orthoginal, 384
periods of, 384
drag force
linear, 351

eclipse, 26
eigenfrequencies, 376
eigenvectors, 376
elastic collision, 176, 177
electric field, 165
electrical force, 165
energy, 157
of harmonic oscillator, 339
units of, 4
energy conservation, 141, 165, 175, 185, 278
application, 161
engineering
radio receiver, 368
equation of motion, 338
equations of motion, 26, 81, 83, 91, 122, 247, 263, 382
equilibrium, 159
error, 6
post hoc ergo propter hoc, 96
conceptual, 6
quantum, 6
random, 6
systematic, 6
escape velocity, 148, 278

estimation, 9, 10
Euler's Identity, 329
Euler's identity, 333
examination, 183
exponential function notation $\exp(ax)$, 333
external force, 133

Ferris wheel, 64, 84
field variable, 388
First Heliacal Rising, 274
First Law, 56, 63, 75, 133
fluxion notation, 220
force
of acceleration, 60
of inertia, 60
magnetic, 157
orthogonal, 157
reaction, 63
time dependent, 101
force components
signs of, 79
force constant, 121
force diagram, 59, 78, 247, 260–262, 268, 381, 387
sppring, 120
forces
conservative, 157
non-conservative, 157
four-bar linkage, 225
Fourier's Theorem, 397
frequency, 399
frequency ν, 330
friction, 95, 175
kinetic, 96
solids, 95
static, 98
Fundamental Theorem of Calculus, 24

gases, 95
general solutions, 376, 378
gravitational acceleration, 275
gravitational constant G, 275
gravitational potential energy, 160
gravity, 59, 157
acceleration of, 59
forces between three masses, 278
general relativity, 277

linear force, 277
Nawton's Law, 277
Newton's Law of, 275
strength of, 276
great scientist
Hooke, Robert, 119
great scientists
Atwood, George, 275
Brahe,Tycho, 274
Claudius Ptolemaeus, 274
Copernicus, 274
Curie, Marie, 20
Curie, Pierre, 20
Euler, Leonhard, 333
Fourier, Jean-Baptiste Joseph, 355
Galilei, Galileo, 57, 122, 133
Gibbs, Josiah Willard, 4, 13
Hamilton, William Rowan, 220
Jacobi, Carl Gustav Jacob, 220
Joule, James Prescott, 4
Kepler, Johannes, 275, 290
Kuhn, Thomas, 275
Leibniz, Gottfried Wilhelm, 23
Maxwell, James Clerk, 103
Michelson, Albert Abraham, 4
Newton, Isaac, 13, 23, 275
Raman, Chandrashekhara Venkata, 400
the Scribes of *Enuma Anu Enlil*, 274
Trimble, Virginia Louise, 301
Weiss, Rainer, 58
Wu, Chien-Shiang, 193
Greek alphabet, 337
group study, viii

hanging mass, 82
harmonic motion, 327, 329
solution, 329
harmonic oscillation, ix, 337
harmonic oscillator
canonical forms for, 339
canonical solutions, 339
complex solution, 351
complex variable method, 341
coupled, 373
damped, 351
damped and driven
average power absorbed, 366

power absorption, 365
response curve, 363
driven, 355
driven and damped, 359
energy of, 339
equation of motion, 339
infinite frequency limit, 362
position of, 338
response curves, 359
transient response, 359
velocity of, 338
harmony, 329
harpsichord, 329
helical motion, 197
Hertz, 330, 392, 399
homework, vii
Hooke's Law, 119

ideon, 175
imaginary number, 331
imaginary part, 331
impetus, 57
indefinite integral, 24
index
kinematics
in circular polar coordinates, 197
inelastic collision, 176
inertia
force of, 78
moment of, 217
inertial reference frame, 73, 124
inertial reference frames, 75
initial conditions, 346
instantaneous velocity, 31
integration
dimensions of, 25
direction of, 25

jerk, 46, 54

kinematics, 23
rotation, 187
kinetic energy, 141
definition of, 143
kinetic friction, 95
Koenig's Theorem, 226

laboratories, 299

laboratory exercises, ix
Large Interferometric Gravitational Observatory, 6
lasers, 392
law of conservation of angular momentum, 208
law of converation of energy, 159
Law of inertia, 57
Laws of Motion, 63
LIGO, 6, 274
linearity, 391
linearization, 343
liquids, 95

magnetic force, 157
magnitude, 13
massless, 164
masslessness, 82, 119
memorization, viii
modes, 376
moment of inertia, 217, 220, 247
momentum, 55, 131, 133
    angular, 205
momentum conservation, 131, 133, 175
Moon
    libration, 287
motion
    at constant acceleration, 23
    helical, 197
motion at constant acceleration, 25

neutral equilibrium, 160
Newton's Law
    applications, 77
Newton's laws of motion, 55
Newton, Isaac, v, 96
non-conservative forces, 157
non-inertial reference frame, 73, 124, 288, 382
non-linear least squares, 346
noninertial reference frames, 75
normal force, 59
normal modefrequencies, 376
normal modes, 376, 378
normalizing factor, 37
notes, viii

ordered list, 13

orrery, 287
orthogonal forces, 157
oscillating spring
    normal modes, 387
    standing waves, 387
oscillating string
    energy in, 392
    period, 392
    potential energy of, 393
    travelling waves, 397
outlier, 6

parabola, 84
partial derivative, 389
pendulum, 233
    angular frequency of, 344
    conical, 348
    double, 381
    energy of, 342
    extended bob, 237
    motion of, 340
    physical, 223
    small-angle approximation, 239
    torque approach, 236, 343
    triple, 386
period, 330, 399
phase, 240, 330
phsyical pendulum
    potential energy, 345
physical basis, 58
physical pendulum
    energy approach, 345
    kinetic energy, 345
    oscillation frequency, 346
    period, 346
planetary orbit, 287
    equal-area rule, 290
    period, 289
position vector, 45
potential energy
    gravitational, 160
    spring, 164
power, 144
    and energy, 144
    and work, 144
power series
    equality of, 125

praxis, 29, 155, 183, 294
principal value function, 332
problem solving, 364, 384
    good notation, 354
projection of a vector, 18
pulley, 82

quadratic equation, 331
quality parameter
    Q, 354
quantization, 391

radian, 5
    dimensions of, 5
reaction force, 63
reading, viii
real part, 331
reduced mass, 288
reference frames, 55, 73
restoring force, 341, 343
right hand rule, 343
    defined, 188
right-hand-rule, 237
rigid body rotation, 223
rod
    moment of inertia of, 225
    non-uniform, 241
roller coaster, 162
rolling cylinder, 226
rolling friction, 95, 102
rolling motion, 217
    without slip, 218
root finder
    may not work, 346
rotation, 187
    around displaced center, 217
    around single axis, 217
    rigid body, 217
rotation vector, 192
rotational motion
    rope and pulley, 218
    with slip, 218

sample examinations, ix
scalar, 13
scalar product, 142, 149
scientific notation, 7

Second Law, v, 55, 63, 75, 81, 90, 100, 142, 165, 186, 249, 263, 373
seismometer, 122
significant figures, 7, 9
simplex method, 346
simultaneous, 58
sketch
    is not a force diagram, 61
Socratic dialogue, viii
Socratic method, 28
solution methods, 100, 124, 165, 177, 190, 240, 264, 288, 290
solution procedure, 30, 77, 82, 92, 200, 227, 248, 361
    checking algebra, 252
    incorrect assumptions, 234
    linearization, 340
    prove contrary is false, 374
solution procedures, 234
special relativity, 75
spring
    forces on, 120
    mass on a, 337
    unstretched length, 121
springs, 119
stability, 159
stable equilibrium, 160
standing waves, 397
static friction, 95, 99
statics, 259
    example
        flagpole, 262
        ladder on a wall, 261
        simple beam, 260
string instruments, 392
studying, vii
subscripts, 80
Sukhoi-27, 88
systematic error, 7

tangential velocity, 47
Taylor series, 125, 333, 343
    convergence, 280
Taylor seriess
    for gravity, 276
tension, 61, 82
tensor product of two vectors

value of, 20
the Global Positioning System GPS, 278
theology, 96
Third Law, 56, 58, 63, 132
three-body forces, 207
times
    meaning of, 4
torque, 206
    components of, 217
    definition, 250
    independent of origin, 209
    units of, 4
torque approach, 250
torque diagram, 233, 247, 260–262, 268
total force, 79, 259
total torque, 259
tractive force, 95, 102, 158
tribology, 95
trigonometric identities, 333

understanding, viii
unit vector, 14
    of **0**, 15
unit vectors
    $\hat{i},\hat{j},\hat{k}$, 15
    $\hat{x},\hat{y},\hat{z}$, 15
    e_1,e_2,e_3, 15
units, 3
    cgs-esu, 4
    furlong, 4
    hogshead, 4
    Newton, 4
    pipee, 4
    SI, 4
unstable equilibrium, 160

vector
    axial, 206
vector equation, 134
vector field, 13, 198
vector product, 187
    anticommutative, 189
    geometric interpretation, 188
    mnemonic for, 189
    not commutative, 192
    parallel vectors, 236
vectors, 13

addition commutative, 14
addition of, 14
basis, 15
basis vectors, 18
component equations, 16
components, 16
define a parallelogram, 14
define a paralllelogram, 188
derivatives of, 45
displacement of, 191
equality of, 13
graphical representation, 14
inner product, 17
magnitude
    from components, 17
multiplication, 17
notation, 13
notation for components, 16
product of a vector and a scalar, 17
representation, 17
scalar product, 13, 17
tensor product, 13
unit, 14
vector product, 13
velocity
    average, 38
    in circular polar coordinates, 199
velocity gradient, 103
velocity vector, 45
viscoelasticity, 95, 102
viscosity, 95
volksraketenwagon, 27
Vulcan
    intramercurial planet, 278

wave equation, 389
wave number, 399
wave vector, 399
wave velocity, 399
wavelength, 399
weak damping, 353
weight, 37
weightlessness, 279
work, 141
    by a spring, 148
    by an internal force, 164
    done by friction, 148

work, gravitational, 146
work-energy theorem, 141, 159, 185
    derivation, 142
    examples, 145
    proof of, 143
worked problems, ix

zero vector **0**, 20

Made in the USA
Monee, IL
28 January 2023

26354977R00236